G.K.Hall
ENCYCLOPEDIA OF MODERN TECHNOLOGY

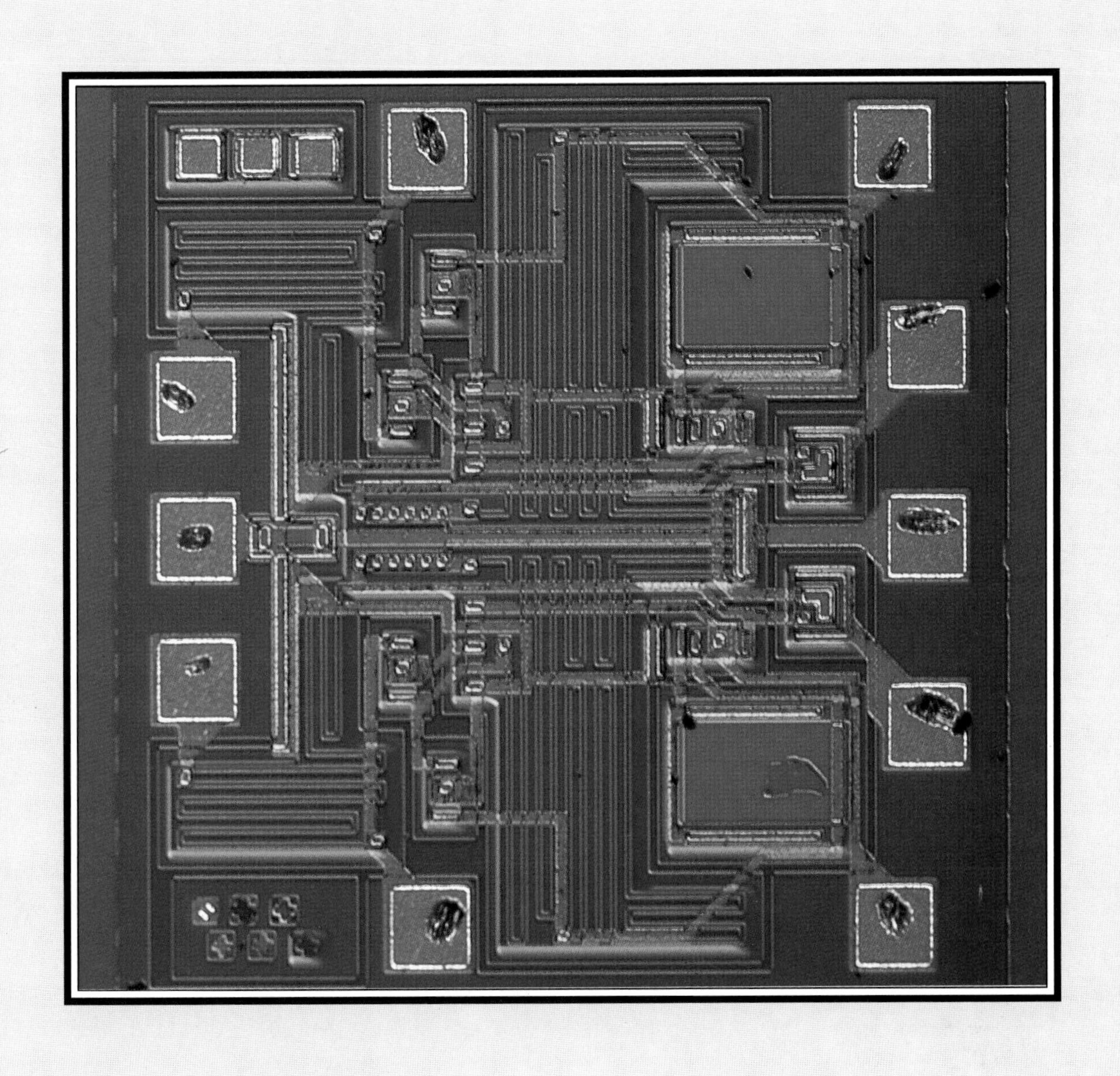

G.K.Hall

ENCYCLOPEDIA OF MODERN TECHNOLOGY

Edited by Dr David Blackburn
& Professor Geoffrey Holister

Editor Peter Furtado
Art Editor John Ridgeway
Assistant Editor Nicholas Harris
Designers Ayala Kingsley, Niki Overy
Picture Editor Mary Fane
Senior Editor Lawrence Clarke
Advisors
Sir Montague Finniston FRS, Chancellor, Stirling University
Dr Erik Quistgaard, European Space Agency, Paris
Contributors
Keith Attenborough (8, 11)
David Baker (31)
David Blackburn (9, 10, 12, 16, 17)
Nigel Cross (22)
David Goreham (24, 25, 26)
Ben E. Hill (23) David Hodges (28)
Geoffrey Holister (18, 19, 20, 21, 27)
Jane Insley (6) Rhys Lewis (13)
Iain Nicolson (7) Denis Ridgeway (26)
Martin Sherwood (14, 32, 33, 34)
Christine Sutton (2, 15)
Graham Warwick (30)
Graham Weaver (1, 3, 4, 5)

AN EQUINOX BOOK

Planned and produced by:
Equinox (Oxford) Ltd
Littlegate House
St Ebbe's Street
Oxford OX1 1SQ

First published in 1987 by
G.K. Hall & Co.
70 Lincoln Street
Boston, Massachusetts
02111

ISBN 0-8161-9056-9

Printed in Spain by Heraclio Fournier SA, Vitoria

Introductory pictures (pages 1-8)
1 Electronic circuit on a silicon chip
2-3 Solar evaporation plant
4-5 Fermilab synchrotron accelerator
7 Scanning tunneling microscope
8 Solar thermoelectric power station

Contents

Introduction

This is not a book about science. Certainly, it deals with the ideas of science, with ways in which understanding of scientific principles can lead to the solution of problems, with notions of experiment and measurement. But the focus here is different. Science is concerned principally with the investigation and exploration of the laws that underlie natural phenomena. The theme of this book is the use of scientific knowledge to create things, to build, to move, to communicate and to control events. This topic we call technology.

Technologists develop models of how things work, or of how they might be made to work. Rarely are these models physically similar to the things they represent. More often they are designs to indicate structure or mathematical equations to describe processes. Such models can be extraordinarily elaborate, but all provide a simplified version of reality which is used as the basis for prediction and action. For this book, it is useful to construct a model for technology itself.

Materials, energy and information technology

In one model, technology is the skills which determine three kinds of flow within society. First is the flow of materials, relevant skills including ore extraction, smelting, casting, purifying or refining, shaping. Second is the flow of energy. This deals with engines, generators, batteries and fuel cells, solar cells, transmission lines for electrical power, windmills and hydroelectric plant. The third flow is that of information; it deals with the devices that collect, broadcast, store and select information, notably in the fields of measurement, printing, transmission of signals, recording, library management, and the control of computer systems.

Early times are often described in terms of materials development. The expressions Stone Age, Bronze Age, Iron Age refer to date by the materials technology of the time. But bronze and iron were not produced without an understanding of how to generate high temperatures, and could not have been produced without the transmission of skills from one generation to another. Materials development was, however, the great technological feat of these times, and this is justly remembered in the name.

Energy conversion is not celebrated as an era by name, but it was increasing skill in energy conversion that led to the Industrial Revolution. First water power, then steam made it advantageous to concentrate work in towns and factories. Growing skill in the techniques of energy conversion led to transport systems. Separate developments led to the electrical distribution systems which now give domestic access to heating, cooling and the power for the small electric motors in widespread use.

Understanding of materials technology is a prerequisite for energy conversion. From earliest days materials have been used at the limits of temperature and mechanical stress, and this will not change. Early steam boilers were prone to rupture and such explosions could be very destructive. Modern energy conversion devices make far higher demands on materials performance, and in such equipment as nuclear reactors there is an additional need for parts to function effectively while being irradiated by a high flux of neutrons.

Only in recent times has information been treated as a commodity. With the advent of telecommunications it became useful to think of the information content of a message and to assign a number to designate its quantity. Such numbers take no account of the nature of the message: it may be a text, a piece of music, a list, an image. This approach provides a structure in which print, disks, tape, photographs and computer records share a common characteristic.

If one word is to encapsulate the activities of our present age, materials technology might suggest *plastics* or *silicon*, while energy-conversion technology might offer *jet-plane* or *nuclear*. But a descriptor in terms of the most rapidly moving technology is the word *information*. This is the age of information.

Technology, measurement and control

In this volume no attempt has been made to catalog the fields in which technology is used, nor to list by instrument or process. The approach rather is to exemplify technological problems and to show the development of some key ideas. Pervasive is the idea of control. Technological work has purpose: it provides food, shelter, travel, communication, defense and leisure facilities. If purpose can be defined then deviations from what is sought can be measured and corrective action taken. As technology evolves, processes become more complex so tasks must be performed with increasing precision and reliability.

Early technology called for personal observation. Gradually systems of measurement were developed which standardized methods of observation and allowed less subjective assessments of quality and precision. Measurement of mass, length and time were primary requirements, and early success is demonstrated by the geometric precision of ancient buildings and in the records of early astronomy. Measurement has extended since that time. Distance measurements of great precision are now made on scales that range from the nuclear to the global. Mass is measured on a scale that can handle sub-units of the atom, while time is measured in intervals linked to oscillations of atomic dimension.

Then there is the range of instruments to show structure. One early example is the optical telescope; a modern example the field-ion microscope which forms images of high magnification. In addition there are devices that produce images of a selective character. Sound may be used to show cracks in solid structures or to form images of underwater objects. Rapidly formed images show the details of explosive events. More sophisticated again are devices that show the distribution of chemicals within living tissues, of crops in a country or the effects of careless or thoughtless industrial processing.

Latest among technological tools are those which process information. Complex activities need measurement systems which gather far more data than can be interpreted directly by human means. Control must be handed to an automated system, human intervention being restricted to assessing system objectives. Here technological skill relates what can be observed to what is desired, and so controls what is done.

The limits of modern technology

Technology is applied to all fields of human activity. The benefits are manifest. More people live longer, are healthier, eat better, have more leisure to pursue their own ideas. More insidious are the negative benefits: toxic chemicals slipping into food chains, the stress of industrial life, global changes of climate mediated by releases of industrial waste, the potential for catastrophic global conflict.

The assessment of technological activity must await the attention of future generations. If the high point of current technological achievement is in the information processing required to control production systems of great complexity, the task for coming generations is much more difficult. It calls for information processing systems capable of assessing technology itself. With the growing scale of industrial operations, a particular technology can no longer be thought of as operating in a closed system without reference to the wider context. The global surroundings can no longer be thought of as an infinite reservoir able to accept unwanted byproducts without degradation. The closed system is now our global one, and particular production systems are subsystems which cannot be allowed to function without reference to adjacent subsystems. This is the primary problem now faced by those working in the fields of technology.

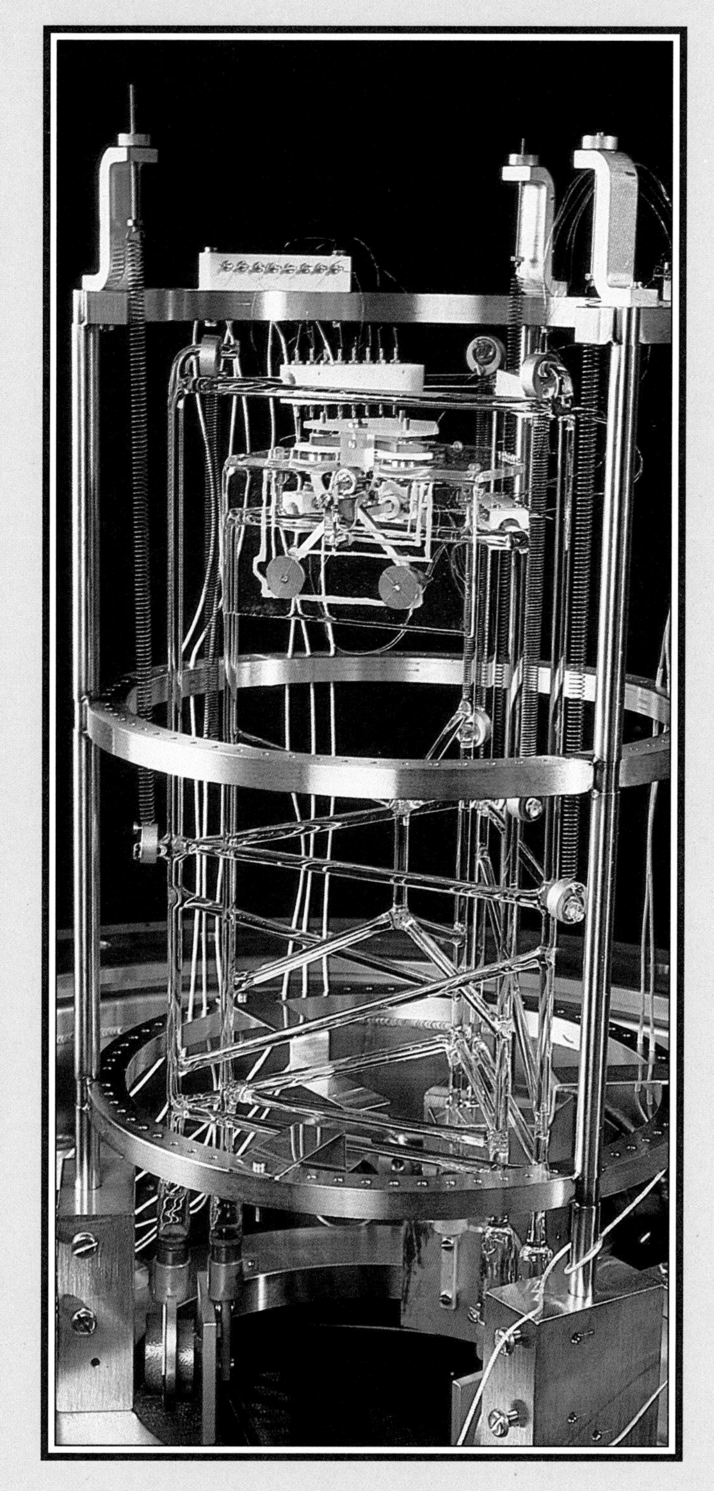

Introduction to Measuring

Counting and measuring...Counting bodily sensations ...The structure of a measuring instrument...The uses of measurement...Establishing standards of measurement ...Measurement and mass-production...Absolute measurements...Automatic measuring machines... PERSPECTIVE...Creating the metric system...Standards, gauges and armaments...The SI system of measurement...Other units

We humans, in common with other animals, experience the world through the senses and, also like them, we have a repertoire of swift, accurate and elegant responses to the things we perceive. However, humans are probably the only animals able to count, and this fact makes many of our actions and perceptions special. *Homo sapiens* might be known as *Homo technologicus*, and we have always understood that we are uniquely able to exploit – even to have dominion over – the natural world. This uniqueness stems from our ability to count, as much as from the ability to speak or to make tools.

Some human actions, such as striking a ball or riding a bicycle, are done without counting, just as cats or birds chase their prey by processing sights and sounds "naturally". Moths, although mathematically ignorant, pursue their prospective mates by following minute gradations in the concentrations of attractive scents. Such movements are controlled by internal judgements which process sensory signals and generate appropriate responses.

Measurement is the process of counting how much of a sensory signal exists. If a language can be devised in which we can express this information, it can then be passed outside our individual bodies. That makes possible three things: communicating to other people information describing what we have seen; contracting on terms of trade or exchange; and controlling systems, with many applications for the pure sciences, technology and adminstration.

▲ *The abacus is one of the earliest and simplest counting devices. It has been used since antiquity throughout the Middle East and Asia for addition, subtraction, multiplication and division. Skilled operators can perform quite complex calculations on an abacus at least as quickly as on an electronic calculator.*

▼ *Counting and measurement are integral to most human activities, not just those now designated as science and technology. We invent many games and sports that depend on counting arbitrary events; the results, such as the exact times for athletes at the Los Angeles Olympic Games of 1984, may be reported with great seriousness.*

Instruments are usually a more reliable means of counting sensations than human bodies, but less versatile in operation

Chang Heng's seismoscope

Direction of seismic waves from earthquake

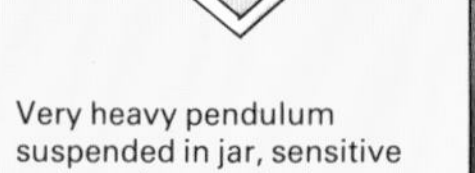

Very heavy pendulum suspended in jar, sensitive to deep, low-pitched tremors

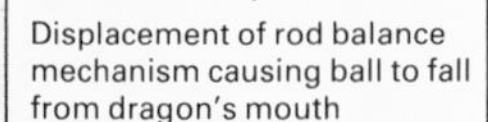

Displacement of rod balance mechanism causing ball to fall from dragon's mouth

8 dragons aligned with points of compass

Direction of tremor shown by location of ball in mouth of appropriate toad

Counting the sensations of our bodies

With our senses we can detect many qualities of the environment: saltiness, redness, loudness, warmth, to name but a few. Athough all such sensations can be set roughly on scales of comparison, few sensations, and none of these examples, can easily be counted. Counting lets us ask "how many" rather than "how much", so that some amounts of quality we sense can be added together to make more. To say "twice as loud" and "half as red" means nothing until we have created special units and techniques of measurement. There are, however, a few sensations which *can* readily be counted and added, and these form the basis of the measurement of many others.

We all live in time and are conscious of its passing (➧ page 23). Days, moons, seasons and years give natural units of time that can be counted, and are units that can be measured objectively (the lunar month is the same throughout the world). The cycle of feeding and becoming hungry makes us aware of time passing in infancy.

We also live in space, and by counting steps or handspans we can size up the world. There are four fundamental spatial qualities that we can measure: distances, areas, volumes and angles can all be calculated using length. Rules of spatial measurement were established by mathematicians and philosophers in many early civilizations, giving definitive standards even though units of measurement were not yet accurately determined (➧ page 37).

The third sensation we can experience directly and count is force. We appreciate this by the quantities of muscular effort involved, and can count and add it by studying, for example, how many stones of a particular size a person can carry (➧ page 31).

There is a fourth quality of matter which we normally detect only in exceptional circumstances: this is its electrical charge. The ability to detect and measure this has allowed the range of sensations that can now be measured to be expanded greatly (➧ page 41).

▲ The pyramids of Egypt, with their careful dimensions and orientation to the points of the compass, show the early association between measurement and the gods.

▲► The Chinese Emperor Chang Heng (2nd century AD) built a seismoscope to detect earthquakes. A heavy weight hung inside a jar on a levered beam. A tremor caused the weight to swing, moving the levers and opening one of the mouths of the dragons around the jar. A ball dropped down to the toad below.

Anatomy of a measuring instrument

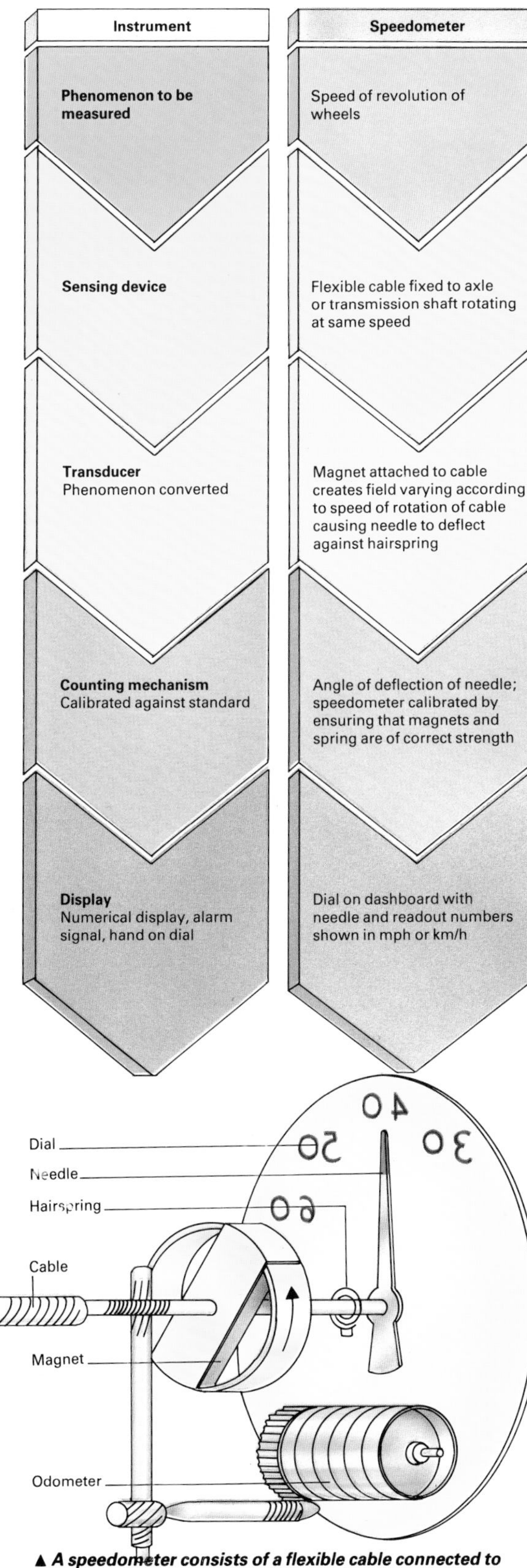

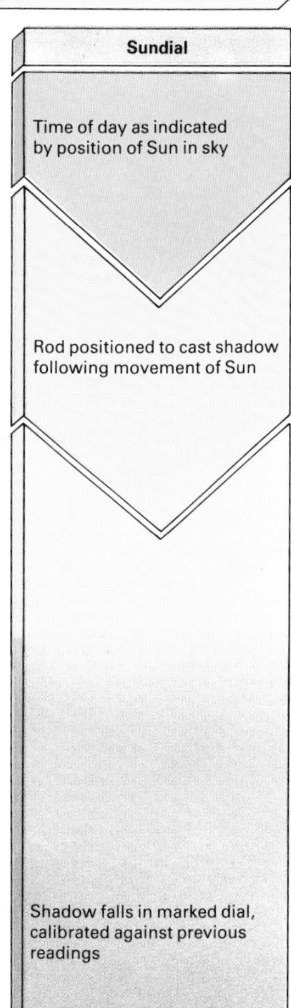

◄ Measuring instruments can all be broken down into the same basic components: sensor, counting mechanism, display, and sometimes transducer.

▲ An 18th-century German sundial, with a compass to ensure that it is aligned due north so that the Sun's rays sweep evenly across the dial at any season or latitude.

▲ A speedometer consists of a flexible cable connected to the drive or transmission shaft by a helical thread and revolving at the same rate. The speed of rotation – the phenomenon to be measured – is transduced by magnetic force into the deflection of a needle on the dial. A magnet is fixed to the cable end and rotates with it. Around the magnet is a deflection cup of aluminum attached to a spindle, and around that a stator cup. The force between the revolving magnet and the stator causes the deflecting cup to move on its spindle against a hairspring. A needle is also attached to the spindle, and this indicates the speed of the vehicle against a numbered dial. The odometer or distance counter works by a simple system of worm threads.

To count sensations it is better to use instruments than to rely on bodily impressions. Instruments can extend the human senses by being more sensitive, expand them by detecting things that humans cannot sense at all, monitor continuously without tiring, respond consistently to a steady input, convert sensation to number automatically, and make comparisons with precisely agreed standards. Against this, humans have the advantage of versatility. The body has a battery of sensors and the ability to switch attention from one to another at will; this could only be matched by a host of instruments, each of which is usually specialized to a particular task.

There are three elements to any instrument: a sensor, a comparator, and a display. The sensor detects the quality to be measured and sends a signal to the next stage. The comparator counts the size of the signal using designed-in units of counting, and the display presents the quantitative information to the observer. The display indicates what the instrument has measured. It is up to the user to decide what to do with the information, including whether to believe the measurement. Displays are often shown in one of our directly countable senses. Angle is used on clocks and other dials. Length analogs of temperature and pressure are found in mercury thermometers and barometers. But where the instrument merely has to show that a danger level has been reached, a flashing light or a loud noise might suffice.

Many instruments go one stage beyond mere analog and provide numbered displays showing how many "agreed units" have been detected. Converting the analog response of the instrument to numbers of standard units is called calibration, and accuracy is the description of the error or lack of it in this counting process (though of course the sensor must also behave consistently, producing the same signal for a constant quantity). The accuracy of an instrument must be distinguished from its sensitivity, which refers to the fineness of detail detectable by the sensor.

Measurement for communication

Measured information is communicated for innumerable purposes in modern society. Scientists, in particular, strive to make accurate measurements so that when they report their observations and experimental results they will not mislead others. Drawings are one technique of transferring numerical information to another person in an accessible form, much used by engineers, architects and planners. Scale drawings especially can inform rapidly and concisely about sizes, shapes and positions. Together with graphic devices such as symbols, color conventions, contours, or numerals, other layers of measured information can be added to it, so turning a drawing into a map.

Such devices can allow many features to be compared directly which could never all have been experienced simultaneously. Thus a map is able to distinguish between vertical and horizontal distance through the calculation of angles; by adopting standard systems of measurement, information from many separate sources can be assembled to form a single image.

Measurement for contract

Trade and industry require agreement about the quantities of goods, many of which may not be countable directly. It may once have been accepted that three sheep were rated as equal to one cow, but land is normally measured by area, work by time elapsed, and precious metals by weight. Again, standard units of measurement are needed, and devices that allow both parties to agree how many units there are in the item being exchanged.

When coinage came into use with the ancient Greeks, commodity markets gave free rein to complicated calculations of all sorts. Early coins were of precious metal and were directly changeable into any

CONTRACT

COMMUNICATION

◄ One role of measurement is to enable contracts to be drawn up such that they can be checked by either party. Modern stock markets and commodity markets (such as the New York Stock Exchange, shown here) offer scope for the most complex and fluctuating calculations. Prices may be quoted in several different currencies, which continually vary in value against one another.

currency. However, as some coinages took on higher and higher fractions of base metal, direct exchange was no longer possible and coinages acquired different relative values against each other. Today, the commodity page of a newspaper shows a confusion worthy of Babel. There are many different commodity units and money units; all the individual values change with time as supply and demand vary, and the relative values of the currencies also fluctuate.

Most commodities are listed by weight. However, oil, at so many dollars per barrel, and grain, listed in cents per bushel, show archaic interest in using volume measurements. Once, these were easy to reproduce and use. It is simpler to calculate the volume of a cart than to weigh it full and empty. Weight measurements were introduced originally for trading in small and irregular shapes of precious metals, such as coins (➧ page 31).

Measurements for control

By counting his subjects and their wealth, a ruler can calculate the taxes he considers his due; by counting physical events, a scientist can make accurate observations of the natural world from which certain principles can be deduced which allow us to control events, perhaps by making a tool that takes advantage of those principles. The stakes can run high. Blind-landing an aircraft calls for the pilot to believe in the measurements shown on the instrument panel. Many pilots have lost faith, overruled their instruments and discovered too late that they were themselves disoriented. To land the aircraft, the critical measurements are height, course, alignment with runway, rate of descent, ground speed, aircraft attitude, and airspeed. At modern airports most of these are measured from the ground by means of radar (➧ page 84), and are radioed to the pilot; the last two depend on airborne measurement.

CONTROL

▲ Plans and blueprints are a method of communicating information about dimensions simply and unambiguously; they can also provide a standard against which the proportions of the building itself can be checked.

◄ The control room of a large radio telescope dish provides the operator with all the information needed to use this huge but highly sensitive instrument.

◄◄ The air traffic control room at Pittsburgh Airport relies on radar-based measurements to track all the aircraft in the vicinity and control their movements through the crowded air space.

Mass-production techniques required a method of duplicating measurements and dimensions exactly, and were pioneered by the armaments industry

Measurement and industry

Although craftsmen have for long been capable of making individual instruments to a high degree of accuracy, until the mid-19th century industrial products were made much more crudely. In the steam engines that powered the Industrial Revolution (➧ page 129) any critical parts that needed to fit accurately were adjusted by hand. Since compatibility of adjacent parts was all that mattered, there was no attempt to measure against standard units.

Ironically, however, these hand-built, non-standardized machines made accurate small-scale measurement a necessity. Their power could permit the mass production of all sorts of goods in factories; but mass production requires components made in different places or at different times to be assembled into a single item. It insists on a dimensional precision, so that parts are made to match before they are brought together for assembly. This means that they have to be measured accurately.

For a long time progress in the field of industrial precision was hampered by a lack of instruments. The calipers and engraved rulers used by craftsmen could offer tolerances no better than 0·25 millimeters. Although it was not until the later 19th century that the now-common hand-held screw-thread micrometers came into general use, the use of uniform, fine-pitch screw threads was the key to short-distance measuring. If one whole turn advances a screw by exactly one millimeter, then it becomes simple to detect a variation of 0·01 millimeter, even though it is too small a distance to be seen unaided. Such mechanical amplification is at the heart of position setting on machine tools (➧ page 19), and ensures that an object of any designated size can be manufactured.

▲ James Watt's micrometer of 1772 is brass, with steel for the anvils. The moving anvil is driven by a thread giving 19 turns to the inch, the larger dial gave 51 turns an inch. Divisions of 0·001 inch are marked.

► Armaments factories led the way in developing techniques of measurement for mass-production. The grindery of the Royal Small Arms factory, Enfield, Britain, is shown here in a print of about 1890.

Making things fit

The first successful application of industrial measurement techniques to mass production was in the firearms factory of Eli Whitney in the 1830s. Hand-made master gauges were made for every critical dimension of a firearm; the gauges themselves did not necessarily have to be measured with great accuracy in inches or millimeters, since what was important was that parts made to match them would themselves fit together.

This system worked well when manufacture was confined within a single factory, but expanding production to another plant required a duplicate set of master gages, and at this stage accurate measurement *was* required. And even within a single factory, the master set of gauges might be too precious for use, and duplicates had to be made. Whitworth made end-gauges of standardized lengths, and accurate screw instruments allowed work and gauges to be compared. Standard plug and ring gauges were also prepared to allow for convenient diametrical measurement of cylinders. By the end of the 19th century, sets of slip gauges were developed by Johansson in Sweden – flat, parallel surfaces that could be "wrung" together to build up defined lengths. A set of 81 blocks allowed the range 0·2 inches to 18 inches to be covered in increments of 0·0001 inches.

All these advances were led by the armaments industry, which pioneered mass-production techniques. Millions of men caried identical guns to the trenches of Flanders and were provided with bullets to fit them. The killing rates of the battles of Passchendaele and the Somme were only achieved through the success of these methods of accurate measurement and mass-production.

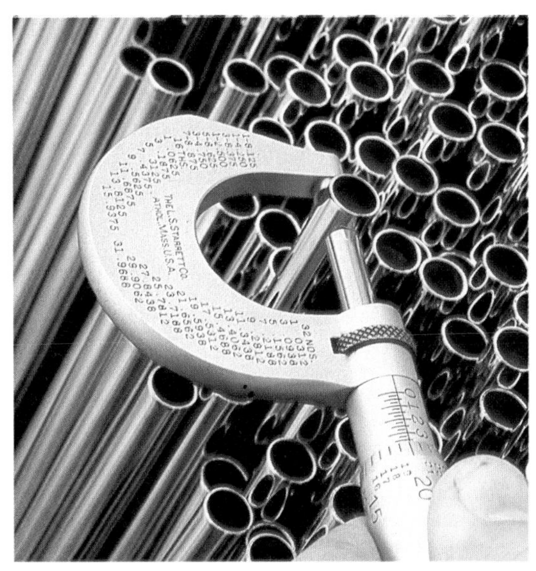

▼ ***The master gauges, including slip gauges and go-not-go gauges, used to manufacture the Royal Enfield rifle of 1853, a standard British infantry rifle for the second half of the 19th century.***

▲ ***A modern micrometer has the same fundamental features as Watt's. The spindle (moving arm) is driven forward towards the anvil (fixed point) by a carefully ground and calibrated screw thread.***

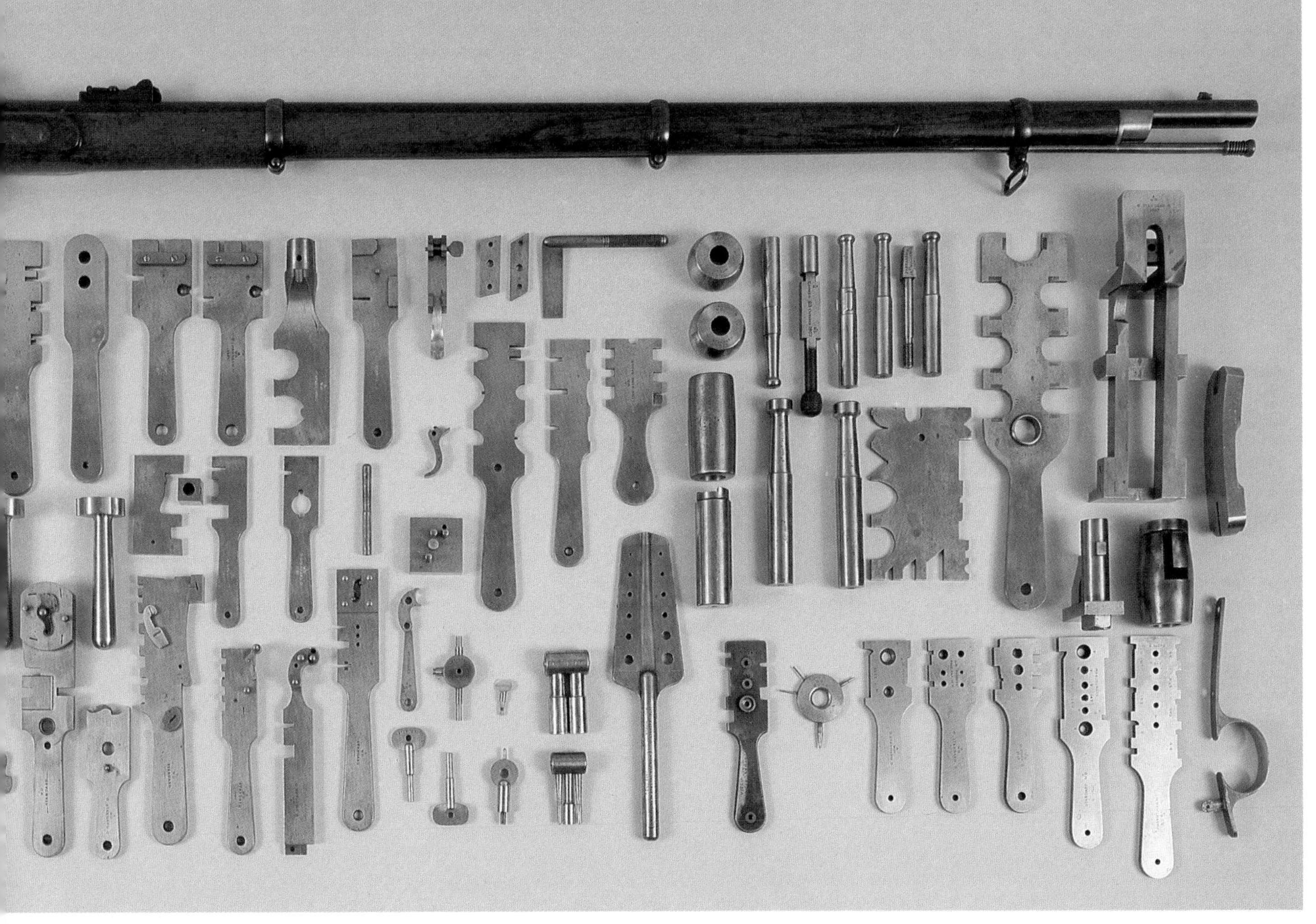

The origins of the metric system

At the time of the French Revolution in 1789, the Académie des Sciences in Paris was an exclusive elite of essentially aristocratic scientists. At that time scientific research needed time and money, and received little support from commerce. The abrupt change of political power surely alarmed the Academicians, but they undertook serious attempts to liberalize their institutional structure. Even so the Académie was disbanded by decree of the National Assembly in August 1793; yet in the four years between Revolution and disbandment it gave its most famous legacy to the world, the metric system.

Units of length and weight in 18th-century France were a confused array of poorly interrelated local standards. Fair trading was not easy, and navigation, surveying, engineering and science were all hampered by the difficulties in communicating data precisely. Propositions for well-defined standards had often been urged, but were always frustrated until reform was dictated by revolutionary ideals. The Académie set up a working committee and in May 1790 a bill was put before the Assembly proposing as a linear standard the length of a one-second pendulum, measured at 45° latitude, half-way from Pole to Equator, and conveniently set in southern France. Great confidence was implied in the definition of the second; this was justified since clocks accurate to a few seconds per day were available. Accuracy of one in 10,000 would define length to about the limits of resolution of the human eye. (The length standard is today once again based on time *▶ page 20.)* *A decimal system of counting was proposed, and weight was to be connected to length by defining a unit based on a standard volume of water at a given temperature; the chemist Antoine Lavoisier (1743–1794) had recently made such measurements.*

A debate ensued. Some argued that counting in 12s was more convenient than a decimal system, and that reform should be limited to standardization of existing units, to avoid sacrificing the convenience of tradesmen to the exigencies of perfectionist science. The scientists' argument for a unified set of standards based ultimately on a "natural" constant, the period of the Earth's rotation, won the day.

With agreement won to a new system, the Académie reconstituted its weights and measures committee. Inspired by the mathematician the Marquis de Laplace (1749–1827), this switched the linear standard away from the second pendulum to a ten-millionth part of the Earth's polar quadrant through Paris. It made sense to couple decimal measures of the right angle (100grad) and a quadrant of latitude so that linear and angular measures on the Earth's surface were interchangeable: 100km along the Earth's surface would equal 1grad along a great circle, saving small corrections of the non-sphericity of the planet.

The new proposal was enacted in a single day, on 26 March 1791. The metric system with its decimal structure was born, and its survival was probably ensured by the new land registry that was made essential by the land reform of the Revolution.

▲ Antoine Lavoisier (1743–1794), the pioneering chemist seen here demonstrating oxygen in air, was instrumental in having the metric system agreed by the Académie.

Prefixes for the SI units

Prefix	Power	Symbol
Tera-	10^{12}	T
Giga-	10^{9}	G
Mega-	10^{6}	M
Kilo-	10^{3}	k
Hecto-	10^{2}	h
Deca-	10^{1}	da
Deci-	10^{-1}	d
Centi-	10^{-2}	c
Milli-	10^{-3}	m
Micro-	10^{-6}	μ
Nano-	10^{-9}	n
Pico	10^{-12}	p
Femto-	10^{-15}	f
Atto-	10^{-18}	a

Units of the SI system

Quantity	Name of unit	Unit symbol
Length	Meter	m
Mass	Kilogram	kg
Electric current	Ampere	A
Temperature	Kelvin	K
Luminous intensity	Candela	cd
Plane angle	Radian	rad
Solid angle	Steradian	sr
Frequency	Hertz; cycle per second	hz
Force	Newton; kg-m per second per second	N
Pressure	Pascal; newton per square meter	Pa
Viscosity	Poiseuille; newton-second per m^2	Pl
Work, quantity of heat	Joule; newton-meter	J
Power, heat flux	Watt; joule per second	W
Quantity of electricity	Coulomb; ampere per second	C
Electromotive force	Volt	V
Electric resistance	Ohm	Ω
Electric capacitance	Farad	F
Magnetic flux	Weber	Wb
Inductance	Henry	H
Magnetic flux density	Tesla; weber per square meter	T
Luminous flux	Lumen	lm
Illumination	Lux; lumen per square meter	lx

► *André-Marie Ampère (1775–1836) was a French physicist who worked on electrodynamics. The amp, the unit of electric current named for him, is defined as the constant current which, if maintained in two conductors placed parallel, 1m apart, would produce a force between them equal to 2×10^{-7} newtons per meter of length (▸ page 42).*

▼► *The German physicist Heinrich Rudolf Hertz (1857–1894) confirmed J.C. Maxwell's theory of electromagnetism and studied electromagnetic waves (▸ page 73). The hertz, defined as the frequency of a phenomenon that has a period of one second, is named for him.*

▲ *The French philosopher Blaise Pascal (1623–1662) carried out original work in science, mathematics and religious thought. His work on barometric pressure and the equilibrium of fluids is commemorated in the SI unit of pressure, defined as resulting from the force of one newton acting over an area of one square meter.*

◄ *The United States physicist Joseph Henry (1797–1878) made advances in electromagnetism, discovering the principle of self-inductance and the principle of the induction motor. The henry is the unit of inductance, defined as the inductance of a closed loop that gives rise to a magnetic flux of one weber for each ampere of current.*

▲ *The English scientist James Joule (1818–1889) studied the equivalence of thermal and mechanical energy and formulated the law of thermodynamics relating to the conservation of energy. The unit of work – the energy equivalent to the work performed as the point of application of a force of one newton moves through one meter – is named for him.*

◄ *The British physicist William Thomson, 1st Baron Kelvin (1824–1907) worked in thermodynamics, and established the absolute scale of temperature named for him. One kelvin is 1/273·16 of the triple point of water (the temperature at which, in a sealed vessel, the three states – liquid, solid and vapor – coexist).*

The International System of Units

In 1960 the scientific community established an complete system of units based upon the meter, kilogram and second standards. The "système international" (SI) also defines units in fields of observation beyond the purely mechanical. Kelvins of temperature, amperes of electrical current and candelas of luminous intensity are examples. Gradually these SI units are penetrating into engineering, but they are still a long way from global usage. The Apollo missions that put men on the Moon still used pounds, inches and degrees Fahrenheit.

The principle of science that asserts that energy can be converted from one form to another is used to connect distinct domains of observation. Electrical instruments measuring ampères and volts (▸ page 41) permit the power consumption of an electric machine to be assessed in watts; the power output of an athlete driving onto an instrumented treadmill can be computed in the same terms, having made measurements of force (newtons) and speed (meters per second).

Measurement and automation

Building complex machines such as automobiles involves making the parts first and assembling them afterwards. The two processes can be done in different plants, or even in different countries. The mass production methods of the 19th century still sought for precision through manual skills: craftsmen handling precision machines tools, checking their work against gauges or using micrometers. Production rates were limited by human capabilities and the availability of skilled machinists.

The breakthrough to much greater production rates came from having machines that could control themselves to do these critical tasks. Such a machine needs "eyes" to detect positions of tool and work; a "brain" to decide what action should be taken; and "muscles" to undertake the action. In modern machine robots, the brain is a computer with a program, and the action is realized by means of electric motors controlled by feedback devices; but it is a measuring problem to provide the "eyes" (➧ page 127). Position sensors are available based on a wide variety of transduction methods (➧ page 41), but sensors using the same principle as the micrometer are among the most common. Linear positions are set up by lead-screws, and the angles they have turned is proportional to distances moved from a predefined starting point. What is then needed is a code suitable for sending data to a computer to replace the engraved lines and numbers that human brains recognize. Optical angular sensors convert angular displacements into electrical voltage pulses using photocells; the rest of the job is then defined in purely numerical terms.

Lead-screws suffer from the problem of backlash which comes from wear, and in some machines sensors which relate directly to the positions of work and tool may be preferred. Systems sensitive to linear displacement include photocells looking at Moiré fringes (➧ page 41), coils sensing alternating magnetic fields, and laser interferometers.

In the industrial environment accuracy is not always the overriding priority. For most applications tolerances of 0·01 millimeter may be sufficient. Reliability, on the other hand, is paramount, since faulty work and shut-down machines cost money.

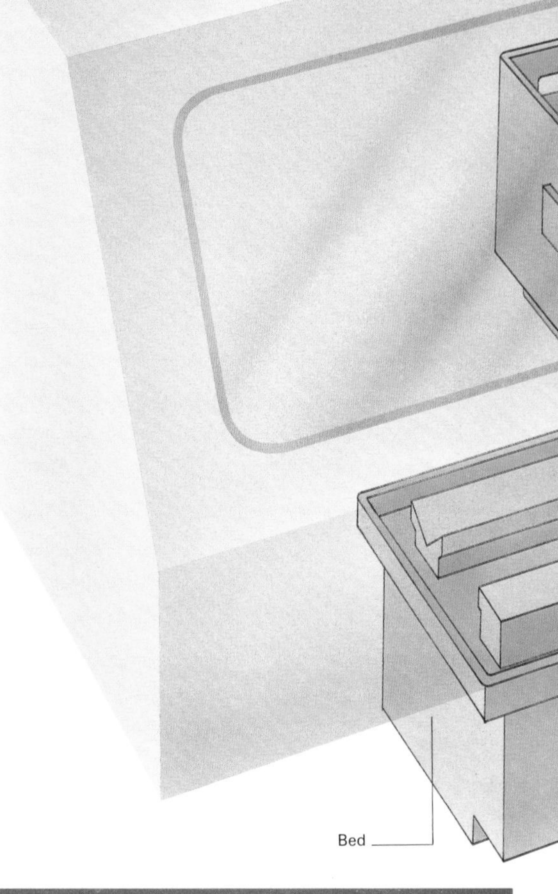

***►** This grinding machine, numerically controlled by computer, is a typical machine tool of the 1980s. The sequence of operations can involve different spindle speeds, grinding-wheel speeds and feed rates. Position sensors inform the computer of progress and dictate the time frame in which the program is executed.*

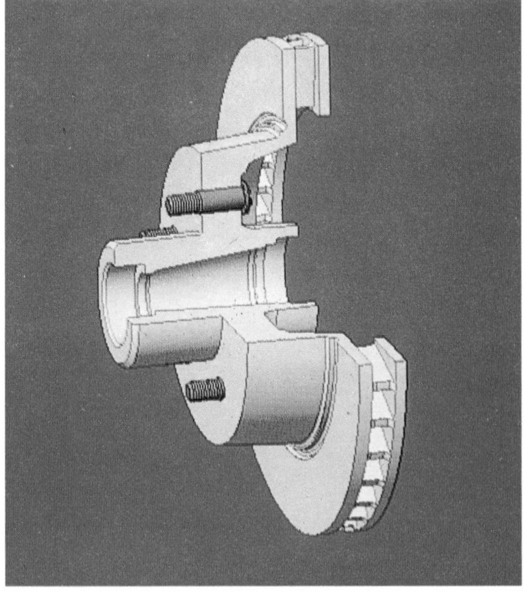

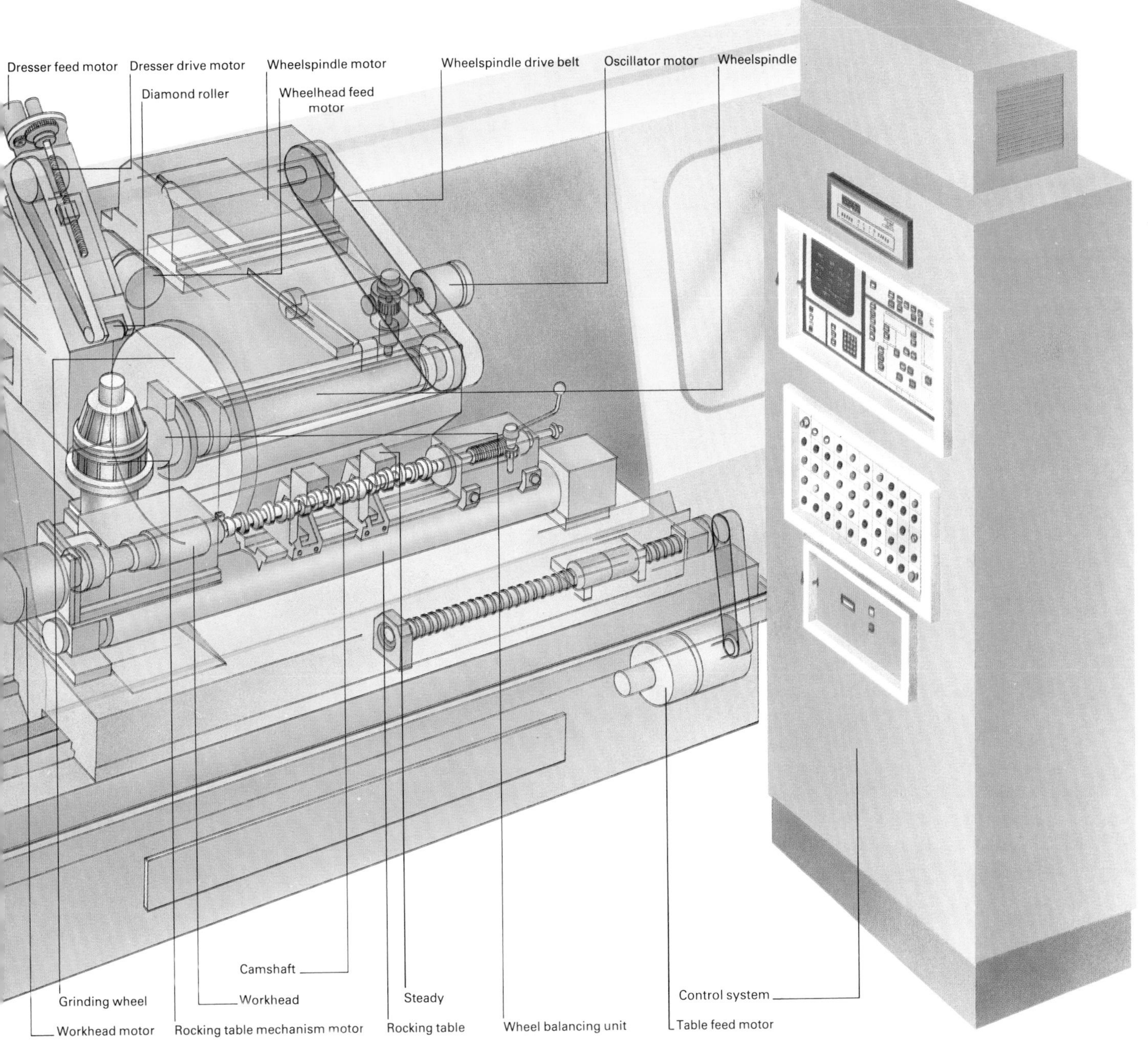

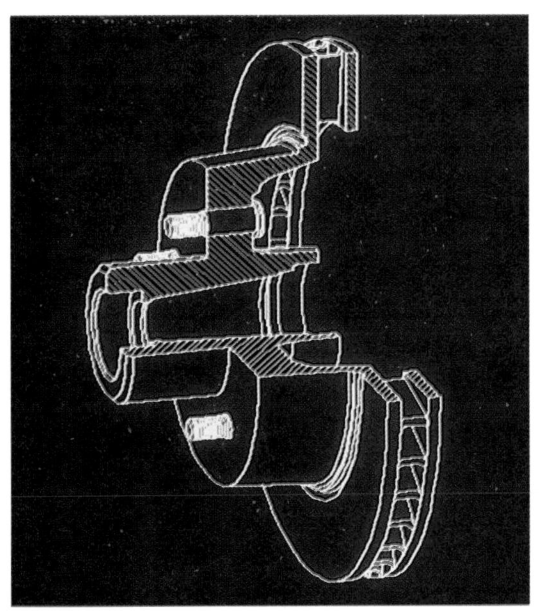

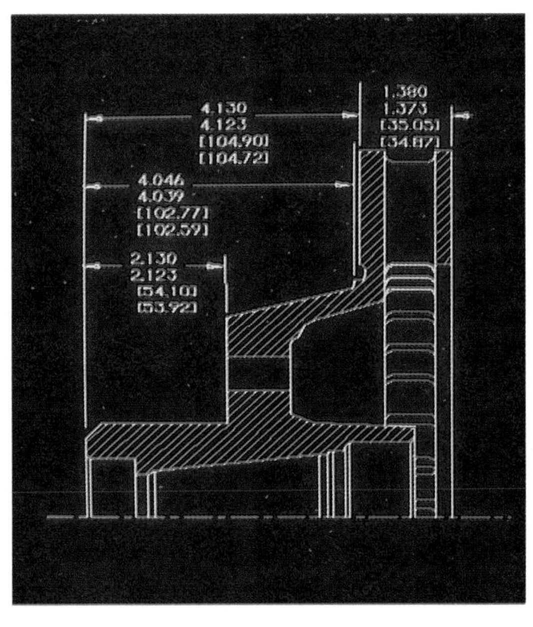

◀ ***Computer-aided designs (CAD) for industrial components, such as this disk brake rotor assembly, can be produced at interactive terminals. The operator first creates a "wireframe" profile of the component and of its internal features and cavities. This model can be rotated on screen in three dimensions, and can be altered simply; a change to one dimension means that all other measurements are recalculated automatically. Special solid modelling software converts the wireframe geometry into wireframe definitions of the facets, and shading is added to make the image more readable. The coordinates for the assembly are stored in the computer, and may serve as the standard to be matched in manufacture.***

Time and length are the two most basic phenomena to be measured and most devices measuring other sensations require reliable standards to be set for these. The achievement of agreed and precise measurements in these fields has been a matter of concern to scientists for centuries, and ever more accurate definitions have been achieved as the technology of measuring has advanced.

Time standards

The solar day and the lunar month afford natural blocks of time. It is easy to count these into larger blocks – weeks, years, – and to devise calendars that are fairly accurate (➧ page 28). Subdividing these units, on the other hand, is not so easy, since it requires a method of determining whether the separate, small, units of time are equal or not. In the early days of clock-making, subjective judgements of regularity had to be used (➧ page 24). Even after the development of the pendulum it was not self-evident that its duration of swing was absolutely constant for a given length. Scientific reasoning offers objective grounds for believing in the regularity or otherwise of phenomena, and modern time standards use these justifications.

For several centuries the basic standard of time was the sidereal day, the time taken for the planet to complete one spin on its axis. Subdivision into standard seconds was measured by carefully controlled pendula. The sidereal day is more constant than the solar day, which varies because the orbit of the Earth around the Sun is elliptical rather than circular. However, the solar day is the familiar calendar interval, and is the usual span divided into 86,400 seconds. The constant pendula used to define seconds had to be calibrated for frequency and regularity against sidereal observations.

Even the sidereal day is not constant. There are tidal effects on the Earth's spin resulting from the motion of the Moon and the perennial variation of the Sun's distance. Also there is a relative motion between the Earth and the reference star. So a new belief of constancy was chosen to be the time standard. An artificially induced change of internal state in atoms of the metal cesium makes them radiate waves of constant frequency (➧ page 27); a second is now defined as the time for 9,192,631,770 of these waves. What has been so defined is a frequency (pulses per second), and electrical waves of exactly that frequency are generated in response to the cesium atoms. Electrical pulses can be counted exactly by electronic devices (➧ page 46), and a frequency can be used to define longer time intervals very precisely. The sidereal day time standard was defined to about one part in ten million; the new standard is fixed to one in a billion.

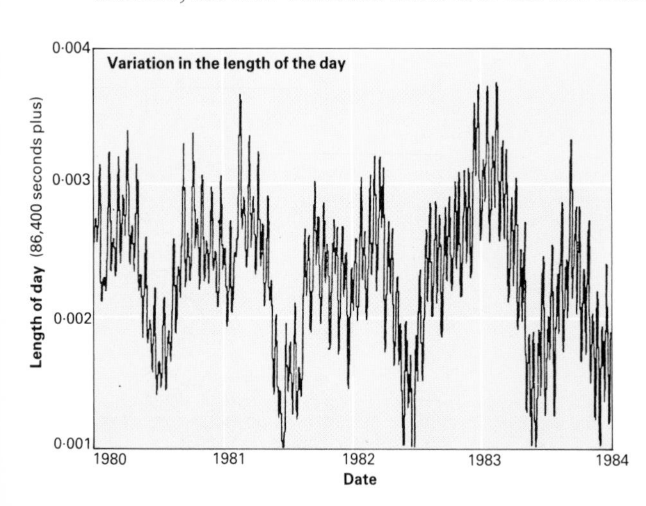

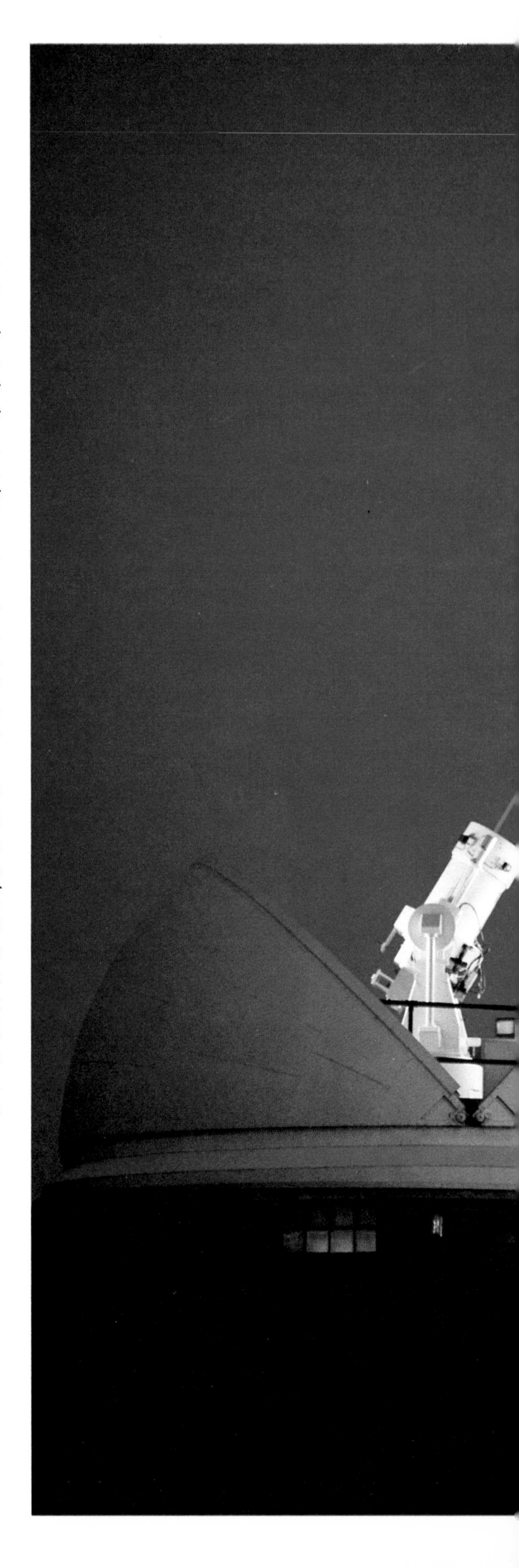

◀▶ ***The length of the sidereal day has been shown to be inconstant owing to variations in the rotation of the Earth. This has been measured by using a laser beam, such as this one at the Royal Greenwich Observatory, Britain, to track the satellite Lageos I, since January 1980. The surface of the satellite is covered in mirrors to reflect the beam. This technique has shown that the length of the sidereal day varies by 0·001 seconds over periods as short as two weeks.***

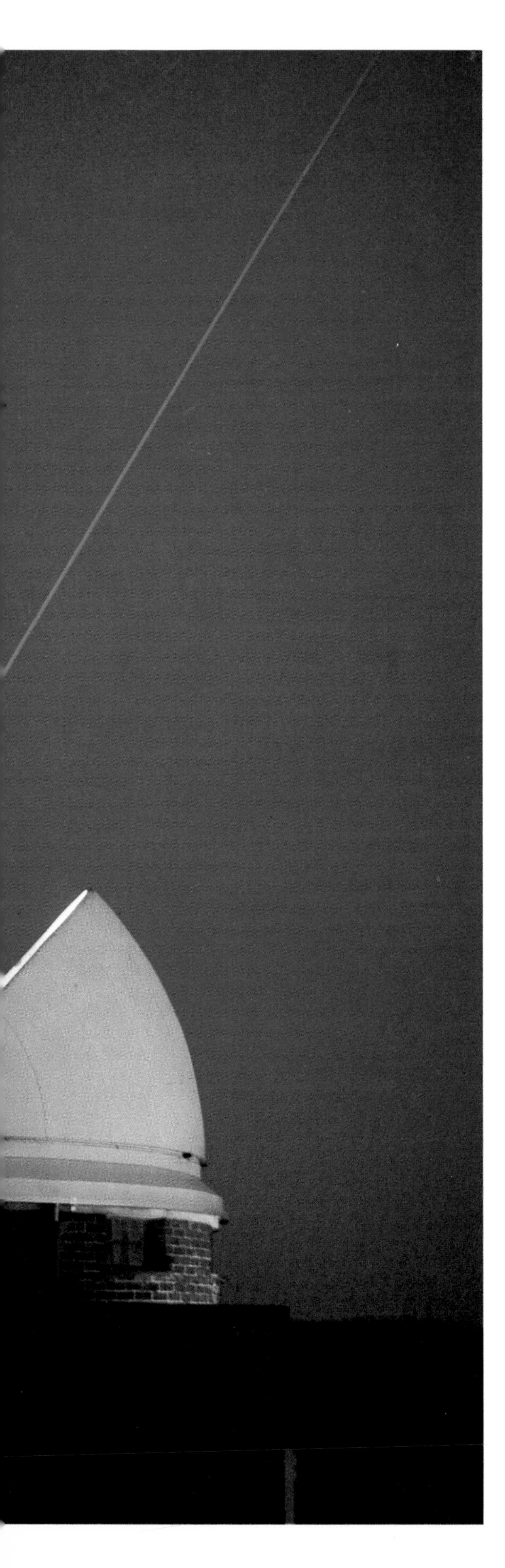

Measuring length

Standards of length, like those of mass (➧ page 31), start as arbitrary choices. As soon as unconstant measures, such as hand spans or paces, became inconveniently variable, a "special stick" for other measures to be calibrated against would be chosen and carefully preserved.

The *Académie des Sciences* in 1793 produced its own "special stick", comprising a rod of platinum, the separation of the parallel ends of which – one meter apart – purported to be one ten-millionth of the polar quadrant of the Earth passing through Paris. It was intended to use the constancy and reproducibility of this astronomical measure to define the standard meter and regard the platinum rod merely as a "state-of-the-art" expression thereof. By 1799, however, the *mètre des archives* itself was adopted as the primary standard.

In the 1870s work was undertaken to reproduce a suite of standard meters for international distribution, and this resulted in 1889 in the adoption at an international conference of a new, carefully designed, standard measure. But as physics advanced it became apparent that the wavelengths of the discrete colors of light emitted by atoms are far more constant than any bar of metal. Moreover, a standard based on a spectroscopic measure (➧ page 106) could be set up cheaply, quickly and accurately in any laboratory. In 1960 the meter was deemed to be 1,553,164·13 wavelengths of a red line in the spectrum of cadmium, as emitted from a lamp built to a standard specification.

Even this excellent standard has now been set aside. The theory of relativity has shown that the velocity of light in a vacuum is an utterly constant property. Since the standard of time is defined to an even better specification than the red cadmium line, the meter can be more accurately defined in terms of the distance traveled by light in a specified interval of time – 1/299,792,458 of a second. Thus the advantages of electrical measurement of time are brought to length measurement.

► An Egyptian yardstick from the 7th century BC.

▼ The experiment to define the meter in terms of the speed of light. The helium-neon laser light is stabilized with methane or iodine vapor, and its frequency measured. From this the distance it covers in a specified length of time is calculated; there is no need to measure the actual distance traveled by the light waves.

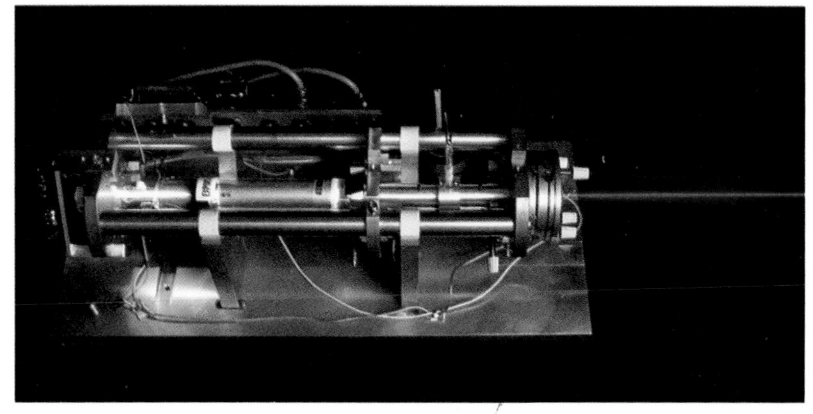

See also
Measuring Time 23–30
Measuring Weight 31–36
Surveying and Mapping 37–40
Measuring with Electricity 41–46
Computer Software 123–128

Units derived from time and length

The dimensions of the world appear to us to number three of space and one of time. Area and volume are combinations of lengths; units for their measurement need a length standard. Strictly speaking, measurement of angle does not, because it is a ratio of lengths. Instead it has its own measures: 90 degrees, 100 grads or $\pi/2$ radians (1·57) making a right angle.

Bringing time into combination with length allows a whole new set of measurements to be made. Speed in kilometers per hour is the most familiar example, with acceleration as kilometers per hour per second. Such "mixed-up" units are often easier for us to imagine than more "rational" ones. Using the agreed standards of length and time, speeds ought to be written as meters per second, and accelerations as meters per second per second.

The rate at which things happen in time often needs to be described, and saying "per hour" or "per second" allows this to be expressed easily. A carpet layer works at a rate of ten square meters per hour; a river flows at a rate of 5,000 liters per second. Even for everyday language these derived units require standards of both length and time. If there were no such standards, a motorist's speed would have to be compared directly with that of a "standard speed" for a speeding prosecution to be sustained: as it is, a calculation from measurements taken with a radar device on the roadside will suffice.

▲ The acceleration of a ball as it bounces can be measured by means of high-speed photography using a strobe (▶ page 88). For larger objects such as submarines or spacecraft, acceleration can be measured by sensing the inertia of a weight suspended in the vehicle.

The units of mass and force

If mass is introduced as a third fundamental standard, more units can be derived. Mass is not so much a dimension of the world as a property of some stuff within it. Mass is the property of matter which has mechanical influences since, to scientists at least, it tells of inertia (a "reluctance to change motion").

The English physicist Isaac Newton (1642-1727) first made the useful connection between the rather abstract quality of mass and the much more countable sensation of force. He showed that mass is force per acceleration. This allows us to define a force. Standards of mass, length and time exist (▶ page 31), so the obvious unit of force is a force which, if it acted on a one-kilogram mass, would make its speed change by one meter per second every second. This is the scientific unit of force, known as the newton (and, accidentally but felicitously, is about equal to the weight of one apple). Weight is therefore a force (the amount needed to effect a pick-up) and reflects just one mechanical situation in which a mass is reluctant to begin moving.

Once force has been established, many new measurements can be made. Energy equals force times distance; power is force times speed; impulse is force times time. The viscosity of a liquid is the shear force per speed gradient, and its surface tension is measured as force per length. Pressure and the strength of a material are both measured as stress, which is force per area, and the rigidity of a beam or bridge is measured as the force sustained per unit deflection.

▼ A Schlieren photograph of the shock waves of a pointed object tested in a wind tunnel. Testing aircraft for drag and turbulence at different windspeeds and air pressures is an essential part of design, with implications for safety and economy.

Measuring Time

2

The earliest devices for measuring time...Mechanical escapements...Pendulum and spring-driven clocks and watches...Electrical and electronic watches...The cesium clock...The calendar...Measuring time backwards: radiocarbon dating...PERSPECTIVE...The marine chronometer...Greenwich Mean Time and Universal Time...Relative time

Knowledge of the correct time is important in many walks of life. It allows the communications industries, from airlines to broadcasting, to coordinate their business; it is vital to navigators, who need to know time precisely to establish their exact positions. Indeed, it was the growth of long-distance sea travel from the late 15th century that led to the development of accurate clocks and chronometers.

The measurement of time depends on the observation of an event that occurs at regular intervals. From the dawn of civilization, time has been linked inextricably with astronomy – the rising and the setting of the Sun and stars, and the waxing and waning of the Moon, are regular phenomena. The sundial, which was in use several centuries BC, gives a measure of daytime, marked by the passage of a shadow across a dial as the Sun moves across the sky.

As more and more accurate clocks have been devised, so the definition of the basic unit of time has become more precise (◆ page 20). The clocks of antiquity, such as the water-clock or clepsydra, were not very accurate. In the 14th century AD, the first mechanical clocks were made, driven by falling weights with a mechanism to transfer their pull slowly and regularly to the clock train. Then in 1581, the Italian mathematician Galileo Galilei (1564-1642) discovered that the natural swing of a pendulum repeats at regular intervals, depending not on the angle of the swing but on the length of the string that holds the bob. A pendulum about 990 millimeters long marks out seconds with its swing. Transferring this time unit to the clock train was the next step.

▲ *The sundial, shown on a 16th-century building in Cambridge, was the earliest form of clock; its limitations are obvious, but portable versions were made, with a compass to aid orientation.*

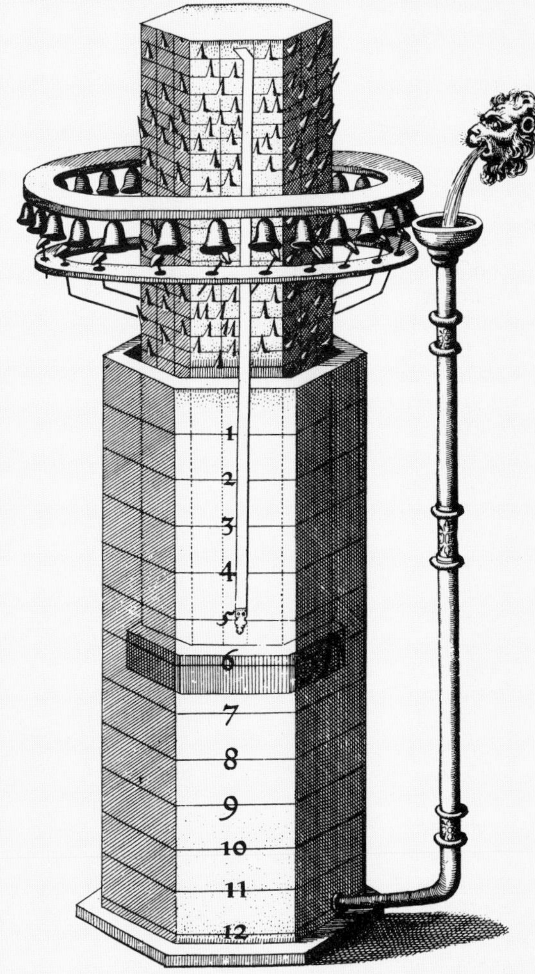

▼ *A 17th-century design for a water-clock. Water enters the tank from the lion's mouth, then escapes; as the water level falls the spiky object drops down, striking the bells to ring the hours.*

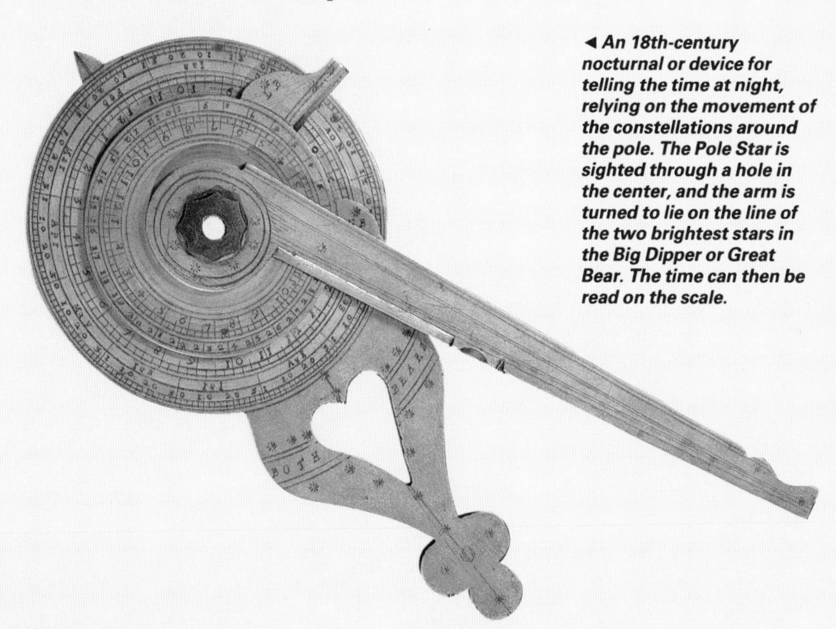

◀ *An 18th-century nocturnal or device for telling the time at night, relying on the movement of the constellations around the pole. The Pole Star is sighted through a hole in the center, and the arm is turned to lie on the line of the two brightest stars in the Big Dipper or Great Bear. The time can then be read on the scale.*

Chronology

3500 BC Earliest sundials used in Middle East
100 BC Tower of the Winds, Athens, contained eight sundials, to indicate time accurately throughout day
1st century AD Romans built waterclocks with pointer moving across dial
1315 Earliest public striking clock installed, in Milan, Italy
1386 Tower clock of Salisbury Cathedral, England, installed; oldest survivng mechanical clock
1500 Peter Henlein, a German locksmith, built spring-driven small clocks
c.1520 Fusée invented by Jakob the Czech of Prague to compensate for drop in tension as the mainspring unwinds
1581 Galileo Galilei demonstrated that duration of pendulum's swing is dependent on its length
1656 Christiaan Huygens applied pendulum to clocks
1660 Anchor escapement invented by Robert Hooke in England, who also devised balance spring for watches
1670 Seconds pendulum introduced by Englishman William Clement
1704 Jewels used in watch movements for first time
1715 Englishman George Graham devised deadbeat escapement for precision clocks
1721 Graham introduced mercury pendulum which compensated for the effect of temperature variation on pendulum length
1735 Englishman John Harrison's first marine chronometer completed
1765 Englishman Thomas Mudge invented lever escapement for watches

The Dutch physicist Christiaan Huygens (1629–1695) developed the first pendulum clock around the middle of the 17th century. The pendulum controls the rate at which a weight falls, or a spring uncoils, by means of a mechanism called the escapement. This releases the weight at regular intervals, allowing energy to escape – hence the name escapement. There are a number of different kinds of escapement, but most consist of a toothed escape wheel and a curved bar bent at the ends to form "pallets" that alternately engage the teeth of the escape wheel. The bar is connected via a spindle to the pendulum, so that as the pendulum swings one pallet releases the wheel and the other engages it again. This allows the escape wheel to rotate one notch at a time, and also transmits an impulse to the pendulum, which keeps it swinging. The escape wheel is connected to the main wheel of the clock train, which is driven by the weight or spring via a series of pinions (gears), spindles and wheels. The train is effectively a gearbox, arranged to pass the drive to the motion work, which converts it into the sweep of the hands around the dials.

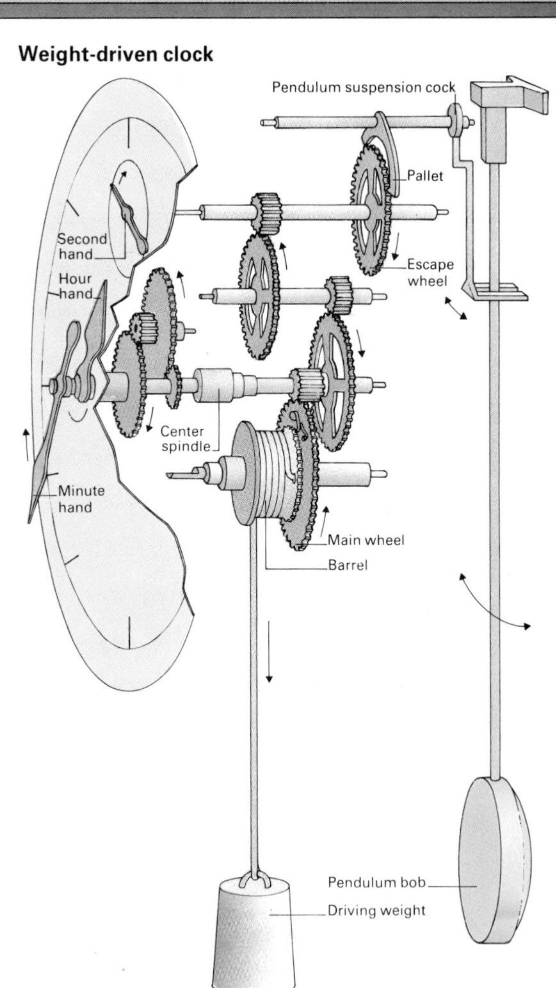

▲ *The clockmaker's skill lay in making parts that ran with little wear. The arbors or shafts and the pinions were usually of steel, and the plates in which they turn were brass, as were the wheels. Wheels were cut on dividing engines.*

◄ *A 17th-century lantern clock. The notched wheel controls the striking mechanism. It revolves with the hour hand and the clock strikes when the locking piece is raised.*

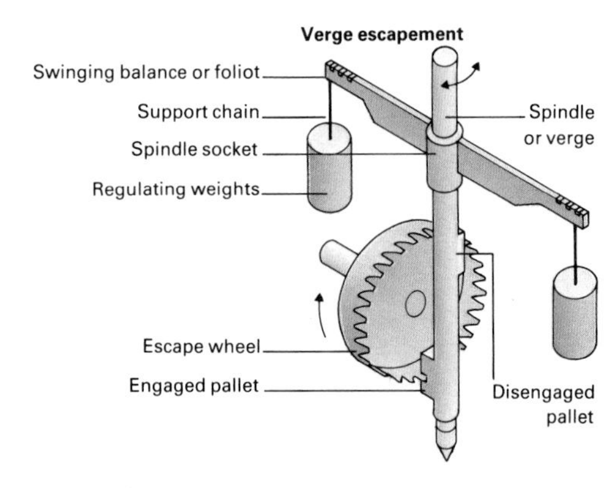

▲ *The verge escapement relies on the swing of a suspended arm, adjusted by movable weights. As it swings pallets alternately engage and release the teeth on the escape wheel.*

1840 Electric clock developed in Britain by Alexander Bain and Barwise
1840 Earliest battery-driven electrical clock, though still with spring and pendulum
1855 Invention of chronograph, able to record time intervals as well as continuous time
1884 Greenwich Mean Time adopted as international time standard
1906 First self-contained battery-driven clock
1918 Mains-electric clock invented by H.E. Warren in USA, counting frequency of AC mains current
1921 Free-pendulum clock built by Briton W.H. Shortt, accurate to one second in 10 years, and powered electromagnetically; used in astronomy observatories
1929 First quartz clock designed by J.W. Horton and W.A. Marrison in USA; accurate to one second in three years
1944 Quartz clock installed at Royal Greenwich Observatory to tell Greenwich Mean Time
1948 First atomic clock, based on ammonia atoms, invented in United States
1953 Electronic watches became available
1953 Possibility of cesium clock investigated in United States
1955 Cesium clock installed at National Physical Laboratory, Britain
1957 Battery-driven electric watch introduced, using hairspring and balance wheel
1962 Hydrogen-based atomic clock built at Harvard University, USA, accurate to one second in 100,000 years
1975 Digital clocks and watches with liquid-crystal displays marketed

The marine chronometer

As European navigators traveled across the oceans, it became necessary for them to calculate their position exactly. From early times the sextant and astrolabe had been used to enable navigators to calculate the height of the Sun, Moon and stars above the horizon. The curve of the Earth's surface means that noon is four minutes earlier or later for each degree west or east of the meridian; this offers a method of calculating longitude if the navigator has an accurate timepiece to compare local time with meridian time. The movement of the ship means that a conventional pendulum cannot be used, and the design has to allow for changes in humidity and temperature.

In 1714 the British Longitude Board offered a handsome prize for the maker of such a clock. Fourteen years later the clockmaker John Harrison (1693-1776) began work on the problem, and in 1735 produced his first marine chronometer, which used a grasshopper escapement and a pair of linked balances terminating in heavy brass balls to counter the rolling of the ship. After successful trials he set about improving the design; his third clock was built by 1759, and a large watch version with a verge escapement and compensated bi-metallic balance appeared two years later.

The watch proved the most accurate of his four designs, and in 1761 it was taken on a five-month trip to Jamaica, during which period it lost less than two minutes, the equivalent to half a degree longitude. Another trial three years later reduced this to 54 seconds in five months. The British explorer James Cook (1728-1779) took a copy of this chronometer to the South Seas, where it proved invaluable. Harrison only received his prize money in 1773, three years before his death.

The pendulum and anchor escapement were the first breakthroughs in making accurate clocks, and from the late 17th century many refinements were introduced to make clocks precise enough for scientific work. In the early 18th century the English clockmaker George Graham (1674–1751) invented the deadbeat escapement and the compensating pendulum; this had a column of mercury on the pendulum, so that when, in hot weather, the metal of the pendulum expanded, thus lengthening the duration of its swing, the mercury column also expanded to keep the same effective length.

Perhaps the last important development to the mechanical clock was devised by a British civil engineer, W.H. Shortt, in 1921. This was the free-pendulum arrangement, and consisted of two clocks, one of which, the slave clock, ran to the frequency of the alternating current of the mains electricity and was used to synchronize the swing of the pendulum of the main clock every 30 seconds. This arrangement was accurate to one second in 10 years, and was widely used for astronomical purposes until the development of quartz and atomic clocks.

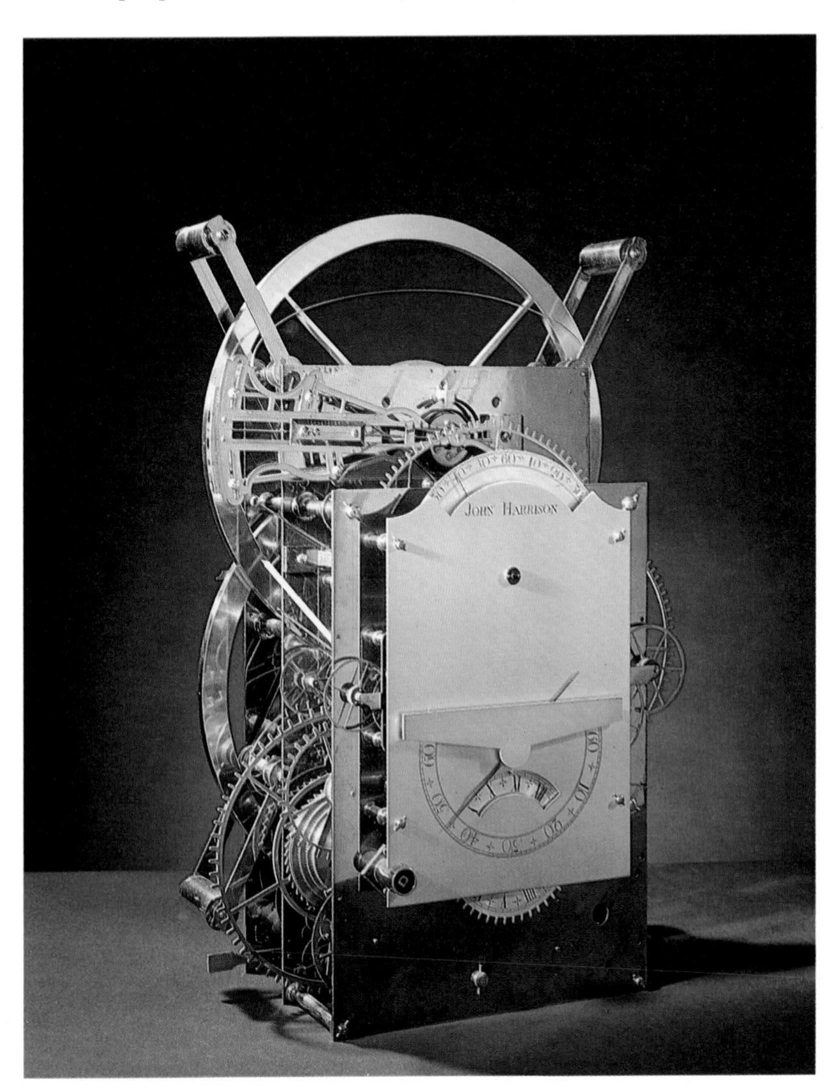

► ***John Harrison's third marine chronometer of 1759.***

Anchor escapement

Suspension cock
Disengaged pallet
Engaged pallet
Escape wheel
Pendulum
Anchor

▲ ***In an anchor escapement the anchor swings with the pendulum. As the wheel turns it pushes one pallet up until the other engages; the curve of the pallet forces the wheel to recoil.***

Deadbeat escapement

Pivot spindle
Dead face
Impulse face

▲ ***In the deadbeat escapement, used for precision clocks, the pallets are carefully designed with a positive locking action so that there is no recoil caused to the escape wheel.***

The mechanical watch

Watches were invented nearly 2,000 years after the first clocks, in the early 16th century. In a mechanical watch, a coiled mainspring provides the driving force. An escapement again controls the rate at which the spring uncoils, but in a watch the escapement is regulated by the combination of a heavy-rimmed balance wheel and a hairspring. The hairspring is coiled within the balance wheel, attached in such a way that as the wheel swings round in one direction the spring uncoils, only to coil up again as the wheel swings back. In this way, the hairspring and balance wheel together act like the pendulum of a clock, the oscillations of the wheel controlling the rate at which the mainspring uncoils.

The quartz clock

Many modern clocks and watches are controlled not mechanically but electronically by a quartz crystal. A crystal of quartz becomes slightly deformed when a voltage is applied across it, and when deformed, a voltage occurs across its face. This phenomenon is known as the piezoelectric effect (➧ page 43), and it allows a quartz crystal to maintain a voltage that oscillates at the crystal's natural vibration frequency, 100,000 times a second. A combination of electrical and mechanical gearing controls a motor so that this oscillation drives the second hand of the watch.

Quartz clocks are accurate to within one second in ten years. Many modern versions are all-electronic; the time appears on a liquid-crystal display, controlled by electrical signals from the circuitry driven by the quartz crystal.

The atomic clock

The most accurate clocks of all are atomic clocks. These do not show the time in hours and minutes, as conventional clocks do. Rather, they set a very precise rate of the passage of time, against which other clocks are calibrated and compared. They depend on the natural vibration of electromagnetic radiation emitted or absorbed by atoms.

Electromagnetic radiation comes in a variety of guises, from radio waves to X-rays and gamma rays (➧ page 73). The radiation is often described in terms of waves of energy pulsating across space at 300 million meters per second, the speed of light. Two properties characterize the waves. These are the wavelength (distance between adjacent crests of the wave) and the frequency (the number of crests passing a point during one second).

Atoms consist of concentric "shells" of electrons whirling around a central nucleus. Whenever an atom emits or absorbs electromagnetic radiation, it is the direct result of an electron emitting or absorbing energy as it moves nearer to or further from the nucleus. These "jumps" made by atomic electrons are very precise, and a particular transition always corresponds to exactly the same energy and hence the same frequency and wavelength of radiation. This fact is exploited in the atomic cesium clock.

In cesium, the outermost electron occupies a shell on its own, but its energy can be one of two narrowly separated values. Electrons spin either clockwise or anticlockwise as they whirl around the atom. The outer electron in an atom of cesium has both these options open to it, but in one case the electron's total energy is slightly higher than the other. So the electron must emit or absorb radiation if it switches from one state to another.

Spring-driven chronometer

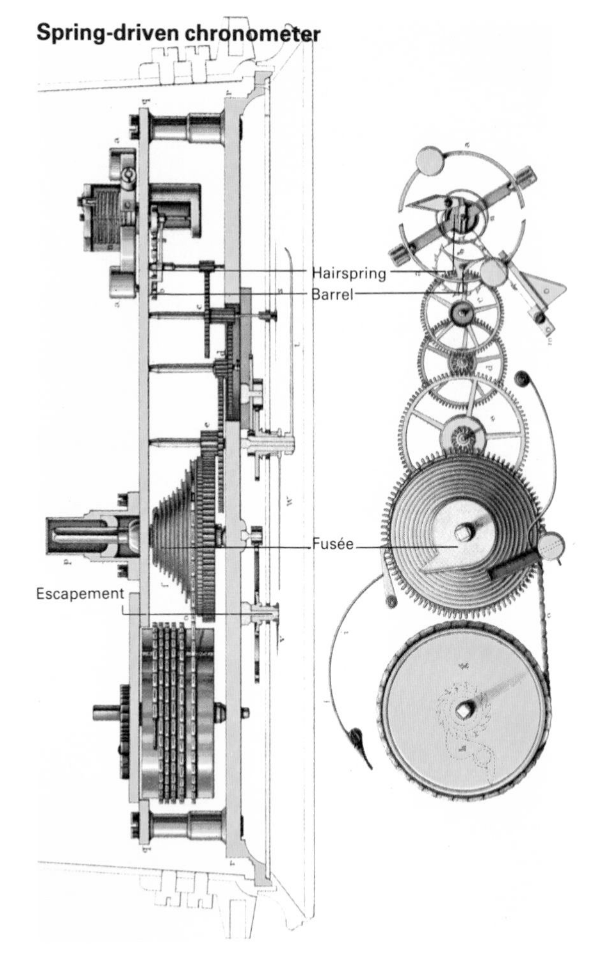

▲ The mechanical watch (like the chronometer shown here) has a balance and hairspring to replace the pendulum of a clock, and a fusée which ensures that the power of the mainspring (analogous to the weights of a clock) does not decline as it unwinds. This cone-shaped drum connects by a cord to the barrel housing the mainspring. As the watch runs the barrel revolves and winds the cord upon itself.

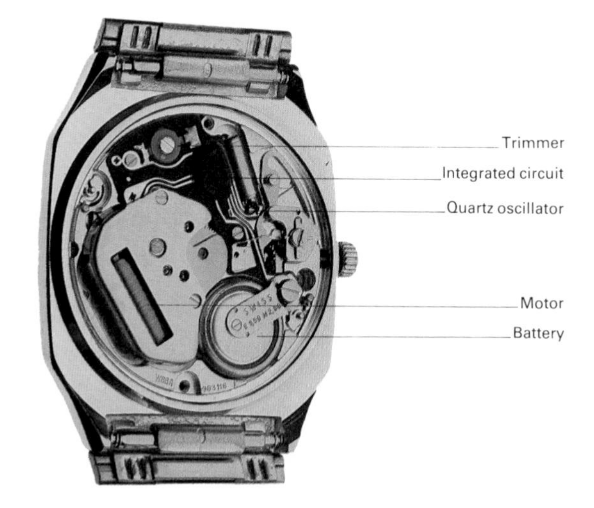

▲ In a modern quartz watch, the battery (bottom right) produces a current which excites the quartz oscillator (resembling a rod, to the top) to vibrate at its natural rate. The integrated circuit divides this to give one pulse per second. This pulse is used, in an analog watch, to drive the motor which drives the hands, or in a digital watch to produce the liquid crystal display (➧ page 46).

▲ *The cesium atomic clock used at Greenwich until 1962.*

A cesium atom clock contains a beam of atoms that issue from a sample of hot cesium metal. The atoms emerge from the metal with their outer electrons in either energy state and they travel down an evacuated tube. Half-way along the tube they pass through a cavity containing microwave radiation vibrating at a frequency set by a quartz clock. Magnets on either side of the cavity ensure that only the atoms that absorb energy in the cavity reach a detector at the end of the tube. The maximum number of atoms reaches this detector when the frequency of the microwaves exactly matches the hyperfine transition of the outer electrons in the cesium atoms, and by adjusting the rate of the quartz clock to maximize the number of atoms detected, the quartz clock is brought into agreement with the cesium standard. The frequency of the cesium radiation, which is microwave radiation, forms the basis for the definition of the second. One second is exactly 9,192,631,770 cycles of the radiation from transitions between two hyperfine levels of the ground state of cesium-133. The frequency can be kept stable to one part in 30 billion, equivalent to one second in 1,000 years.

The cesium atomic clock

▼ ***A cesium clock consists of a vacuum chamber. At one end cesium atoms are heated, pass into the chamber and are focused by magnets into the main tube. In the center is a slit resonating at a rate determined by microwave radiation; the atoms then pass through another magnet to a detector. Unless the resonator vibrates at the precise frequency of cesium, many atoms will not reach the detector.***

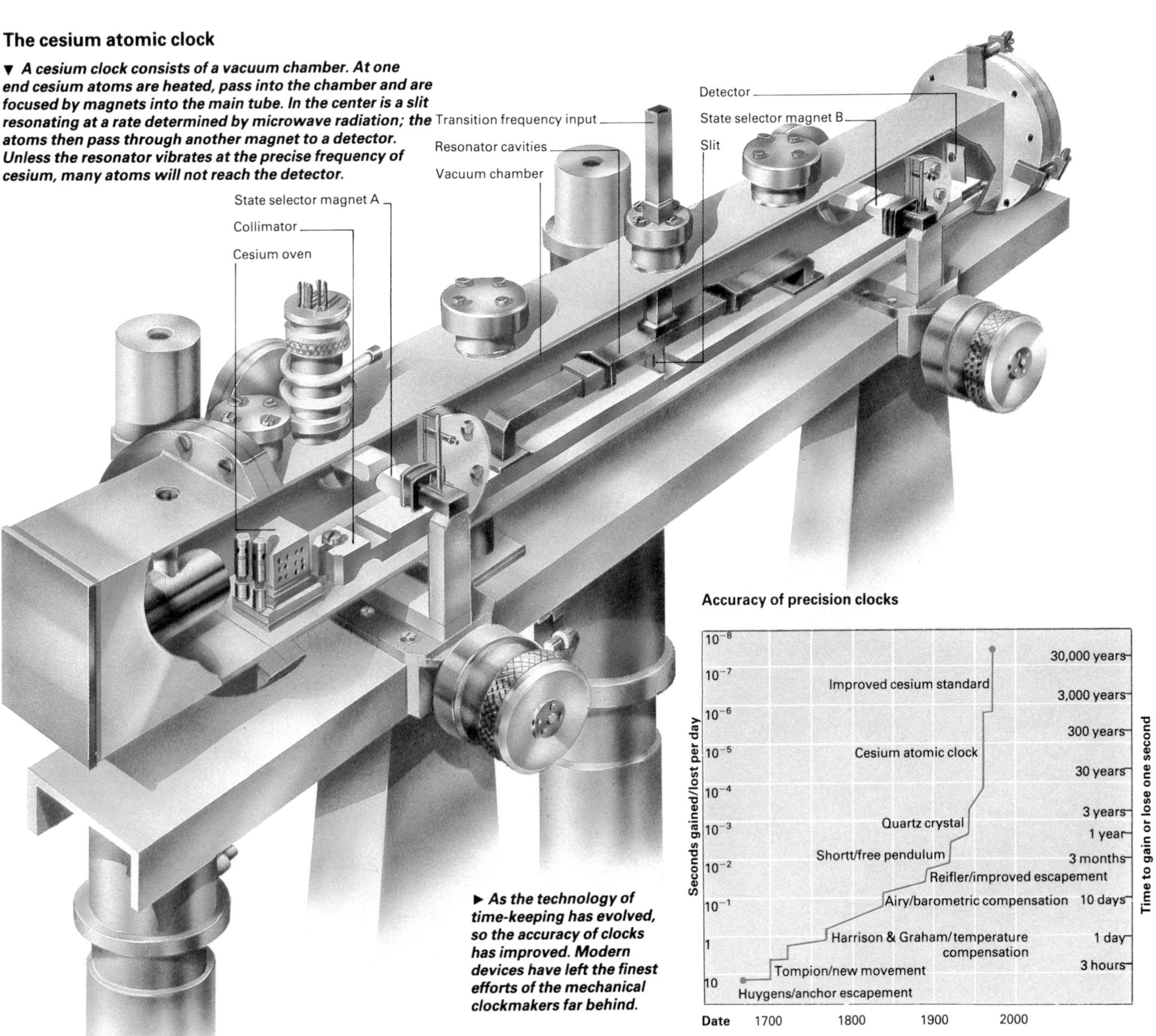

► ***As the technology of time-keeping has evolved, so the accuracy of clocks has improved. Modern devices have left the finest efforts of the mechanical clockmakers far behind.***

Until relatively recently, different countries maintained different prime meridians and time standards

Time standards

Since 1884 Greenwich Mean Time (GMT) has been the main international time standard. Before that time, there were many different time standards even within Britain, a fact that brought confusion when railway timetables were first drawn up. GMT is based on solar time, but it refers to a fictitious "mean sun", as opposed to the real Sun. The concept of the mean sun is necessary because the Earth's orbit about the real Sun is elliptical, not circular. This means that the apparent speed of the Sun across the sky varies through the year, which in turn gives rise to variations in the length of the solar day. For GMT the meridian at which the Sun's passage is measured passes through Greenwich, on the south bank of the river Thames east of London. This was the original location of the Royal Observatory founded by Charles II (r. 1660-1685) in 1675. Its original purpose was that of providing accurate positions of the stars for navigation.

Almost a century later, in 1766, Nevil Maskelyne (1732-1811), the fifth Astronomer Royal, published the first edition of the Nautical Almanac. *This is a set of tables listing the Moon's position against a background of stars, given in terms of the time at Greenwich. A navigator east or west of Greenwich uses a sextant to observe a different relationship between the Moon and stars, and treats it as if transposed to an earlier or later time. In other words, longitude (position to the east or west of the meridian) is equivalent to time; a rotation of 15 degrees around the globe corresponds to one hour. Longitude therefore specifies angular position west or east of a meridian and may be calculated from the* Nautical Almanac, *now known as the* Astronomical Almanac.

By the 1800s, there were many prime meridians, as different countries based the zero of longitude on their own cities. The resulting confusion was resolved at an international conference in Washington, USA, in 1884, when delegates from 25 countries agreed on the Greenwich Meridian as the prime meridian for the entire world. From that time, GMT gradually became accepted as the main international time scale.

*In 1935, GMT became known in scientific circles as Universal Time. Today, international time signals provide Universal Time Coordinated (UTC). The rate of UTC (the precise length of the seconds) is provided by a number of atomic clocks at various locations around the world. These are more regular than the rotating Earth (**♦ page 20**), but many people, especially navigators and astronomers, still need to relate time to the Earth's rotation. To account for tiny irregularities in the Earth's motion, UTC is adjusted by the insertion of leap seconds at the end of June and December. This procedure keeps the atomic version of Universal Time in step to within one tenth of a second a year.*

The calendar

Clocks record the passage of time and allow observers to measure periods of time. But it is often important to deal in much longer timespans than can be measured by clocks. For this purpose, people use a calendar. The calendar now used in many countries was introduced in 1582 by Pope Gregory XIII (r. 1572-1585). Before that time, a calendar introduced by the Roman Emperor Julius Caesar (100-44 BC) in 46 BC was commonly used in Europe. The Julian calendar was based on a year (the time taken for the Earth to complete one orbit of the Sun) of 365·25 days. In practise, each year was 365 days long, and an extra day was inserted every four years to give leap years. But the year is more nearly 365·2422 days long, and the Julian calendar eventually became out of step, being one day too long every 128 years. By Pope Gregory's day, the calendar was ten days out.

A new calendar was worked out by the Bavarian astronomer Christopher (Clavius) Schussel (1537-1612). This Gregorian calendar is three days shorter than the Julian over a period of 400 years. Normally years divisible by four are leap years, but only one year in four divisible by 100 is a leap year. In this calendar, a year is on average 365·245 days long, and it will take 3,300 years for it to become one day out of step with the solar year.

► Airy's transit circle lies on and is used to define the line of the prime meridian, or 0° longitude, passing through the Greenwich Observatory in London. A transit telescope is used to measure the time and height of stars at their zeniths so that the sidereal day can be calculated. It must be aligned exactly north-south and its rotation axis must be horizontal and east-west.

◄ *A quartz sphere, said to be the most perfectly round object ever made, intended to be sent into orbit and spun as a gyroscope. It will be used to test one of the predictions of Einstein's theory of general relativity, that there is a drift in the plane of the Earth's orbit.*

The sidereal day

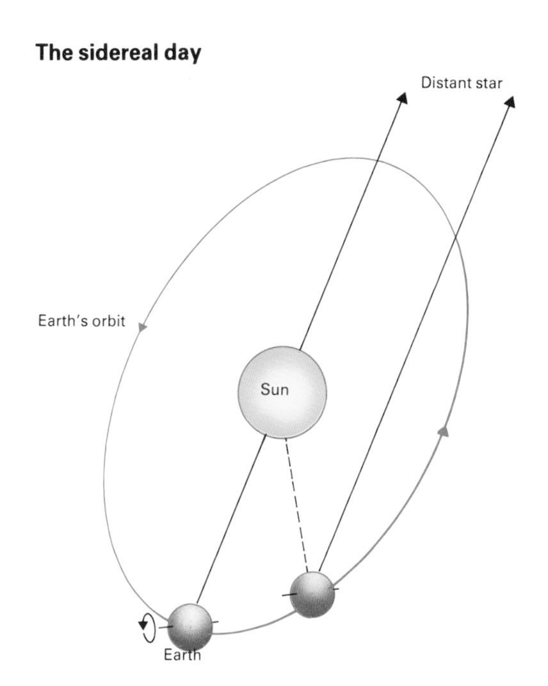

► *The movement of the Earth on its orbit around the Sun means that a day measured purely in terms of the position of the Sun (a solar day) is a little more than one revolution of the Earth on its axis. Hence the sidereal day, calculated by means of observations of the height and passage of much more distant stars across a transit telescope, is more accurate. A solar day is 3 minutes 56·5 seconds longer than a sidereal day; the year comprises 366·24 sidereal days, one more than the number of solar days.*

Time and relativity

The notion of absolute time was discredited by the theory of relativity developed in the early 20th century by the German physicist Albert Einstein (1879-1955). Einstein's theory comes in two parts. The first, known as special relativity, deals with motion and measurements by observers moving relative to one another. One of its predictions is that, to an independent observer, a moving clock appears to run more slowly – the closer to the velocity of light the clock gets, the slower it runs. The second, or theory of general relativity, deals with gravitation; here Einstein showed that a clock runs more slowly when gravity is greater.

In 1971 two American physicists, Joseph Carl Hafele and Richard Keating, took atomic clocks around the world on jet aircraft, flying westwards on one journey and eastwards on the other. Comparison with clocks that had remained on Earth showed that the clocks that had traveled eastwards had run slower by about 59 billionths of a second. The westward-moving clocks had, on the other hand, gained relative to those on Earth by about 273 billionths of a second. Hafele and Keating had in fact observed two different effects caused by the two kinds of relativity. The eastward-moving clocks "lost" time because their total velocity was faster than the Earth's own eastward rotation and time for them passed more slowly. The westward-moving clocks gained since their total velocity was less than for a clock on Earth. In addition, both sets of clocks gained because the aircraft flew at high altitudes where the Earth's gravitational force is weaker.

► ***An 8th-century AD calendar of the central American Mayan civilization. The passing of time was counted in blocks of 20 days; a year comprised 18 such blocks plus 5 "unlucky" days. It was known that this was out of synchrony with the solar year by less than a quarter of a day. The Mayans also counted cycles of 260 days. A day named in both systems was thus unique within a 52-year span.***

See also
Introduction to Measuring 9–22
Measuring Weight 31–36
Measuring with Electricity 41–46
Seeing beyond the Visible 73–86
Analyzing the Atom 109–114

Measuring the past

Atoms can be used to measure very long periods of time. The technique of radiocarbon dating is used by archeologists and geologists to assign dates to rocks or ancient artefacts by studying the radioactive decay of their atoms. The technique requires a fairly reliable "clock" within the materials themselves, which can be used to reveal how much time has elapsed since the objects were formed. Several elements provide such clocks, but the best-known example is the isotope carbon-14.

Radioactivity occurs when the nucleus of an atom decays to another type of nucleus, while at the same time emitting some form of radiation. The process is random, in the sense that there is no way of knowing when a particular nucleus will decay. But in a sample of many nuclei, radioactivity provides a useful clock because the time for a certain proportion of the nuclei in any sample to decay is always the same for a particular isotope. Radioactive nuclei are generally described in terms of their half-life – the time taken for half the original number of nuclei to decay. The most useful isotopes for dating purposes have very long half-lives, up to billions of years.

In 1945, the American scientist Willard Libby (1908–1980) realized that tiny amounts of a radioactive form of carbon, namely carbon-14, are being produced continuously in nuclear reactions high in the Earth's atmosphere, where high-energy cosmic rays from outer space bombard the nitrogen in the air. This carbon is quickly trapped in carbon dioxide and can enter plants during photosynthesis, and then enter the food chain. As a result carbon-14 is found anywhere that ordinary carbon (carbon-12) is found, although in very small quantities. By comparing the amounts for an old specimen, say a fragment of bone, with a modern sample, an archeologist can calculate the relative age of the specimen: the older the object the smaller its proportion of carbon-14.

Timing radioactive decay

Carbon-14 decays to nitrogen-14, with a half-life of 5,730 years. This is the common form of nitrogen, so it is not possible to estimate how many decays have occurred in a sample by measuring the amount of nitrogen-14. Instead scientists must measure the amount of carbon-14 that has not yet decayed in proportion to the amount of carbon-12.

The amounts of radioactive carbon are minute – only a single atom of carbon-14 for every thousand billion atoms of carbon-12. The measurement of such small quantities involves some very intricate techniques. The main method is to monitor decays of the carbon-14 atoms remaining in a sample, by detecting the electron emitted each time a carbon-14 nucleus changes into nitrogen-14. A gram of modern material produces only 14 electrons per minute; a sample 5,730 years old of the same size emits only half this number, and so on, with the number halving for each 5,730 years back in time. Most laboratories do not date objects more than 60,000 years old with this method, although sophisticated techniques based on the use of particle accelerators (▸ page 109) are beginning to extend carbon dating still further back in time, and require smaller samples to study.

Other radioactive nuclei that are useful in dating over very long periods include rubidium-87 and potassium-40. Rubidium-87 decays to strontium-87, while potassium-40 produces argon-40, in both cases with half-lives of billions of years. Measurements based on the proportions of these nuclei have shown rocks on Earth to be over 2·5 billion years old. The same technique has been applied to rocks collected from the Moon on the Apollo space missions and has shown them to be as much as 4·6 million years old.

◀▶▼ Radiocarbon dating can be done with particle accelerators. The sample (about one milligram) is excited in an ion source, and the beams are steered into the accelerator by an injection magnet (left). They are fed into the analyzer magnet, which acts as a mass spectrometer (▸ page 107), bending each isotope differentially, allowing only the carbon-14 to pass through the final filters (right) to the detector.

Radiocarbon dating

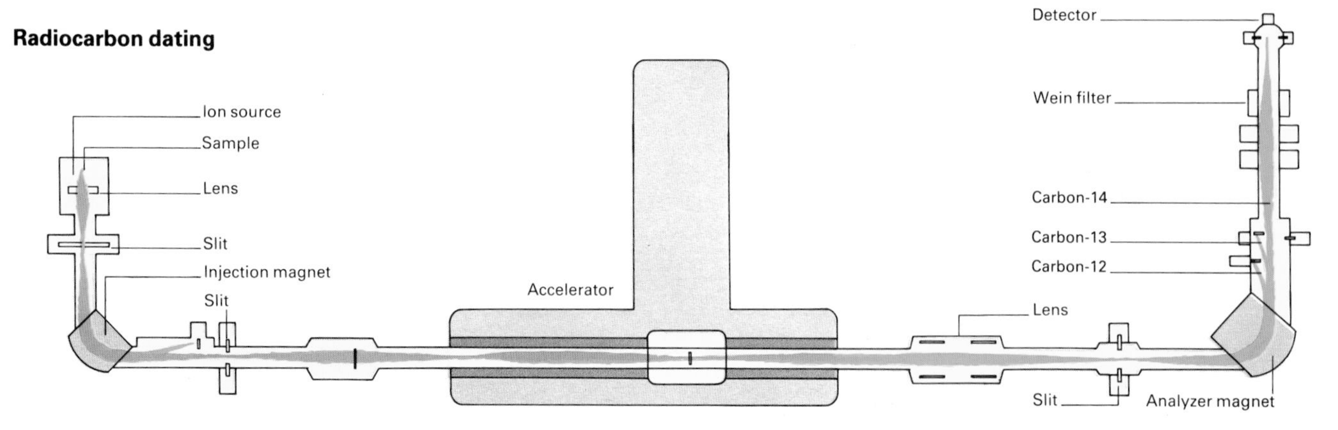

Measuring Weight

3

Early weight measurement...Balances...The chemical balance...Transducing weight measurement...Electronic weight measurement...Strain gauges...Weighing and industry...PERSPECTIVE...Defining the kilogram... The Ramsden balance...Problems of calibration

Instruments for weighing were quickly needed in every society where the exchange of goods played a significant role. Valuable commodities were generally traded by weight, and people could not be trusted to make accurate enough weighings by sensory perception alone, nor be trusted to be fair. "Objective" balances and steelyards are ages old and are still in use.

Arguably, though, such balances do not measure anything at all. They are "null detectors", in so far as they are used to sense when the turning effect of the standard weights piled on one pan equals the turning effect of the unknown weight placed in the other. When they balance, the "signal" is zero (◆ page 12). At this point the measurement part of the operation comes into play, as the weighing is done by counting the standard weights used. The "display" is the numbers stamped on the weights – not a very automatic instrument, but an effective one.

Commercial dealing, domestic purchases and the work of an apothecary are activities that call for very different ranges of weight. It is hard to achieve accurate comparisons between them, yet this is necessary if all weights are to be referred back to a single standard. Once a definitive standard of weight such as the kilogram had been defined, the next task was to duplicate it and then subdivide it as accurately as possible so that any range of weights could be measured in terms of the new units. These tasks called for sensitive and reliable balances, and it was not until the 18th century that instruments of sufficient quality were built and a universally accepted standard set.

▲ The simple, or equal-arm, balance, used for millennia and seen here in a market in India. The sensitivity of the balance depends primarily on the length of its arms and the quality of its central pivot; it would be significantly increased if a pointer were added to measure the angle of displacement of the arms when the pans held slightly unequal weight and if this angle could be accurately calibrated.

▲ The standard kilogram as defined in 1899, made of 90 percent platinum and 10 percent iridium. It is kept at the International Bureau of Weights and Measures at Sèvres, Paris, in several bell jars to protect it from corrosion and from changes of temperature, pressure and humidity.

Defining the kilogram

Standards of weight or mass have always been arbitrary. In earliest times a particular stone would have been chosen. The standard kilogram was originally defined by the Académie des Sciences in Paris in 1791, and is still kept there. It is a small cylinder of platinum iridium alloy (about 40mm high and wide), estimated as equal to the mass of one liter of pure water. It would have been inconvenient to use water itself as a standard, since it would have had to be contained in something, and water is notoriously hard to purify since it contains traces of so many different substances; so an equivalent amount of a solid was chosen. The alloy has a low expansion as temperature rises, so that weighings are not affected by changes in air buoyancy, and it is immune to corrosion. Nevertheless this piece of metal is still a unique and arbitrary choice of standard – a sophisticated version of the primitive stone. It cannot be reproduced in another place by following a recipe; copies of it have to be made and tested in Paris to be distributed to other countries. Elaborate precautions are taken to preserve the master standard from damage.

To make the first standard half-kilogram masses, two pieces of metal had to be filed down carefully, ensuring, with the help of a good balance, that they stayed equal, until eventually the two together balance the standard whole kilogram. This must have been a painstaking procedure until a comprehensive suite of sub-kilogram standards was established. There was much room for doubt. Were the apparent differences real or the result of errors in the construction of the balances? How small a difference could the balance detect? An accuracy of one part in 100,000 was required, and this was a severe challenge.

Recently metrologists have recognized that a standard of force or energy could be used instead of a mass standard, and coupled through electrical standards (◆ page 42) to the time standard. Such a definition would remove the arbitrariness of the kilogram, since time measured on atomic clocks (◆ page 29) can easily be duplicated. No everyday advantages flow from the new measurement.

Scales for use in the marketplace and chemical balances both rely on counting the value of the standard weights placed in the pan. This method is rather crude, and it is more convenient and may be more accurate to use a transducer (◆ page 11). The tilt angle of a balance can be measured directly and interpreted as a weight difference between the pans. Most modern balances use weights to measure to the nearest gram, and then sense the tilt by an optical device for higher levels of sensitivity.

The sensitive balance

A balance is sensitive if a small weight difference between the pans produces a noticeable tilt. Several features of instrument design determine how much tilt results from how much weight, and it took instrument makers some time to get everything right. Aiming for high sensitivity meant building a balance that took a long time to settle, was sensitive to tiny geometrical distortions due to temperature changes and responded unwontedly to the faintest air currents.

The design has to ensure that there is an equilibrium tilted position when slightly different weights rest on the pans. This will only happen if the center of gravity of the beam lies below the plane of the pivots. As the heavily loaded pan moves downwards, the center of gravity shifts to the other side of the pivot. The balancing point is found when the turning effect of the weight of the beam matches that of the small extra weight in the heavier pan.

The beam has to be light, yet stiff enough to carry the capacity load in each pan without significant distortion. At the same time it must be possible to see the beam tilt when a minimal weight difference exists between the pans. These aims conflict because a large tilt comes from long arms, but the longer the arms, the less stiff they can be made. Deep vertical sections of a trussed design are commonly used. When the pans are loaded, both ends of the beam will deflect downwards to some extent; the central pivot and the pan suspension points must then be coplanar. To ensure that the center of gravity of the beam is close to, but below the central pivot, some instruments have an adjustment consisting of a small weight on a vertical screw.

Adjustment is the key to accuracy as well, because the "exact" engineering implied in this desciption is impossible. Slight inequalities in the lengths of the arms can be compensated by interchanging the loads from one pan to another, and a screw adjustment can be used to compensate if the pan and beam on one side is slightly heavier than on the other.

◀ *The modern chemical balance is all-electronic, and is usually readable to 0·0001g. These machines rely on strain gauges (◆ page 35) rather than on mechanical principles, and are but little affected by changes in temperature, humidity or height. The balance may take as little as two seconds to stabilize. It can also be linked directly to a computer.*

The chemical balance

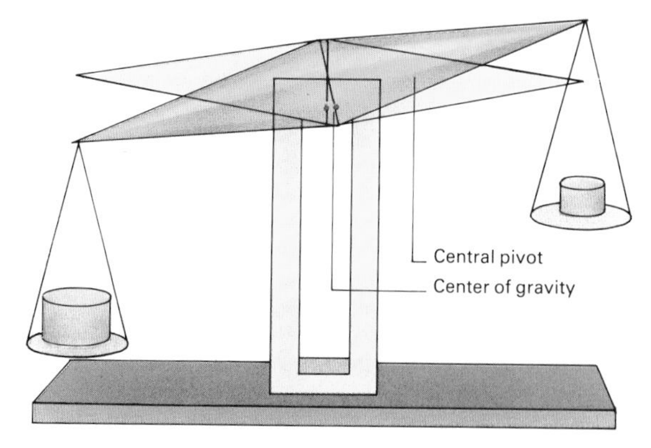

▲ ***One feature of the Ramsden balance was the screw adjustment above the central pivot on which a small weight hung. This adjusted the center of gravity of the beam and kept it just below the pivotal point. When one pan was pulled down, the center of gravity shifted to the other side and countered the weight difference. The sensitivity of the instrument could be adjusted as balance was approached.***

Jesse Ramsden – instrument maker

Accuracy was the passion of Jesse Ramsden (1735–1800), perhaps the foremost instrument maker of the 18th century. He was an inventive and imaginative yet practical engineer, who constantly sought to find fault in his own designs and improve them – sometimes to the extent of missing delivery dates, by as much as 23 years in one instance.

The son of an innkeeper in Halifax, England, Ramsden received a mathematical education from a local parson but until the age of 23 worked in the clothing industry. He then became apprenticed to a London instrument maker and in the 1760s married the daughter of the lensmaker John Dollond.

The key to much of his accuracy was an early invention which improved the constancy of screw threads along their length. His telescopic mountings for astronomical and geodesic surveying were in great demand (◆ page 38). He could give reproducible settings of angular position to better than 1 arcsec. His workshop of some 60 craftsmen produced barometers, pyrometers, sextants, standard-length survey rods and chains, balances, electrical machines, levels, manometers and drawing instruments.

Ramsden's precision balance

The accuracy goal for copying mass standards in the late years of the 18th century was around one part in a million – an error of one milligram was allowed in "standard" kilograms. The English instrument maker Jesse Ramsden was one of the select few to build a balance to this challenging specification; his balance made in 1787 was intended to measure the density of spirits to assess their alcohol content for duty.

The balance arm is 300mm from the central pivot to the pans' suspension points. The pointers on the ends of the beam extend 400mm from the pivot, so that a minute angle of tilt can be detected as the pointers move across engraved scales.

Ramsden made use of hollow cones to resist bending without excessive weight. This shape was

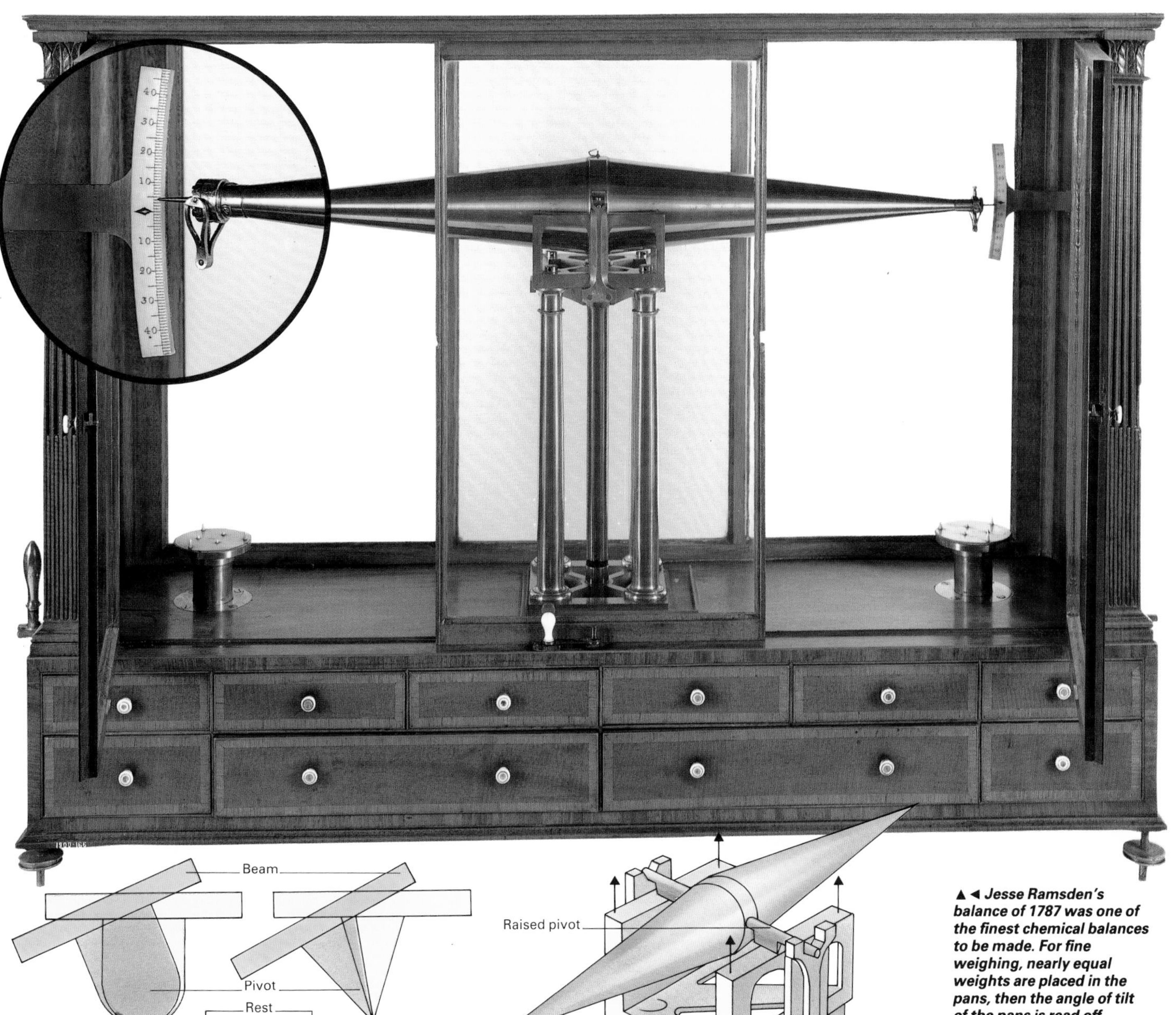

▲◄ Jesse Ramsden's balance of 1787 was one of the finest chemical balances to be made. For fine weighing, nearly equal weights are placed in the pans, then the angle of tilt of the pans is read off against markers at the end of the beams and calculated as a difference in weight. Ramsden adopted the knife-edge pivot, which may suffer from crumbling or digging into the rest, rather than the U-shape which rolls on the supporting plate and thus makes the balance arms unequal. When the balance is ready the beam is lifted off the Y-shaped rests, by raising the supporting agate beam structure, to come into contact with the steel knife edge.

easy to make, by panel-beating a sheet around a mandrel and brazing the seam together. The beam is two cones joined base to base, and carries a load of 2kg with little deflection. Nevertheless, some displacement can be detected, caused by bending which adds to the tilt displacement on the heavy side and reduces it on the light side. It was necessary to examine the displacements of both ends of the beam so that the bending effects could be cancelled.

Another concern was the equality of the arm length. A small nut on a screw at the end of one arm allowed the zero of the balance to be set, but equal weights on unequal arms still produce some tilt. Changing the weights from pan to pan allowed this error to be eliminated.

A difficult design feature on all precision balances was the mechanism for lifting the beam off its delicate knife-edge supports and lowering it again without disturbing the zero setting. Ramsden made particularly good knife-edge bearings for the beam. The edges were of hardened steel and bore on ground and polished flat agate plates. The edge needs to be a very abrupt angle with no rounding so it is highly stressed even with a modest load. To maintain the sharp angle requires the steel to be very hard but to carry the stress it must be strong. Hard steels tend to brittleness, so a compromise is needed on the temper of the steel and the geometry of the edge. The accuracy of the balance and the survival of the edges depend on exact alignment of the parts; not easy to ensure in the 18th century when steelmaking and precision measurement techniques were very limited.

Strain gauges provide an accurate method of measuring the weight of anything from a pinch of pepper to an aircraft

▲ A spring balance used to weigh the day's catch at a salmon fishery in Scotland.

► In a spring scale the object to be measured extends the spring, and the instrument measures this extension by a rack and pinion arrangement that moves a pointer on a dial.

▼ A letter balance similarly converts force into a deflection to be measured, but no spring is involved.

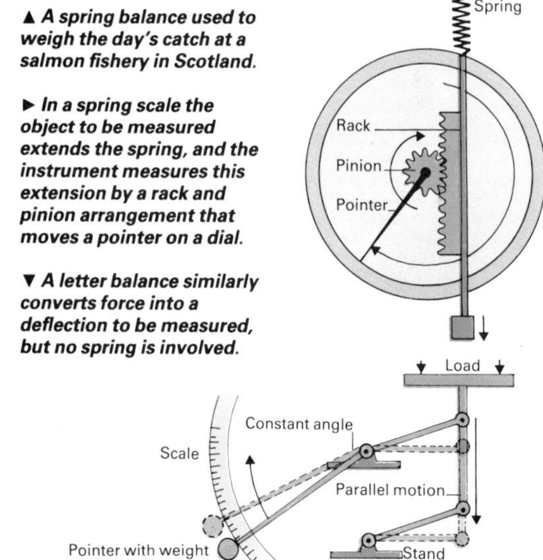

In letter balances there is no attempt to balance two equal arms; the balance is pivoted such that placing a letter on the pan causes the other arm to move across a dial; the angle of deflection consititutes the signal to be measured. In spring balances the entire responsibility for weighing is given to a transducer (the device which converts the force to be measured into a different, more readily measurable, phenomenon). The force of the weight in the pan stretches a spring and the extension of that spring is the signal to be counted. A pointer moves over a scale and by marking scale positions when standard weights are put in the pan, the scale lengths are calibrated. This is how most domestic scales work, though usually without much engineering sophistication. Shop scales are required to be more accurate.

Sensitivity on a spring balance is decided by the stiffness of the spring. The deflection of a platform resting on several thick coil springs could weigh an aeroplane (useful information immediately before takeoff), while a delicate hairspring can be sensitive to quantities as small as a milligram.

▶▼ *Strain gauges are used to measure heavy weights of all kinds. A public weighbridge for trucks usually operates on the principle of the steelyard, in which one arm of a balance is longer than the other, permitting a relatively light weight in the pan on the long arm to balance a much heavier weight on the shorter arm. However, this is relatively inaccurate and calculating the weight and center of gravity of an aircraft during manufacture and before takeoff has been made simpler with strain gauge load cells linked to microprocessors. Aircraft are weighed every couple of years to check for changes in weight from paint, dust, moisture and modifications to equipment.*

▲ *A strain gauge consists of a narrow metal ribbon attached to the surface on which the strain is to be measured. The pattern of the ribbon depends on the stresses in the material. The changes in resistance in the ribbon are measured.*

▼ *A strain gauge load cell, used industrially in the manufacture of aluminum.*

Strain gauges and electronic weighing machines

Weight is the force that gravity exerts on a mass, and is only one of many forces that engineers have to take into account when designing a structure. Tensions in suspension bridge cables, wind loads on buildings, engine thrusts or forces stressing structures from aircraft to oil rigs all have to be designed for. The strain gauge is an electrical transducer for measuring force. A ribbon of metal changes its electrical resistance when stretched. Associated electric circuits can detect a resistance change very sensitively and interpret it as a change of length; the force which caused the change in length can then be deduced. This form of transducer is a sort of spring sending out electric signals about its length.

Modern electronics has now so guaranteed the accuracy, convenience and reliability of electrical measurement that strain gauges are taking over from springs as the commonplace transducers for weighing. Everything in modern shop scales is electrical; we place the goods on the pan and instantly a digital display of the weight is shown.

See also
Introduction to Measuring 9–22
Surveying and Mapping 37–40
Measuring with Electricity 41–46
Chemical Analysis 105–108
Air Transport 217–224

Designing for use

Every weighing machine, like every other kind of measuring instrument, is designed to meet its own particular needs. The sensor can be designed for high sensitivity (as in the case of the chemical balance) or it can be designed for a wide range (a shop scale may measure from 100 grams to 10 kilograms). It may be important to follow a rapid variation of the measurement or to detect an average over an extended time span. The scales used to monitor packet filling on an automatic production line need to respond quickly, for example. It is clearly convenient that the calibration of an instrument should be constant in time, but this may not help if the sensor responds to something other than what is to be measured. Temperature compensation is used in strain gauges to get round the problem that electrical resistance changes with temperature. The limit of measurement by a sensor is set by how much its signal fluctuates when it is supposed to be at exact and steady zero. This fluctuation is called noise, and is inherent to any instrument. A most effective, and surprising, improvement in measurement results if many measurements are averaged, since the random signals of the noise tend to cancel, leaving the underlying signal clear (➧ page 170). The "signal to noise" ratio of a sensor needs to be large if ultrafine limits of measurement are to be reached.

▲ A weighing machine used to check the contents of cans of paint. Such machines may also be used on an automated production line: when the can is full the flow stops and the next can is brought into position.

Fair trading and weighing

The idea of fair trading is that the seller should never cheat the purchaser by giving less than the stated mass. This gives rise to some interesting problems.

The gravity force on a 1kg mass is 0·5 percent less at the Equator than at the poles, due to the spin of the Earth and its slightly flattened shape. A maker of spring scales at the Equator may calibrate the scales accurately by putting a mark on the dial where the pointer rests when a 1kg standard weight is put into the pan. But if the scales are sold to a customer from a very high latitude, north or south, when the customer takes the scales home and puts the standard measure on them they will indicate a weight of more than 1kg. A little must be taken off the pan to give a reading of exactly 1kg; the scales will not give a fair trade. In this instance the transducer is sensitive to place, and must be calibrated where it is to be used. A balance does not suffer from this problem since its true kilogram weights travel with it.

Every measurement is subject to some error, although it may be quite small if the instrument is accurate and gives consistent responses. A factory packaging flour may aim to put exactly 1kg into each of thousands of bags, but because of fair trading laws it must "never" put less in. It is a problem for the instrument makers, and also one for the factory managers, who must decide how much more than the stated weight should be put in to ensure that no more than, say, one bag in a thousand is under weight. Their calculation depends on the accuracy of repeated fast measurements as the flour flows into the bags. And they must also decide whether to spend extra money on accurate instruments or on flour given away unnecessarily.

◄ Electronic weighing machine used in a cheese warehouse.

Surveying and Mapping

Ancient surveying techniques...Ruler-and-compass methods...Triangulation...PERSPECTIVE...Jesse Ramsden, instrument maker...Photogrammetry and aerial surveying

In 1693 Louis XIV (r. 1643-1715) of France discovered that his kingdom was about 40,000 square kilometers smaller than he had thought. A new method of measuring his boundaries, based on astronomical observations, set the south and west coasts of France noticeably closer to Paris than had previously been thought.

Measuring the "lie of the land" has been a preoccupation of governments since ancient times. The civilization of the Egyptians depended on the river Nile annually flooding and depositing a new load of fertile silt upon its delta, and farms had to be laid out again as the waters receded from that featureless landscape. The Chinese built the Great Wall to defend themselves from barbarians. The Romans created roads running straight across western Europe, and planned towns in accurate checkerboard patterns. Land measurement is necessary in any society in which people live together closely.

Many of these ancient surveys were achieved with surprising accuracy considering the primitive instruments available. But they were all essentially "flat Earth" surveys. Mapping large areas in which the curvature of the Earth is a significant factor is much more difficult (➧ page 39). Replacing the imaginary world maps of early times with measured descriptions of the planet is an achievement of post-Renaissance technology. Even so the English Royal Engineers took eight years to map Scotland after the Jacobite Rebellion of 1745, using measuring tapes for distances and magnetic compass bearings for angles. These methods have intrinsic errors. Stretchiness, temperature variation and placement errors always afflict long-distance measurement by any kind of "ruler", and compasses are not reliable in a world that is full of iron. Today, radar-based survey instruments make highly accurate long-distance measurement commonplace and aerial stereophotography readily provides information about relief, irrespective of the wildness of the terrain, the vagaries of the weather or the hostility of the inhabitants (➧ page 40).

▲ A portolan map of the Mediterranean Sea, made in the early 17th century. Compass-bearings were noted from point to point, and a picture of the coastline was built up. The map was covered in a network of geometric lines to assist the mariner in setting a course. Precise distances were of less importance, and could not be accurately measured. North is to the right of the map.

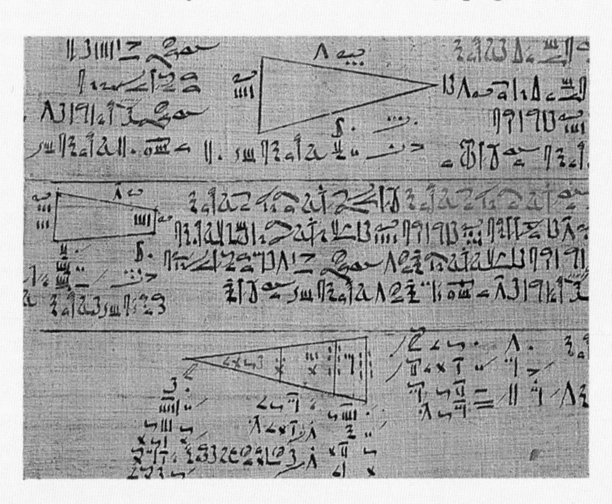

◄ The geometry of triangles was important to the ancient Egyptians who had to reclaim land after the annual floods. However, they had no means of measuring angles precisely. This treatise on measuring land areas is from 1600 BC.

► Not all civilizations have had a similar need for accurate mapping. The medieval maps of the world, such as this one from Hereford Cathedral, England, places towns symbolically rather than realistically.

Triangulation

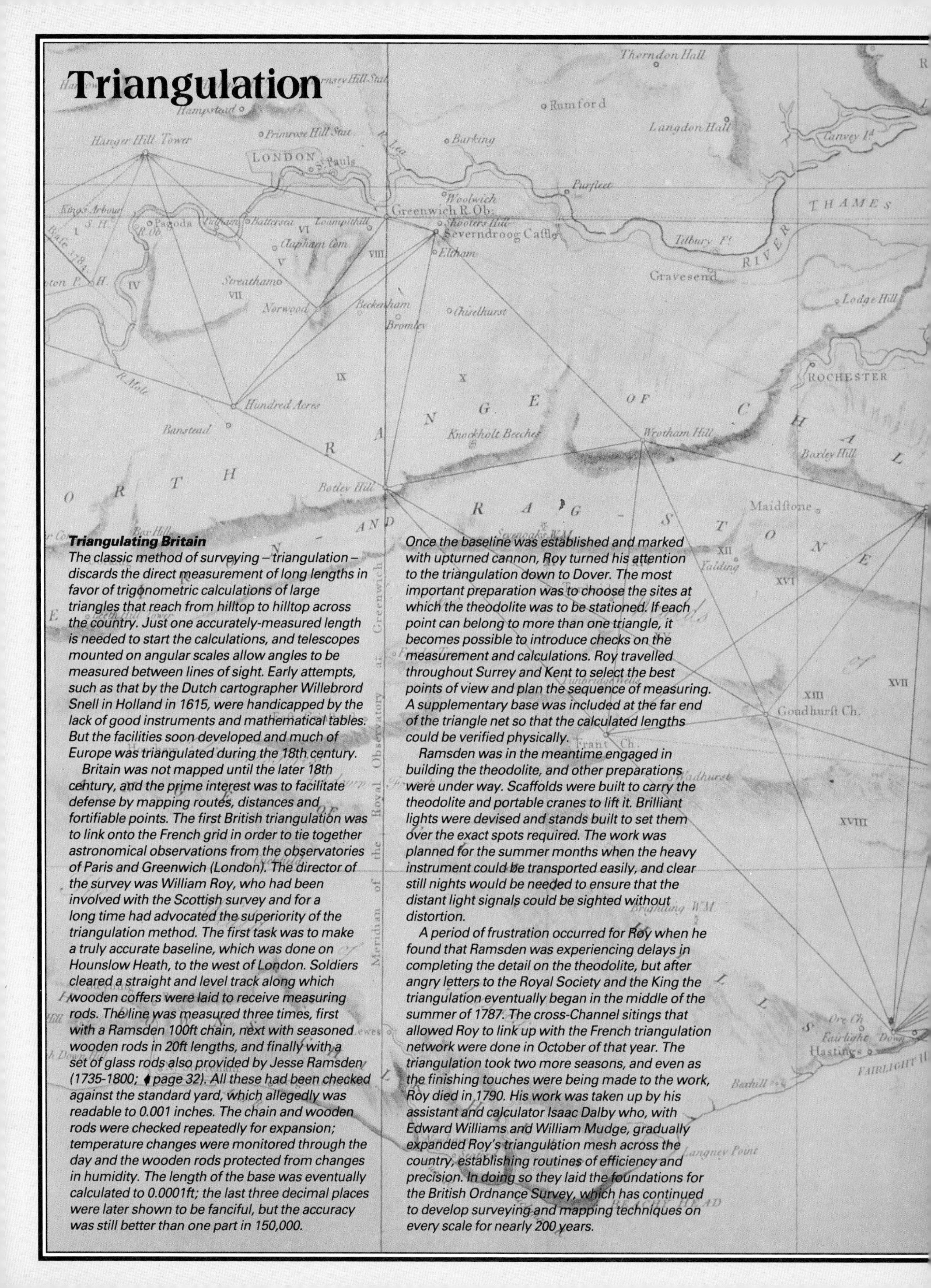

Triangulating Britain

The classic method of surveying – triangulation – discards the direct measurement of long lengths in favor of trigonometric calculations of large triangles that reach from hilltop to hilltop across the country. Just one accurately-measured length is needed to start the calculations, and telescopes mounted on angular scales allow angles to be measured between lines of sight. Early attempts, such as that by the Dutch cartographer Willebrord Snell in Holland in 1615, were handicapped by the lack of good instruments and mathematical tables. But the facilities soon developed and much of Europe was triangulated during the 18th century.

Britain was not mapped until the later 18th century, and the prime interest was to facilitate defense by mapping routes, distances and fortifiable points. The first British triangulation was to link onto the French grid in order to tie together astronomical observations from the observatories of Paris and Greenwich (London). The director of the survey was William Roy, who had been involved with the Scottish survey and for a long time had advocated the superiority of the triangulation method. The first task was to make a truly accurate baseline, which was done on Hounslow Heath, to the west of London. Soldiers cleared a straight and level track along which wooden coffers were laid to receive measuring rods. The line was measured three times, first with a Ramsden 100ft chain, next with seasoned wooden rods in 20ft lengths, and finally with a set of glass rods also provided by Jesse Ramsden (1735-1800; *◆ page 32**). All these had been checked against the standard yard, which allegedly was readable to 0.001 inches. The chain and wooden rods were checked repeatedly for expansion; temperature changes were monitored through the day and the wooden rods protected from changes in humidity. The length of the base was eventually calculated to 0.0001ft; the last three decimal places were later shown to be fanciful, but the accuracy was still better than one part in 150,000.*

Once the baseline was established and marked with upturned cannon, Roy turned his attention to the triangulation down to Dover. The most important preparation was to choose the sites at which the theodolite was to be stationed. If each point can belong to more than one triangle, it becomes possible to introduce checks on the measurement and calculations. Roy travelled throughout Surrey and Kent to select the best points of view and plan the sequence of measuring. A supplementary base was included at the far end of the triangle net so that the calculated lengths could be verified physically.

Ramsden was in the meantime engaged in building the theodolite, and other preparations were under way. Scaffolds were built to carry the theodolite and portable cranes to lift it. Brilliant lights were devised and stands built to set them over the exact spots required. The work was planned for the summer months when the heavy instrument could be transported easily, and clear still nights would be needed to ensure that the distant light signals could be sighted without distortion.

A period of frustration occurred for Roy when he found that Ramsden was experiencing delays in completing the detail on the theodolite, but after angry letters to the Royal Society and the King the triangulation eventually began in the middle of the summer of 1787. The cross-Channel sitings that allowed Roy to link up with the French triangulation network were done in October of that year. The triangulation took two more seasons, and even as the finishing touches were being made to the work, Roy died in 1790. His work was taken up by his assistant and calculator Isaac Dalby who, with Edward Williams and William Mudge, gradually expanded Roy's triangulation mesh across the country, establishing routines of efficiency and precision. In doing so they laid the foundations for the British Ordnance Survey, which has continued to develop surveying and mapping techniques on every scale for nearly 200 years.

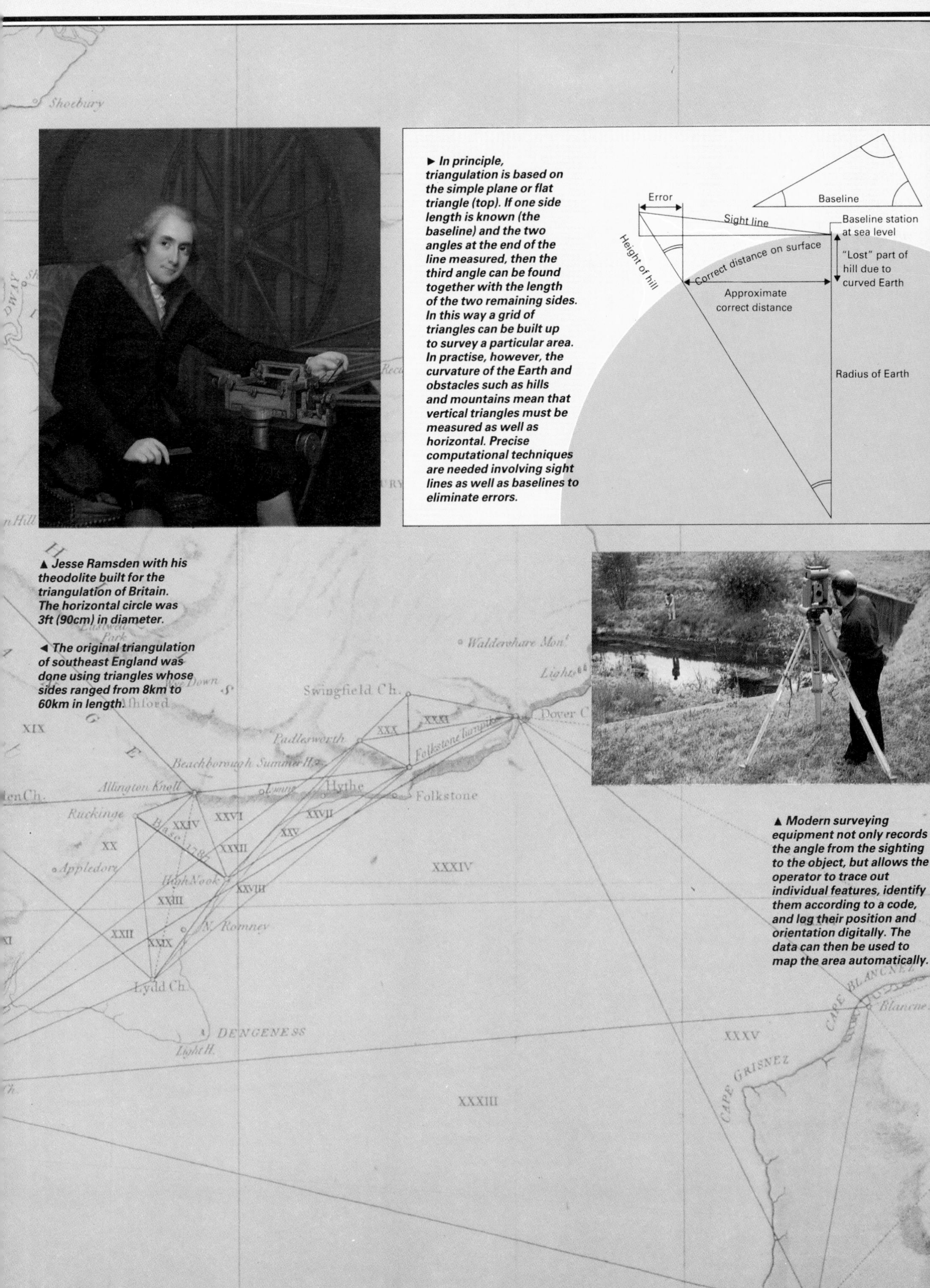

► *In principle, triangulation is based on the simple plane or flat triangle (top). If one side length is known (the baseline) and the two angles at the end of the line measured, then the third angle can be found together with the length of the two remaining sides. In this way a grid of triangles can be built up to survey a particular area. In practise, however, the curvature of the Earth and obstacles such as hills and mountains mean that vertical triangles must be measured as well as horizontal. Precise computational techniques are needed involving sight lines as well as baselines to eliminate errors.*

▲ *Jesse Ramsden with his theodolite built for the triangulation of Britain. The horizontal circle was 3ft (90cm) in diameter.*

◄ *The original triangulation of southeast England was done using triangles whose sides ranged from 8km to 60km in length.*

▲ *Modern surveying equipment not only records the angle from the sighting to the object, but allows the operator to trace out individual features, identify them according to a code, and log their position and orientation digitally. The data can then be used to map the area automatically.*

See also
Introduction to Measuring 9–22
Measuring Weight 31–36
Measuring with Electricity 41–46
Photography 87–90
Computer Software 123–128

Mapping from the air

Although large-scale surveying is still done on the ground, the majority of smaller scale work, and particularly mapping of difficult terrain, is now done with the technique of aerial surveying, also known as photogrammetry. A light aircraft flies up and down in strips over the region to be mapped, taking a sequence of photographs with a camera pointing directly downwards. The flight path, focal length chosen and frequency of shooting gives images that overlap considerably, so that each spot on the ground is covered by at least two photographs. The change in position of the aircraft when these are taken provides a slightly different angle of view for each feature on the ground.

When two such photographs of a single area are viewed together through a stereoviewer, a three-dimensional image of the terrain results. The operator matches identifiable points on this image against points of known latitude, longitude and height above sea level, and, using these as the base, plots the positions of all other features on the photographs. By moving a cursor across the stereo image the position and orientation of the features can be logged digitally in three dimensions. The automatic drawing machines then convert these data into a map to the required scale and density of information.

Aerial surveying

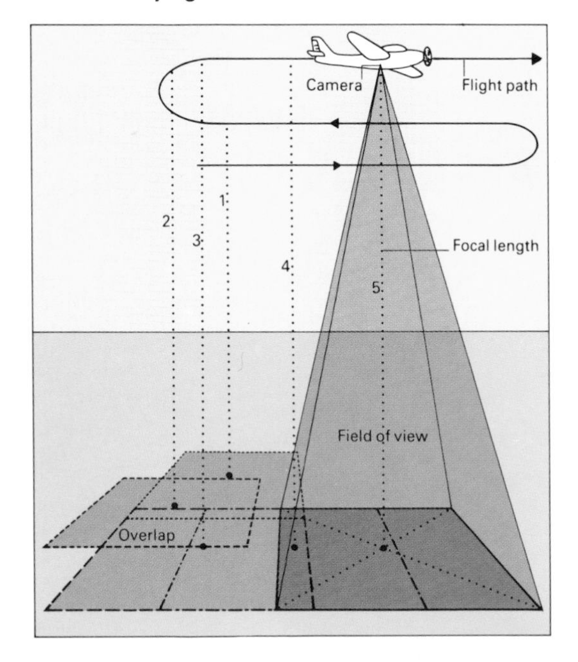

▲ When planning the flightpath and camera angles for aerial surveying, the individual photographs in a strip are arranged to overlap by about 60 percent, and the photographs of adjacent strips overlap by 25 percent. In this way, every feature is covered twice. Aerial photogrammetry is normally done when the Sun is high in the sky to ensure that shadows do not obscure important details.

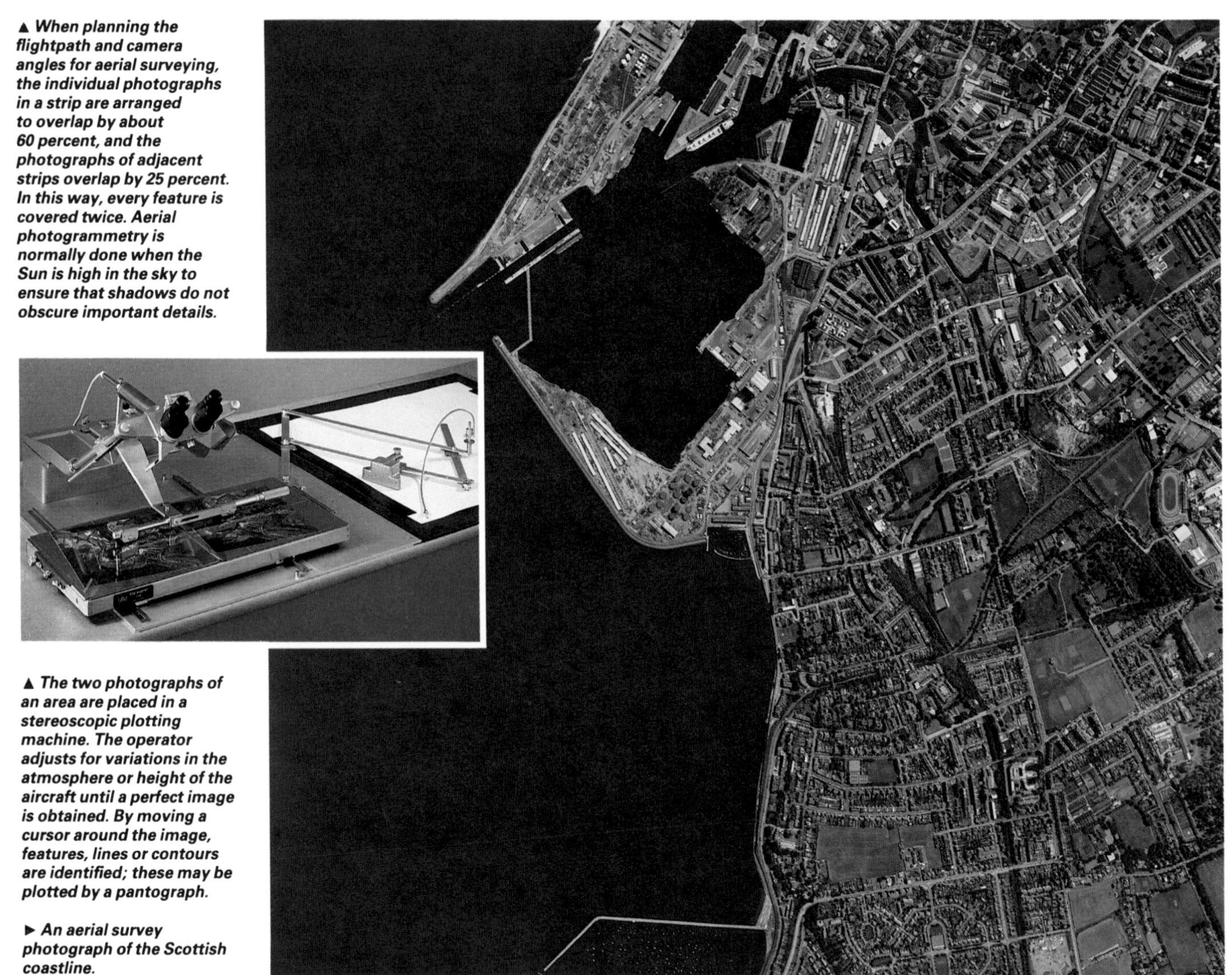

▲ The two photographs of an area are placed in a stereoscopic plotting machine. The operator adjusts for variations in the atmosphere or height of the aircraft until a perfect image is obtained. By moving a cursor around the image, features, lines or contours are identified; these may be plotted by a pantograph.

► An aerial survey photograph of the Scottish coastline.

Measuring with Electricity

5

Electrical charge and current...Measuring electricity: ammeters and voltmeters...Electronic measurement techniques...The chart recorder...Very precise techniques: the flash converter...PERSPECTIVE... The elements of electricity...Ampère and the current balance

The evolution of technology has been dramatically speeded up by the developing mastery of electricity. Electricity gives us power at our fingertips (➧ page 135), and equally through myriad forms of transducer (➧ page 43) allows accurate and speedy measurement of many phenomena that could previously barely be detected at all. The measurement *of* electricity is at the heart of all measurement *by* electricity. The elementary principles of what electricity is must therefore be understood.

The electrical quality of matter is called charge. It is regarded by scientists as one of the fundamental qualities of nature, more to be recognized by its effects than explained in terms of anything else. Like mass, length and time, is should be regarded as simply "there". Just as masses exert gravitational forces on one another even though separated, so do charges; and electrical forces are much stronger than gravity. Whereas gravity holds solar systems and galaxies together, electrical forces hold atoms together. They are, in fact, some 10^{40} times stronger than the gravity force at atomic distances.

Unlike gravity, which can only attract, electrical forces can both attract and repel. This is explained by recognizing two distinct kinds of electrical charge, labelled as "positive" and "negative" for the convenience of counting the net charge in assemblies of charged particles such as atoms.

The structure of the atom suggests why electrical phenomena are so many and varied. Most physical and chemical explanations of matter are couched in terms of the interactions between the clouds of electrons in close-together atoms. Electrical current in a metal wire, for example, is regarded as a flow along the wire of electrons which have broken free from their atoms. The fact that the wire becomes hot is seen as a consequence of electrons crashing into the fixed atoms; and a pressure or voltage is needed to drive them along because of these continual collisions. The idea of current as charge flow, and of voltage as a pressure which can drive the flow, are crucial to understanding the techniques of measuring electricity.

It is useful to recognize the relationship between the current and the voltage for a metal wire. The wire offers a "resistance" to the current flow which is dependent on its dimensions and materials. This resistance is defined as the "pressure per unit flow", or voltage divided by current. This relationship can be exploited in measuring devices, since carefully measured wires with a known resistance provide a means of transducing currents into voltages and vice versa.

Electrical currents also produce magnetic effects, and conversely magnets act on wires carrying electrical current. These electromagnetic effects are at the core of all electrical machinery (➧ page 136). They are also used to sweep a spot across a TV screen (➧ page 188), and are prominent in the field of electrical measurement.

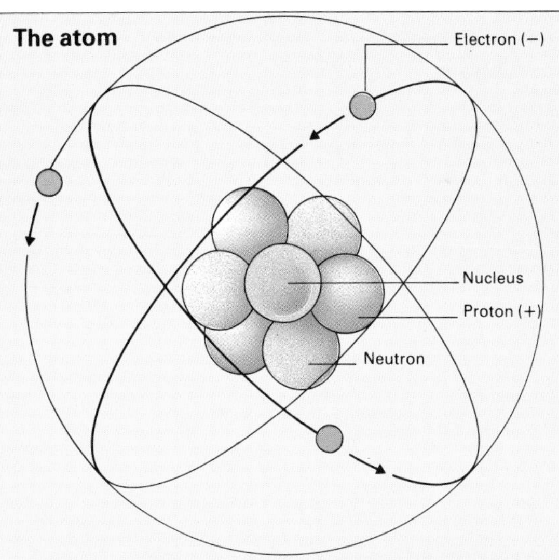

▲ ***The atomic nucleus is positively charged, electrons negatively. They are usually in balance but an ion (atom with the "wrong" number of electrons) is electrically charged, and may cause a current, or flow of electrons, around it.***

The elements of electricity

An atom consists of electrons orbiting around a nucleus. When an electromotive force (EMF), measured in volts, is applied to it, electrons may move away to another atom. The net movement in one direction is the electric current or flow, measured in amps. Atoms which hold some electrons weakly are good conductors, whereas those with strong bonds allow virtually no movement and are non-conductors. The flow of current down a wire resembles the flow of water in a pipe, and the EMF is like the pressure needed to move the water. The flow of water may be affected by the resistance of the pipe (a small diameter means high resistance), and a wire similarly has resistance, measured in ohms, to the current. A small wire has higher resistance than a thick one, and, as in plumbing, the longer the wire the more force is needed to push the current along.

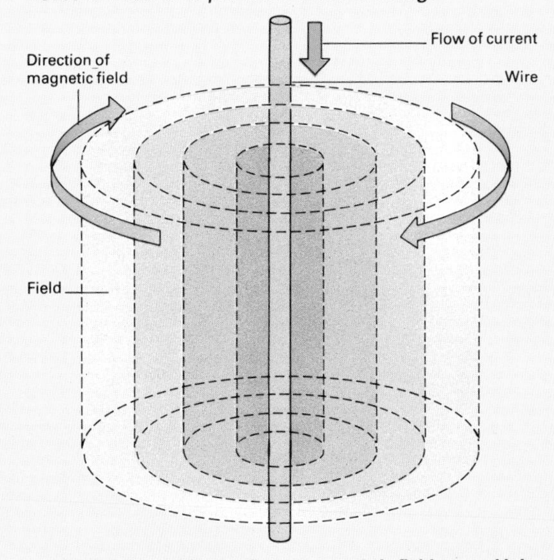

▲ ***A current in a wire produces a magnetic field around it in a plane at right angles to its flow. If this field interacts with that of a permanent magnet and the deflection is measured, it can be used to measure the strength of the current.***

Instruments to measure electric current are designed with a range of possible sensitivities

Moving coil electrical instruments

Beneath the dial and pointer display of the familiar electric meter is a miniature current balance (current measuring instrument). The current to be measured interacts with a magnetic field to produce a force. This distorts a spring until the spring exerts a balancing force, and a pointer moves through an angle proportional to the current. The current is therefore transduced twice, first to force and then to deflection; if a voltmeter is required, there are three transductions: voltage to current, current to force, and force to deflection.

The design of electric meters is directed to several goals because their applications are widespread. Satisfactory performance is likely to require an instrument to be sensitive to a chosen level (from 1,000 amps to 0·001 amps or less depending on needs); accurate to a speciied precision of calibration; and stable (it is important that the characteristics do not alter, springs tire, bearings stick or magnets fade). It should be responsive, with the needle moving quickly and unwaveringly to its correct indication; linear so that equal steps of current produce equal deflections through the range of the instrument; and robust (it may suffer excess current, may be installed in a vibrating machine or subjected to temperature and humidity changes).

Measuring voltages

When a voltage is measured in a moving coil meter, some current is lost at the meter itself in order to get the balance of mechanical forces. This is avoided by using a potentiometer. The potentiometer balances voltages directly, and only uses the moving coil meter as a null detector (◀ page 31). A steady current is passed through a piece of resistance wire, so that the voltage steadily declines along it (in other words, the current in transduced into a uniform voltage gradient). The test voltage is set in an opposite direction across the resistance wire, using a sliding contact which is moved along until a current meter on the text circuit indicates that no current is flowing: at this point the voltages in the two circuits are equal and opposite. The voltage drop down the wire is calibrated as a standard so centimeters of wire can be interpreted as volts, and the voltage in the test circuit calculated from the length of resistance wire across which it is connected; since it is driving no current its full value can be assessed.

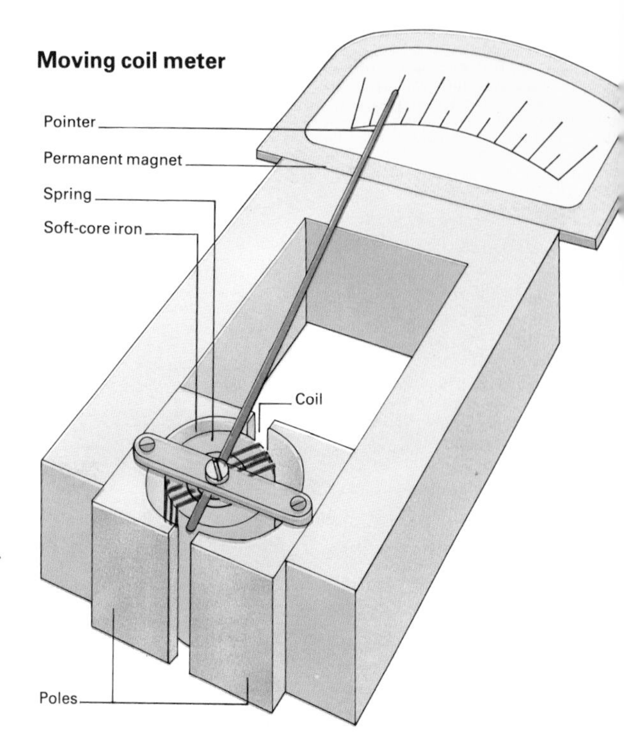

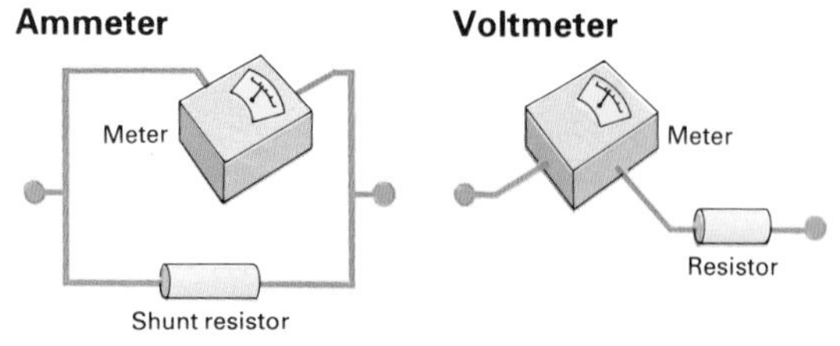

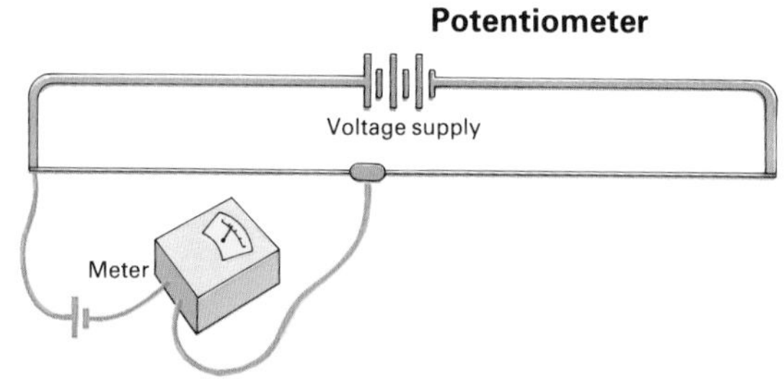

A standard current

The ampere is named for the French physicist André-Marie Ampère (1775-1836) – one of the pioneers of electrical measurement. It is the unit of electrical charge flow rate. The ampere or amp is defined in terms of the forces between wires carrying equal currents lying in a standard geometry, "straight parallel conductors, one meter long and one meter apart". In practise this arrangement is not easy to set up without error, and the wires do not exert a convenient force on each other when one amp flows. As a result a "current balance" uses an arrangement of coils which brings very long lengths of parallel conductors close together in a compact geometry. The relation between this arrangement and the standard can be calculated. Two coils are mounted on the arms of a balance and two are fixed; the forces are measured by lifting standard weights.

◀ ***Replica of Ampère's current balance.***

◄ In a moving-coil meter, a coil carrying the current to be measured is placed on a cylindrical iron core which rotates between the poles of a magnet. A pointer is attached to the coil spindle, restrained by a spring. When a DC current passes through the coil, its magnetic field interacts with that of the permanent magnets, and the coil turns on its spindle against the spring. The deflection indicates the strength of the current. If an AC current is to be measured, it must first be converted to DC by a rectifying diode.

► An industrial multimeter offers a range of electrical measurements; different models may be suited to requirements that range from electronic and laboratory applications to power engineering. A multimeter measures current (amps), pressure (volts) and resistance (ohms), in AC and DC.

◄ An ammeter can be used to measure different ranges by using a shunt from the main circuit; varying the strength of the resistor allows different magnitudes of current to be measured. A voltmeter is wired up in serial with the resistor. A potentiometer consists of a standard voltage put through a known resistor; the test voltage is applied across the resistor in the opposite direction, by means of a sliding contact. When no current flows, the voltages are balanced and so the test voltage can be calculated.

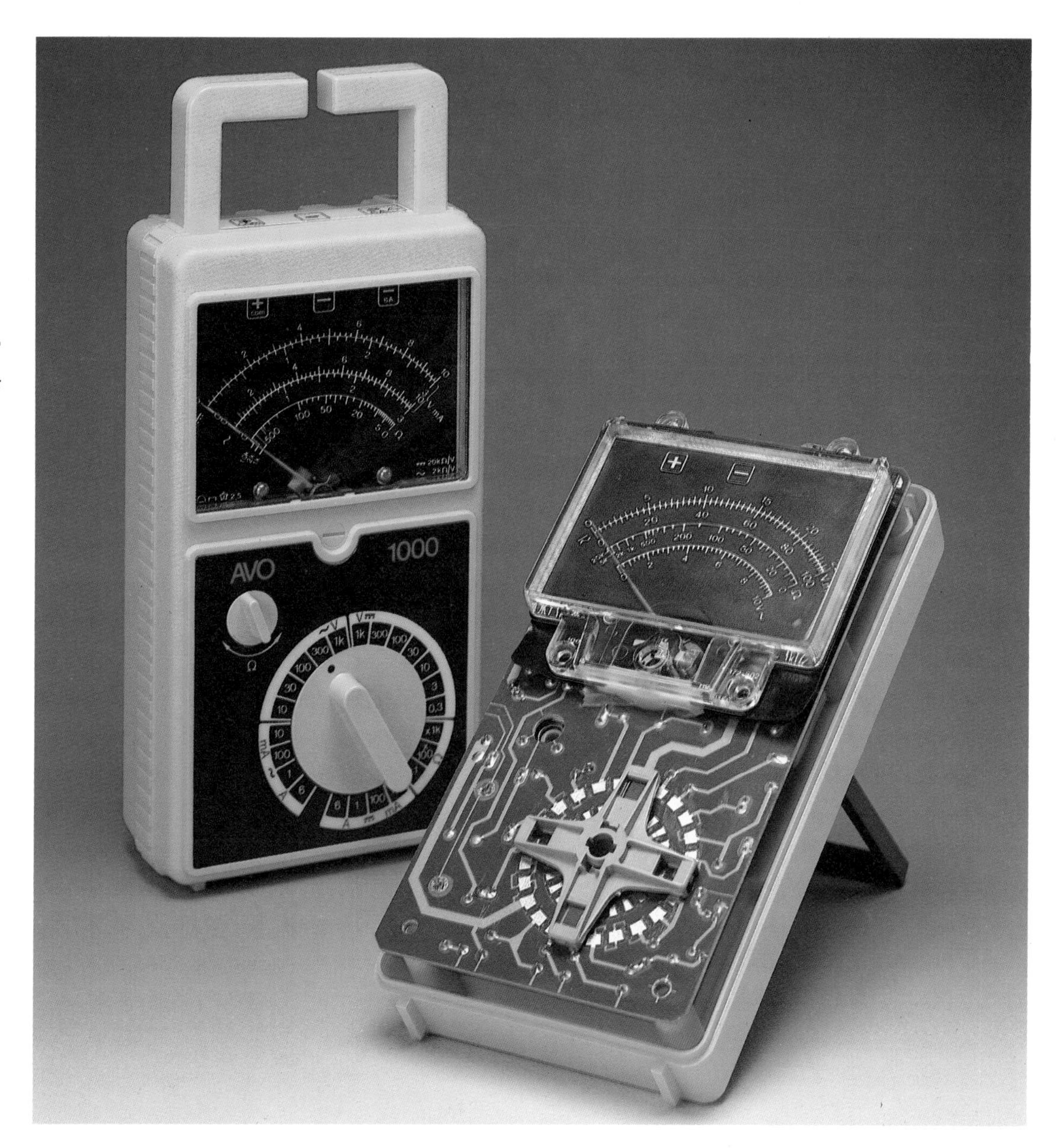

Transducers

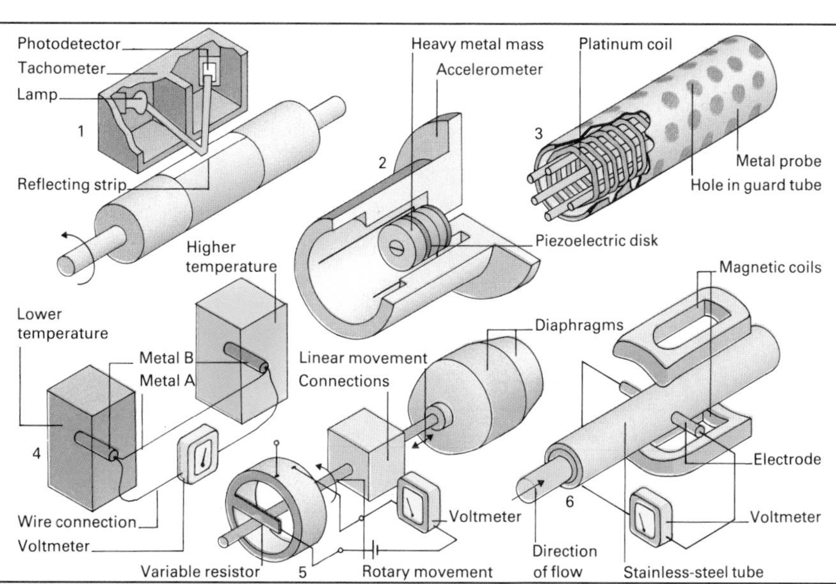

► Many different devices can be used to transduce different types of signal into an electric current, thus allowing the signal to be measured accurately. A tachometer may use a photoelectric cell (1), a semiconductor that generates a current when struck by light. An accelerometer may be one of many devices to employ the piezoelectric effect for transduction: here crystals become variably charged when they are subjected to changes in pressure (2). A metallic resistance thermometer (3) takes advantage of the changing resistance of a metal, in this case platinum, at different temperatures. In the thermocouple device (4) dissimilar metal wires are joined and one junction is heated. An electric potential is formed at the junction. A pressure meter or microphone may exploit the variable resistor, a substance that changes its resistance when pressure is applied to it (5). A flow meter uses the principles of electromagnetism (6); the conductive substance to be measured flows through an insulated tube into which electrodes are fitted, and a magnetic field is set up at right angles to the flow by a magnetic coil.

Electronic techniques of measuring electricity have revolutionized measurement in many fields

Electronic measurement techniques

Measurement techniques have been revolutionized by electronics. Whatever is to be measured, every instrument maker will now find a way to transduce it to a voltage or frequency, and display the signal digitally. Even angles, which for centuries provided the final stage of accurate measurement (◀ page 38), are now displayed in this manner. The collapse of the craft of mechanical instrument making is most evident in the field of time-keeping (◀ page 26), but is found in many other fields.

Electronic circuits accept voltage signals while drawing virtually no current from them, and they can perform several processes with speed and precision. Signals can be amplified by controlled amounts; they can be added or subtracted, averaged over time or compared with standards. Pulses can be counted and very rapid changes (billions of cycles per second) followed faithfully. New fields of measurement such as radar (▶ page 84) depend entirely upon electronic signal processing for their very existence.

There are several methods of measuring voltage using electronics. One of the most common is the cathode ray oscilloscope (CRO), on which the display is a bright spot caused by the impact of a narrow stream of electrons on a suitable screen. As in a television receiver (▶ page 190), the electrons travel at very high speed through a vacuum and can be steered by voltages applied to electrodes or coils set into the vacuum tube. Quite high voltages are required for this steering function, so the machine must incorporate stable amplifiers to magnify small signals to effective levels.

As each electron passes between a pair of electrodes or coils it receives a force impulse for an instant. This diverts it upward, so that the spot on the screen moves by an amount proportional to the

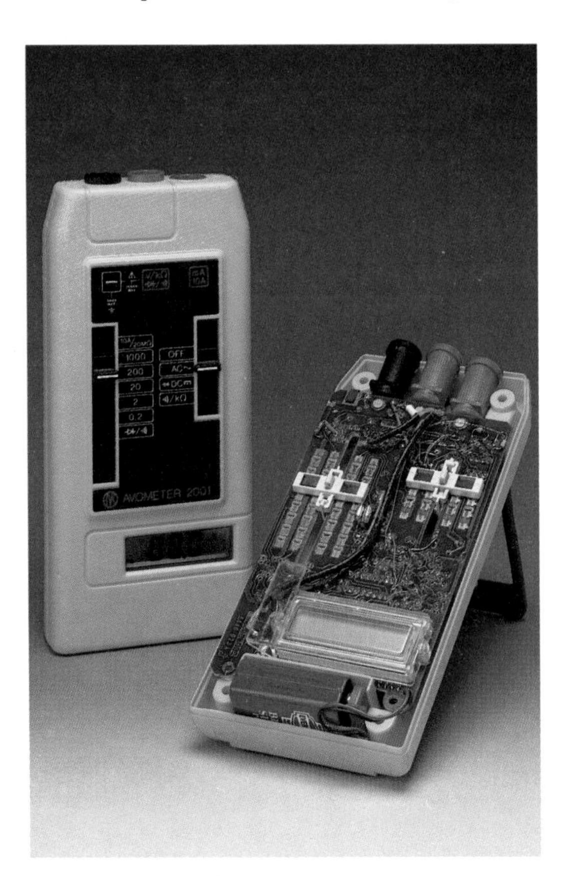

◀ ***Electronic techniques, as well as conventional moving-coil methods, may be used for multimeters, offering a similar range of features but with a liquid-crystal display rather than a dial. Such devices are used in the laboratory and in field service, including testing vehicle electrical circuits.***

Oscilloscope

Signal
In form of voltage repeating at regular intervals

Vertical amplifier
Amplifier input signal to create varying magnetic field in CRT deflection coils.

Cathode ray tube
Electron beam directed at screen, deflected by magnetic fields. Vertical deflection is signal and horizontal deflection is time.

Trigger
Starts sweep generator at certain point on signal

Horizontal amplifier
Amplifier sweep signal to create varying magnetic field to drive spot sideways.

Sweep generator
Produces steadily increasing signal voltage against a set timebase

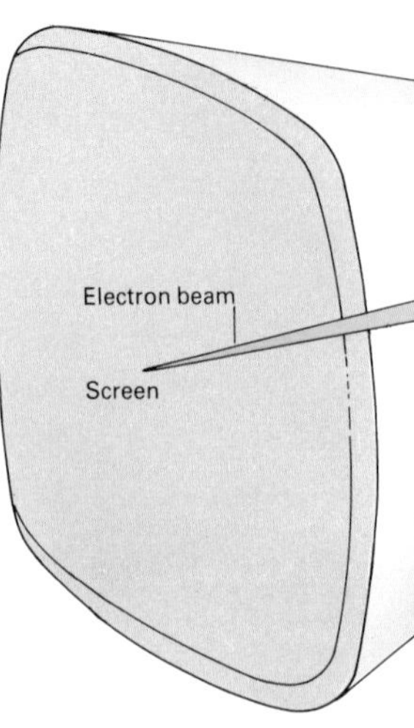

signal voltage. Because the duration of this impulse is so short, extremely rapid variations of voltage can be followed on the screen.

Usually the screen is also swept at chosen speed from left to right across the screen by a repeated ramp of voltage (the "timebase") applied to another pair of plates or coils. This spreads out the signal variations so they can be seen. The waveform of a repetitious signal can be displayed by always starting the timebase sweep at the same signal level. The spot then repeats its up-and-down pattern on the same parts of the screen every sweep, and a persistent image can be generated – hence the name oscilloscope.

The chart recorder

A second method of electronic measurement, and one which offers a permanent record of the display, is by use of the chart recorder. Given a transducer with a voltage output, the system is an automatically balancing potentiometer.

A person using a manual potentiometer has to decide which way to move the slider and how far, in order to obtain a zero reading on the ammeter. This can be done automatically by using the out-of-balance voltage to drive a small electric motor that moves the potentiometer contact; at balance there is no voltage so the motor stops. This principle takes advantage of an "error signal" to make a "correction", and is known as feedback. It is the key of all automatic control, and is used in chart recorders. The motor drives a pen which moves across the chart, while another motor advances the paper at a selected rate. Transducer outputs are rarely sufficient, either in voltage or power to run these motors, so electronic amplification is necessary. Switching in more or less amplification as required permits different sensitivity ranges to be selected.

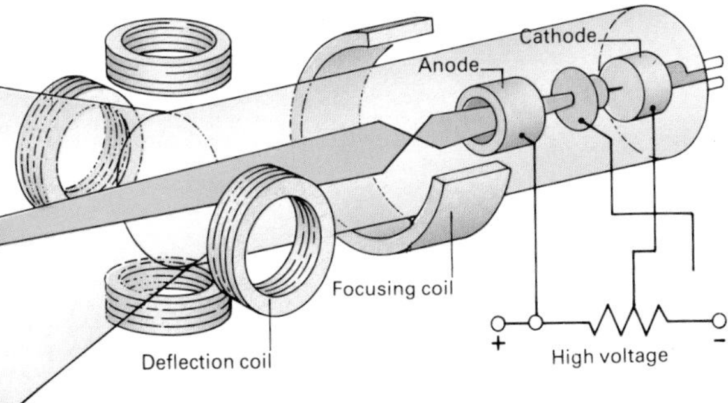

◄▲ An oscilloscope amplifies the current to be measured and passes it through a magnetic coil or electrode in a cathode ray tube to cause a deflection in the beam of electrons. The deflection is displayed as a change in the position of the dot on the screen.

► The chart recorder amplifies the signal voltage, and uses it to drive a servo motor. This compares the strength of the signal with the previously received signals and moves a pen forwards or backwards according to the relative strength of the signal.

See also
Introduction to measuring 9–22
Measuring Time 23–20
Seeing beyond the Visible 73–86
Electricity 135–140
Radio and Television 185–192

Liquid-crystal display

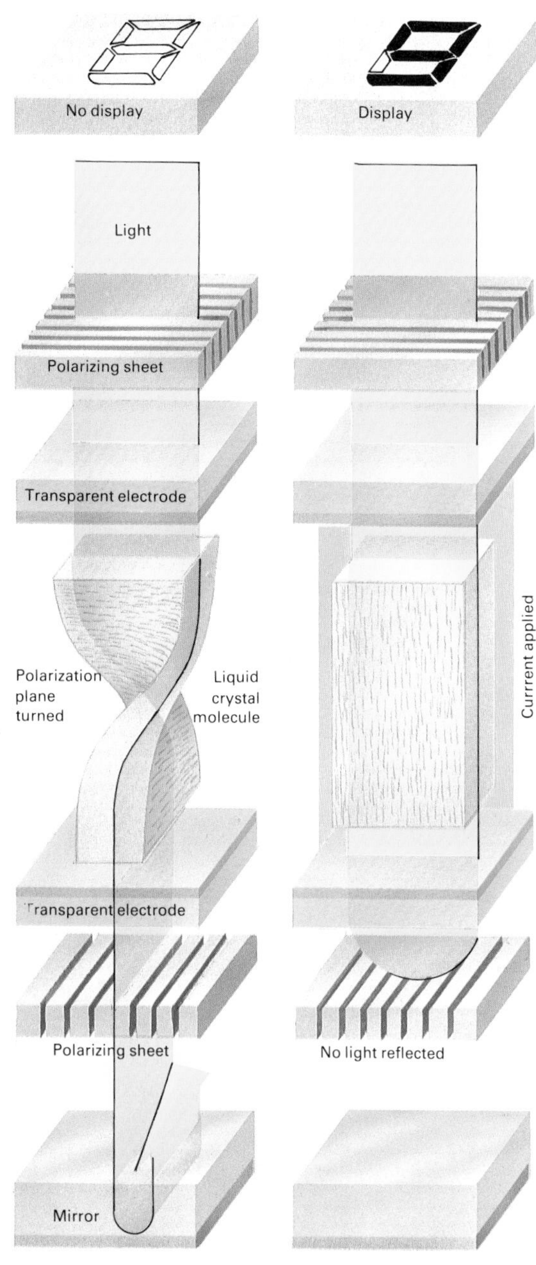

▲ ***Liquid crystals – organic compounds with rodlike molecules – change shape when a charge is applied. In an alphanumeric display, the crystals lie between sheets of glass whose polarization planes lie at right angles to one another. In their normal state the crystals are slightly twisted: light is turned to pass through both sheets and reflect off the mirror. The crystals change if a charge is applied; the light is blocked and the display shows black.***

The flash converter relies on electronics to establish and maintain precise and stable reference voltages. This is an electronic equivalent of the potentiometer wire, and offers a new way of measuring the voltage.

The advent of complex integrated circuits on silicon chips (➧ page 116) offers the opportunity of several hundred reference circuits, closely spaced. A signal is presented momentarily and simultaneously to all the reference circuits and is compared to each of them individually. If the signal is larger than the reference, the comparator outputs a binary code one; if smaller, it outputs a zero. The answers from all the reference circuits constitutes a particular set of numbers which correspond with a certain voltage. Electronic logic converts this number into a set of instructions to light or not to light on the digital display, so that the discovered voltage can be read easily.

Meters using this principle are rapid to read, draw virtually no current from the signal source and are very reliable. They are also comparable in price to moving-coil meters of similar performance which rely on delicate mechanical engineering.

One great advantage of these instruments is that the numerical storage of information allows computers to work on it. The possibilities that such devices open up are very large. It is conceivable, for instance, to use a photocell of television camera to transduce light intensity to voltage with a very rapid response, and connect these to a flash converter which could keep up with the task of enumerating the ever-changing electrical signal of the image. Thus a cruise missile is able continually to compare the terrain over which it is passing with the map it carries in its memory, and make fast course adjustments.

Electronics also allows voltages to be measured in terms of time – the most accurate method of measurement available to us (➧ page 20). The principle again depends on comparing a test voltage with a reference voltage, and the electronics provides a reference "ramp", or voltage increasing at a known steady rate.

To begin a measurement a trigger starts the ramp and simultaneously switches pulses from a stable high frequency oscillator to a counter. The counting continues until the ramp has risen to equal the test voltage, and at the instant the output of the comparator reads zero the count is stopped. If the frequency of the oscillator and the rate of the ramp are calibrated against standards, the number of pulses means a certain value for the test voltage, which can be calculated and displayed. Finally the logic circuits reset the ramp and counter to zero ready for a new sample of test voltage.

A standard voltage is sometimes used to check the accuracy of the calibration

Measuring voltage in terms of time

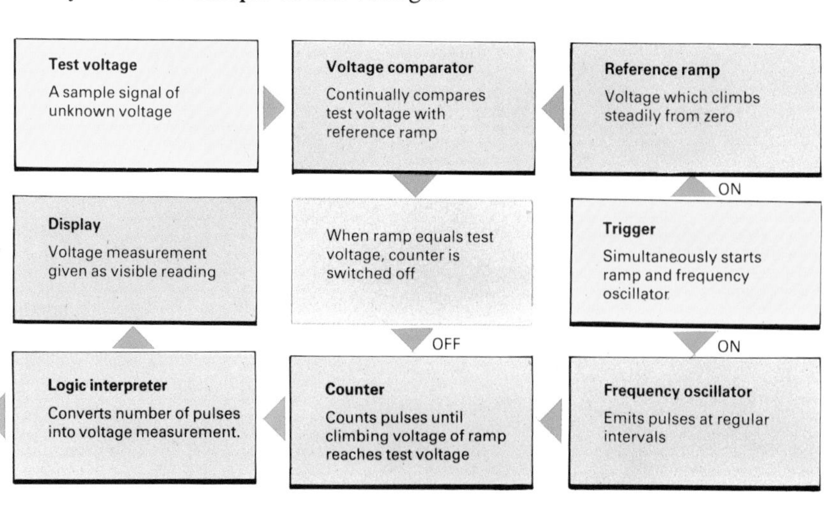

Reset commands: put counter and ramp back to zero and admit new test voltage sample

► ***A reference ramp is the most accurate method of measuring voltages.***

Microscopes

6

Origins of the microscope...Principles of compound microscopy...Phase-contrast microscopy...The electron microscope...Scanning tunneling microscope... PERSPECTIVE...Chronology...Optical microscope... Polarized light and fluorescence microscopy... Interference contrast microscopy...Scanning electron microscope...Emission microscopy...X-ray microscopy

The microscope is a tool for observing small objects or phenomena from a very short distance. Its antecedents lie in ancient times, and an Assyrian lens dating from about 700 BC has been discovered, carved out of natural crystal. The usefulness of magnified images for biologists, chemists, mineralogists, geologists, materials scientists and doctors has led to the development of various instruments and techniques, many of which go beyond the visible spectrum.

The principle of compound microscopy, by which two lenses are aligned so as to multiply the magnification of each individually, was first conceived in about 1600 by Zacharias Jannsen of Middelburg, Holland. Problems of distortion and aberrations were so considerable, however, that despite the compound microscope's great potential, magnifying glasses using expertly-ground lenses were to make the first advances.

The pioneer of modern microscopy, the Dutch biologist Anton van Leeuwenhoek (1632-1723), developed an effective type of simple microscope, with a single glass bead lens set between two brass plates in front of a pointer on which the specimen is placed. He was able to produce a magnification of more than ×300, and is credited with being the first to see bacteria and spermatozoa. The English physicist Robert Hooke (1635-1703) made early studies with a compound microscope, but the field was not fully developed until the 19th century when advances in lens manufacture made it possible to eliminate spherical and chromatic aberration in lenses.

▲ *A characteristically ornate microscope of the late 17th century, made by Christopher Cock of London.*

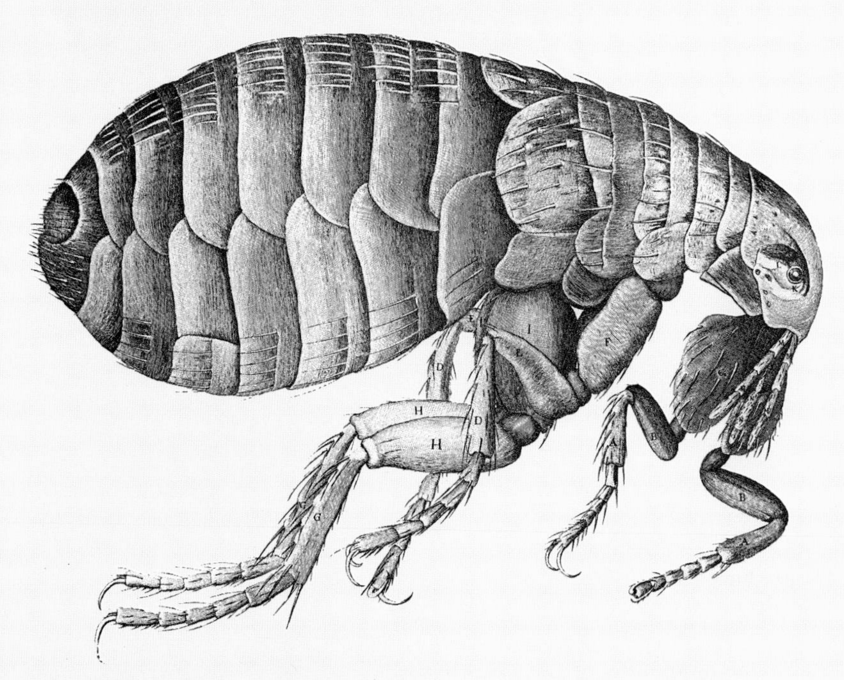

◄ *The invention of the compound microscope in 1600 was of immeasurable significance to science: previously invisible organisms and minute detail came under human scrutiny for the first time. True microscopy began in the latter half of the 17th century with the publication in 1665 of "Micrographia" (drawings of small things) by Robert Hooke, the earliest major collection of microscopical sketches. Using a glass globe and spirit lamp to give a constant and intense illumination of the specimens under his multiple-lens microscope, Hooke produced a number of sketches, including this detailed pitcure of a flea, that caused a sensation throughout Europe.*

Chronology

c1550 Simple microscope invented
1590 Compound microscope invented in Italy
c1600 Zacharias Jannsen perfected compound microscope design
1610 Galileo Galilei independently designed compound microscope
1665 *Micrographia* published by Robert Hooke in England; first major collection of microscopical sketches
1673 Simple microscope improved by Anton van Leeuwenhoek, magnifying up to ×300
1830 Joseph Lister's paper to British Royal Society revolutionized microscope objective design
1830 Achromatic microscope, with aberration-free lenses, invented
1850 Oil-immersion objective lens invented by Gianbatista Amici
1851 Single-objective binocular microscope invented
1872 Microscope objectives manufactured using precise mathematical formulae
1878 Ernst Abbe's modern microscope described
1897 Stereomicroscope for use in microsurgery produced by Karl Zeiss in Germany
1903 Ultramicroscope developed to observe particles of diameter less than 10 percent of the wavelength of visible light
1904 Dark-field microscope adapted for use with ultraviolet light by August Köhler in Germany
1919 Ultraviolet microscope developed, magnifying up to ×2,500

The principle of the microscope

The eye's ability to perceive fine detail is termed its resolving power, and the smallest separation between two points that it can detect is its limit of resolution. Several factors are involved. First, an object cannot be seen if its image is smaller than the distance between two light-sensitive nerve endings on the retina. The image may be made larger by bringing the object closer, but the eye cannot magnify it further by bringing it closer than about 25 centimeters, a distance known as the near point. Second, aberrations in the eye lens may distort the image, and, third, the wave nature of light may impose its own limitations.

If an image can theoretically be resolved, there must also be enough light to illuminate the object, which must appear significantly different from its background.

The basic function of the microscope is to produce a magnified image of a small object by focusing light projected onto or through it with lenses. An ordinary magnifying glass can produce an image of a distant object, such as the Sun, on a piece of paper held behind the lens and at its focal length. If an object is moved closer to the lens than the focal length, the image is upright and larger than the object, but it cannot be formed on a surface. This form of "virtual" image is the key to the simple microscope.

In the compound microscope, the magnification is achieved in two stages. The objective lens produces an upside down magnified real image of the object; this is formed at the focal distance for the eyepiece, which acts as an ordinary magnifying glass to magnify the image. This final image can be viewed with a relaxed eye, accommodated to infinity. In photomicrography, a different eyepiece is used to focus the final image on the camera photographic plate.

The resolving power of a microscope is directly proportional to the amount of light that can pass into the lens system, and the measure of this is the numerical aperture. This determines the amount that can be seen. In a great deal of microscopical work, the numerical aperture is increased by using special lenses with a drop of oil between the objective and the slide as oil bends light more than glass. This is called immersion microscopy.

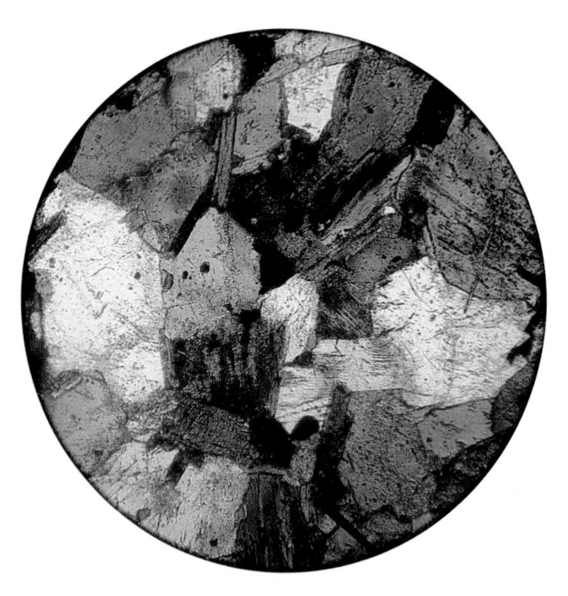

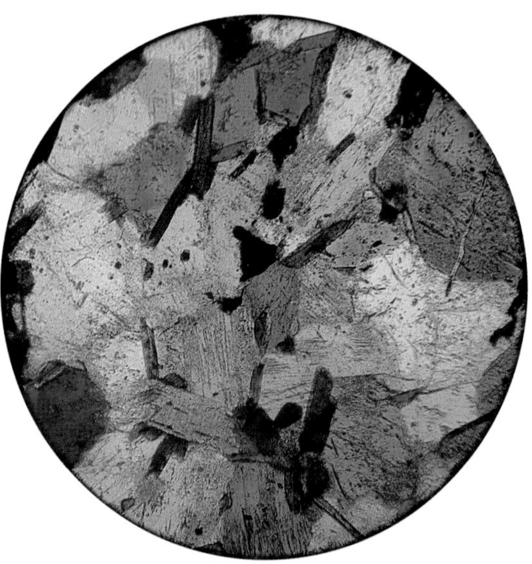

▲ **Polarized light microscopy is valuable for the study of crystalline materials. These images are of the same thin section of granite, seen in ordinary bright (top) and polarized light (above). Amorphous parts show black, while crystalline structure is revealed in colors that identify the composition.**

Polarized light, ultraviolet and fluorescence

A polarizer may be used in the illumination system, rather like a color filter, and a second, called an analyzer, in the eyepiece lens system. The polarizer only allows light to pass whose vibrations lie in a cetain plane, and the analyzer is oriented at right angles to that. If the specimen does not affect the light, the result is darkness. If it does, then regions of the microstructure which are optically active can be revealed. This is particularly useful in metallurgy and geology, and the identification of organic crystals.

As the resolving power of the microscope depends directly on the wavelength of the light used, it can be improved by using radiation of a shorter wavelength than visible light, such as ultraviolet light. Normal optical glass will not transmit it, so lenses of quartz or synthetic fluorite must be used. This method was first described in 1904 by August Köhler (1866-1948), but it has largely been overtaken by advances in electron microscopy. One important application, however, is in fluorescence microscopy. Certain substances are fluorescent: they absorb ultraviolet light and convert it into light with a longer wavelength which can be detected by the eye. Many natural materials, botanical and mineralogical ones in particular, are fluorescent, and this can be used to identify them. Other substances which are not themselves fluorescent can absorb fluorescent dyes in a selective manner, and this can be used to reveal their microstructure. This is particularly useful in medicine, and has been applied to the identification of chromosomes.

1924 Louis de Broglie demonstrated that electron wave motion is similar to that of visible light
1927 Ability of magnetic fields to focus electrons demonstrated by H. Busch
1931 Prototype electron microscope built by M. Knoll and E. Ruska
1933 L-stand microscope construction introduced by Zeiss
1934 Phase-contrast microscope invented by Fritz Zernicke
1936 First electron microscope entered commercial use
1938 Prototype scanning electron microscope built by Von Ardenne
1940 Prototype phase-contrast microscope built by Zeiss
1944 Soviet physicist N.D. Sokolov proposed idea of acoustic microscopy
1947 Electron-probe analyzer developed to determine identity of chemical constituents at high resolution
1955 Automatic camera microscope produced by Zeiss
1956 Field-ion microscope developed by Erwin Mueller, with magnification of more than × 2,500,000
1965 Scanning electron microscope commercially available
1973 First scanning acoustic microscope built by a team at Stanford University, California
1974 Holographic electron microscope invented at University of Texas to visualize electrons
1980 Scanning electron microscope used to make microscopic microchips
1984 Scanning tunneling microscope developed by Binnig and Rohrer in Zürich

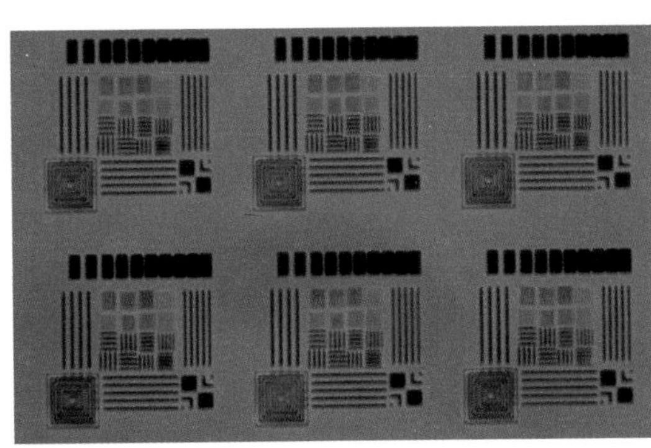

▲ ***Modern fluorescence microscopy most often uses non-ultraviolet light to excite fluorescence in a specimen. Here residual photoresist on a semiconductor chip is excited by visible green light to fluoresce in red.***

Compound microscope

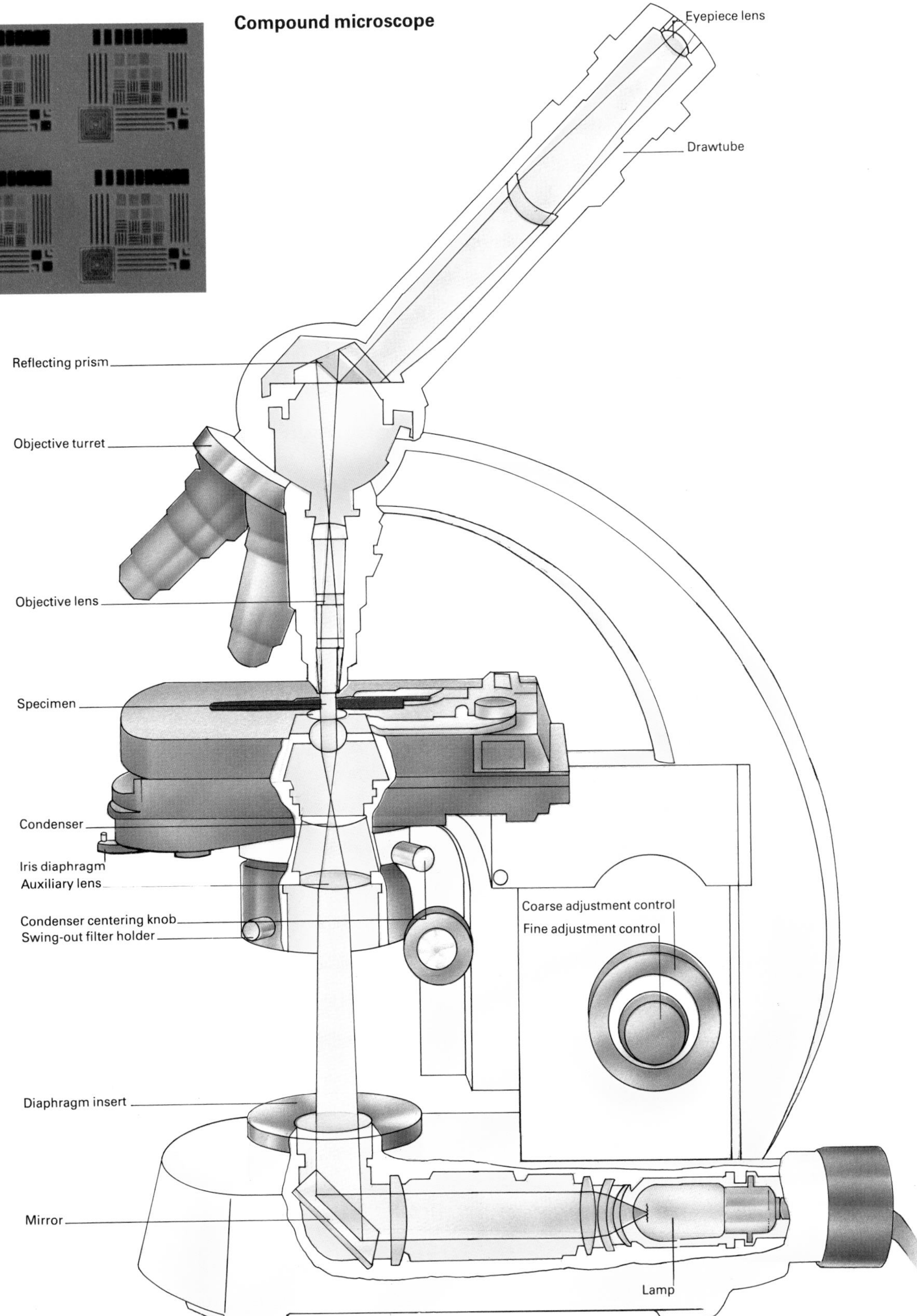

► ***A specimen mounted on a slide is supported by a stage which has a central hole through which light passes. It is viewed through a tube carrying two lens systems: the objective and the eyepiece. Beneath the stage a third system, known as the substage condenser, is used to illuminate the specimen. The objective is the most important of the three, and adjustments to the microscope are provided to maximize its efficiency. Rotating nosepieces carry two or more objectives, allowing for a ready switch in magnifications.***

Splitting and then recombining the microscope's light beam can give striking relief effects on the subject

Phase contrast microscopy

The principle behind the phase contrast microscope was developed in 1934 by the Dutch physicist Fritz Zernike (1888-1966), although it was not until about 1942 that the technique became widespread. Zernike was awarded the Nobel Prize for Physics in 1953 for his invention.

The phase contrast microscope allows the observation of details in unstained preparations (and therefore also in living cells). In a conventional optical microscope contrasts are produced in the image because different parts of the object absorb light to differing extents. Those parts of the object which differ in refractive index but not in absorption of light appear equally bright. The phase contrast microscope exploits the fact that a material's refractive index determines the speed at which light passes through it – a high refractive index slows light more than does a low index. When light passes through an object with a range of such indices, some areas transmit the light more rapidly than others and the emergent light is said to differ in phase.

To produce a visible image, a ring-shaped diaphragm is placed between the light source and the condenser. The condenser and the objective form an image of the ring behind the objective, where a glass plate with a black ring of the same size as the ring image is placed. If there is no object under the microscope, all the transmitted light is screened off by the ring. But if there *is* an object under the microscope, light passing through material of differing refractive index falls beside the ring, not on it, and a bright image on a black background is formed. This is made up of light, which has been slowed by differing amounts, resulting in an image that is difficult to interpret. Using a gray ring instead of a black one behind the objective gives a more natural appearance.

Another improvement is to enhance contrast by grinding the glass plate to form a ring which is either thicker or thinner than the rest of the plate. This causes another phase difference, between the light passing through the two thicknesses of glass. In the ultimate formation of the image the two groups of light waves interfere in such a manner that the differences in refractive index in the object are seen as differences in brightness.

If the plate is ground with the phase ring thinner than the rest of the plate, parts of the object with a low refractive index appear bright against a gray background, and those with a high refractive index appear darker. This is called positive phase contrast or dark contrast. If the phase ring is thicker the reverse is true; this is called negative phase contrast or bright contrast.

Applications of phase contrast microscopy

A variation of phase contrast microscopy is used in metallurgy for the study of specimens which have been lightly deformed, but its most extensive application is in the biological sciences, where it is an essential tool for most serious work. Since structures need not be stained, dynamic events such as cell division can be studied, as can the action of drugs and other physical and chemical processes. It is also possible to measure the refractive index of a substance by comparing its image contrast with a set of standard mounting media.

Although the limit of resolution of the optical microscope is only about 200 nanometers (millionths of a millimeter), the presence of features such as surface steps in metal of only 5nm can be detected. If two such features are present they need to be separated by a distance greater than the limit of resolution in order for both to be seen.

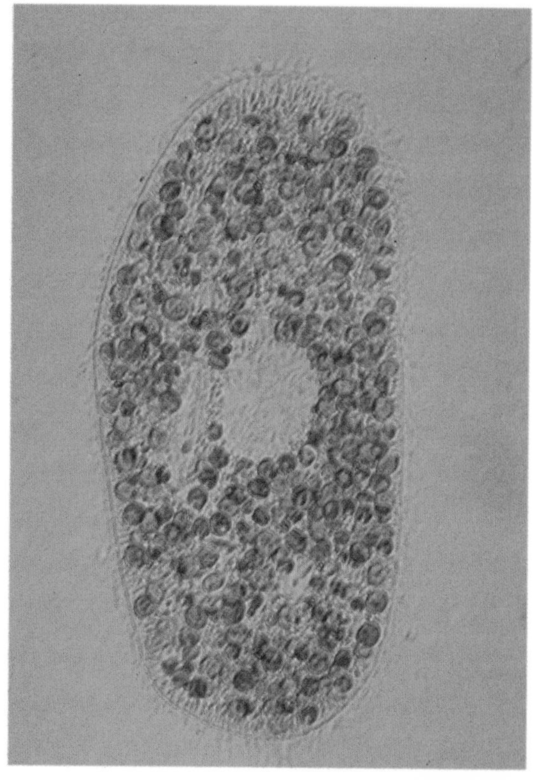

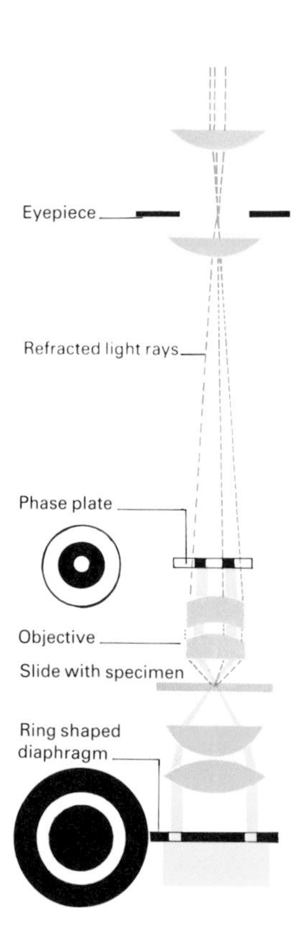

▲ *In a phase contrast microscope, the light rays from the specimen which form the image are those refracted by the specimen, avoiding the opaque ring on the phase plate.*

◄► *The invertebrate* **Paramecium bursaria** *seen at a magnification of ×150 in the bright field of an ordinary microscope (left), in phase contrast (right) and interference contrast (far right).*

▼ *In whatever mode a microscope is set, photomicrographs are most readily obtained from photographic equipment with automatic exposure control integrated within the instrument itself. A modern camera microscope such as the model seen in operation below is fitted with a screen for general viewing. One of the objectives is designed for phase contrast work. A slider installed beneath its stage contains an interchangeable diaphragm for phase-contrast, polarized-light or simple bright-field microscopy.*

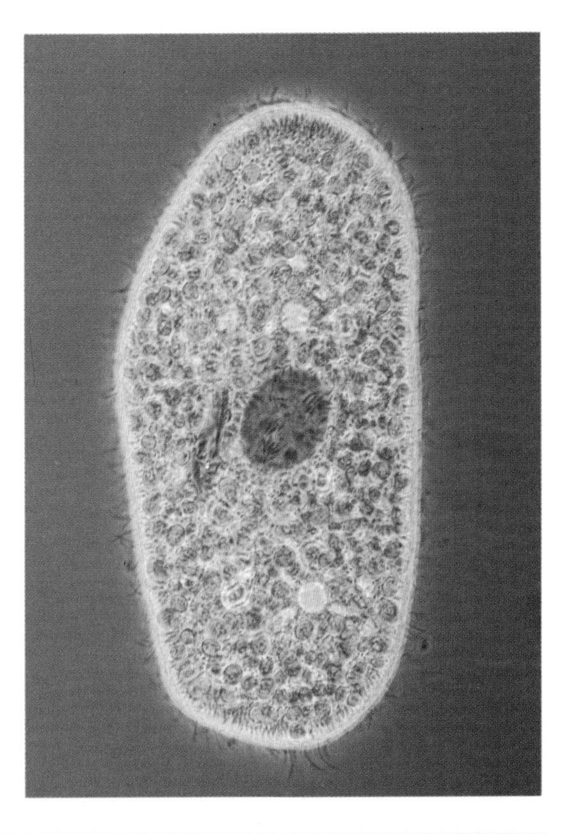

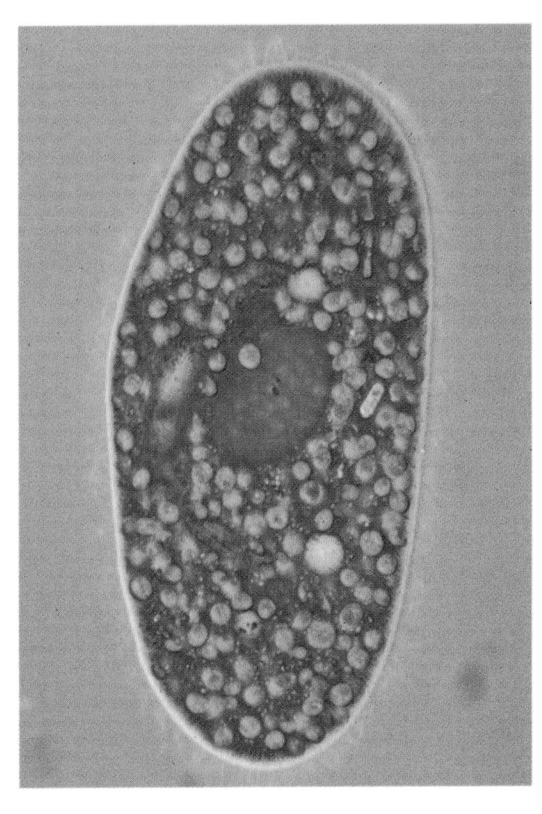

Interference contrast microscopy

Like phase contrast, interference contrast microscopy uses a beam of light which is split and then recombined. The essential difference is that whereas in phase contrast the light is divided by fine detail in the object, in an interference contrast microscope the light is divided by a beam splitter such as a half-silvered mirror. A direct beam is allowed to pass through the specimen, while a reference beam, whose phase may be controlled, is brought by a separate path to form a final combined image with the reference beam. This system permits adjustment to give the best contrast. Quite striking relief effects may be produced; the image is sharper than in phase contrast, and there is no bright halo around revealed features. The technique is more sensitive to gradual changes in microstructure, and can show surface tilts as well as steps. Contrast-enhancing techniques in photomicrography can further improve the quality of an image of the specimen in interference contrast work.

A simple example of a method of producing interference contrast is by a two-beam interferometer. Monochromatic light (that is, light of only one wavelength) is split into two equal beams at right angles to one another. One beam passes through the objective to be reflected from the specimen surface in the normal way; the other is focused through an identical objective onto a metallized optical flat mirror. The beam reflected from this reference surface is reunited by the beam splitter with that from the specimen to form the image. The phase of the reference beam can be adjusted by controlling the distance of the optical flat mirror from the beam splitter and may vary across the image by incorporating a small tilt.

The electron microscope

In 1878 the German physicist Ernst Abbe (1840-1905) proved that the optical microscope is limited by, and dependent upon, the wavelength of the light used – the shorter the wavelength, the greater the resolution and therefore the more details that can be seen. However, even when ultraviolet light is used, the smallest detail that can be distinguished is of the order of 100 nanometers.

Resolving finer detail depended on two discoveries. First, in 1924 the Frenchman Louis de Broglie put forward the theory that a moving electron was characterized by a wave form very similar to that of light, and this was verified experimentally in 1927. Second, the German physicist Hans Busch discovered an analogy between the focusing effect of a convex (converging) lens on a light beam and that of a magnetic coil on a beam of moving electrons. Since the wavelength of the electron is very much shorter than the shortest wavelength of light, an electron beam lens and a magnetic coil lens can be used to form an image of objects far smaller than those visible with optical microscopes. The first commercial electron microscope appeared in 1936; since then many types of instrument have been developed.

Passing the beam through the specimen

The transmission electron microscope (TEM) consists of an electron "gun" and an assembly of magnetic lenses, enclosed within a column evacuated to prevent the scattering of the electrons by collisions with molecules of gas. The electron beam passes through a specially thinned specimen and is then focused to produce a magnified image.

The electron source is usually a pointed tungsten filament heated to about 2,500°C. The filament is kept at a high negative potential (−40 to −100kV), and the electron beam is accelerated through a small hole in the earthed anode before being focused on the specimen by a system of condenser lenses. The wavelength of the electrons is determined by the accelerating voltage.

The possible magnifications are increased by having more than one projector lens – a magnification of ×200,000 might be divided between the objective (×100) and two projector lenses of ×20 and ×100. Variations in magnification can be achieved by altering the current in the coils of the magnetic lenses.

Individual lenses take the form of coils whose common axis is the axis of the microscope. The magnetic fields caused by small currents passing through the coils are concentrated by a soft iron casing around the coil. If very strong magnetic fields are required, accurately machined pieces of soft iron called pole pieces are inserted in the center of the coil. Apertures of different sizes can be placed in the lens system. The condenser aperture controls the illumination of the specimen, and the objective aperture is important in controlling the contrast of the final image. Specimens are placed in the vacuum chamber through a vacuum airlock; other attachments tilt the stage, or allow thermal, chemical or mechanical treatments to be carried out under continuous observation.

Transmission electron microscopy is used extensively in metallurgy, in particular to study crystal defects and their interaction under mechanical stress, and also to investigate phase transformations. An early biological application was in the study of plant and animal viruses; other work has been carried out on the internal structure of a wide variety of plant and animal cells.

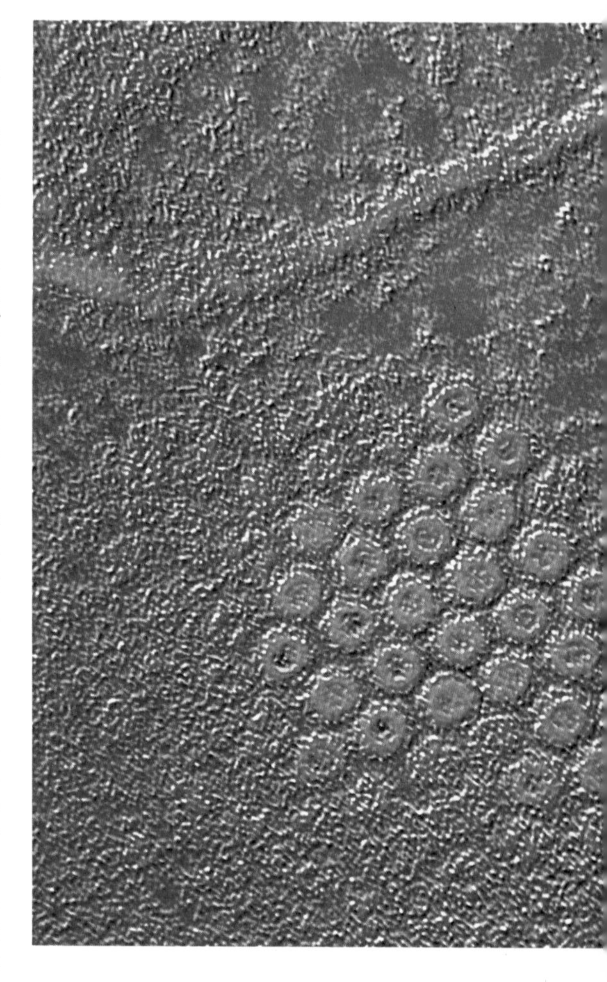

Scanning electron microscopy

*The scanning electron microscope produces its image in a very similar way to the cathode ray tube of a television set (**▸ page 188**). A "gun" fires a beam of high-energy electrons through a series of magnetic fields which focus it to a fine spot; the specimen is placed at this point. Other magnetic fields deflect the spot to follow a raster, a pattern of lines like that on a TV screen, and a linked cathode ray tube is controlled so that its illuminated spot traces the same pattern. The ratio between the sizes of the rasters traced on the face of the tube and on the specimen determines the level of magnification.*

When the high-energy electron beam strikes the specimen, secondary electrons are emitted, the number of which depends on the shape and nature of the specimen at that point. These secondary electrons are collected and provide a signal that controls the brightness of the cathode ray tube beam. The light intensity at the tube face thus varies in synchrony with the emission of electrons from the specimen surface and forms an image of the specimen. This image has the advantage of being three-dimensional.

Scanning techniques have a general applicability. As with transmission electron microscopy, X-rays and electrons of selected energy can be used with different kinds of collector to produce X-ray maps. They may also be used to analyze the chemical elements present in the surface.

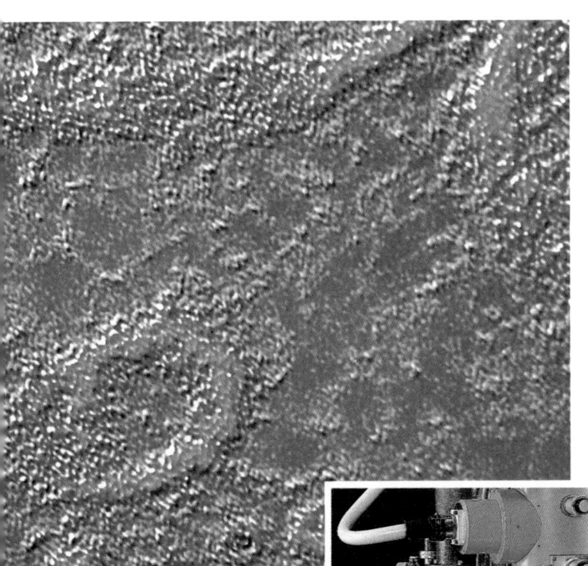

► *Section through an electron microscope.*

◄ *The high resolution of a TEM permits analysis of microorganisms, like the hepatitis A virus. Magnification ×60,000.*

▼ *In a conventional TEM, the objective lens focuses electrons scattered as they pass through the specimen. In this Zeiss EM 902, an imaging spectrometer, an array of mirrors and prisms analyzes the scattered electrons to improve image contrast.*

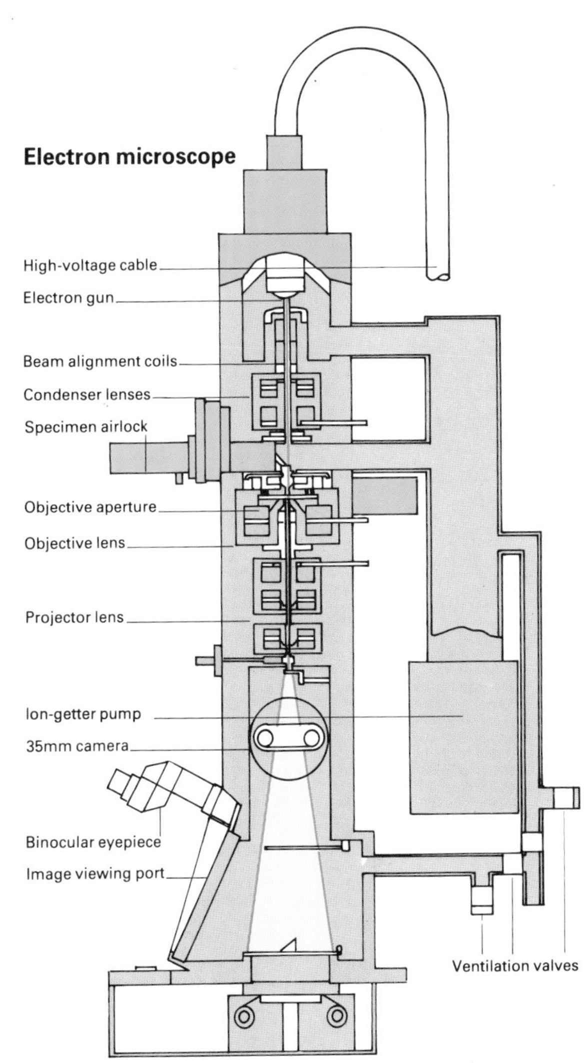

▼◄ *The scanning electron microscope, which appeared in the mid-1960s, enabled microscopists to study three-dimensional objects by probing their surfaces with fine pencil beams of electrons. Typical of the results obtained from SEM equipment is this image of a bird cherry aphid next to its recently cast-off exoskeleton.*

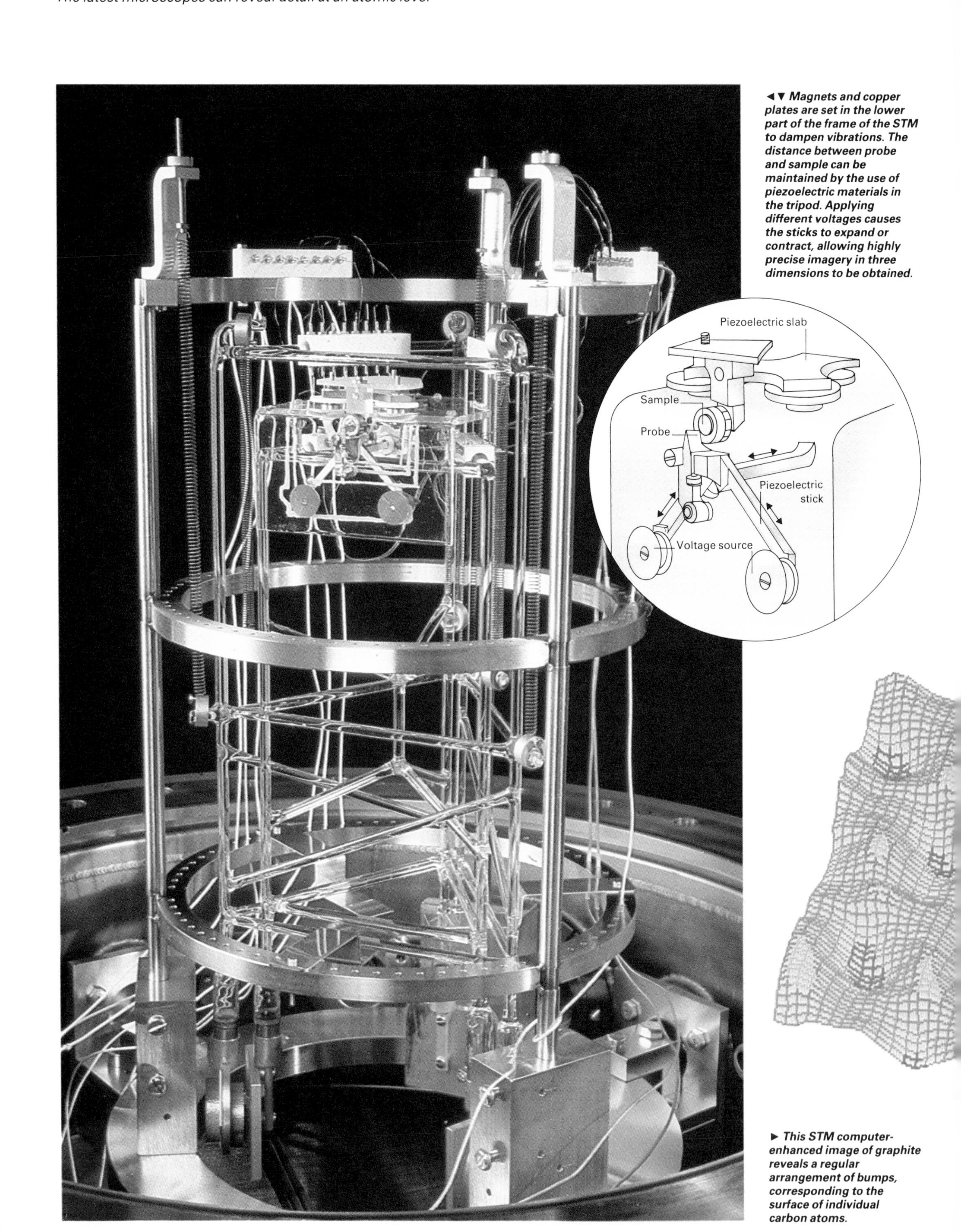

◄▼ Magnets and copper plates are set in the lower part of the frame of the STM to dampen vibrations. The distance between probe and sample can be maintained by the use of piezoelectric materials in the tripod. Applying different voltages causes the sticks to expand or contract, allowing highly precise imagery in three dimensions to be obtained.

► This STM computer-enhanced image of graphite reveals a regular arrangement of bumps, corresponding to the surface of individual carbon atoms.

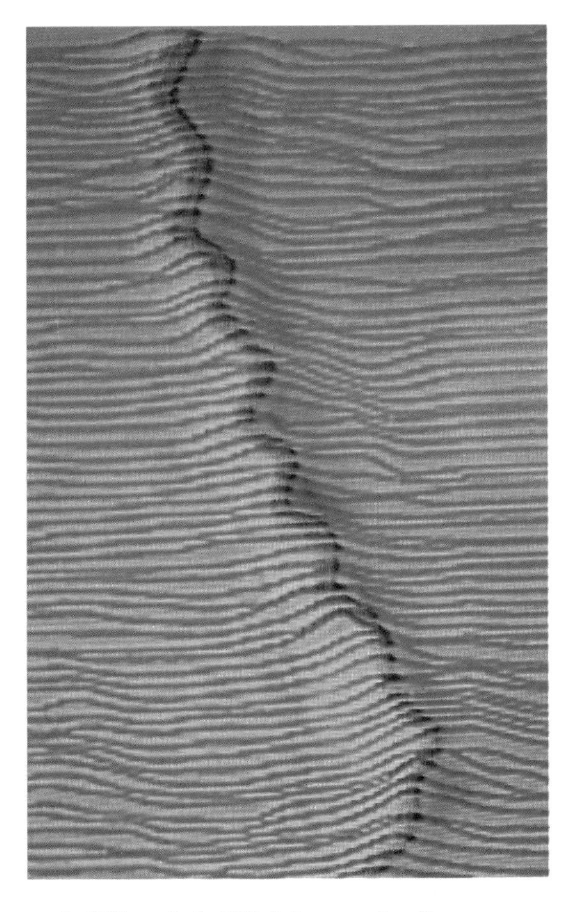

▲ *An STM graph of a DNA chain on a carbon film.*

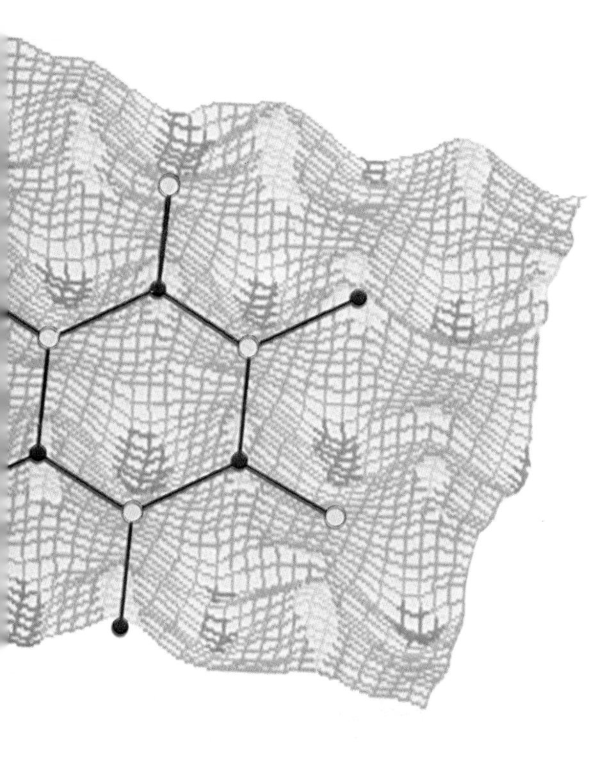

The scanning tunneling microscope

A remarkable experiment developed by Gerd Binnig and Heinrich Rohrer at the IBM Research Laboratory in Zürich and first reported in 1984 shows surface profiles and permits identification of superficial atomic species. The instrument is a scanning tunneling microscope (STM), and it operates with a precision well below the interatomic spacing. The image is formed by moving a fine point just above the specimen surface and recording the height of the point as it moves.

The key feature of the tunneling microscope is the system of position detection. It depends on the fact that, on an atomic scale, the surface of a material does not correspond to an immediate change from a solid to, say, an adjacent gas. The outermost part of a material is the electron cloud surrounding the atomic nuclei of the surface layer and the density of electrons in this cloud falls off progressively with distance out from the surface; a given increase in distance corresponds to a fixed factor in density, though this fall-off is very rapid.

When the search point of the microscope is brought towards the surface, electron clouds begin to overlap significantly at a separation comparable with the interatomic spacing. Although the materials are not fully in contact, electrons may move from the point to the surface and vice versa; this form of exchange is called tunneling. If a small potential difference is maintained between the point and the surface, a current will flow, and its magnitude, which depends on the overlap of the electron clouds, is an indication of the separation between the point and the surface.

The search point is mounted on a framework of piezoelectric sticks, to keep it one interatomic spacing from the surface. Such sticks vary slightly in length when a potential is applied across them, so by matching the current between the point and the specimen to the voltage controlling length, a chosen separation may be maintained. An image is formed by recording the voltage which controls the separation as the search point scans the surface.

Making a search point of atomic dimensions is simpler than might seem likely. Fine tungsten carbide points may be produced by grinding, and one atom will normally have a position more prominent than the others. To turn this into a single atom probe, a voltage is applied between the tungsten carbide point and the specimen which, with some luck, will transfer an atom from the specimen to the probe. This is then used as the probe tip.

The STM is unique in avoiding the wavelength limitation to which conventional microscopes are subject. No free particles – photons or electrons – need to be focused to form the scanning tunneling image so the wavelength limitation does not apply. The high resolution of the STM has been used to describe the surface of silicon and to show the detailed symmetry of its dimpled surface. Potentially of greater interest are studies of oxygen adsorbed on nickel; since the tunneling current depends on both the tunneling distances and the electronic structure of the surface, atoms of different elements may be identified and their distribution over the surface measured.

When applied to biological studies the technique shows another useful feature. It may be used in air and water, so specimens are not necessarily destroyed in the course of examination. Few biological studies have yet been done, but it has been used to scan the surface of the nucleic acid DNA and has resolved a pattern corresponding to its helical structure. The technique has also been used to study viruses.

See also
Measuring with Electricity 41–46
Seeing with Sound 65–72
Seeing beyond the Visible 73–86
Photography 87–90
Chemical Analysis 105–108

X-rays in microscopy

Another area of recent progress in microscopy has been in the use of X-rays. The advantages of using X-ray technology in producing images of internal structure, to reveal crystalline structure or chemical composition, have long been understood (▶ page 74); but until recently scientists were unable to focus the rays successfully. Another major difficulty was that of the damage that might result to the sample.

Effective X-ray microscopy has, however, become available in recent years; synchrotrons (▶ page 110) can now provide a highly concentrated source of X-rays which is focused by the use of precision-etched Fresnel zone plates. These allow only certain sections of the beam to pass through, then deflect them them at small "grazing-incidence" angles (▶ page 74) in such a way that they add up constructively in step at a common focal point.

In the scanning X-ray microscope, the microfocused waves of differing frequencies are scanned over the sample (by moving the sample and not the beam, thus minimizing the danger of damage). Detectors pick up the transmitted X-rays, and permit analysis of different absorption patterns.

Another new microscopic device that uses X-rays is the scanning proton microscope. A focused beam of high-energy protons is scanned across the specimen; this induces the atoms to emit X-rays. A computerized image color-codes the different wavelengths which identify specific elements within the sample.

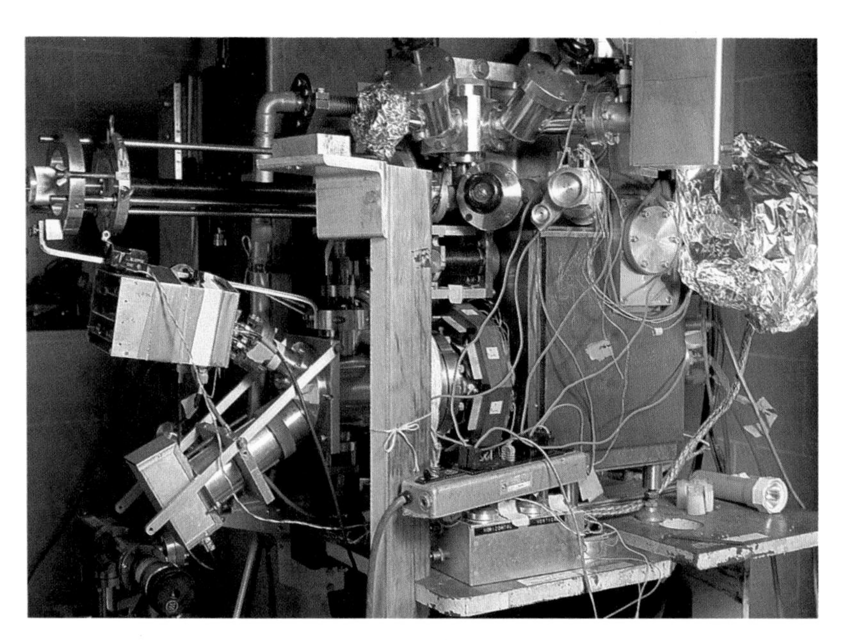

▲ The field emission electron microscope was a forerunner of the scanning tunneling microscope. It was the first device to produce micrographs showing individual atoms, in 1970-71.

Emission techniques

In some forms of microscopy the sample is not illuminated. Instead the sample is made to emit particles by the use of high temperatures or large electric fields. Examples of images formed by this method include thermionic emission, field emission and field ion microscopy.

In thermionic emission microscopy, electrons in the specimen are supplied with energy to allow them to escape from the surface. They are then accelerated through a series of objective and projector lenses which focus an image of the surface on a fluorescent screen or photographic plate. The magnifications normally obtained are in the range of ×100 to ×12,000.

To release electrons from the surface the sample must be heated to temperatures in the range 1,200–2,000°C. Many metals melt at these temperatures, but if the surface is coated with a thin layer of cesium or barium the temperature required to produce electron emission is reduced to about 500°C.

Field emission microscopy uses an electric field to cause the release of electrons from an unheated specimen. The electrical fields required to do this are enormous but feasible if the metal sample has the form of the tip of a needle, and the electrons can be registered on a curved screen around the tip, forming a magnified image of the specimen surface.

Magnifications of up to ×200,000 can be obtained in this way and are used to show reactions at solid surfaces, such as adsorption of gases, oxidation and corrosion, and the growth of thin films formed by vapor deposition.

Field ion microscopy calls for yet higher electric fields of opposite sense. These strip ions from the specimen and allow yet higher magnification.

◀ To obtain this field-ion micrograph, a needle of iridium was placed in a gas-filled chamber and a positive potential applied to its tip. Gas atoms became ionized and were repelled from the tip to strike a fluorescent screen. A magnified image of the atomic structure of the tip was then formed; the dots are the locations of individual atoms, the ring-like patterns are facets of a single crystal of the metal.

Optical Telescopes

7

Refractor and reflector telescopes...Modern telescopes: mounting, siting and control...Astronomy and photography...Photoemissive detectors...Observatories ...Widening the aperture...PERSPECTIVE...Early refractors ...Early reflectors...Chronology...Image Photon Counting System...Charge-coupled device...Hubble Space Telescope...Spherical telescopes

◄ **Using refracting telescopes fitted with a finely-ground objective lens, good-quality images were being obtained by the late 19th century. The instruments used were often of a size containable within an observatory. This 60cm-aperture refractor was built by the American Alvan Clark (1832–1897) and installed in Lowell Observatory at Flagstaff, Arizona.**

Optical astronomy is concerned with the collection, detection and analysis of light from the remote depths of space. Light is collected by telescopes, of which there are two basic types – the refractor and the reflector. A refractor employs a lens (the objective) to collect light rays and bring them together to form an image at its focus or focal plane, while a reflector uses a concave mirror (the primary) to achieve the same result. In practise telescopes usually contain additional lenses or mirrors.

Large refractors suffer from optical and mechanical problems which are complex and expensive to deal with, and for this reason all large modern telescopes are of the reflecting type. In very large reflectors the observer or the instrumentation can be placed in a "cage" at the prime focus – in front of the primary – without obscuring too much of the light traveling down towards the main mirror. In other systems, smaller secondary mirrors reflect light to the side of the tube or back through a hole in the primary, as in the widely adopted Cassegrain design. The surface of a Cassegrain primary has a parabolic curve, but better images are obtained if the curve is made hyperbolic, and the system is then called a Ritchey-Chrétien. In the coudé design, a system of mirrors reflects light to a fixed position – often in a separate room – regardless of the direction in which the telescope is pointing. Catadioptric systems use a combination of lens and mirror to form an image, a classic example being the Schmidt – a wide-angle telescope that can photograph areas of the sky six degrees wide (compared to half a degree or less with other systems).

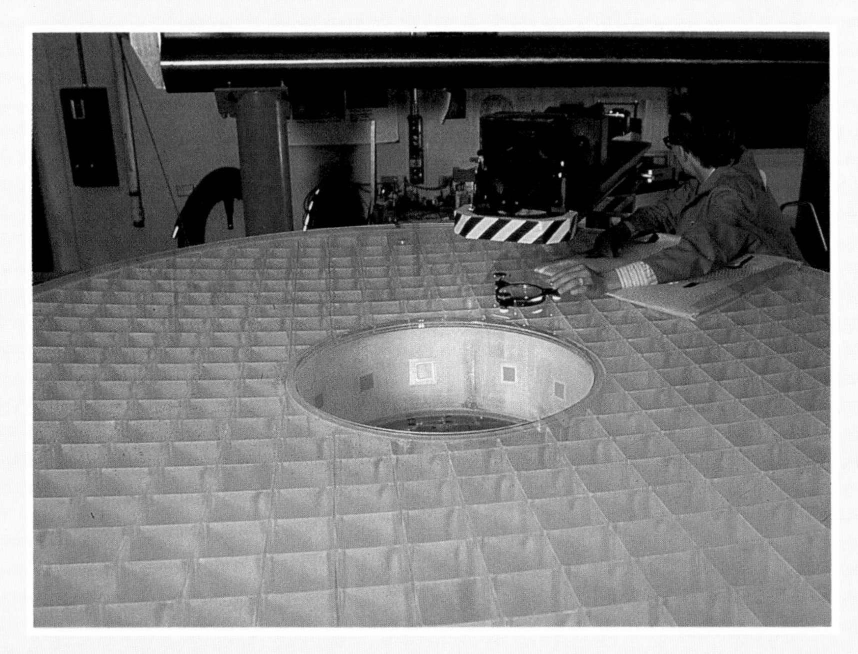

◄ **Highly-accurate mirrors are an essential part of the modern reflecting telescope. For the NASA Hubble Space Telescope, new technology features in the design of the 2·4m primary mirror, seen here under construction. The lightweight mirror is made up from two sheets of ultra-low expansion titanium silicate glass sandwiching a fused glass honeycomb structure.**

Early refracting telescopes

Despite rumors that crude telescopes existed a little earlier, most historians of science believe that the telescope was invented in 1608 by the Dutch instrument maker Hans Lippershey (c.1570–c.1619). Certainly the first person to use the telescope seriously for study of the heavens, and to appreciate the significance of what he saw, was the Italian scientist Galileo Galilei (1564–1642). Although the telescopes he made were the best of his time, they were of poor optical quality and magnified no more than about 30 times. Even so, when he began observing in 1609-10, he was able to see craters and mountains on the Moon, spots on the Sun, the phases of the planet Venus, the four moons of Jupiter, and resolve the Milky Way into a myriad of stars. His observations revolutionized humanity's perception of the universe.

Early refracting telescopes suffered from two major defects – spherical aberration and chromatic aberration. As a result, star images were blurred and surrounded by a fuzz of color. These aberrations could be reduced by making the curvature of the lens as slight as possible, but this gave the lens a very large focal length. As a result absurdly large "aerial" telescopes – up to 60m long and slung from tall masts – were built in the 17th and early 18th centuries. A famous example was constructed by Johannes Hevelius (1611–1687) at his private observatory in Danzig; he used it to help him draw lunar maps.

In 1674 the English glassmaker George Ravenscroft (1618–1681) invented flint glass, a very hard, clear glass containing lead. Half a century later Chester More Hall (1703–1771) realized that crown and flint glass spread out light by differing amounts. He designed a compound achromatic lens, comprising a crown lens followed by a flint one; by this means he succeeded in canceling out the color spread to a large extent. By the late 1750s, the optician John Dollond (1707–1761) could make relatively short refractors in which both chromatic and spherical aberrations were much reduced.

The development of large refractors culminated in the commissioning of the 1m aperture refractor at Yerkes Observatory in Wisconsin in 1897. Still in use, this giant refractor has never been surpassed.

Chronology

c.1570 Earliest refracting telescopes invented
1609 First astronomical studies performed by Galileo using a simple refracting telescope
1611 Convex eyepiece lens proposed by Johannes Kepler
1645 Eyepiece adapted for terrestrial telescopes by Anton Schyrle von Rheita; erector lens plus field lens produced wider field and erect image
1663 Reflecting telescope proposed by James Gregory, but not built until after Newton's instrument
1668 First reflecting telescope, using metal alloy mirror, constructed by Isaac Newton with a magnifying power of×30; Cassegrain independenty invented his own version
1729 Achromatic telescope invented when Chester Moor Hall devised achromatic lens – comprising concave lens of flint glass and convex lens of crown glass
1733 First compound lens constructed by George Bass to Hall's design
1740 High-quality concave mirrors for Gregorian telescopes produced with introduction of James Short's new grinding process
1753 Achromatic telescope lens, compensating for color scattering, built by John Dollond
1789 Speculum alloy mirrors with 60 percent reflectivity incorporated by William Herschel into two telescopes
1845 Largest-ever speculum mirror used by Lord Rosse in his reflecting telescope to discover spiral structure of nebulae
1856 Mirrors made from deposition of silver on glass,

The development of the reflecting telescope

The first reflecting telescope was built in 1668 by the English scientist Isaac Newton (1642–1727). He realized that, since all wavelengths of light are reflected equally, a curved mirror should not suffer from chromatic aberration. From his day to the later 19th century, mirrors were made of an alloy known as speculum metal, which was hard to work with, reflected poorly, tarnished quickly and was greatly affected by temperature changes. Newton's speculum metal was a mixture of copper, tin and arsenic; it reflected only 16 percent of the incoming light.

The art of mirror-making was improved by the German-born English astronomer William Herschel (1738–1822). He devised new speculum metals which reflected up to 60 percent of light when freshly polished, and eventually he constructed a 1·2m aperture telescope housed in a 12·2m long tube. The largest metal mirror of all was the 1·8m speculum made for the 3rd Earl of Rosse (1800–1867) and installed in his giant telescope at Birr Castle in Ireland in 1845.

Silver-on-glass mirrors were first made for astronomical telescopes in 1856. Glass provided a better base, was easier to work and less affected by temperature, while freshly deposited silver reflected up to 95 percent of light. Mirrors of this type opened the way for the reflector to become the dominant type of telescope. The first such large mirror was the 1·5m installed at Mt Wilson, in California, in 1908. This was followed in 1917 by the epoch-making 2·5m instrument at Mt Wilson. A 5·1m giant installed at Mt Palomar in 1948 remained the largest optical telescope until the 6m Soviet instrument was installed at an observatory in the Caucasus mountains in 1976.

▲ ***Isaac Newton's reflecting telescope of 1668.***

improving reflectivity to more than 90 percent
1897 Yerkes 1m telescope commissioned, the world's largest refractor
1930 Camera telescope invented by Bernard Schmidt combining best aspects of reflecting and refracting telescopes, and invaluable for viewing whole sky
1931 Reflectivity improved still further when John Strong developed vacuum method of depositing aluminum onto astronomical mirrors
1932 Naturally-occurring radio emission from sky discovered by Karl Jansky, allowing development of radio telescopes
1934 Electron telescope invented independently by Vladimir Zworykin and Manfred von Ardenne
1937 First radio telescope built by US engineer Grote Reber
1946 Photomultiplier tube introduced for detecting and enhancing faint stars
1946 Radio interferometers developed in Britain and Australia
1946 5·1m reflector established at Mt Palomar
1949 Rocket-borne radiation counters measure X-rays from Sun – the first X-ray astronomy
1969 Multichannel spectrometer begins operations at Mt Palomar observatory, allowing multiple simultaneous observations
1972 Satellite Landsat 1 launched by NASA to relay images of Earth
1978 International Ultraviolet Explorer Satellite launched
1983 Exosat X-ray telescope launched
1983 Infrared Astronomical Satellite (IRAS) launched

From an engineering point of view, the simplest way to mount a telescope is the altazimuth system. In this the telescope can pivot around two axes, one vertical and one horizontal. Traditionally, most large telescopes have been mounted equatorially, with the "vertical" axis tilted over to be parallel with the Earth's axis of rotation. The motion of a star across the sky can then be followed – in principle – merely by turning this axis at a rate of one revolution per day. Equatorial mounting suffers from changing stresses and balance, and hence deformation, as the orientation of the telescope changes. Modern computer control allows both axes of an altazimuth to be driven precisely to follow the motion of a star across the sky, and several large, recently-constructed telescopes have adopted this system.

From the astronomer's point of view, the most important functions of a telescope are its ability both to collect light and to resolve fine details in images of distant objects. The amount of light which a telescope collects depends on the surface area of the objective or primary mirror and is proportional to the square of the aperture. Thus a 2m aperture telescope will collect four times as much light as a 1m telescope and will reveal fainter, more distant objects. In theory, the larger the aperture, the better the resolving power or resolution. A telescope of 0·15m aperture should resolve details as small as one arcsec in apparent diameter (an arcsec is a second of angular measurement, 1/3,600th of one degree) – roughly equivalent to the height of a human seen at a range of 400 kilometers. In practise, turbulence in the Earth's atmosphere causes images to wobble about and smears out a star's image into a "seeing disk" seldom smaller than one arcsec across and often larger than that. Although astronomers try to minimize the obscuring and turbulent effects of the atmosphere by siting their telescopes on carefully-selected mountain locations, often arid, cold environments, large Earth-based telescopes cannot attain their theoretical resolving powers.

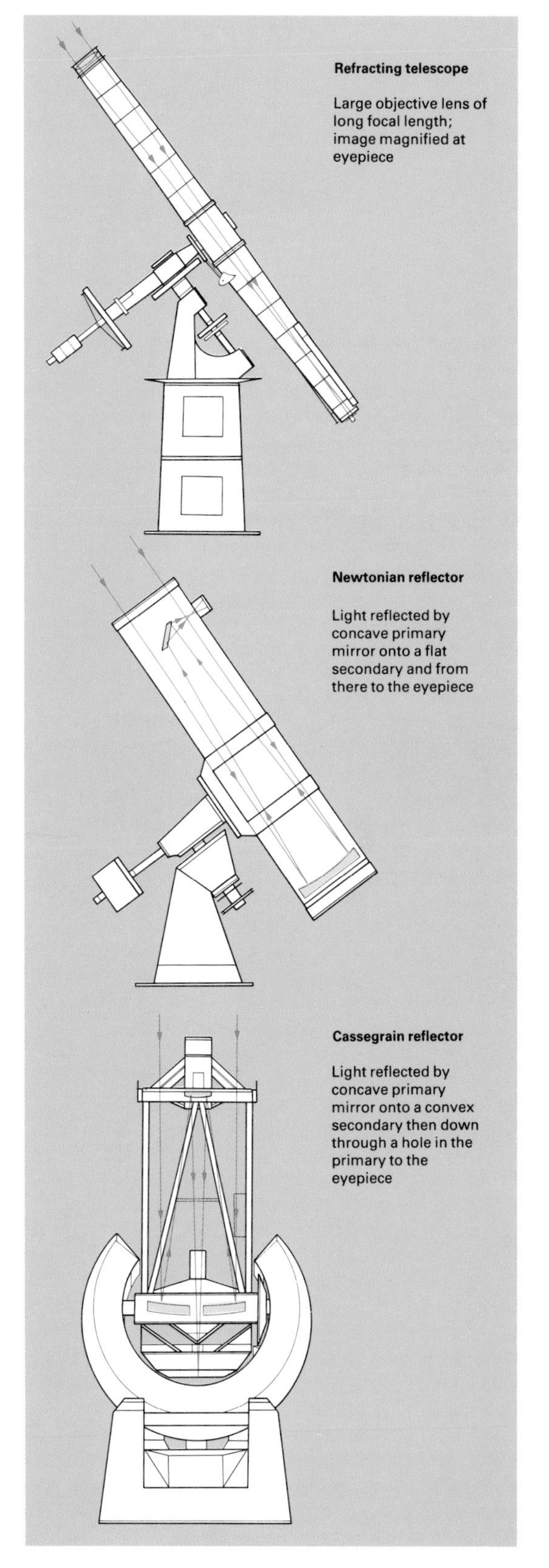

◀ The La Palma observatory in the Canary Islands houses the 2·5m "Isaac Newton" telescope (shown here). On the same site is the 4·5m "William Herschel" telescope. With its superior optics, computer control and instrumentation, it should become the most powerful telescope on Earth, despite being only third in terms of sheer aperture.

▶ Modern telescopes, such as this one at Siding Spring in Australia, permit the astronomer to work in comfort in the control room. Computers are used to store the data, and image-enhancement techniques are provided to isolate minute variations in the wavelength or intensity of the radiation and present them in vivid color.

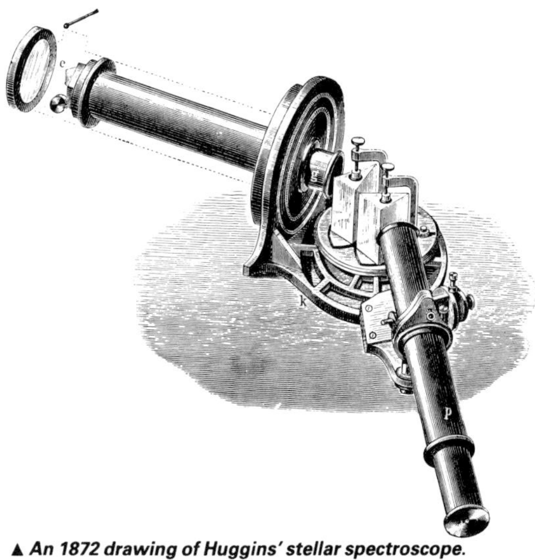

▲ An 1872 drawing of Huggins' stellar spectroscope.

Analyzing the stellar spectra

One key instrument in the armory of the modern astronomer is the spectrograph. In 1802, about a century after Newton showed that sunlight could be split into a rainbow band of color by a glass prism, the English scientist William Wollaston (1766–1828) built a spectroscope into which light was admitted through a narrow slit. This improved the sharpness of the spectrum and revealed the presence of dark lines at particular wavelengths in the colored band of light (▶ page 107).

By 1814 the German physicist Joseph von Fraunhofer (1787–1826) had mapped the wavelengths of more than 500 lines in the solar spectrum. By 1856 Gustav Kirchhoff (1824–1887) and Robert Bunsen (1811–1899) had established that while a hot solid, liquid or dense gas emits a continuous spectrum of all wavelengths, a low-density gas emits only at a few wavelengths, so giving an emission-line spectrum. They also showed that if a continuous spectrum passes through a rarified gas, light will be absorbed at certain wavelengths to give a pattern of dark absorption lines. Since this happens in the atmosphere of a star, and each element gives its own unique pattern of lines, analysis of the spectrum can reveal the star's chemical composition, temperature and density, as well as its speed of approach or recession.

Spectroscopy is the foundation of astrophysics. One of the first to exploit its potential was the British astronomer William Huggins (1824–1910), who in 1864 identified chemical elements in stars and showed that some nebulae (luminous patches in the sky) were composed of glowing gases. Four years later he measured the radial velocities of some stars.

In a modern spectrograph light is spread out either by prisms, or by a diffraction grating, a glass plate on which more than 500 lines per millimeter have been ruled like furrows on a field. The best spectrographs can resolve spectral lines narrower than 0·001 percent of the wavelength of visible light. The spectrum is then recorded photographically or electronically for subsequent computer processing and analysis.

▲◀ Intensified images suffer from spurious spots of light or "noise". An Image Photon Counting System (IPCS) detector, mounted on the end of this telecope, scans the intensified image by computer, rejects unwanted noise and counts every genuine photon registered during the exposure. Blips on the output scan are caused by photons hitting the tube; the smaller wiggles are electron noise.

Photography and electronic devices have replaced the human eye for practically all forms of professional observation. Photographic emulsion, unlike the eye, integrates the light which falls on it so that, within limits, the longer the exposure, the more light is registered, so that objects can be revealed which are too faint to be seen visually. Photographs provide permanent records of large numbers of images which can be analyzed and measured at leisure. Images can be scanned electronically and processed by computer to emphasize specific features. For example the computer can assign different colors to subtly different levels of brightness and so bring out clearly, in false-color images, features which are inconspicuous or undetectable to the naked eye.

In an hour's exposure a Schmidt plate can easily register a million star images. Months of laborious visual measurement of plates such as these can be eliminated by devices such as the automated photographic measuring machine (APM) at Cambridge, England, which, by means of its computerized laser and television scanning system, can measure to within one micrometer the positions of all these stars in an operating run of eight to ten hours' duration.

Photographic emulsion, however, is not an efficient collector of light; even the best emulsions register only three or four of every hundred photons which fall on them. Electronic detectors have light-detecting efficiencies 10 to 100 times better than average emulsions. The basic principle of a photoemissive detector is that incoming photons expel electrons from a sensitive surface (photocathode). Other detectors, such as the photodiode, utilize the ability of light to

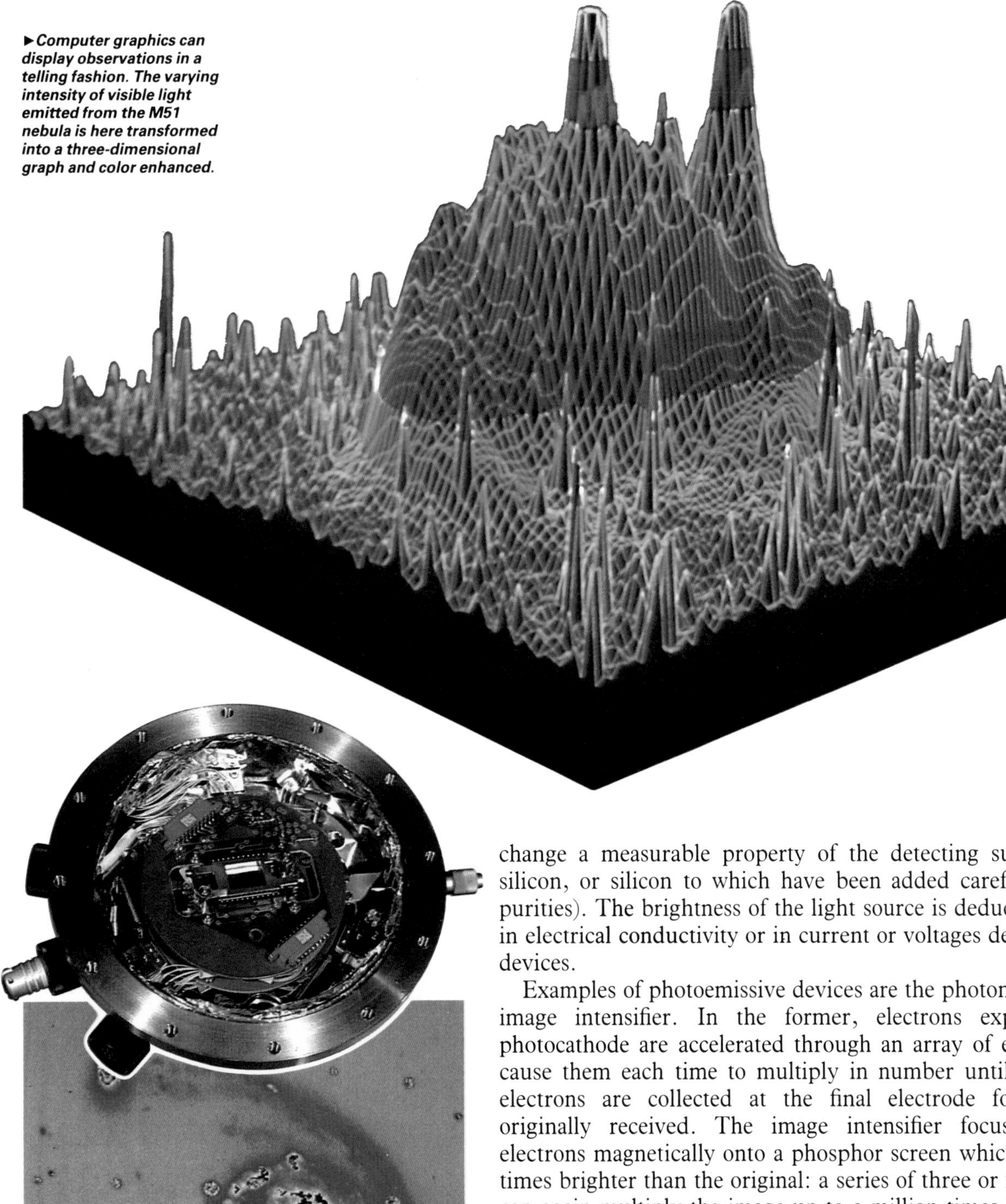

▶ *Computer graphics can display observations in a telling fashion. The varying intensity of visible light emitted from the M51 nebula is here transformed into a three-dimensional graph and color enhanced.*

▲ *A pair of binary galaxies such as NGC 7752 and NGC 7753, in orbit around each other, would scarcely be visible in an ordinary photograph. This false-color image of the larger was taken with a CCD camera, the instrument seen face-on in the picture above. The CCD itself is the small rectangle fitted in the center of the specially cooled camera jacket designed to attach to a telescope.*

change a measurable property of the detecting substance (usually silicon, or silicon to which have been added carefully selected impurities). The brightness of the light source is deduced from changes in electrical conductivity or in current or voltages developed by these devices.

Examples of photoemissive devices are the photomultiplier and the image intensifier. In the former, electrons expelled from the photocathode are accelerated through an array of electrodes, which cause them each time to multiply in number until up to a million electrons are collected at the final electrode for every photon originally received. The image intensifier focuses the released electrons magnetically onto a phosphor screen which glows about 50 times brighter than the original: a series of three or four such devices can again multiply the image up to a million times brighter than the original light received.

Images of extended objects such as galaxies can be recorded by arrays of photodiodes, or by a charge-coupled device (CCD). A CCD consists of a silicon chip little more than one centimeter across, divided into a grid of several hundred thousand elements or pixels. Photons entering the silicon liberate electrons, which are stored in the pixels so that each builds up an electrical charge proportional to the amount of light falling upon it. At the end of an exposure the charges are read off systematically and the image reconstructed by computer. CCDs can detect up to 70 percent of incoming photons and are so efficient that a one-meter telescope using a CCD is as effective as a five-meter telescope using the best photographic plates.

Astronomers detect light with the aid of photographic emulsions or with much more sensitive electronic devices. The light is analyzed by a range of instruments including polarimeters, photometers (which measure brightness) and spectrographs. Polarimeters analyze the way in which incoming light is vibrating and can reveal information about how the light was produced, or what has happened to it on its journey through space.

Not even the best telescopes and instrumentation can operate to their full potential under the turbulent, polluted atmosphere, or in the glare of artificial lighting. To minimize these problems, observatories have been established on remote mountain sites far from cities. Today the major instruments are clustered on a limited number of favored localities. Telescopes in the 3–4-meter aperture range are to be found, for example, on Kitt Peak, Arizona (where the world's largest solar telescope is located) and, to view the Southern skies, in the Chilean Andes or Australian sites such as Siding Spring. Arguably the best of all ground-based locations are mountain peaks on isolated islands where the smooth flow of air contributes to better visibility. Mauna Kea in Hawaii (altitude 4,200 meters) and El Observatorio del Roque de los Muchachos on La Palma in the Canary Islands (2,400 meters) are both sites of expanding international observatories for this reason. At such sites, however, difficulties of access and physical discomfort can hinder study. Computer control and television viewing from a console in a heated room can help, and remote control and observation from thousands of kilometers away is increasingly practiced over existing telecommunication links.

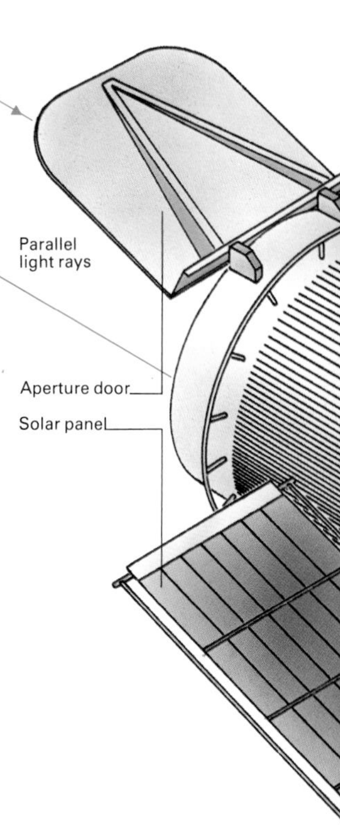

***►* The Space Telescope consists of three elements: the Support Module, made up of the satellite body, solar panels, computers and telemetry equipment; the Optical Telescope Assembly – the telescope itself plus fine guidance and optical control sensors; and the Scientific Instruments, including two cameras, two spectrometers and a photometer.**

***◄* The telescope, newly in operation on La Palma, of the Royal Greenwich Observatory, is one of the most advanced in the world. It can be controlled via satellite from Edinburgh in Scotland.**

***▼* Siding Spring Observatory, housing the Anglo-Australian telescope, stands on a mountain top in New South Wales. The location was chosen for its distance from city lights and the clear atmosphere.**

Space telescope

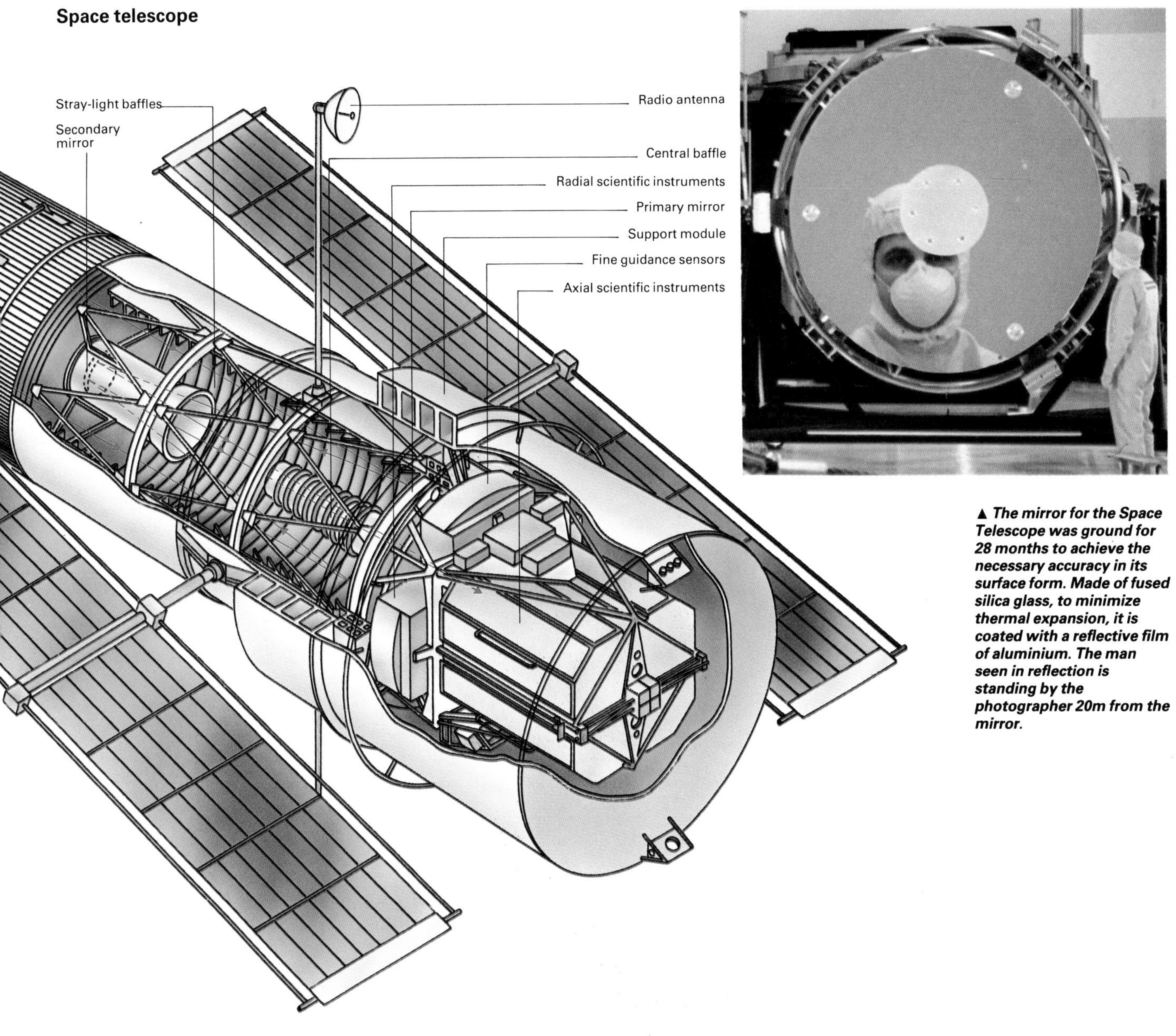

▲ The mirror for the Space Telescope was ground for 28 months to achieve the necessary accuracy in its surface form. Made of fused silica glass, to minimize thermal expansion, it is coated with a reflective film of aluminium. The man seen in reflection is standing by the photographer 20m from the mirror.

The Space Telescope

Satellite telescopes have mainly concentrated so far on those radiations which are wholly or largely obscured by the atmosphere – X-rays, ultraviolet and infrared. Nevertheless, telescopes and arrays of telescopes in space offer the best prospects for visible light grasp and resolution. The "Edwin Hubble" Space Telescope was planned to be placed in a 600km orbit by the Space Shuttle in 1986 (▶ page 225), but the explosion of the Challenger early in 1986 caused the launch to be postponed. It will herald the dawn of a new era of space-based optical astronomy and will surely revolutionize the subject. Its state-of-the-art instrumentation should detect stars 50 times fainter and up to seven times more remote than the best ground-based telescopes can manage. It should also attain its theoretical resolution of about 0·05 arcsec, and easily resolve details ten times better than can be seen from the ground.

See also
Introduction to Measuring 9–22
Microscopes 47–56
Seeing beyond the Visible 73–86
Photography 87–90
Space Transport 225–228

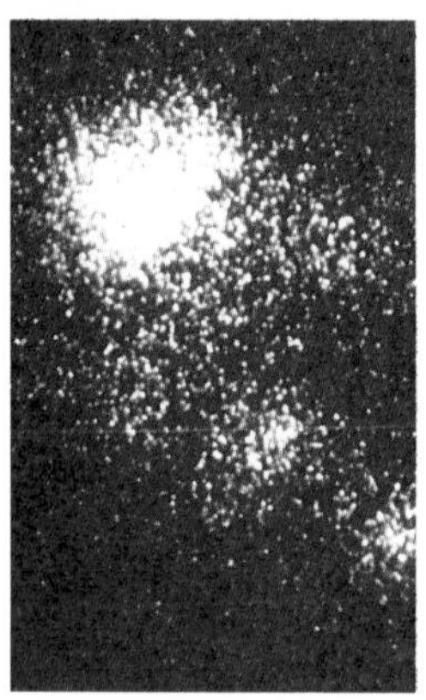

▲ The Multiple-Mirror Telescope on Mount Hopkins, Arizona, is one of the world's largest. With laser beams giving continuous alignment checks, its six mirrors are automatically adjusted by a computer. The building which houses the telescope is designed to rotate with it.

◄ The center of R136 nebula as observed using the technique of speckle interferometry. Astronomers were later able to resolve this image into eight separate stars.

Now that detectors are so efficient, the only way to gain substantially more light and thus probe even deeper into the universe is to build larger telescopes. While single mirror telescopes of 8–10 meters aperture are actively being considered, a cheaper alternative approach is to use several smaller mirrors aligned to combine their light at a common focus. The largest current example is the Multiple-Mirror Telescope (MMT) in Arizona, which uses six 1·8-meter mirrors to give the same light-grasp as a single 4·5-meter mirror.

One way of improving results is to minimize the adverse effects of atmospheric turbulence. "Speckle interferometry" is a technique which allows this effect to be very much reduced. An exposure of a few milliseconds "freezes" the disturbances produced by large numbers of little air cells, and the image of a star then consists of a large number of spots or speckles. Resolution approaching the theoretical limit of the telescope can be achieved by mathematically combining numerous speckle exposures.

Much greater improvements in resolution can be achieved by interferometry techniques which combine the light received by quite widely separated telescopes. The intensity interferometer at Narrabri in Australia was devised specifically to measure the diameters of stars, and has measured apparent diameters as small as 0·0005 arcsec. Aperture synthesis – a technique used to great effect in radio astronomy (➧ page 83) – is now being developed for optical astronomy. Two or more telescopes mounted on a track feed light to a common point. The spacing of the telescope is varied and the Earth's rotation is used to change the orientation of the baseline (the line joining the telescopes) relative to the sky. The data acquired in a series of runs can be combined to give the same resolution as a single dish equal in aperture to the maximum separation of the telescope. Following the success of a smaller prototype, the *Centre d'Etudes et de Recherches Géodynamiques et Astronomiques* (CERGA) in southern France is building a system which eventually will be equivalent to a 300-meter aperture.

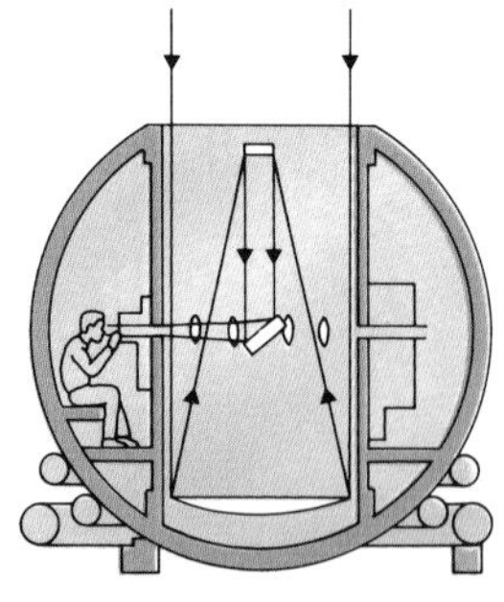

▲◄ Spherical "boule" telescopes, made of concrete and each with 1·5m apertures, comprise the CERGA optical interferometry station in southern France. At any one time, two or more boules can be moved along rail tracks to follow the motion of a star, and can be rotated continuously to direct the light they capture into a central laboratory. An astronomer can rotate each miniature observatory by "walking" it with his or her feet and using a computer for fine adjustment.

Seeing with Sound

8

Sound waves in air...Sound in liquids...Sonar...Sound in solids – seismic surveys...Ultrasound...PERSPECTIVE... Echoes from the atmosphere: sodar...The sofar channel...Sonar side-scan...Imaging the seabed... Seismic refraction...Seeing the fetus with ultrasound... Acoustic microscopes

Sound propagates through air as the result of the oscillations of air molecules away from a vibrating surface that represents the source of the sound. As the surface vibrates to and fro, the molecules alternately bunch together and move apart, creating a series of compressions and oscillations as wave fronts that spread out and move away from the source like the ripples from a stone thrown into a pond.

An important characteristic of sound is its frequency, the measure of the number of complete oscillations or cycles per second, measured in Hertz (Hz). Pitch is the subjective impression of a sound's frequency. Frequency doubles each time the pitch rises by an octave; thus the frequency of middle C on a piano is 256 Hz, while that of the highest C on the keyboard, three octaves higher, is 2,048 Hz. Another important characteristic of sound is its wavelength. Literally the distance from crest to crest of adjacent sound waves, the wavelength is defined technically as the velocity of sound divided by the frequency; the longer the wavelength, the lower the frequency. Audible sound wavelengths lie between 2 centimeters and 15 meters.

We are able to use our ears to identify the direction from which a particular sound is coming, even if we cannot see the source. From antiquity, fishermen navigating near a coast in dense fog have made loud noises and listened for echoes to help them guess their distance from known points, thus taking advantage of the relatively low velocity of sound.

Modern acoustic ranging techniques use horizontal or vertical arrays of microphones rather than ears. The direction and distance of the sound source are determined by sophisticated signal-processing techniques (◆ page 46) from the difference in arrival times at each microphone. Indeed acoustic detection and ranging of aircraft is enjoying a new lease of life as modern aircraft become more successful at flying below radar screens (◆ page 170).

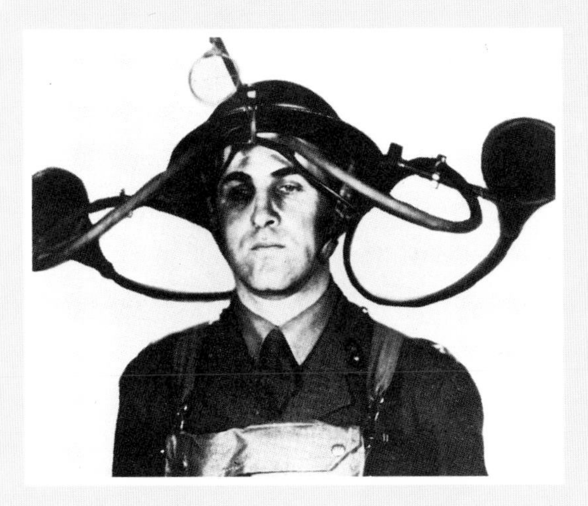

◄ ***We can detect the direction of sound wavelengths equal or shorter than the width of our heads, by sensing the differences between the sounds at the two ears. These Second World War ear trumpets effectively increased ear separation and enabled the wearer to locate lower frequency sounds such as artillery fire.***

Sounding the atmosphere

The first accurate estimation of the speed of sound was made by Englishman William Derham in 1708, from an experiment in which gunfire reports were timed from a distant church tower. It became obvious during the 19th century, however, that in the open air the accuracy of sound speed determination, particularly when made over the long base lengths necessary to ensure adequate time resolution, was limited by uncertainties regarding the temperature and wind velocity.

John Tyndall (1820–1893) reported clearly audible echoes being returned immediately following cessation of a fog signal directed out to sea, even when no fog was present. He was able to show that this was not reflection back from the irregular sea surface and concluded that the reflection was from some atmospheric disturbance. This disturbance is now known to be turbulence caused by friction between moving air and the ground. During his experiments, Tyndall tried directing his sound source upwards and found that echoes persisted for a much shorter time. This caused him to deduce correctly that the most turbulent regions extended for only a relatively short distance above the ground surface. Tyndall's experiments on this subject may be regarded as originating the concept of acoustic sounding of the atmosphere (or SODAR as it is sometimes called, an acronym for SOund Detection And Ranging). Today the echo-sounder provides a unique tool for probing atmospheric heterogeneity and immediate application is found for meteorological conditions, such as inversions, likely to lead to unacceptable levels of atmospheric pollution in urban areas. A typical frequency range for SODAR is from 1,000 Hz to 4,000 Hz.

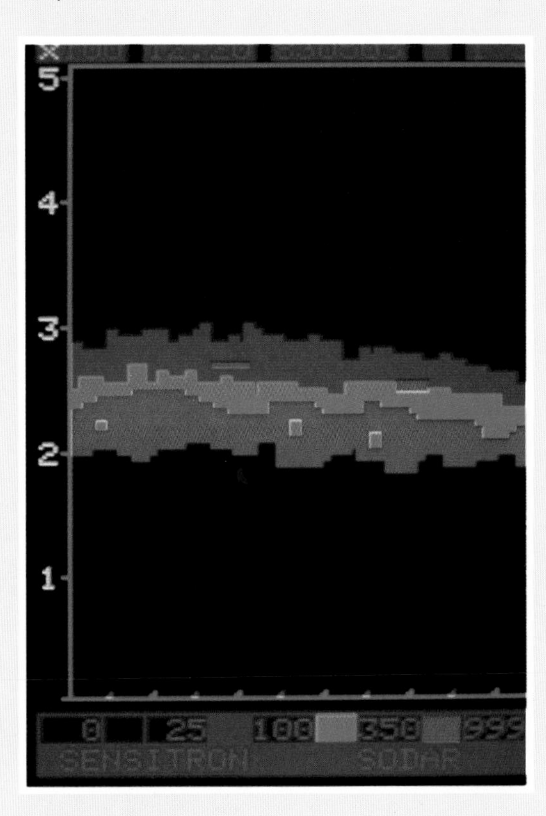

► ***Atmospheric data collected by sodar can be processed by computer to produce images such as this profile of air humidity, with readings plotted by altitude.***

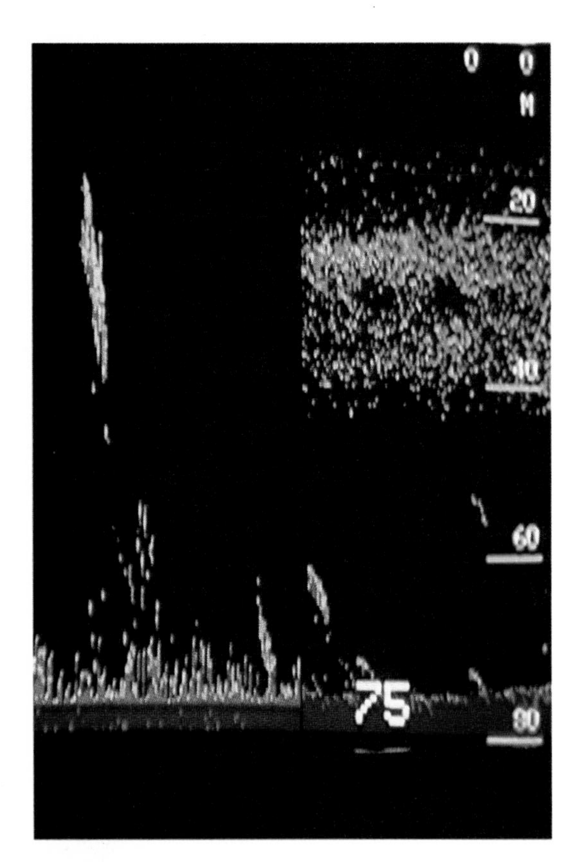

▲ *These two adjacent undersea profiles of shoals of fish were produced by a high-resolution echo-sounding system. The images feature deepening color, from blue through yellow to red, denoting increasing density.*

The sofar channel

Sound in liquids, like sound in air, is affected by temperature and pressure. An increase in temperature or pressure means decreasing density and hence increasing velocity. The surface layer of the ocean contains water mixed by turbulence and warmed by the Sun. Below this the temperature, and thus the velocity of sound in water, drops with depth. In what is known as the deep isothermal layer, the temperature remains more or less constant but the sound speed increases as a result of the increasing pressure. So there is a minimum sound velocity at the top of the deep isothermal layer, about 1,000m below the ocean surface in middle latitudes. This is significant since in a layered system sound waves bend in the direction of the lowest velocity. Sound directed upwards or downwards, therefore, from the region of minimum velocity, is bent back towards this region and effectively trapped, forming a deep-sea sound channel. The worldwide SOFAR channel (SOund Fixing And Ranging) of underwater sound detectors (hydrophones) and signal processing stations was built to facilitate rescue work at sea and exploits this channel extensively. Anyone seeking rescue at sea may drop a small depth-charge overboard, set to explode in the deep-sound channel. The location of the explosion can be determined by calculating the difference in the transit time of the sound to listening stations at widely separated sites.

Sound in liquids

Many liquids and solids are opaque to both radio and light waves (➧ page 73), and sound may be the only source of waves that can conveniently be harnessed to examine their structure. The velocity of sound through the material is sensitive to structure, since in any medium it is directly proportional to the square root of its stiffness, and inversely proportional to the square root of its density. Liquids and solids are denser than air but they are so much stiffer that the velocity of sound in most liquids and solids is higher than in air. In water at 20°C with a density of 1,000 kilograms per cubic meter, the velocity of sound is approximately 1,500 meters per second; in a solid material of density 7,500 kilograms per cubic meter, the velocity of sound is about 6,000 meters per second. In gases and liquids, the stiffness depends on pressure; in solids, on elasticity and rigidity. As sound travels through a liquid, it is affected by layering, turbulence and scattering from obstacles; through solids it is affected by layering, cracking and changes in elastic properties.

Sonar

In the sea, sound is used in both active and passive SONAR (SOund NAvigation and Ranging). The underwater acoustician uses hydrophones (essentially microphones designed for underwater use) to detect and measure the velocity of local water movement associated with the passage of an underwater sound wave. Passive sonar, arrays of hydrophones permanently fixed on or near the bottom of the sea, can be used for ranging and detection, but active sonar systems are more effective. These generate underwater signals and receive signals reflected and scattered from the targets of interest, like the echo-sounders used on SODAR. Deployed on or near ships, echo-sounding sonar systems scan the surrounding water by the movement of transducer arrays which transmit a sound beam downward. A profile of the ocean's depth or of the distribution of fish in shoals can be obtained as the ship moves forward. The frequency range typically used is between 100 kHz and 10 MHz. Newly developed high-resolution sonar can not only give the distribution of sound reflectors and scatterers in space, but can form an acoustic image of an individual scatterer. High-resolution echo-sounders with narrow acoustic beams can show details of the structure of a shoal of fish, even distinguishing individual fish. Resonance frequency analysis of the returned signals assists in estimating the size of the fish, and can even determine their species.

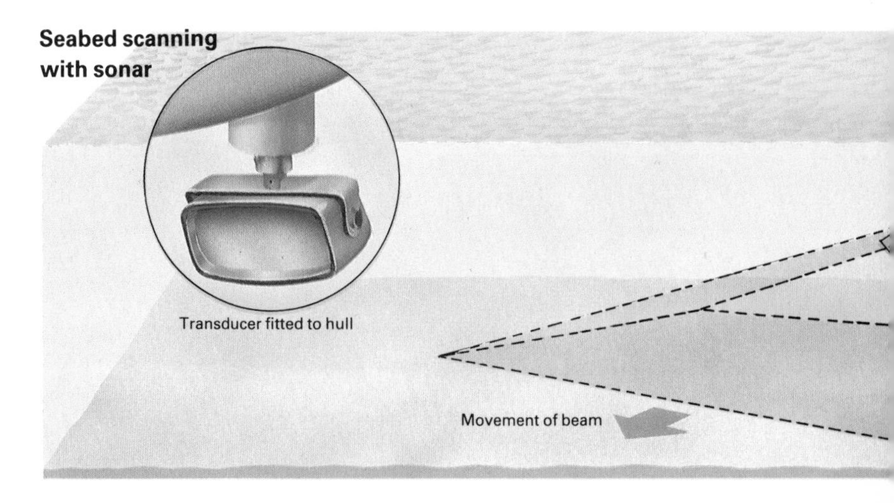

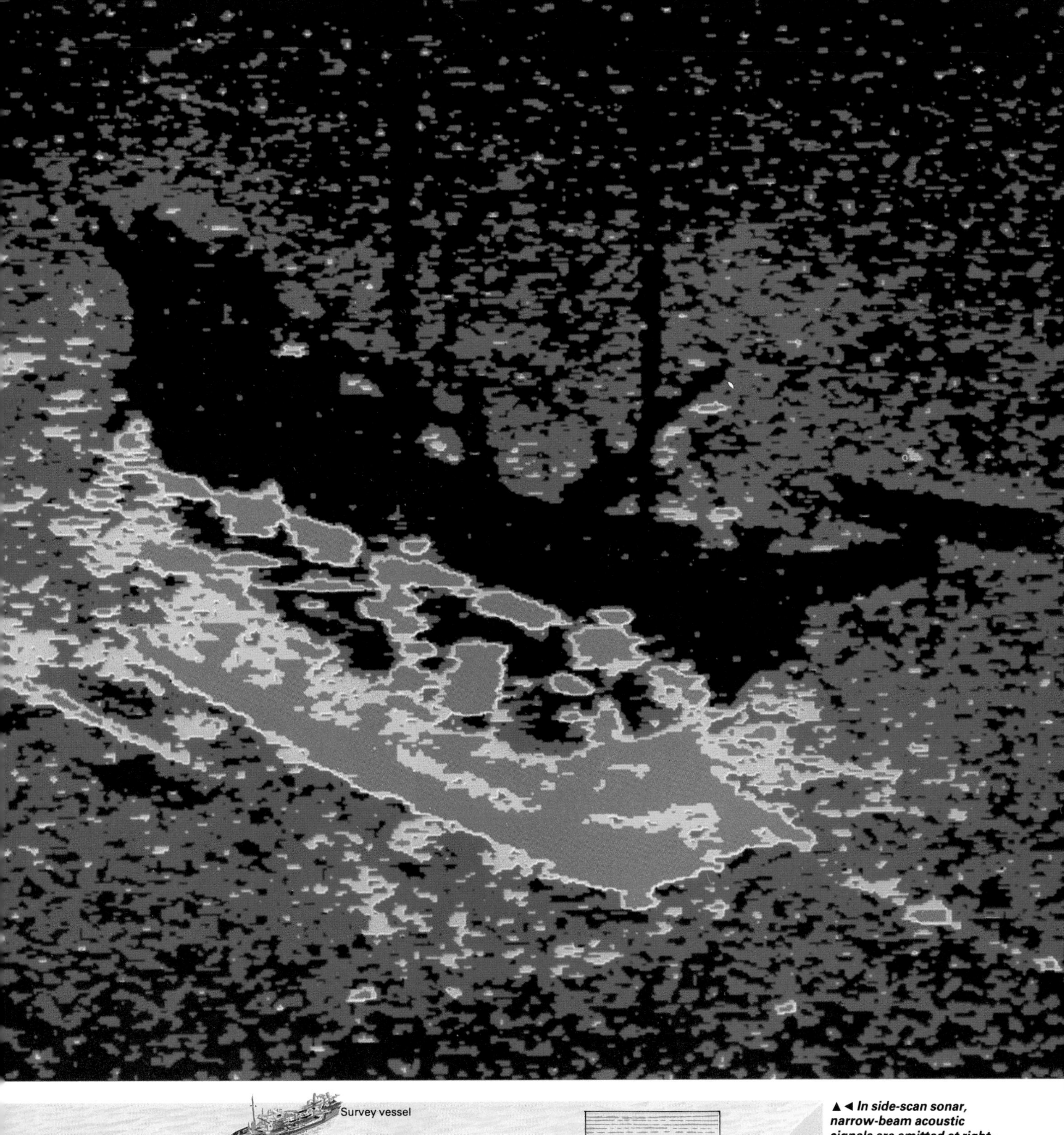

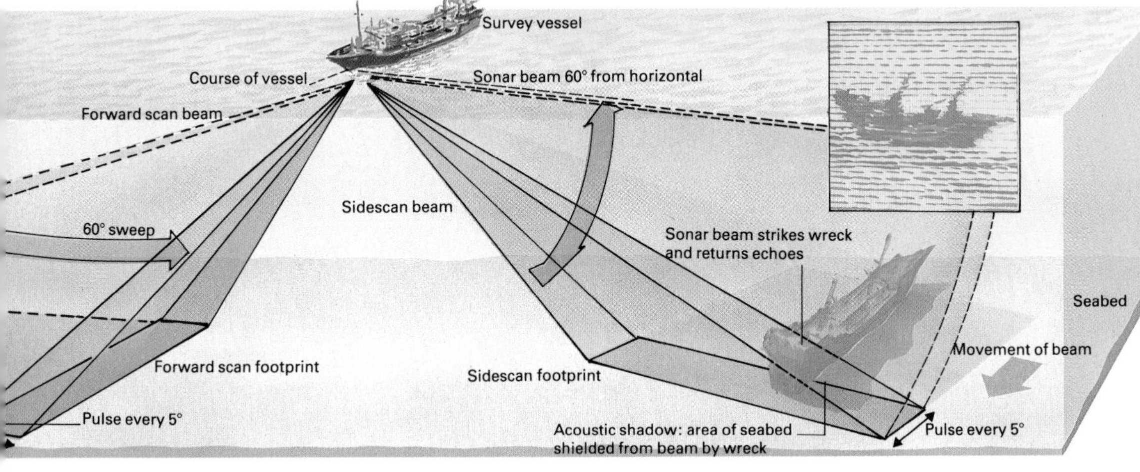

▲◄ In side-scan sonar, narrow-beam acoustic signals are emitted at right-angles to the ship's track in a continuous sweep, while the returning echoes are recorded line by line. A typical image obtained, such as this one of a sunken schooner (above), is processed by computer, with depth of color indicating intensity of echo. An acoustic "shadow" thrown by the wreck where the sea bed is hidden from the acoustic signals reveals details of the object in elevation.

Seismic surveying can be used to prospect for oil deposits, or to check on the quality of the topsoil

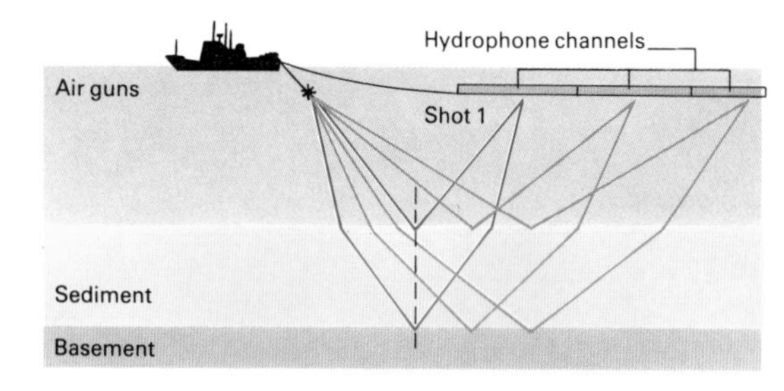

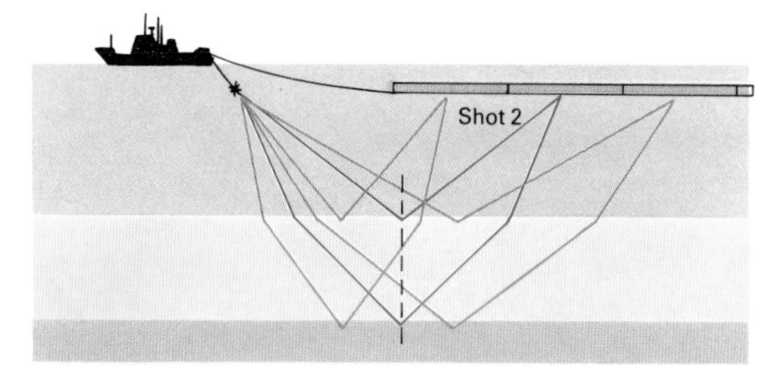

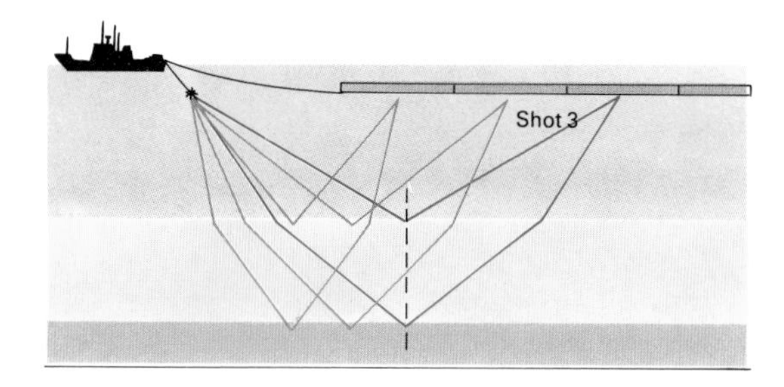

◄▼ Multichannel seismic profiling is a technique to give enhanced images of the Earth's crust beneath the sea bed. As with conventional surveys, a research vessel tows air guns and a streamer of hydrophones. Sounds from shots fired at the sea bed are reflected, either directly or via underlying rock layers, then recorded as signals. With this new technique, the hydrophones are grouped in receiver channels: reflection points sampled by the first channel on one slot are sampled again by the second on the next shot and so on, until all channels have sampled the same points. The four diagrams to the left illustrate how the sampling process is achieved through the coordination of the vessel's speed and the air gun's firing intervals. In processing the signals, recordings from a common depth point are identified by computer (1), brought into phase (2), then placed side by side with successive readings (3), which go to build up the final image (4).

Seismic reflection

During a seismic refraction survey the source and receiver may be sufficiently close together that a seismic reflection results. In this case there is no path from source to receiver (other than the direct one) that takes less time than the path involving direct reflection at the first sub-surface layer.

Using this process, a picture of the various reflecting rock profiles beneath the sea bed can be built up. Shots fired by underwater air guns are reflected both from the sea floor and the underlying rock strata. The signals received by hydrophones can be timed and deductions about the depth of rock layers made from the results.

More sophisticated images can be obtained by the technique of multichannel seismic reflection. By coordinating the firing interval of the air gun and the speed of the research vessel, the echo signals can be phased in such a way that reflections from the same points are recorded by each channel in succession. Once channel trace records are matched to particular points, arrival times for the shots at different channels can be brought into phase by correcting for the time lapse involved. The resulting composite signal indicates amplified peaks where layer discontinuities are struck. When thousands of such signals are placed side by side to create a continuous image, a profile of the Earth's crust immediately beneath the sea-floor can be built up.

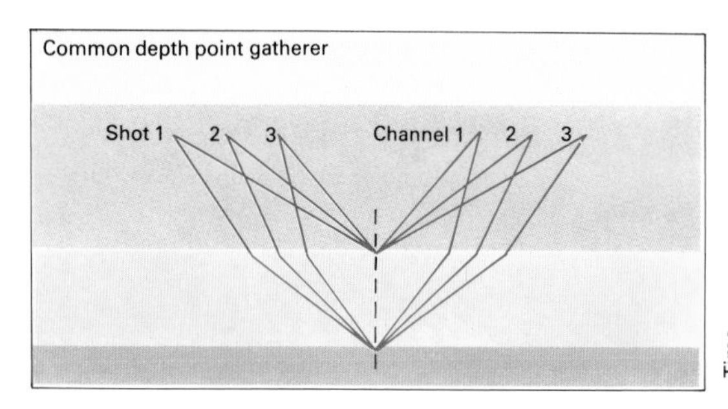

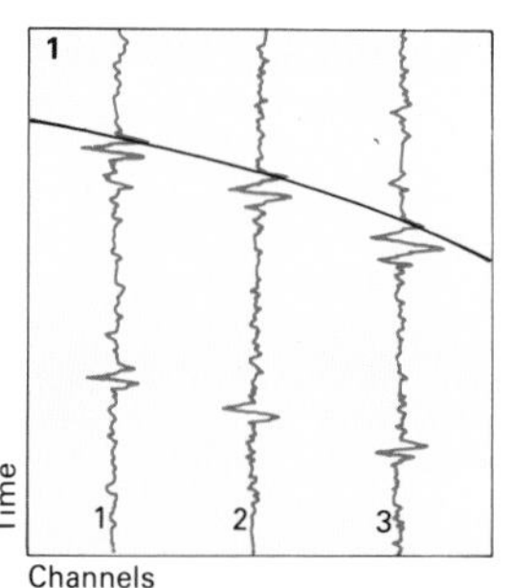

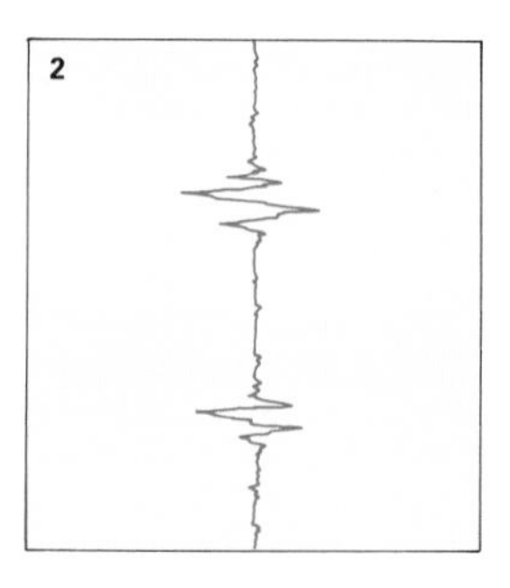

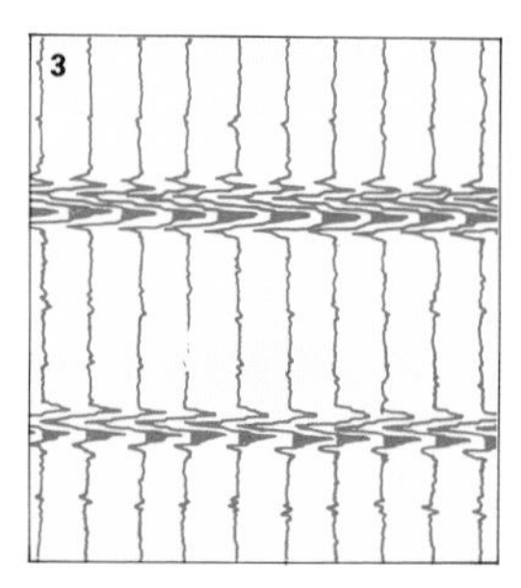

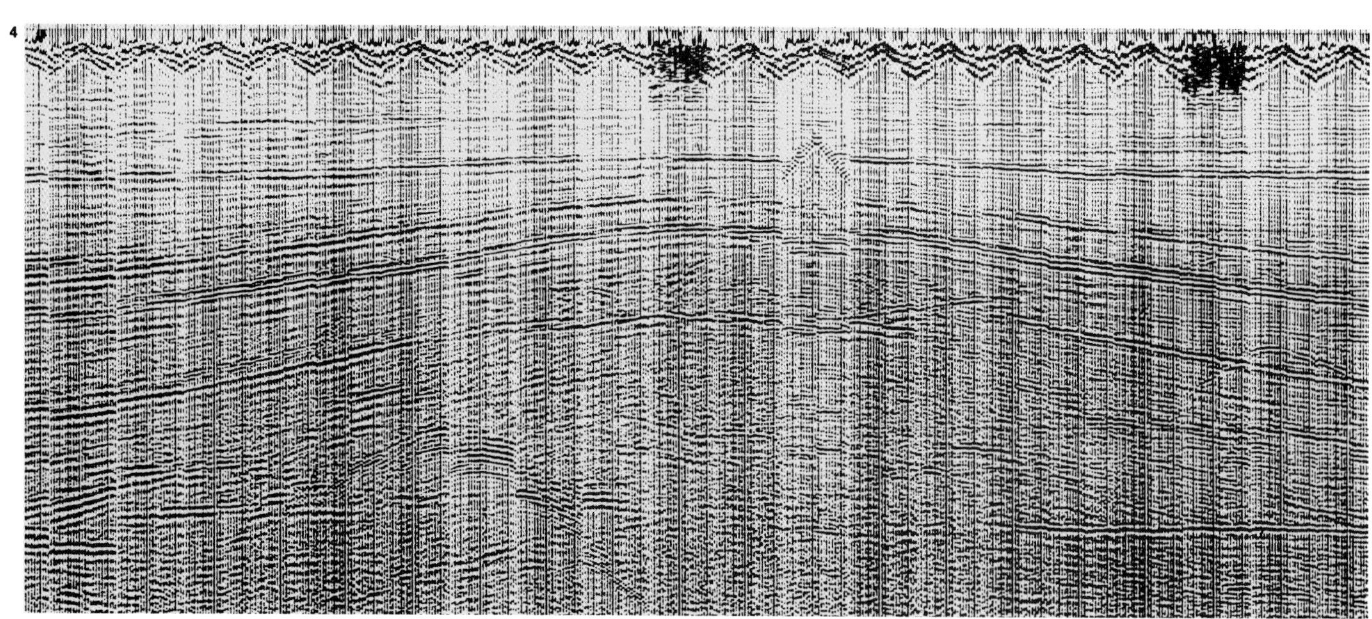

▲ ***Images obtained from seismic reflection profiles are of interest to oil and gas companies. Here geophysicists are analyzing the results of a seismic survey, working from both a printout showing the strata under examination in cross-section, and a three-dimensional computer-processed image displayed on the screen, produced from the same data.***

► ***Seismic profiles, whether by refraction or reflection, are compiled from techniques that involve low-frequency sound waves being sent through the ground. For land-based surveys, trucks can be fitted with equipment which transmits the seismic pulses in a more controlled manner than explosions.***

Sounds in solids

Sound or seismic waves in the solid earth are observed for a variety of reasons. Earthquakes generate waves on the grandest scale, and studying them systematically has implications for the safety of humanity, as well as for our interest in the structure and evolution of the Earth. Information gleaned from artificially-generated seismic waves is indispensable to the search for oil and gas. This applies not only to the exploration for new resources, but also the evaluation of the fields whose existence is already known. Geophysicists determine properties of rocks penetrated by oil wells by observing at various depths seismic waves generated by a distant explosion or a sound source nearby in the same well. Oil prospectors appraise and develop their drilling constantly by integrating knowledge from seismic data and bore-hole information. Engineers use shallow seismic refraction surveys to give information about the rigidity of rock layers into which building foundations are to be sunk.

Marine geologists also employ seismic profiling as the basis for sub-bottom profiling. The sea bed reflects most sound waves directed at it, but a significant portion of sound at sufficiently low frequencies (such as sound emitted by small air guns) will penetrate the sea-floor sediments as refracted waves and be reflected from the strata below. Acoustic investigation of the sea bed for engineering purposes, or for the detection of cables and pipes, is becoming increasingly popular and effective. To see deep into the sea bed, long streamers of hydrophones towed behind a ship receive and store these acoustic signals. However, information about layers thousands of meters beneath the sea bed, which is of increasing interest in the search for oil and gas reserves, is obtainable only from seismic refraction techniques.

The possibility of using acoustic reflection techniques to survey soil conditions has recently been investigated. Cultivated soils need to be porous to allow air to penetrate to growing roots and carbon dioxide formed during photosynthesis to escape into the atmosphere. The extent to which sound can penetrate the soil is also determined by the degree of porosity. Acoustic measurements of the reflection or the transmission of sound in the audio-frequency range at the soil surface may thus be used to measure soil porosity.

Seismic refraction

The principle of the seismic refraction method is to initiate seismic waves at one point and then determine how long it takes for the waves to reach a series of observation points. In traveling from the source to the detectors, the waves will be refracted and reflected at the various discontinuities that lie in their path. The first waves to arrive at a detector, known as the head waves, will not have followed the shortest path if there is a high-velocity deeper layer. Once the wave from the explosion reaches that layer it is refracted so that it travels parallel to the interface, forming secondary waves that are themselves refracted back through the low-velocity upper layers towards the detector. If an appreciable portion of the path lies in the high velocity layer, then such waves are able to travel to the detector faster than those that travel straight through the upper layer to reflect from the interface. By recording the arrival times at the seismograph of both direct and refracted waves, the thickness of the upper layer can be calculated from the difference in their velocities.

Ultrasonic tests are as useful to find faults in mechanical equipment as they are to check the development of an unborn child

Ultrasonic body scanning

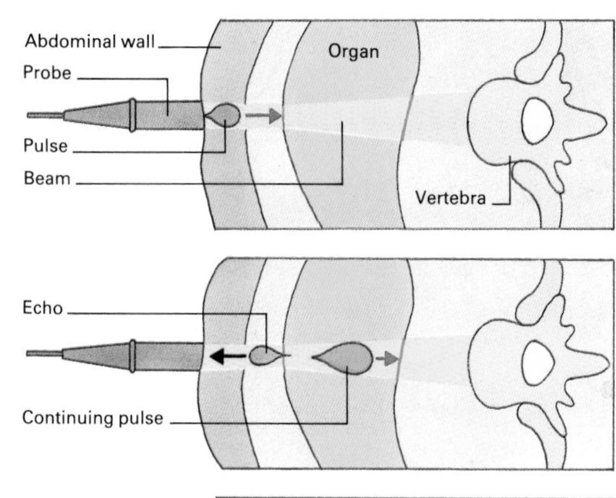

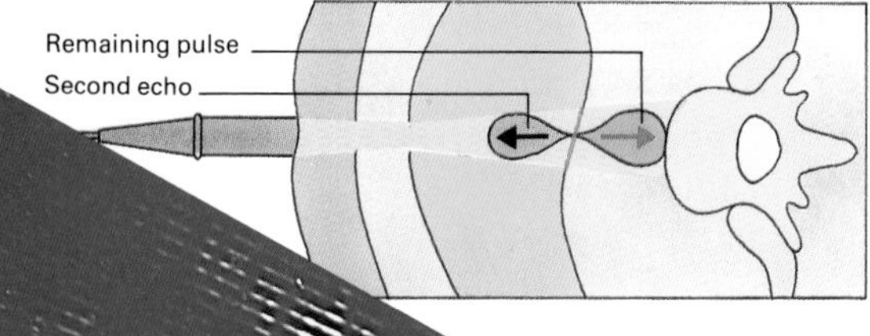

Ultrasound

In contrast to light waves and X-rays (➧ page 74), sound waves are attenuated only weakly in steel, concrete and certain plastics. So if vibrations are set up within a solid component such as a machine part, at ultrasonic frequencies between 1 and 20 MHz, we can acquire information about its internal structure from the variations in the signals received. Ultrasound is a sensitive detector of fractures, so this technique of non-destructive testing (NDT) is particularly useful in industry for inspecting the vital parts of machinery during their service life – for instance welded joints in pipelines and storage tanks, heat exchangers in nuclear power plants and support beams in offshore structures. Any flaws or material defects can be detected before they result in a major catastrophe.

The higher the frequency used in ultrasonic flaw detection, the better the ability of the ultrasonic wave to reveal details and detect small defects. On the other hand, the rate at which the signal deteriorates increases with frequency and the range at which flaws and defects can be seen is much reduced. Thus the operating frequency of flaw-detecting equipment is selected carefully according to its intended use.

In its simplest form, a high-voltage pulse (up to about 600 volts,

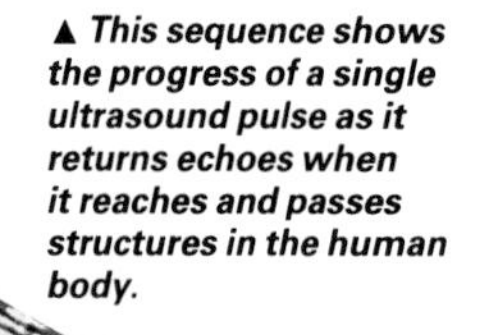

▲ This sequence shows the progress of a single ultrasound pulse as it returns echoes when it reaches and passes structures in the human body.

Imaging the unborn child

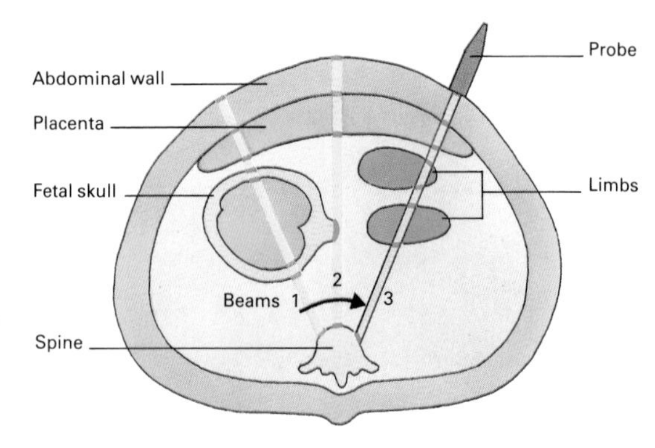

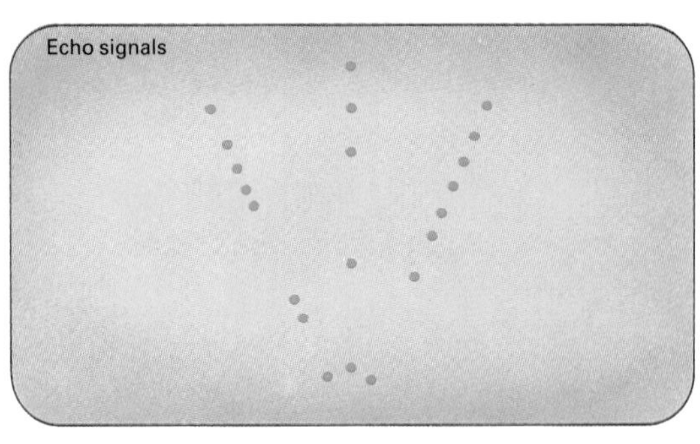

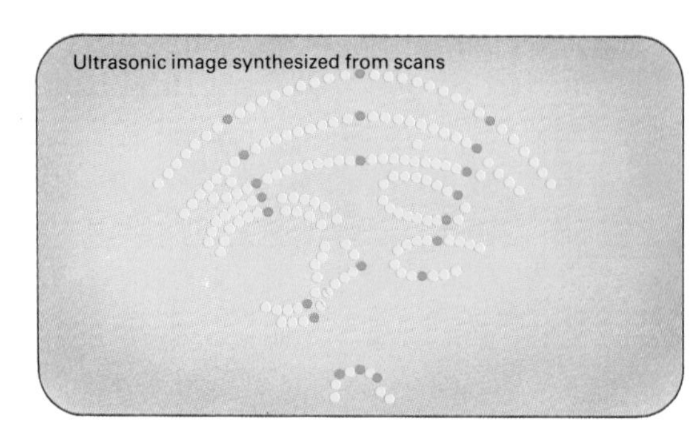

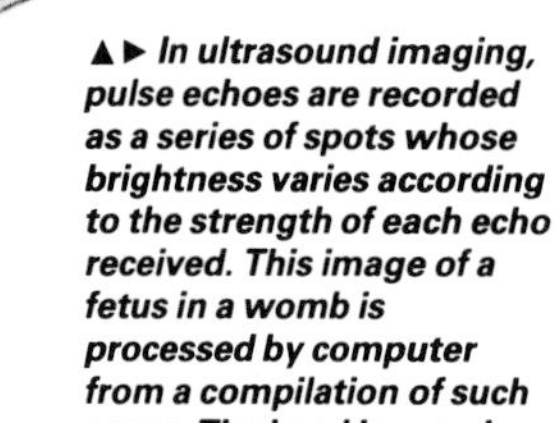

▲► In ultrasound imaging, pulse echoes are recorded as a series of spots whose brightness varies according to the strength of each echo received. This image of a fetus in a womb is processed by computer from a compilation of such scans. The head is seen in profile above.

and lasting only a fraction of a second) is applied repeatedly to a transducer which produces a short pulse of sound waves directed into the object under test. These waves are reflected back to the transducer by any defective interfaces or surfaces which are encountered. The returned sound waves are converted back into an electrical signal in the transducer and this can be displayed on an oscilloscope. The same basic technique is used for medical ultrasonic investigation, routinely used to examine fetuses in their mothers' wombs without the risk of damage to tissue by potentially harmful X-rays.

Images from ultrasound

Ultrasonic systems can also be used to provide images of features and defects within objects, by scanning with a single transducer or by using a phased array. Echo sources are pinpointed in two dimensions from the successive plotting of two coordinates – the transducer's position and the distance from it of the defect – allowing for the known angle of the beam through the material. As the transducer travels across the component, a continuous picture is built up. The range of echo strengths can be recorded digitally and represented as different colors on screen, enabling any faults to be identified and characterized with greater confidence.

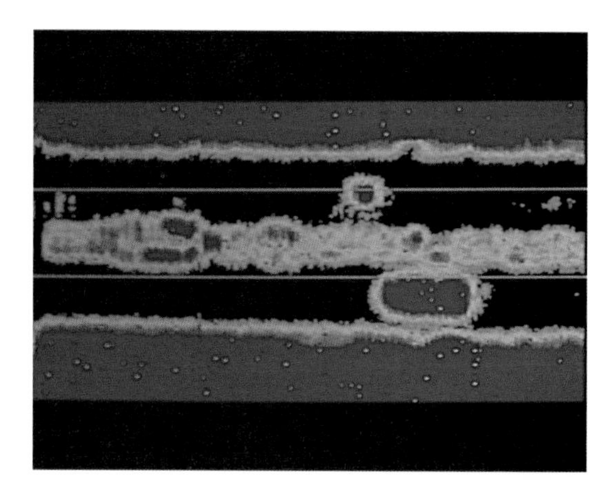

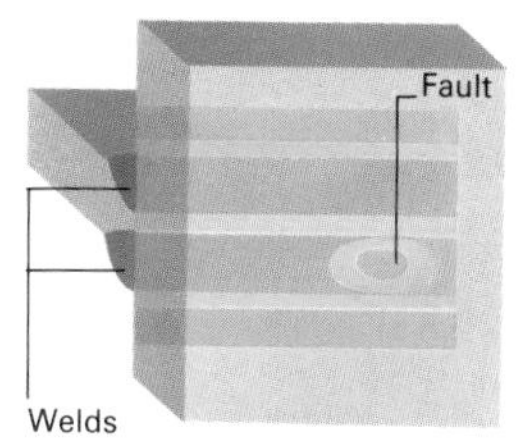

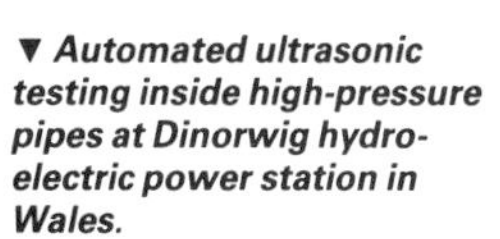

▲► An ultrasound probe linked to a computer can image faults such as these in a welded joint.

▼ Automated ultrasonic testing inside high-pressure pipes at Dinorwig hydro-electric power station in Wales.

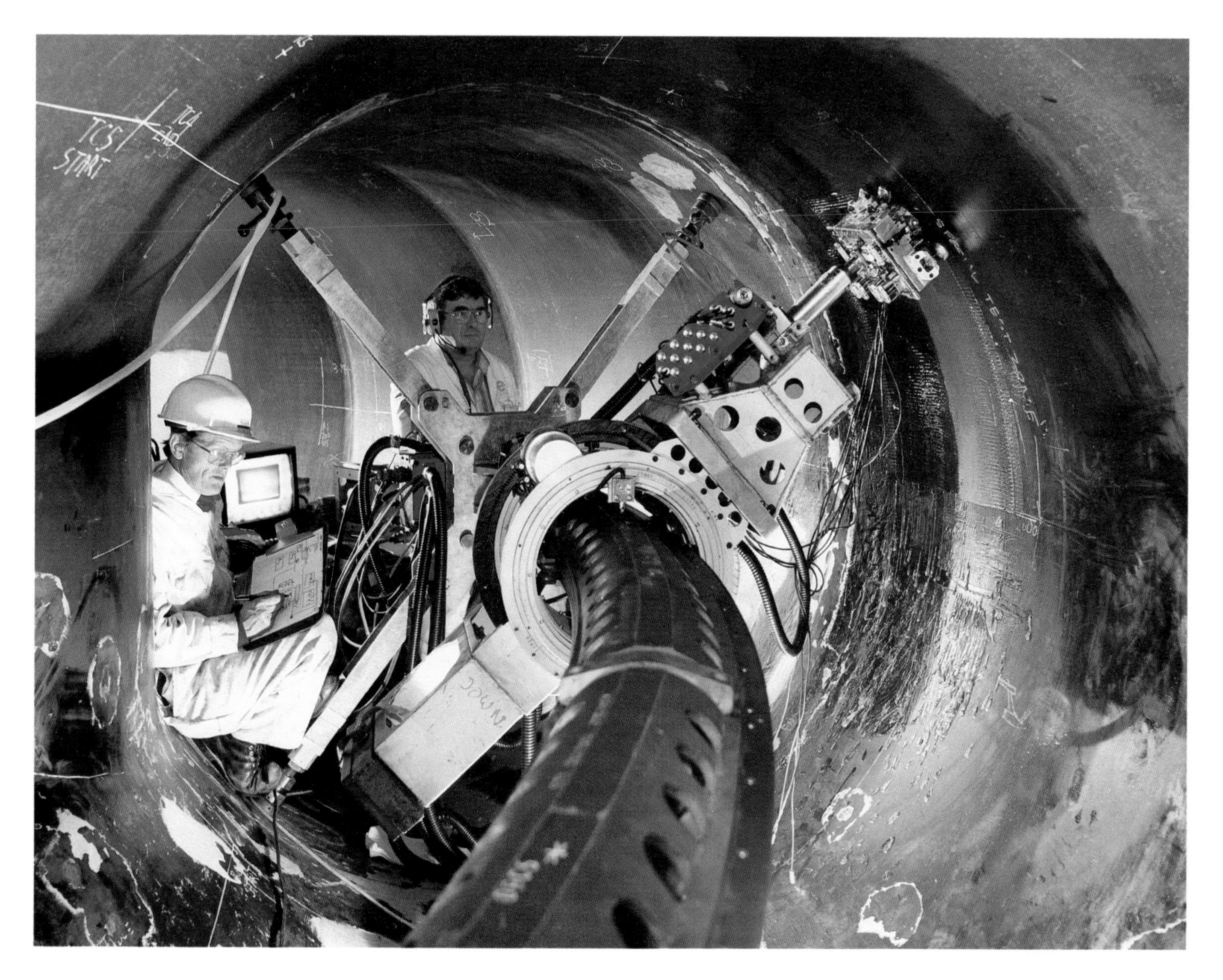

See also
Surveying and Mapping 37–40
Measuring with Electricity 41–46
Microscopes 47–56
Seeing beyond the Visible 73–86
Recording on Tape and Disk 91–96

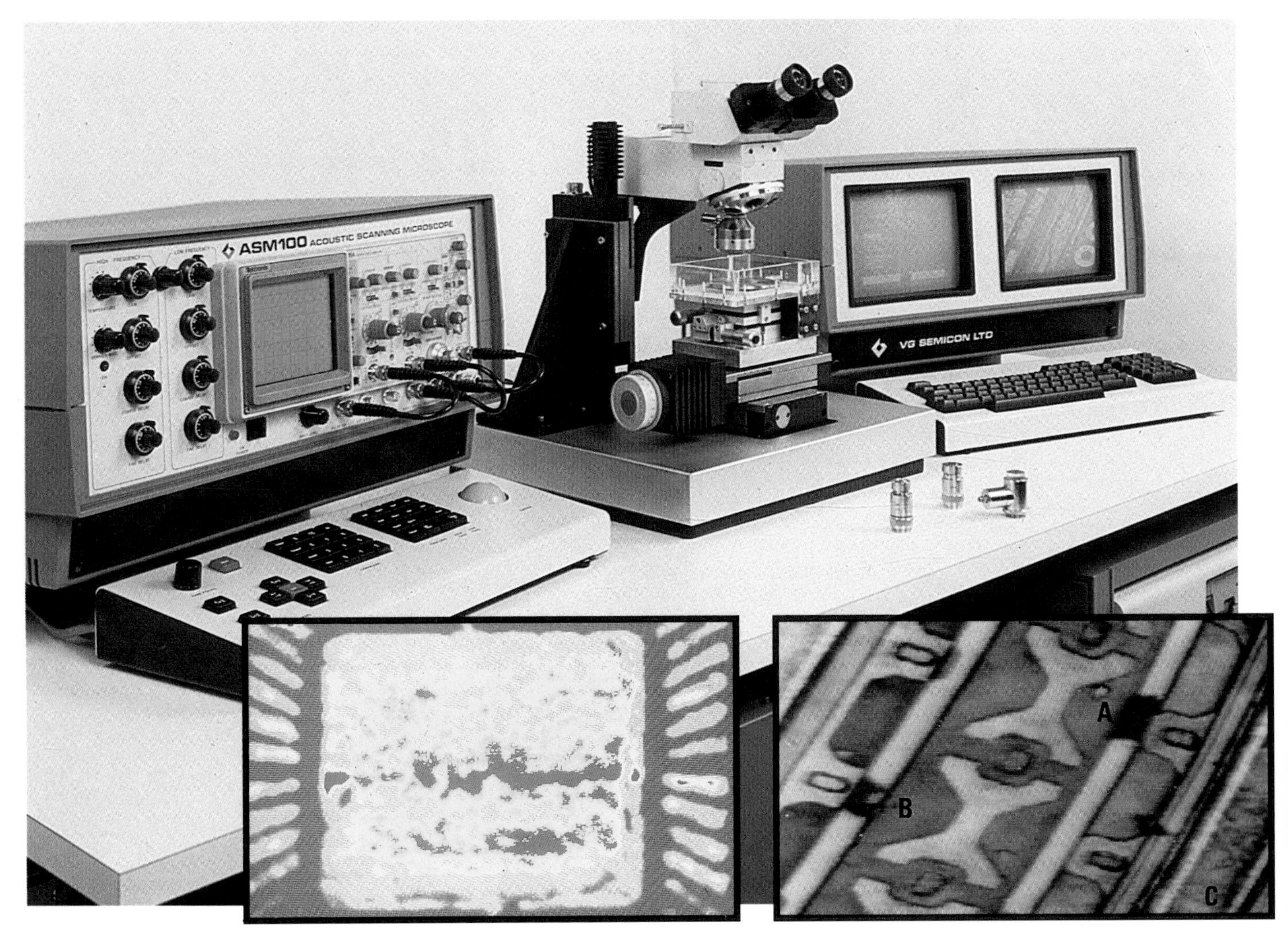

▲ **In this acoustic scanning microscope, the instrument itself is flanked on the left by a device that controls electronically both the frequency and the pulse timing of the acoustic signal, and on the right are visual displays linked to a computer. Design of lens, choice of frequency and length of pulse all affect depth penetration in the specimen and final image resolution. Inset are typical results obtained from examination of integrated circuits, showing moisture ingress (left) and, at greater magnification, faults in the circuitry (right).**

Acoustic microscopy

The original idea for an acoustic microscope was put forward in 1944 by the Soviet physicist N.D. Sokolov, who observed that the wavelength of sound in water at a frequency of a few gigahertz (billion cycles per second) was comparable to that of visible light. Sokolov suggested that a device which could focus high-frequency ultrasonic waves should be able to resolve images in the same way as the optical microscope does (◆ page 48). The first scanning acoustic microscope was built in 1973 by a team at Stanford University, California, led by Professor C. Quate. By 1983 the group had achieved a resolution of 0·09 micrometers, very close to the theoretical maximum.

The acoustic equivalent of the lens is a tiny spherical depression ground into the end of a sapphire rod (which supports a high velocity of sound waves with low internal losses). At the other end of the rod is a transducer which sets up within it a controlled vibration or acoustic wave. The depression at the tip focuses the wave at a spot roughly one wavelength in diameter. As the sound dissipates quickly in air, the lens is held in indirect contact with the specimen by a coupling medium such as water. The high-frequency waves are reflected back from the specimen, and the strength of the returning signal depends on how much acoustic energy is absorbed by the sample at that spot. An image is built up by scanning the focal point over the surface of the specimen and converting this acoustic signal into a cathode-ray oscilloscope picture which is displayed on a screen.

One area of application of the acoustic microscope is in the study of biological systems, as the process of observation does not appear to harm living organisms or tissues in the way that preparation by staining or holding a vacuum often does. Another is in examining and forming images within optically opaque materials. The technique has been used to detect flaws in silicon chips and other microelectronic assembling (◆ page 122), and has the potential to be used in other areas of the electronics industry, such as probing microcrystalline grain structure, properties of alloys, and the surface texture of materials used in electronic components.

Acoustic microscopy also has considerable potential for dentistry. By enabling studies of zones of demineralization on teeth and improving understanding of the natural remineralization process, the acoustic microscope is laying the foundation for the production of chemical methods of restoration which may eventually supplant the common practise of drilling and filling.

Seeing beyond the Visible

9

Electromagnetic radiation...X-rays...Infrared... Infrared astronomy...Landsat imagery...Radio astronomy...Radio interferometry...Radar... Magnetic resonance inspection...PERSPECTIVE...The electromagnetic spectrum...CAT scanning...Industrial radiography... X-ray astronomy...Ultraviolet astronomy ...IRAS...Image processing...Origins of radio astronomy ...Microwave astronomy...Seasat

"A picture is worth a thousand words". Experimental observation and detailed calculation cannot reveal the truth of the saying, but pictures, like words and numbers, may convey valuable information for the scientist as well as the artist, and each method of communication is valuable in its own context.

In technological work, pictures are often described as images since, unlike those produced by an artist, they are constructed according to a process that may be defined and repeated. Their function is to present information about events or structures not open to direct or common human experience. Often these images simply extend experience, for example by depicting structures that are small (◀ page 47) or events that are rapid (▶ page 88). Other images may show phenomena for which no human sense is apposite (◀ page 9), such as the concentration of a chosen atomic species on the surface of an industrial component, the condition of a crop in some remote part of the world, the nature of astrophysical events or objects in distant space, or even the internal structure of a living creature.

Imaging electromagnetic radiation

A number of technologies have been developed to detect such invisible phenomena as images that are open to direct inspection. The use of sound waves to build sonar images (◀ page 65) is one such example; another, more general, is to image electromagnetic radiation emitted in regions of the spectrum to which the human eye is not sensitive. In fact the region of human sensitivity represents the merest chink in the huge wall of this radiation, which to existing knowledge extends from the very short wavelengths of gamma-rays to the long ones of radio waves.

Humans can see a spectrum of colors extending from violet to red. Beyond the violet end, other electromagnetic radiation of shorter wavelength can be detected in a progression named: near ultraviolet, far ultraviolet, X-rays, gamma-rays. Beyond the red a progression to increasing wavelength is given the names: near, middle and far (or thermal) infrared, microwaves and radio waves. Observations can be made in each of these wavelengths.

If an object emits radiation naturally, it may be imaged by detectors sensitive to the appropriate frequency. If it does not, an image may be formed by illuminating the object and observing what comes back. Analysis of the signal may be complex. In radar detection it may be enough to note the position and speed of a single point representing an aircraft (▶ page 84); in magnetic resonance inspection (▶ page 86) a three-dimensional representation of chemicals in body tissue may be sought. Both cases may call for computer calculations, with the image formed only in the final stage, often with the information color-coded for convenience.

The electromagnetic spectrum

In 1831 the British physicist Michael Faraday (1791-1867) demonstrated that an electric field could be produced from a fluctuating magnetic field. In doing so he raised the question of how electrical and magnetic effects might cross empty space. Faraday himself noted the possibility that such electromagnetic forces might be like "vibrations upon the surface of disturbed water". Then in 1864 the brilliant Cambridge mathematical physicist James Clerk Maxwell (1831–1879) published a complete theory of electromagnetism: according to this, the oscillatory motion of an electric charge would produce energy in the form of an electromagnetic field radiating from this source as waves traveling at constant velocity, that of the speed of light. Maxwell concluded that light was therefore a form of electromagnetic radiation, and that radiation of shorter and longer wavelengths than light should exist. The theory was dramatically confirmed by the German physicist Heinrich Hertz (1857–1894); in his experiments (▶ page 185) electromagnetic waves were generated and detected electrically. When he measured their wavelength, they were a million times greater than that of visible light.

The way in which energy is related to the different wavelengths of radiation remained a mystery until the German physicist Max Planck (1858–1947) postulated that energy is radiated in discrete particles or quanta, each of an energy proportional to the wavelength of the radiation. Different kinds of radiation were classified according to the range of energy of these quanta (termed photons by Einstein), or arranged along the electromagnetic spectrum according to wavelength or frequency.

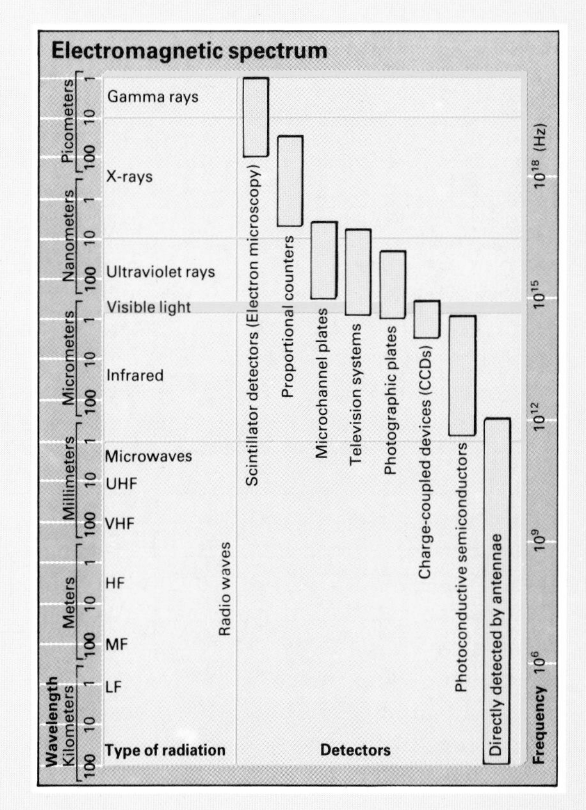

Imaging with X-rays

X-rays were discovered in 1895 by the German physicist Wilhelm Röntgen (1845-1923). At first he did not understand what they were; but he did establish that they were unaffected by magnetic or electric fields and could not be reflected or refracted, as can visible light. He also discovered that they penetrated apparently opaque material, that the amount of absorption depended on the density of the material, and that photographic emulsion would register them. The value of X-rays for photographing the inside of a living body quickly became obvious.

In 1912 it was established that X-rays are a form of electromagnetic radiation of a wavelength much shorter that visible light, around $0{\cdot}1 \times 10^{-9}$ meters. This is about equivalent to the space between the atoms in a solid, and as a result X-rays can be strongly diffracted by crystalline structures, in which the atoms are regularly spaced. The study of diffraction patterns of X-rays of known wavelength can reveal the structure of unknown crystals – a technique known as X-ray crystallography – and, conversely, crystals of known dimensions can be used to identify unknown X-ray wavelengths – X-ray spectroscopy. X-rays can also be used in microscopy (◆ page 56), where they offer a better ultimate resolution than light because of their short wavelength, but a less good one than electrons (◆ page 52). Once an image is obtained, the image contrast can show the presence of microstructural features through differential absorption; it can also show their chemical composition. Because they penetrate quite deeply into thick objects, X-rays are used to reveal the internal details of biological and metallurgical specimens. The increased depth of field which can be achieved also allows a complete three-dimensional view of the structure to be obtained.

Generating X-ray images

X-rays are produced when any material is bombarded with very high energy particles or radiation. Electrons from a heated cathode are focused on a tungsten anode and accelerated using a high voltage. A fraction of the energy of the electron beam is converted into X-ray photons, which are emitted in all directions; those to be used for imaging are permitted to escape through "windows" in the tube housing. In the technique known as radiography, the sample is placed between the X-ray source and a photographic plate or fluorescent screen. A simple shadow image of the specimen is formed, with different regions giving rise to different degrees of absorption.

In X-ray tomography (also known as computer assisted tomography, CT or CAT scanning), the combination of scanning techniques with sufficient computing power to calculate the density of an image point by point enables the construction of a new range of images. A modern tomograph machine consists of a movable frame in which an X-ray tube produces a fan-shaped beam of radiation 10-15 millimeters thick, the intensity of which is measured by an array of detectors on the opposite side of the frame. The subject lies inside the frame in the path of the X-ray beam; the detectors measure the attenuations of the signal reaching them and thus the amount of absorptive material the beam has passed through. By rotating the frame and taking measurements every few degrees, it is possible to build up an image of the distribution of X-ray absorbing material in the plane cut by the beam, and therefore to obtain an X-ray "slice" through the subject.

CAT scanner

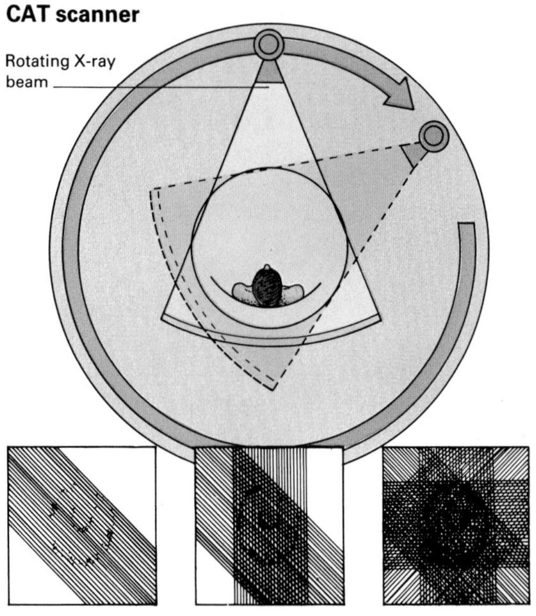

▲ Modern CAT scanners can obtain an X-ray slice through a subject in a few seconds, storing up to 1·5 million absorption point readings, which may be taken with the patient lying either horizontally or at "tip-up" angles. A computer builds up an image first by producing a complete tomograph from superimposed radial X-ray strips, and then color-coding it for display.

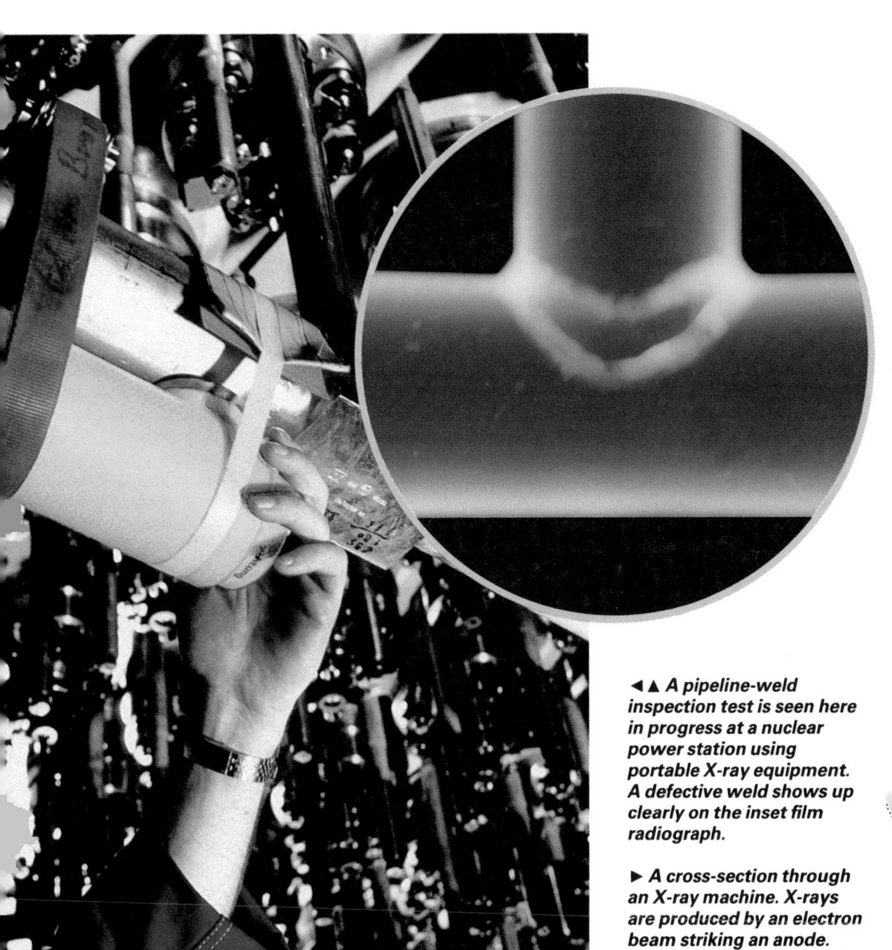

◄▲ A pipeline-weld inspection test is seen here in progress at a nuclear power station using portable X-ray equipment. A defective weld shows up clearly on the inset film radiograph.

► A cross-section through an X-ray machine. X-rays are produced by an electron beam striking an anode. To prevent overheating the anode is either rotated or, as here, liquid-cooled.

◄▼ In the Exosat telescope, X-radiation is brought to focus via a diffraction grating and filter wheel on the face of two detectors, which register the strength and position of arriving photons. Images can be compiled from such data, and color-coded for increasing intensity. The example shown reveals the central region of the galaxy M82.

Exosat X-ray telescope

X-ray optics

Filter wheel

Grating

Star tracker

Pre-collimator

Radiography in materials testing

Ionizing radiation is used in many industries as one of the most advanced methods of examining materials or equipment for subsurface defects. The principles employed are similar to those used in medical radiography or airport security systems. Defects in items show up as shadows on the radiographic image. Interest is growing in replacing film by special screens which can be viewed through a television system. These give a "real-time" radiograph, a direct picture of the object as it is being viewed, which can be recorded on video tape. Such systems are increasingly attractive to industry now that image processing, microfocus X-ray tubes and detection systems can give results comparable to those on film. In most cases far higher levels of radiation are possible than in medical radiography. Gamma-rays may be used instead of X-rays, and neutron radiography has been employed to see inside jet engines while they are running, so that the movement of the parts at different engine speeds can be identified and followed.

X-ray tube

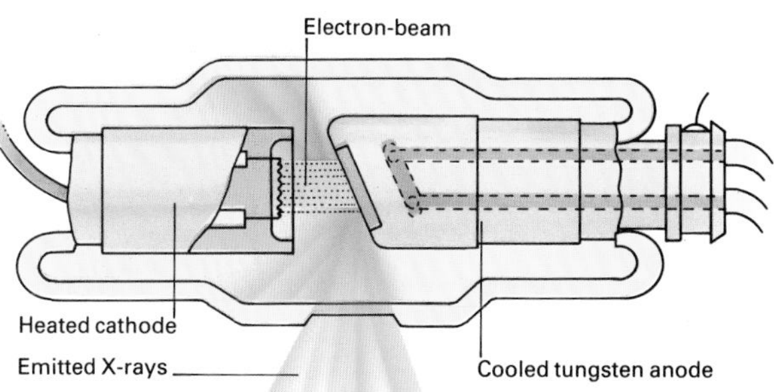

X-ray astronomy

Instruments designed to detect X-rays enable astronomers to discover information about intergalactic gases at multimillion degree temperatures. However, since the Earth's atmosphere absorbs incoming X-radiation, observation can only be made with the help of rockets and satellites. The first X-ray satellite, "Uhuru", launched in 1970, detected X-rays with a collimator – a grid placed in front of the detectors to limit the field of view – and a proportional counter – a chamber of argon gas which absorbs the incoming electrons and converts them into an electric current proportional to their energy.

The absence of an X-ray telescope – no substance could act as a lens or mirror to focus X-rays without absorbing or transmitting them – meant that astronomers still lacked the means to image the structure of cosmic X-ray sources. Then, on the Skylab space station in 1973, the use of "grazing-incidence" mirrors, shaped like the insides of a slightly tapering metal cylinder, showed that X-rays could indeed be reflected and focused if they hit or "grazed" a metal surface at a very shallow angle. Images transmitted from the Einstein Observatory, a satellite launched in 1978, and its successor Exosat (1983) confirmed the greatly-improved operational sensitivity and resolution now available to X-ray astronomy.

Astronomy in the ultraviolet waveband

Ultraviolet astronomy has been dominated by the study of hot stars and interstellar gases. Conducted in wavelengths of between 1050 and 3000 Å, it is beset by problems of absorption and "noise". The source of the noise lies in the omnipresent atomic hydrogen of interstellar space, which dominates all other emission beyond 912 Å.

Since ordinary glass lenses simply absorb ultraviolet light, aluminized mirrors coated with lithium fluoride or magnesium fluoride are used to reflect it. The detector may take the form of a familiar photographic emulsion, though without its normal gelatin surface which would be all too efficient as an ultraviolet absorber. The very first glimpse of cosmic ultraviolet radiation was obtained in this way in October 1946. Other detecting devices include photomultipliers, image intensifiers (◆ page 61) but especially vidicon tubes (◆ page 93).

A vidicon camera was the preferred detector for the International Ultraviolet Explorer, launched in 1978. It comprises an ultraviolet-to-visible converter coupled by fiber optics to a secondary electron conduction (SEC) tube. Preparation of the camera, exposure and reading of the image are all performed for the astronomer in "real time" by a sequence of commands from one of the ground stations, in Madrid or Maryland.

▲► This industrial component undergoing testing by magnetic particle inspection is sprayed with a fine magnetic powder while in the presence of a magnetic field. The particles are attracted to surface cracks which are made to fluoresce under ultraviolet light (right).

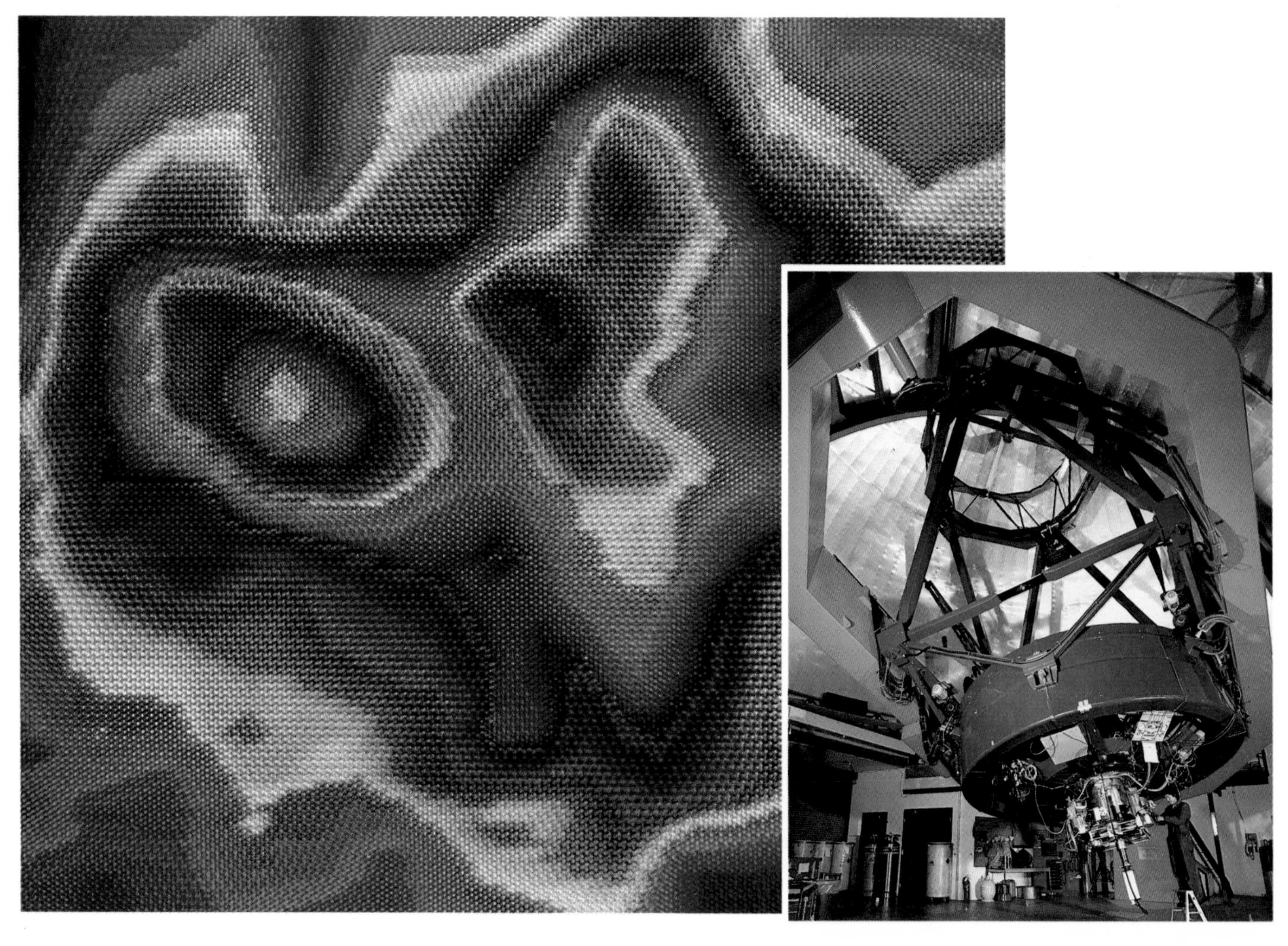

The Infrared Astronomy Satellite

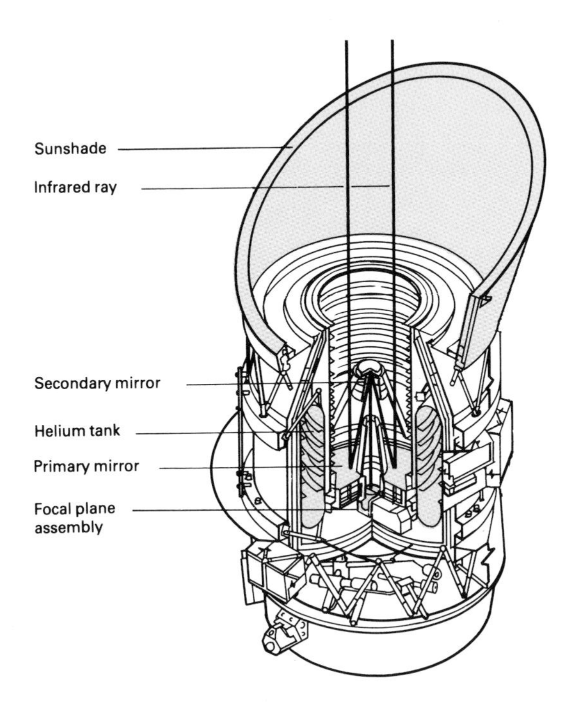

The Infrared Astronomy Satellite
From its launch in January 1983 to the following November when its refrigeration system failed, the Infrared Astronomy Satellite (IRAS) conducted an all-sky survey pinpointing over 200,000 infrared sources. This satellite-borne reflecting telescope, sponsored jointly by the United States, Britain and The Netherlands, had a beryllium primary mirror, with an aperture of 57cm. At the heart of the IRAS was a battery of instruments in the focal plane of the telescope. Since its chief task was to map the infrared sky, the array of 62 detectors had to screen observed signals to guard against cosmic rays and magnetospheric particles of high energy. Of the 15 or 16 detectors used for each of four infrared wavelength bands from 8 micrometers to 120 micrometers, at least two observed the same segment of sky, and only radiation detected by both was accepted as a true "signal".

Infrared radiation

In 1800 the German-born astronomer William Herschel (1738–1822) discovered by chance that energy from the Sun was radiated beyond the red end of the spectrum. Invisible to human eyes, this infrared radiation in its "thermal" band is detected by our senses as heat. Sensitive devices can detect minuscule heat emission and produce images displaying differential values. This technique, known as thermography, has proved useful in medicine (where it may reveal disorders including arthritis and tumors) and manufacturing. Infrared imaging also has applications in satellite photography, astronomy, and military detection systems.

Infrared astronomy

The infrared radiation from the skies nearest to the wavelengths of visible light is easily detectable from the Earth, and the far infrared is accessible from mountain sites. Most of the wavelengths in between are blocked by atmospheric absorption and must be detected from satellite observatories. Even for the few transparent "windows" in this wall, the infrared astronomer is beset by the overwhelming presence of radiation from terrestrial sources, since the soil, flora and fauna and buildings, having absorbed sunlight during the day, reradiate heat by night.

Ground-level infrared astronomy requires bolometers and photoconductors. A bolometer is a thermal detector which measures the change in temperature induced by the absorption of an incoming photon, by means of a sensor of germanium or silicon. Photoconductors, which are more sensitive for the shorter wavelengths of radiation, measure the change in voltage or current produced when an incoming photon liberates electrons on the surface.

Several strategems have been adopted to image cosmic infrared sources despite the ambient thermal "noise". Detectors are cooled to reduce background noise from the telescope itself, especially the metal struts of its supporting structure. British and United States astronomers have sought the shorter wavelengths from mountain-top telescopes: the UK Infrared Telescope on the volcanic peak of Mauna Kea in Hawaii is a flux collector of 3·8-meter aperture on an extremely dry site. Images from this telescope can now rival the quality of optical images. Astronomers are continually developing new techniques, such as Fourier transform spectroscopy, which has permitted the study of planetary atmospheres, and the application of the charge-coupled device (♦ page 89).

◄ The United Kingdom Infrared Telescope (UKIRT) is the world's largest instrument specifically designed for infrared astronomy. It is equipped with a remote control facility and a 3·8m mirror sufficiently accurate even to detect submillimeter waves normally collected by radio telescope antennae.

◄◄ Infrared telescopes are important to the study of nebular emission. Image-processed contour maps such as this one of the Orion nebula can reveal information about the birth of stars.

► Patterns of heat loss in a material can locate internal faults in thin metals, composites or at coating interfaces. In the technique known as pulsed video-thermography, a composite material consisting of a carbon-fiber skin bonded to an aluminum honeycomb (♦ page 246) is subjected to pulses of heat from a xenon flash tube. Infrared cameras record a frame-by-frame sequence of heat diffusion through the object which is displayed on screen. Defects show up as deviations in the expected patterns for the material being tested.

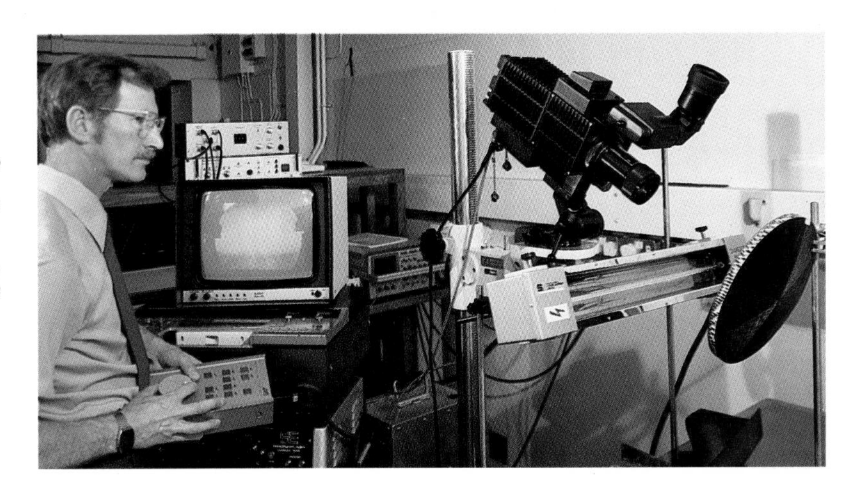

Studying the Earth from space

Different areas of the Earth's surface, whether arable land, urban sprawl, forest, desert or sea, reflect sunlight with varying intensity along the visible-to-infrared band of the electromagnetic spectrum. Looking down at the Earth from space with instruments sensitive to these reflectance values allows certain information to be gathered that would be difficult to obtain from ground surveys. By converting these findings into image form, maps can be produced that are particularly useful to the assessment and analysis of a wide range of geographic phenomena.

In July 1972, NASA launched Landsat 1, the first of a series of highly successful satellite systems specifically designed to acquire data about Earth resources. The orbit of the Landsat satellites – keeping pace with the Sun's apparent westward progress across the Earth – is planned so that, over the 20 visits made to the same "scene" each year, data are collected at the same time of day to allow accurate comparison across time and between scenes at any one time. Data are collected by a device known as the Multispectral Scanner (MSS), its sensors tuned to different wavelengths of the electromagnetic spectrum. Two channels are sensitive to bands in the visible spectrum (green and red), the other two to bands in the near infrared.

In each of the four bands, the scanner uses six detectors to record variations in the intensity of Earth-reflected sunlight, encoding the values in digital form. An oscillating scan mirror permits a scan of six lines at a time, each line 185 kilometers long, containing 3,240 separate picture elements (pixels), each of which covers an area of the Earth's surface 79 by 56 meters. With 2,340 such lines making up a single Landsat scene (185 kilometers square), each band records more than 7,500,000 pixels per scene in 25 seconds; the four bands in total thus provide some 30 million observations.

The data are transmitted as electrical pulses to one of several Earth stations. A computer produces an image by reassembling the array of pixel data in correct position in the scene. Three of the four wavelengths are assigned a primary color of the photographic process (red, blue and green) and are then combined and reproduced in varying strengths across the image according to the reflected brightness measurements received from the pixel array. The result is known as a false-color image – the strongest readings are those from healthily growing vegetation, which is normally depicted in Landsat imagery as a bright red color.

Conversion from digital data to imagery offers considerable scope for modification and processing by the computer: the many permutations available can be tailored to suit analytical purposes, such as geological mapping, measuring soil moisture, differentiating vegetation species, with no information loss from the bank of primary data.

In July 1982, the second generation of NASA's Landsat series was initiated with the launch of Landsat 4 followed, after its failure, by Landsat 5 in March 1984. In addition to the MSS carried by other Landsats, these were equipped with a Thematic Mapper which recorded data in seven wavelengths, including the middle and thermal infrared bands. Landsat 5 scans from a lower orbit and with much higher resolution than the earlier Landsats, and provides much sharper imagery. Similar satellites are being launched by other nations, including the French SPOT system (1986), which has the innovative capacity to observe the Earth's surface obliquely and thus make stereoscopic imagery available.

▼ Landsat 5 is equipped with a Multispectral Scanner and a high-resolution Thematic Mapper. Orbiting the Earth at an altitude of 805km, the satellite covers an 185km-wide swathe of the surface below as it passes overhead in a north-south direction. Its instrumentation is powered by solar energy.

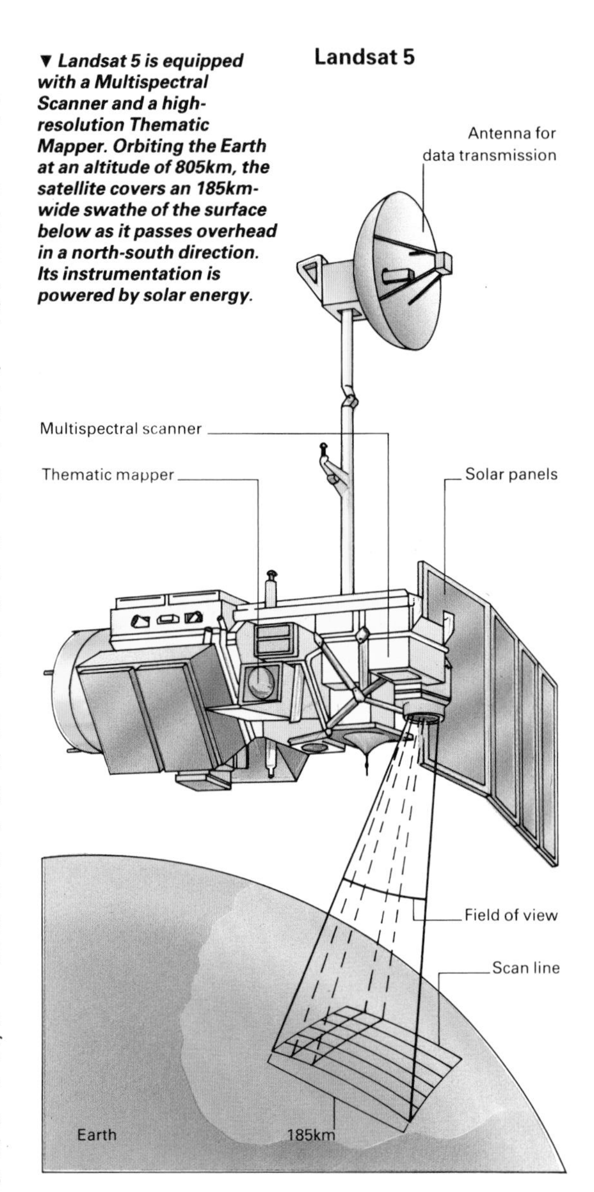

▲► Images of Abu Dhabi as recorded by the Landsat 5 Thematic Mapper in April 1984 and processed by different techniques. False-color composite (1), combines the reflectance data from three bands. On the contrast stretched and edge enhanced image (2) reflectance intensity values are displayed in maximum contrast with high frequencies further enhanced to discriminate between areas with similar values. Maximum likelihood classification (3) classifies areas with similar values to identified known features (16 land-cover types are classified in this image). In the principal components analysis (4) data from correlated bands are transformed to an uncorrelated set to enhance certain details, such as relief, which is otherwise invisible.

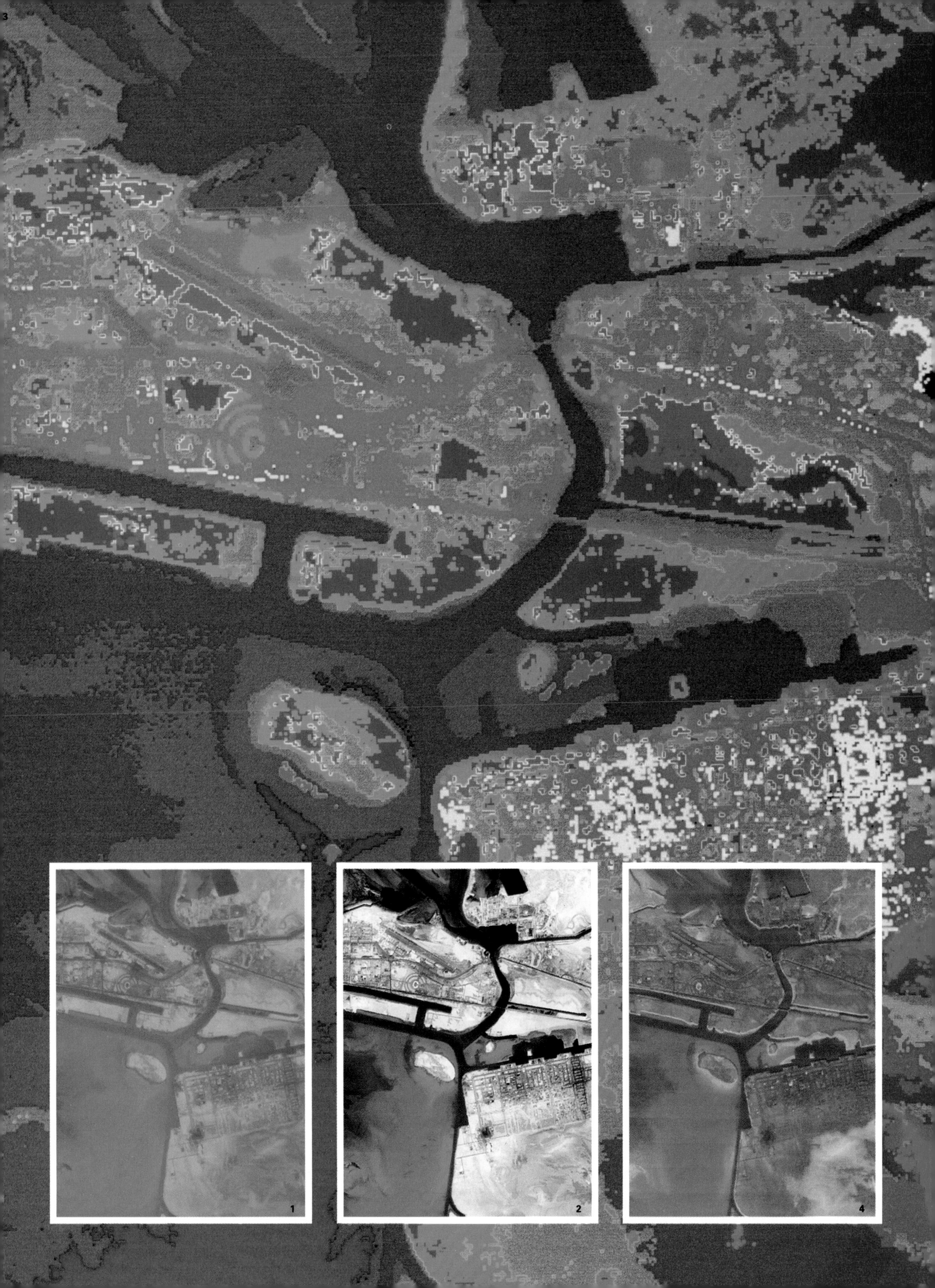
3
1
2
4

The largest steerable radio telescopes are 100 meters in diameter

◀ Karl Jansky with his rotating aerial array at Holmdel, New Jersey.

▶ With a dish of 100m diameter, the Effelsberg Radio Telescope near Bonn, West Germany, is the world's largest fully steerable radio telescope. A mesh rim dissipates wind pressure which would distort the dish, and the valley site protects it from radio interference.

▼ This false-color image of the Andromeda Galaxy was made at 11cm wavelength by the Effelsberg telescope. Increasing emission intensity is coded from purple, through blue, green and yellow to red at the galactic center.

The origins of radio astronomy

Although the pioneering work of Hertz and Marconi made available the means for detecting radio waves (⧫ page 185), it was not until 1932 that the American radio engineer Karl D. Jansky (1905-1950) recognized that some radio "noise" had an extraterrestrial origin. From the fluctuations in its intensity Jansky deduced that the source lay towards the galactic center in the constellation Sagittarius. This radiation is 20 million times longer in wavelength than the rays that a human eye can see. Another American, Grote Reber (b. 1911) took up the subject and mapped the radio contours of the Milky Way.

The contrasting instruments employed by Jansky and Reber foreshadowed the nature of "radio telescopes" for some years ahead. Jansky's instrument was an array of aerials some 30m long, only twice the wavelength of the radiation he was studying. This anticipated the technique of using an array of multiple aerials taken up again by radio astronomers two decades later. Reber constructed the first proper radio telescope, a paraboloid dish 9·4m in diameter working initially at a wavelength of 9cm, then 33cm and finally 1·87m.

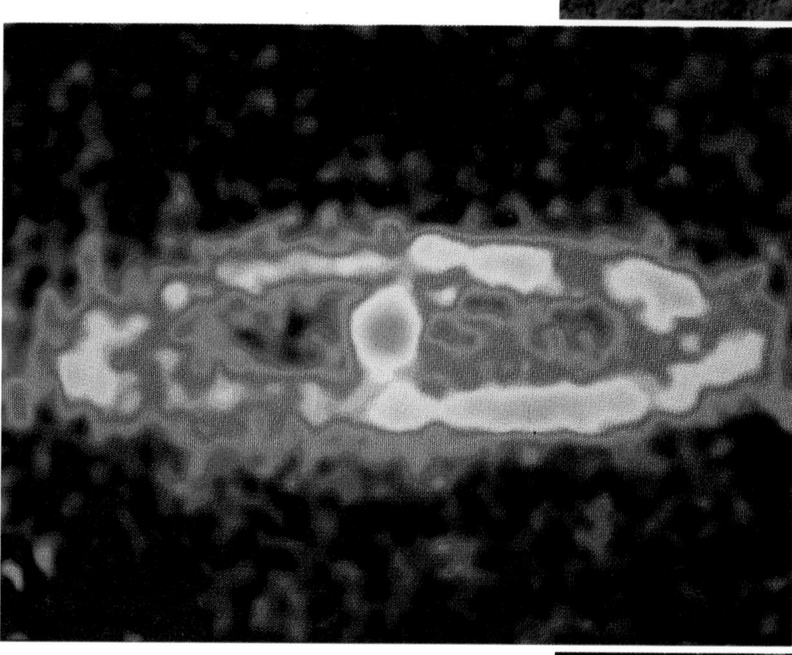

Radio astronomy

Today it seems surprising that astronomers did not predict that radio waves would be emitted from the stars, but at one time there seemed to be good reason why the cosmos would be radio-quiet. The stars burn at temperatures so high that their maximum emission lies in the optical region, or even the ultraviolet. The form of radio emission that *was* predicted, its source being the most abundant form of matter in the Universe, was atomic hydrogen. In 1944 the Dutch astronomer H.C. van de Hulst predicted that hydrogen emission would be detected at a wavelength of 21 centimeters, but seven years passed before advances in radio-telescope construction technology enabled astronomers to "listen" to cosmic hydrogen between the stars.

The radio telescope

The principle of the radio telescope is simple: radio waves are reflected from the paraboloidal dish to the focus, where a small aerial supported by struts feeds the signal to the receiver. For a given

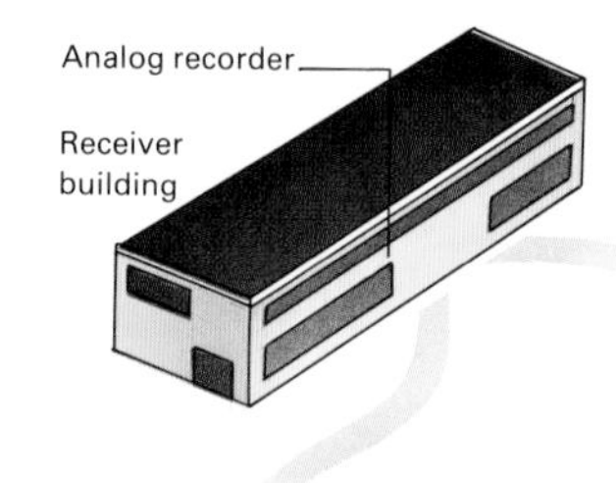

▶ The radio telescope is made up of a number of components, not just the antenna. A steerable dish reflects and focuses incoming radiation to the feed antenna, where it is converted into electrical signals and transmitted to a receiver station. The signals are very weak voltage fluctuations: a series of amplifiers is required to boost them by a factor of 10^{15}. The output is then stored in a computer where it is processed and displayed as an image, usually with measures of brightness coded by color layers.

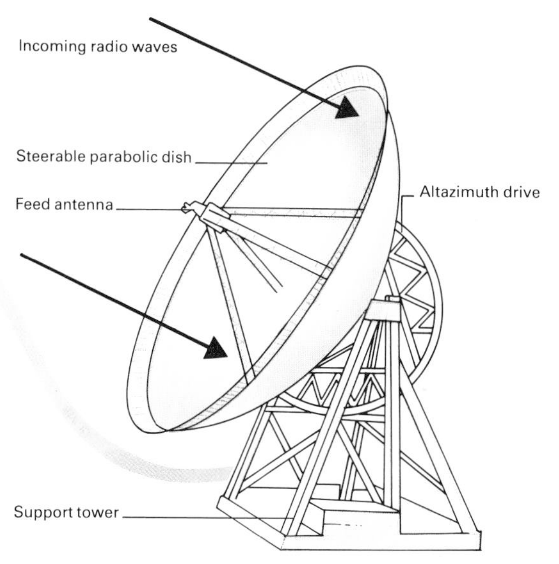

wavelength, resolution is inversely proportional to the diameter of the dish: the larger the radio telescope, the finer the detail observable. In radio astronomy, the wavelength of radiation received can be a sizable fraction of the telescope diameter, although a practical limit to the size of the dish is soon reached: above a diameter of 100 meters a fixed dish must be used (with its attendant disadvantages) or a new approach found. Such an alternative arrived in 1946 when British astronomer Martin Ryle and Australian J.L. Pausey independently built a radio interferometer, a pair of radio telescopes whose separate signals were combined in a common amplifier.

A radio source produces signals whose interference changes over the period of observation thus yielding information about its size and position. By using an interferometer instead of a large dish, however, some degree of sensitivity is lost in exchange for resolution. The first discrete source of radio emissions from the sky, detected in the constellation Cygnus, was pinpointed with an interferometer, and detection of other such sources followed.

Microwave astronomy

The detection of radio wavelengths at the very short end of the spectrum, centimeter, millimeter or submillimeter waves – collectively known as microwaves – allows astronomers to study interstellar clouds and distant exploding galaxies known as quasars.

*Emission of electromagnetic radiation accompanies changes in energy levels that occur or can be made to occur in atoms and molecules. At very short wavelengths, these changes are very small indeed: masers are used to boost the emitted microwave signals by the use of stimulated emission in exactly the same way that lasers amplify light (**▶ page 101**).*

Radiation at wavelengths of 1mm down to 0·3mm can be observed from the Earth's surface providing absorption by water vapor is minimized. Dry, high-altitude sites such as Kitt Peak in Arizona or Pico Veleta in the Sierra Nevada of southern Spain are most favored. To work at short wavelengths, telescopes need to be enclosed for protection from the wind and rain, and have a very smooth dish surface of highly accurate configuration to focus the waves efficiently.

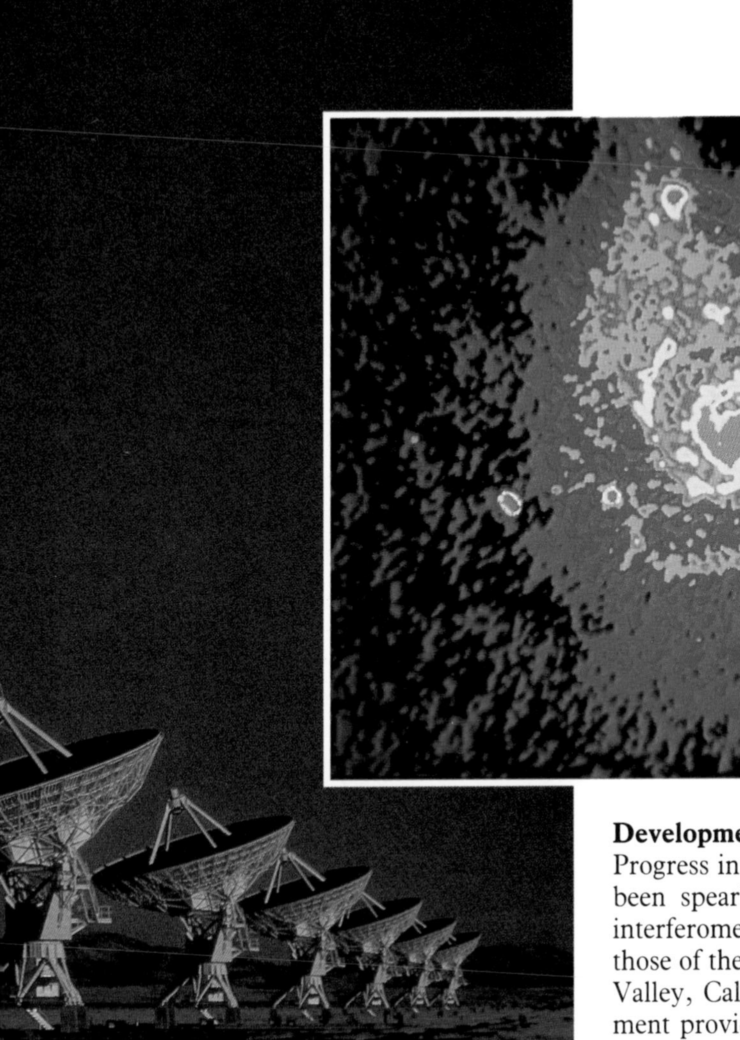

◀◀ *The Very Large Array (VLA) telescope at Socorro, New Mexico, was built to apply the technique of aperture synthesis (now also being used in optical astronomy ◆ page 64). Composed of 27 movable dishes, each 26m in diameter, positioned on a Y-shaped railroad track, the interferometer can synthesize a single dish 27km in diameter.*

◀ *The well-defined spiral arms of the Whirlpool (M51) Galaxy are clearly visible in this image produced by the New Mexico VLA telescope. The galaxy, 20 million light years away, emits radio waves at 21cm wavelength. The computer-powered color-coding runs from red for highest emission, through yellow, green and blue to purple for lowest.*

Developments in radio telescopes

Progress in radio telescope technology over the last three decades has been spearheaded by multiple elements evolving out of the early interferometers. By 1957 the advantages of the interferometer and those of the steerable dish were combined in the instrument at Owens Valley, California, the largest interferometer of its day. This instrument provided a sufficiently accurate position for a source originally detected at Cambridge to permit astronomers to home in on a star which eventually proved to be one of the first quasars. Five years later the British radio telescope at Jodrell Bank was linked by a microwave radio relay to a pair of radio telescopes at the Royal Radar Establishment, Malvern, a baseline of over 100 kilometers. At Cambridge the "one mile" telescope was constructed with two fixed dishes and a mobile one on a rail track. This interferometer was seminal both technologically and scientifically. It provided the testbed for the technique of "aperture synthesis" by which interferometric data from a set of smaller instruments are combined to simulate a single very much larger one (the basis of the Very Large Array, billed as the "world's biggest telescope", at Socorro, New Mexico). Furthermore, it permitted rival cosmological theories to be tested by counts of faint radio sources.

The notion that interferometers of longer baseline yield radio maps of higher resolution suggested that positioning their elements continents apart, separated even as far as the diameter of the Earth, would produce optimum results. A technique called Very Long Baseline Interferometry (VLBI) makes fresh demands on the instrument designer. To maintain the phase information about the incoming wave front – information crucial to the technique – its arrival time must be determined precisely. Wires linking the closely spaced interferometers of the early days were superseded by radio links at larger separation; for VLBI the distances are too great for either method. Instead, time signals from atomic clocks (◆ page 29) synchronize the radio signals. VLBI, although laborious, rewards radio astronomers with resolutions down to one hundred millionths of an arcsecond.

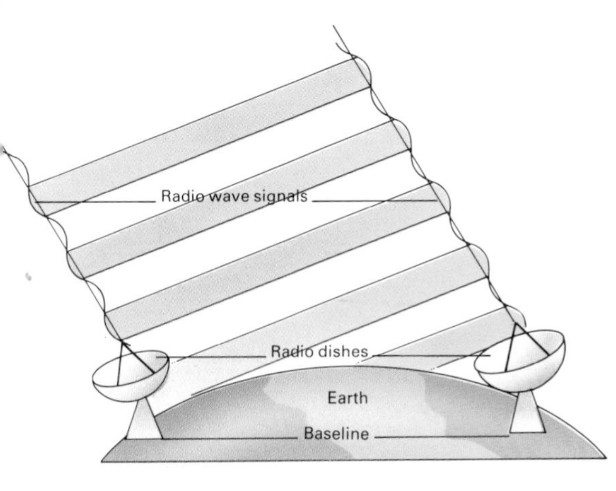

◀ *The James Clerk Maxwell Telescope on Mauna Kea, Hawaii, is the largest in the world for submillimeter wavelength observations. Roof shutters, doors and a transparent Teflon membrane covering its viewing aperture protect the dish from high winds.*

▲ *In VLBI, a baseline separating the two radio antennae of the interferometer can be as long as the diameter of the Earth. Signals are recorded from a common radio source, the arrival times synchronized and the data combined.*

Radar

Since its origins in war as a technique for the early detection of enemy aircraft, radar (RAdio Detection And Ranging) now has many civil uses, and it is used by spacecraft studying the surfaces of planets, including the Earth. In principle radar is simple, but its practise is far more complex. Fundamentally, a radar system sends out a beam of radio waves (frequencies of 3–30 GHz on the electromagnetic spectrum), and extracts its information from reflections, just as a torch allows you to see in the dark. Traveling, like all electromagnetic waves, at the speed of light, these waves move much faster and farther than sound waves. They are also able to penetrate clouds and atmospheric layers and, according to the frequency transmitted, can be made to reflect from a number of surfaces, including metallic objects, planetary surfaces or the zone of the Earth's atmosphere, known as the ionosphere.

The radar receiver has to allow the operator to interpret the reflections it has detected. The waves are precisely controlled so that measurements can be carried out on the returning signals as fully as possible. The spread of the beam must be minimal so that direction can be pinpointed exactly; large aerials are used to produce a parallel beam and to pick up the weak return signals. Radar sets usually transmit pulses of waves, rather than a steady beam; this allows more

The early days of radar

In 1935 the British physicist Robert Watson-Watt (1892-1973) wrote a paper on the "Detection and Location of Aircraft by Radio Methods" for the British Air Defence Committee. The basic principle of transmitting and receiving radio echoes from metal objects was well known, and in 1904 a ship had been detected by this technique at the range of 1·6km. Watson-Watt believed that its application was feasible in long-range detection of aircraft. He explained how to measure distance, bearing and height of an aircraft, and stressed the urgency of development: active research was already in hand in Gemany and the United States.

The argument was accepted and the team worked at a furious pace. They chose to build a system using the known technology of 10m radio waves rather than pursue research into potentially more effective shorter waves. By September 1938 the first section, covering the approaches to London, of the "Chain Home" had been erected – a network of radar stations watching the eastern coast of Britain for signs of airborne attack. It could spot bombers over 150km away, which gave a 30-minute warning. During the Battle of Britain in 1940 "Chain Home" ensured the efficient deployment of fighters and fuel, and gave time for civilians to take shelter.

"Chain Home" was a practical triumph and the technological breakthrough to modern precision radar using waves 100 times shorter (10cm) soon followed. The magnetron generator that was needed to transmit these microwaves was invented by two British physicists, J.T. Randall and H.A.H. Boot of Birmingham University, in 1939 and soon moved into production. By 1941 it provided the basis of the aircraft on-board radar systems which gave eyes to nightfighters and provided long-range bombers with adequate navigational capability.

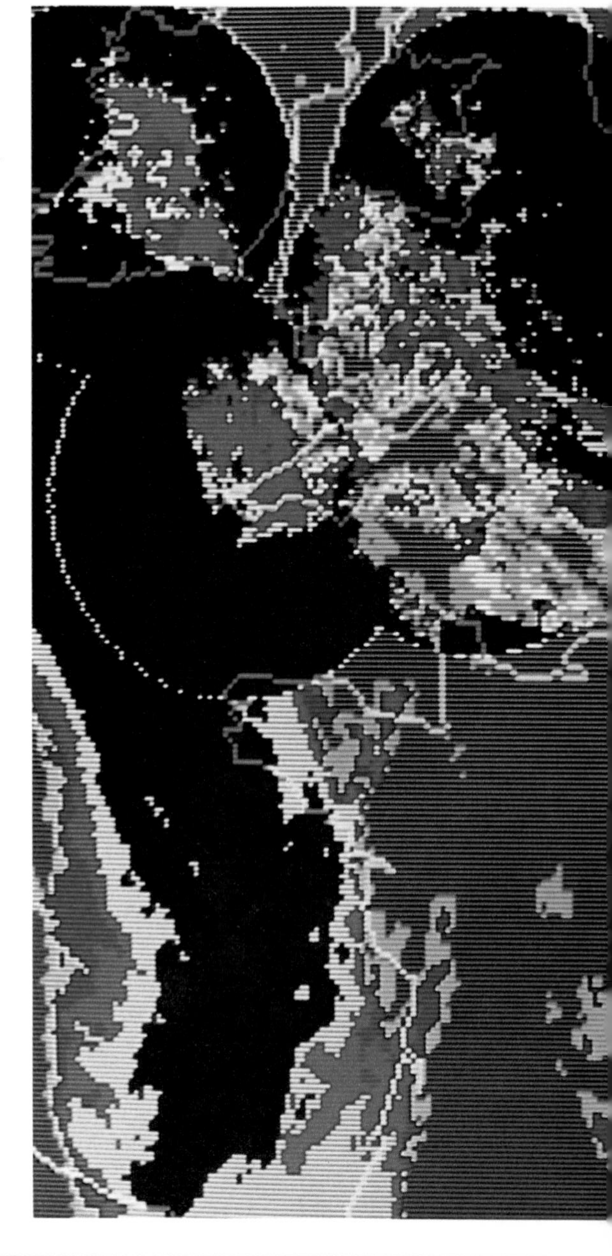

▶ This composite image shows a front of warm and humid air across the British Isles. A satellite radar system can record variations of water vapor in the air by reflecting a series of pulses at controlled frequency and plotting the different strengths of the return signals.

▼ The EISCAT project (European Incoherent SCATter facility) uses this huge compound radar system at Tromsø in Norway to plot the density of free electrons in the ionosphere by detecting echoes scattered back from its high-frequency radiowave transmitters.

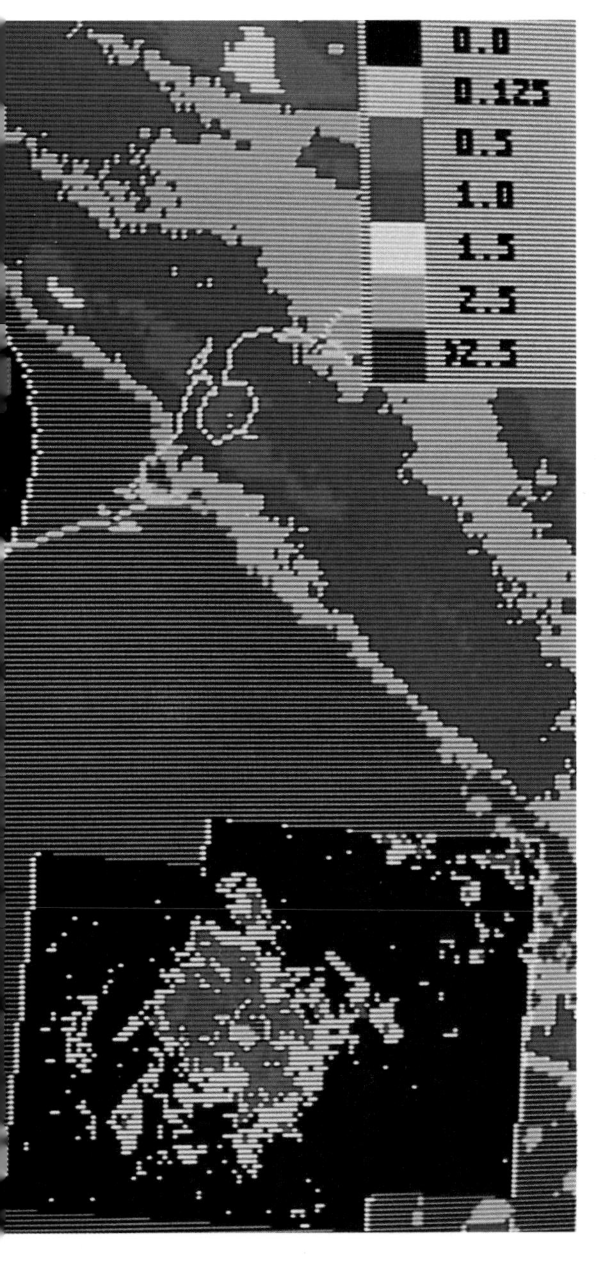

intense illumination, and allows the range of a reflector to be deduced by measuring the time taken for a pulse to go out and return. The frequency is either kept constant or is changed by a controlled amount over a short time span, a factor which allows the range or speed of reflecting objects to be calculated more easily.

When waves bounce off a moving target they are either pushed closer together or spread out, depending on whether the reflector is moving into the wave train or away from it. This effect, known as the Doppler effect, is the same as occurs with sound waves and is commonly experienced when a siren on a car drops in pitch as it passes and moves away. In radar, the echo from a moving object has a slightly different frequency from the transmitted wave. If this change is measured, the velocity of the reflecting object with respect to the radar set can be calculated.

Some continuous wave radars rely on Doppler shift to avoid swamping their receiver circuitry with transmitted signals. Their circuits are designed to ignore all echoes within some small frequency range of the transmitted wave; this means that they detect only objects moving above a chosen threshold speed. Such a "moving target indicator" is used in many aircraft detection systems and landing control, where echoes from fixed objects, buildings and hills could obliterate the signal from a low-flying aircraft.

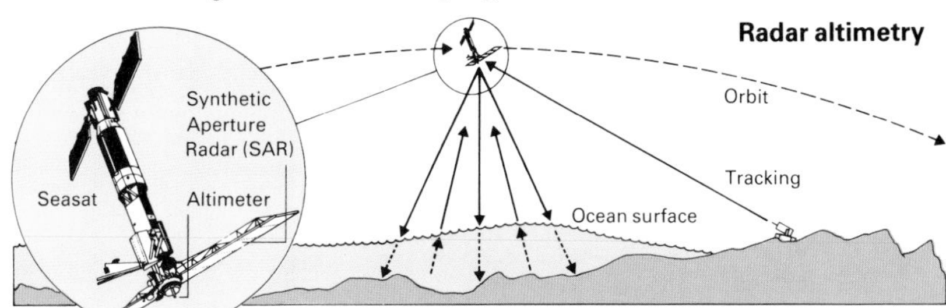

▼ ***Seasat had a synthetic aperture radar, enabling it to record side-scan images of the type pictured below. Surface roughness was recorded by measuring signal loss through scattering and plotted against positional data obtained by use of the Doppler effect.***

Studying the ocean from the sky

The satellite Seasat was launched in 1978 for oceanographic research. One of its instruments was a radar altimeter, an instrument pointed directly downwards. It measured the reflectivity of the ocean, significant wave height and the mean level of the ocean – averaged out over an area large enough to cancel out the effects of "choppiness". Its primary purpose was to map the features of the ocean floor by determining deviations in the height of the sea surface. Seamounts have a high gravity field, resulting in a "pile-up" of water over them; the opposite happens in trenches. During the satellite's 106 days of active life, it sent back to Earth some 60 million spot heights, sensitive to variations as slight as 10cm.

It was not straightforward to translate Seasat's data into useful information. The fact that the Earth is not perfectly round had to be taken into account for the seabed measurements. In addition, the satellite's own orbit was elliptical: its rate of rise or fall therefore had to be detected by Doppler shift of the signals, allowing the onboard computer to calculate the appropriate correlations. To construct a map it was necessary to compile data taken at different times; moving features such as wind and tides had to be eliminated by equating data from repeated points. Calibration to connect variations in the sea surface with the topography of the sea bed was done in terms of known soundings.

See also
Microscopes 47–56
Optical Telescopes 57–64
Seeing with Sound 65–72
Digital Recording 97–100
Radio and Television 185–192

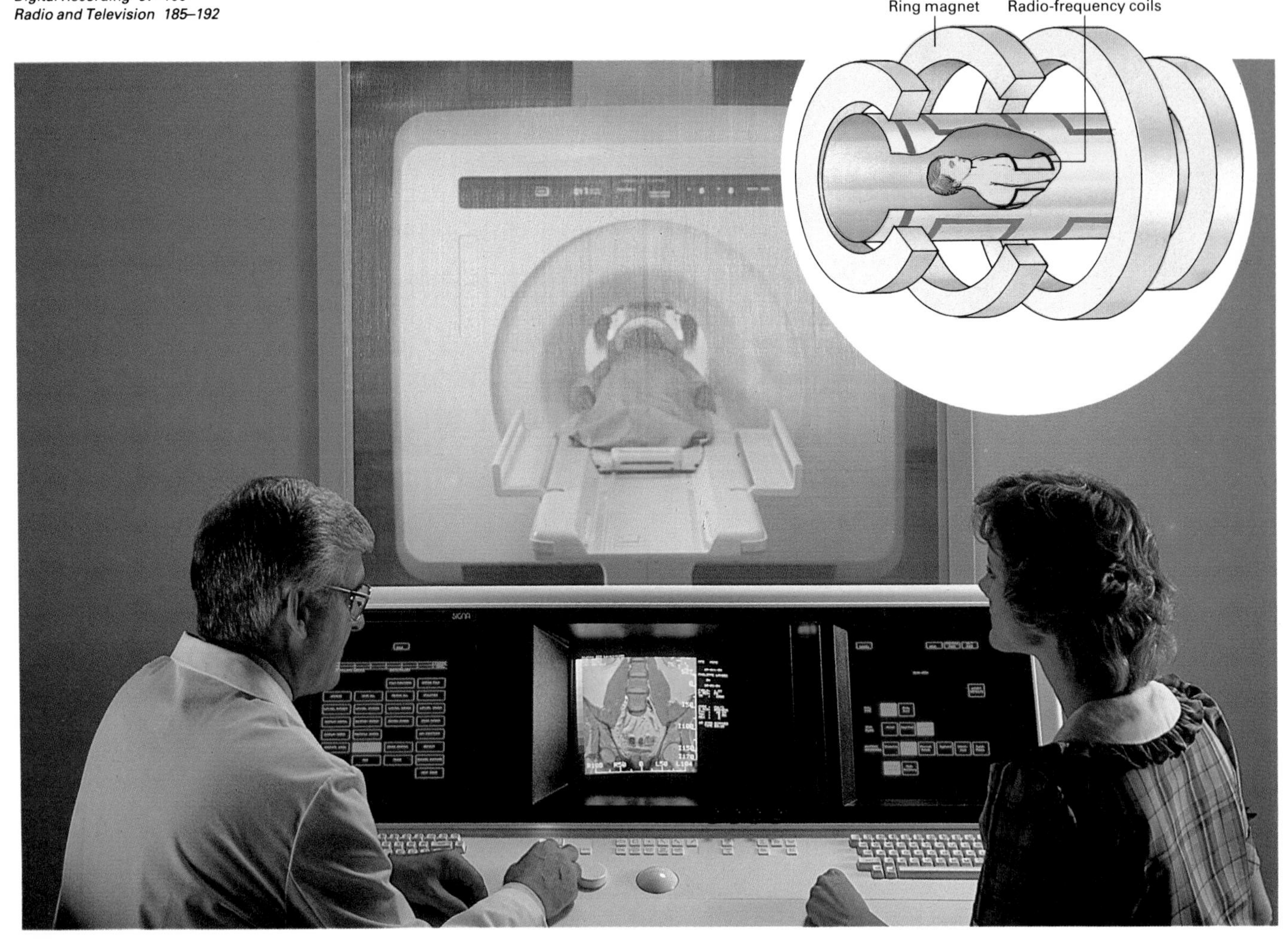

Magnetic resonance inspection

The technique of magnetic resonance inspection (MRI), formerly known as nuclear magnetic resonance (NMR) was developed in the 1970s as a method of non-invasive imaging of the soft material of the human body, a technique derived from chemical analysis and with other applications in industrial non-destructive testing.

A subject is placed in a large cylindrical magnet that produces a very powerful magnetic field. Rather than X-rays which may be damaging to living tissue, harmless radio waves, at high or very high frequencies, are directed at the subject. The nuclei of certain atoms within the subject, induced by the magnetic field to behave as tiny magnets themselves, absorb the radiation – their propensity to do so is known as "resonance" – and then emit radio signals when the radiation source is switched off. For a given strength of magnetic field, the frequency of resonance varies from one element to another, hydrogen responding particularly strongly. By altering the frequency of the radio waves applied, the distribution of concentrations of particular elements, such as hydrogen, sodium, potassium and carbon, along a cross-section of the subject can be imaged. A computer reads the strength of signals received point by point across the subject and produces a display in which different colors encode different intensities.

Magnetic resonance inspection is also used in the non-destructive testing of materials containing organic chemicals including plastics and oils. The technique is still at the experimental stage, but one potentially attractive application is to provide a method of monitoring the flow of oil through running motors.

▲ In MRI equipment, such as this medical body-scanner, radio-frequency (RF) coils control radiation frequency, and magnetic gradient coils control field strength. This arrangement allows operators to isolate the most common elements in the tissues to be investigated and to detect subtle variations in their concentration.

▼ The English 16th-century warship "Mary Rose" was raised in 1982 and MRI was used to monitor the penetration of conservation chemicals into her timbers. The presence of hydrogen (and therefore of water) indicates regions of degradation, and is seen in this cross-sectional scan. White areas have the highest intensity.

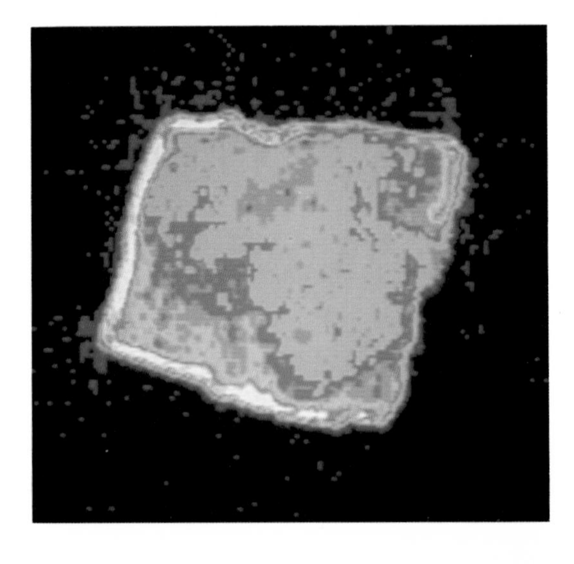

SAVE FOR

Jill ?

phy

ages...High speed photography
logy...The image converter
phy...The single-lens

◄ British physicist William Fox Talbot invented the negative-positive process. His calotype was a paper-positive process derived from a negative image on light-sensitive paper.

ow to form an image. Using the camera nt object could be projected through a wall of a darkened chamber. Obtaining more difficult process. A French army 1765-1833) sought for more than ten ould hold an image, and in 1826 he was liest known surviving photograph was of his family home at Chalon-sur-Saône he adjoining buildings. The image was d it required an exposure of eight hours.

partnership with another Frenchman, . Together they managed to put their only after Niépce's death in 1833 did Daguerre bring up ... tivity of the plates to the degree that he could take a photograph of a human being. Known as daguerrotypes, these silver-coated copper plates were treated with mercury vapor to develop the image, which was fixed with salt.

Multiple copies of images, prints, were made possible by the development of the negative by William Fox Talbot (1800–1877) in 1839. Portrait photography was naturally an early commercial goal. The first professional studio was opened in New York in 1840. It had large reflecting mirrors illuminating the subjects with sunlight, and the usual exposure time was one to four minutes. The experience was not a comfortable one for the subjects since head and body supports were required and the glare was unpleasant.

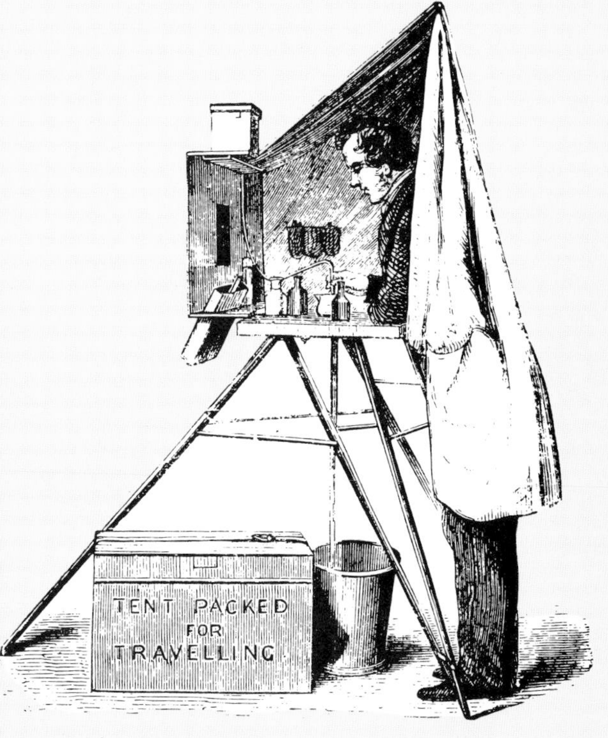

▲ The wet collodion process, perfected by Frederic Scott Archer in 1851, rapidly overtook the calotype and daguerrotype in popularity by combining reproducibility with fine quality. As the collodion-coated glass plate had to be exposed wet, a traveling photographer had to carry his equipment with him. This portable darkroom folded out of a suitcase.

◄ The long exposures demanded by the early photographers meant that their subjects had to be pefectly still. Daguerre's view of the Boulevard du Temple, Paris, taken in 1833, included the image of a man having his boots cleaned, who stood in one place long enough to be recorded. He was the first person ever to be photographed.

Chronology

1725 Discovery of light-sensitive properties of silver nitrate by Johann Heinrich Schulze
1802 Unfixed prints by light action made by Thomas Wedgwood in Britain
1816 Negative images recorded on paper by Joseph Nicéphore Niépce in France
1826 Earliest photograph made by Niépce
1835 First photographic negative made by William Henry Fox Talbot
1839 Two important photographic processes announced: Jacques Daguerre's daguerrotypes, which used silver-coated copper sheets, and Fox Talbot's calotypes, a negative-positive process.
1851 Wet collodion process invented by F. Scott Archer
1853 Dry plates introduced
1884 Orthochromatic plates introduced, using emulsion sensitive to different light intensities
1888 First roll film made by Eastman Kodak
1889 Film made of celluloid and coated with silver bromide film introduced by Eastman Kodak
1891 Motion pictures introduced with the Kinetoscope, developed by Thomas Edison and William Dickson
1891 First permanent color photographs made by Gabriel Lippmann
1892 The "Photo-Jumelle" precision camera designed by Jules Carpentier
1894 First use of high-speed photography by Etienne-Jules Marey in Paris
1900 First appearance of the inexpensive Kodak "Brownie"
1904 Panchromatic sensitizers

High speed photography

Specialized photographic equipment is required to study events that occur too quickly to be seen by the eye and to freeze their rapid motion on film. The illumination of the object must be of high intensity, since light falls on the film for only a short moment of time, and a trigger arrangement must be incorporated that can time the exposure to coincide with the event to be photographed.

Modern conventional cameras with fast shutter speeds and very light-sensitive film can effectively freeze normal motion. A reasonably strong tennis player may serve a ball at speeds approaching 160 km/h. On a print from a conventional camera taken from an exposure at one-thousandth of a second, showing the player about 5cm high, the ball will have moved about 1mm. If the image of the ball is 2·5mm in diameter, it will appear clearly identifiable, but blurred. Many fields of research, however, particularly in materials science, require a more exact image. To examine the way a golf ball is deformed when struck by the head of a club, or how a crack propagates through a sheet of glass, or the progress of impact damage when a steel ball hits a steel surface, the exposure time must be much shorter.

Standard cine cameras move the film intermittently, and the shutter is only opened when the film is stationary. Acceleration is hard on both film and mechanism, and the technique cannot be used for more than about 1,000 images per second. Specialist high-speed cine cameras move the film continuously and use a rotating prism to shift the image in synchronism with the film during exposure. With this approach, speeds of 10,000 images a second are possible, although magazine containing 30 meters of film is used up in about half a second. Clearly cameras of this sort cannot be triggered by hand. Normally the event itself is used as the trigger: for example, the head of a swinging golf club might interrupt a light beam and activate an electronic switch as the club approaches the ball.

The fastest mechanical cameras use a rotating mirror. An image is formed on the mirror and re-imaged through a series of fixed lenses onto a fixed film held in an arc at a distance of about 1 meter from the mirror axis. The camera is run up to speed in a darkened room, then the event and a xenon flash tube are triggered together. The flash is so rapid that the light is extinguished before the mirrored image makes a second pass across the film. On such machines the mirror can be driven to about 20,000 revolutions per second by an air turbine, so the light may cross 25 lenses, each of which provides a 25mm image, in just five microseconds. This technique is very useful in explosives work.

To study rapid motion, a flash tube must emit a high number of light pulses per second. In one such device, known as a stroboscope, sparks resulting from the ionization of gases subjected to a high electrical field last no more than 10 nanoseconds (10×10^{-9}s) and will repeat at rates of 20,000 per second. These lamps can be used with high-speed cameras to give very sharp images of moving objects. Alternatively, they can be used to provide continuous observations of repeated phenomena.

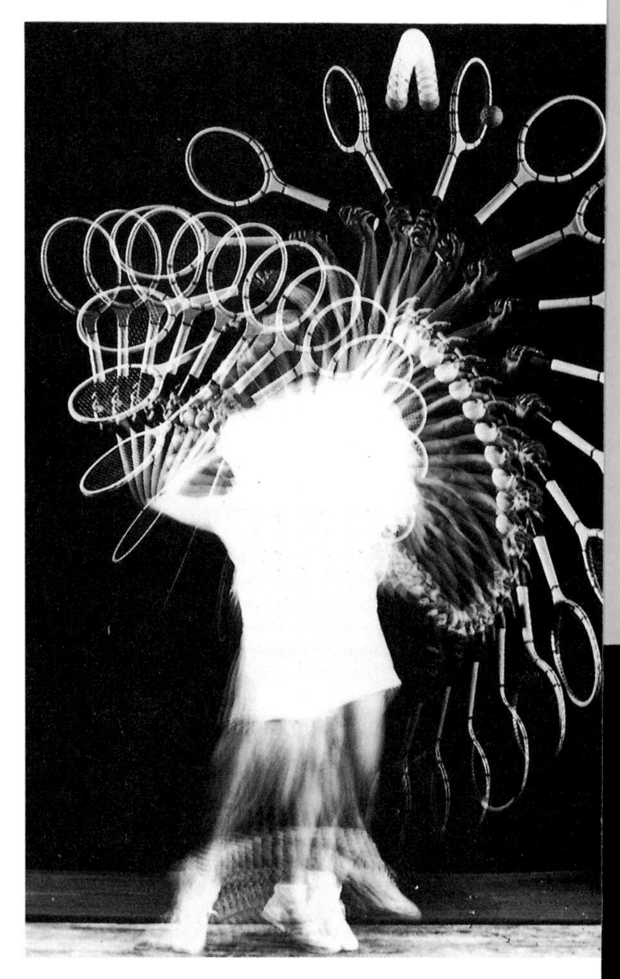

The image converter
Most rapid of all photographic devices is the image converter. In this, the light forms an image on a photocathode, a surface which emits electrons when light falls upon it. These are accelerated and brought into focus on a second surface which is coated with a luminescent material that produces light when struck by electrons. A conventional camera, commonly using a Polaroid film, records this second image.

Deflection of the complete electron beam allows successive images to be placed at different parts of the final photograph and the interval between images can be reduced to about 1 nanosecond. This is the equivalent of forming one billion images per second.

A useful feature of the image converter camera is that when the electrons are accelerated from the photocathode they are given energy and so may release more photons from the final image than were contributed to the initial one. Because of the very short exposure time this gain is important and image converter cameras are sometimes used with an extra stage, called an image intensifier (◆ page 60), which gives a large gain in light output.

developed (emulsion sensitive to all colors)
1907 Autochrome process launched by the Lumière brothers, first marketable color film
1924 First commercially successful precision miniature 35mm camera, the Leica, produced by Leitz
1925 First flash bulb developed
1928 Rolleiflex, first rollfilm twin-lens reflex cameras marketed by Franke & Heidecke
1932 First successful photoelectric exposure meter developed
1932 Rotating prism high-speed camera first used
1934 Image tube, forerunner to the image converter, invented.
1936 Kodachrome produced, first commercial color film with three emulsion layers
1937 First 25mm single lens reflex camera, the German-made Exakta
1939 First electronic flash developed by Harold Edgerton
1942 Color roll film launched by Kodak and Agfa
1944 Rotating mirror framing camera first used
1947 Polaroid one-step printing process invented by Edwin Land in USA
1963 Kodak Instamatic instant-print color film developed
1972 First single-lens instant camera marketed by Polaroid-Land
1976 Microprocessor-controlled automatic exposure introduced by Canon
1978 Konica introduced first automatically focusing camera
1982 All-electronic camera introduced by Sony
1982 Instant transparency films introduced by Polaroid

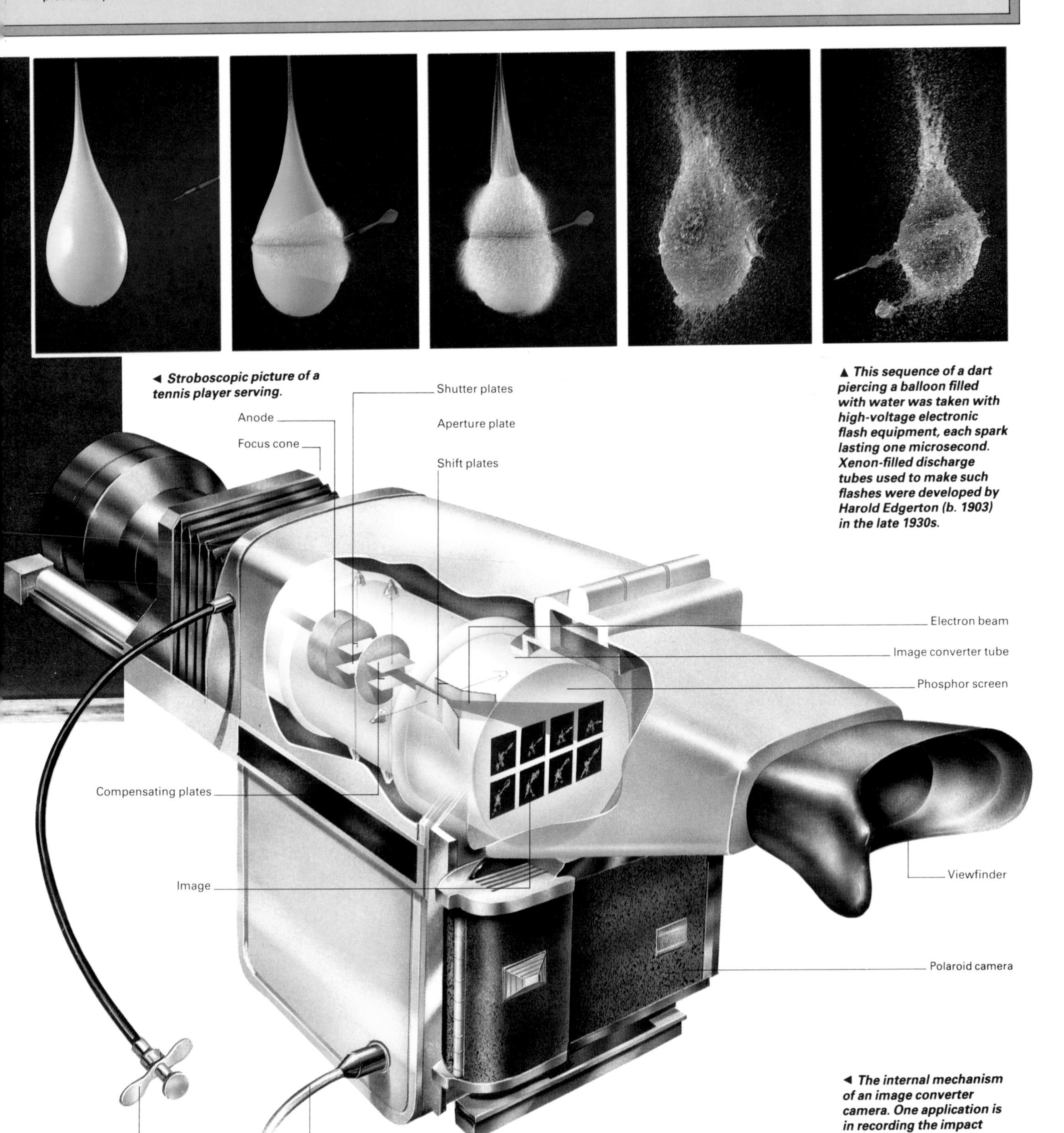

◀ ***Stroboscopic picture of a tennis player serving.***

▲ ***This sequence of a dart piercing a balloon filled with water was taken with high-voltage electronic flash equipment, each spark lasting one microsecond. Xenon-filled discharge tubes used to make such flashes were developed by Harold Edgerton (b. 1903) in the late 1930s.***

◀ ***The internal mechanism of an image converter camera. One application is in recording the impact made by projectiles in armor. Acoustic or fiber-optic triggers coordinate illumination and the electron deflection system.***

See also
Measuring with Electricity 41–46
Optical Telescopes 57–64
Seeing beyond the Visible 73–86
Recording on Tape and Disk 91–96
Printing 171–176

Color photography

A photographic emulsion is a preparation of silver halide crystals suspended in gelatin coated onto a base of transparent cellulose acetate. Modern color film comprises three superimposed layers of emulsion, each sensitive to light in the different primary colors – blue, green and red. By adding together different proportions of these colors, any other color can be produced. During exposure, the crystals are affected by light falling on them – the more intense the light, the more grains are affected – so that a latent image forms. The film is then processed with a developer which reduces the silver halide to black metallic silver; the image is then made permanent by fixing. The image is blackest where the original was brightest, and is thus a negative image.

For a color transparency, the developer reacts with dye-forming chemicals contained within the emulsion to form dyes. The dyes formed in each layer are "complementary" to the primary colors in the negative image to a degree that corresponds to the amount of light received. The areas with least dye in each layer correspond to the brighter parts of that particular color. In this case any other color is produced by subtracting, rather than adding, the dyes – cyan (blue-green), magenta (blue-red) and yellow (▸ page 174). Thus where the original color was blue, on the film there is no complementary yellow dye, and only magenta and cyan dyes show through. When light passes through the transparency for viewing, magenta and cyan layers subtract their complementary red and green lights, to leave the original blue. In the same way, the original green and red areas are also reproduced.

▸ Before the advent of flexible roll film, inventors sought ways to improve the collodion process. In this camera of 1865 the glass exposure plate was sensitized and processed within the camera itself by passing chemicals through the bulbed pipette on the top.

▼ In the single-lens reflex camera, the image is projected onto a focusing viewfinder by a movable mirror and pentaprism. Lens aperture and shutter-speed control exposure for the lighting conditions detected by the photocell. When the shutter button is pressed the mirror flips upwards, exposing the film.

The single-lens reflex camera

Flash tube
Battery
Automatic flash electronics
Self-timer switch
Flash control
Operating button
Shutter-speed selector
Film-advance lever
Liquid crystal display
Microcomputer control panel
Data-imprint switch
Power connection
Rewind crank
Film speed setting
Pentaprism
Photocell, measuring ambient light
Photocell, controlling automatic flash meter
Aperture-control ring
Iris diaphragm
Focusing ring
Frame counter
Film-advance selector
Automatic film-advancing mechanism
Battery
Mirror
Rear lens group
Front lens group

Recording on Tape and Disk

11

Recording electrical signals... Videotape recording... Microphones and loudspeakers... Assisted resonance... PERSPECTIVE... Early days of sound recording... Chronology... The video camera... The vidicon tube... Storing sounds on tape and disk... Mono, stereo and Dolby stereo... Recording-studio consoles

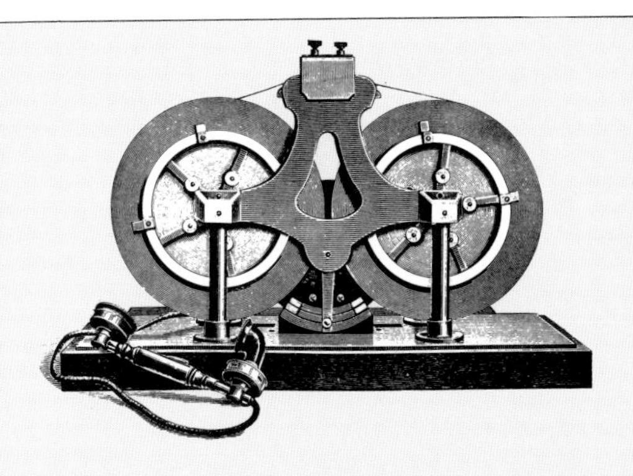

▲ **Poulsen's telegraphone that recorded on a magnetic wire.**

Electrical recording began to supersede mechanical recording in the late 19th century. Originally applied only to sound, subsequent electronic inventions in the 1900s (➧ page 186) showed that the technology for reproducing sounds and pictures would differ only by degree.

An acoustic or visual signal arrives in wave form at the detector (a microphone for sound, a camera tube for pictures) where this signal is converted or transduced into an electric current. The information is held by this current in the form of varying strengths and frequencies, and can be preserved either by imitating these varying levels electromagnetically (analog recording) or by converting the information into a series of binary codes (digital recording). Magnetic tape or disk can be used for storage with either technique. Naturally, a moving picture involves a much higher level of information per second than does sound and so introduces many practical difficulties, but the principle of storing them (video recording) is exactly the same.

The sound recording industry has made many recent developments towards convenience of operation and higher fidelity of reproduction – cassette tapes, digitization, Dolby stereo, compact disk – and all these have been passed on to video recording, despite the problems of packaging the complex equipment necessary in a form cheap enough to reach the domestic market. The results are only just beginning to be realized: all-digital audio-visual systems, using a single tape or compact disk (➧ page 97), played back through digital amplifiers and loudspeakers, with high-definition television pictures (➧ page 190), will soon become commonplace.

◀ **An exhausted Thomas Edison poses with his phonograph after five days and nights continuous work improving it. Its name was used for recording machines long after the disk system replaced the cylinder.**

The development of sound recording

The first skirmish in the curious technological battle between recording on tape or disk began with the very first experiments in recording sound, in 1877. The American inventor Thomas Edison (1847-1931), one of the most creative engineering designers of the age, was working on a device to repeat Morse coded messages sent by telegraph (➧ page 177). When using a waxed paper tape on which the message was written by a stylus, he noticed that if he pulled an already inscribed tape past the stylus it produced a note. He reasoned that controlled use of the stylus should allow him to record the notes of his choice and to replay them at will. Edison's phonograph did just that. In its earliest form the stylus followed a spiral groove on a drum of tinfoil, and sound was recorded as indentations in the groove. The indentations were made by speaking against a diaphragm whose consequent motion was transmitted to the stylus. In replay the motion of the stylus activated the diaphragm. The content of Edison's first two recordings is well known. At the first attempt the word "Halloo" was just recognizable; from the second the words "Mary had a little lamb" could be picked out with some certainty.

By 1899 the Danish engineer Valdemar Poulsen (1869-1942) had a competing system running which he demonstrated at the Paris Exhibition of 1900. He also used a drum with a spiral groove, but instead of indenting the groove, he bound into it a steel piano wire. An electromagnet free to slide along a rod set parallel to the axis of the drum made contact with the wire and followed it as the drum rotated. Current from a microphone energized the electromagnet and magnetized the wire with a strength proportional to the current. For replay the same components were used. Movement of the magnetized wire through the magnet induced a current in an earphone which reproduced the original sound.

From these devices of Edison and Poulsen two separate recording industries developed. Almost to the present day they have remained in competition, with the question always being whether to base sound reproduction on the physical movement of a stylus in a groove or on the currents induced in an electromagnet by a moving magnetized medium.

Chronology

1857 Frenchman Léon Scott invented phonoautograph
1877 American Thomas Edison and Frenchman Charles Cros independently invented phonograph, first talking machine
1877 Microhone patented by Emile Berliner, after principles of "variable resistance of loose contact" had been elucidated previous year
1885 Graphophone demonstrated by Alexander Bell and Charles Tainter
1888 Gramophone demonstrated by Emile Berliner
1897 First black shellac-based gramophone records made
1989 Magnetic sound recording begun with the telegraphone, patented by Valdemar Poulsen of Denmark
1901 Emile Berliner introduced black disk with spiral groove; sound vibrations were recorded as sideways deflections
1917 Condenser microphone invented by E.C. Wente in the USA
1927 Magnetic recording improved when W. Carlson and G. Carpenter patented the AC bias method of producing more signal and less noise
1925 Electrical recording introduced, with sounds recorded by microphones, transduced into electric current and converted into mechanical form to cut the master disk
1929 Plastic tape with magnetic coating developed by Dr Fritz Pfleumer of Germany
1932 Lapel microphone developed
1933 Alan Dover Blumlein pioneered stereophonic recording; two sound tracks recorded within a single groove,

Videotape recording

One advantage currently held by tape over disk is its facility for home recording. In an audio tape recording using conventional equipment, an amplified electric signal energizes an electromagnet contained in the record head, which orientates in imitation of the original waveform iron-oxide particles on the tape drawn across it. Operating an erase-head to demagnetize the tape prior to recording permits reuse. On playback, the new magnetic field pattern on the tape induces the reverse process with the electric signal fed via an amplifier to loudspeakers.

Audio systems with the highest fidelity are those that capture the widest range of frequencies: a bandwidth of 20kHz is about the maximum that can be accommodated. A monochrome television signal is about the equivalent of 150 such channels running simultaneously, a color one about 275. The development of a domestic video recording system therefore required considerable extra sophistication.

In the mid 1950s, the British Broadcasting Corporation (BBC) demonstrated its Vision Electronic Recording Apparatus (VERA) in which large reels of tape ran at high speed past a stack of stationary record heads. This technique of video recording was almost immediately surpassed by the American Ampex Corporation's introduction of rotating record heads. A 50mm wide tape ran slowly across a drum with four video heads rotating at 250 revolutions per second. Each head scanned the tape transversely, "writing" approximately 16 slightly sloping lines of picture-information across the width of the tape before switching to the next head. This system, known as Quadruplex, was capable of very high picture quality, but its size, cost and rate of tape consumption made it unsuitable for domestic use.

The domestic video became a reality with the evolution of the helical scan technique in 1959. The tape is wrapped in a skew path around the rotating drum in such a way that the two video heads lie in oblique tracks across the tape. With the drum rotating at only 25 revolutions per second and using a standard 12·65-millimeter tape, the amount of tape required was considerably reduced, an improvement which enabled Philips to bring a cassette recorder out in 1972.

The next aim was to eliminate the guard bands between the video tracks, thus reducing tape consumption further and prolonging playing time. By the "tilted azimuth" technique, adopted in the mid-1970s, the adjacent tracks were laid side by side but at slightly different angles to one another. Interference between tracks was cancelled out since the heads only read their own tracks.

As home video recorders increase in popularity, further refinements are under way. Systems incorporating stereo, and miniature video recorders using 8-millimeter tape have been launched. Digital video recording is the most significant step to come. In the system pioneered by Sony, each of the 525 or 625 lines per television picture frame is sampled at 720 points or pixels; each sample is given an 8-bit code (with error-correction bits added to protect against blemishes ➧ page 100). At this rate 216 million data bits are needed to record each second, and there is virtually no loss of picture quality.

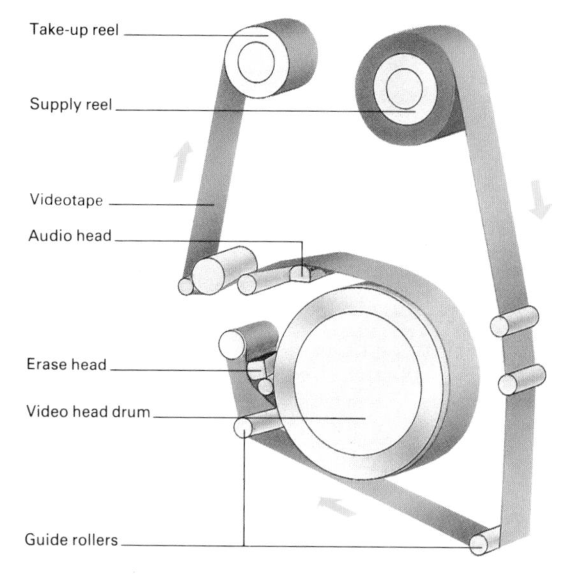

▲ ***Tape cassettes, used in both domestic television recorders and cameras contain a series of rollers that guide the tape past the video and audio recording heads at 2cm per second.***

▼ ***A close-up of videotape shows the diagonal arrangement of video tracks, each around 25 micrometers wide. Audio signals are recorded across the top of the tape.***

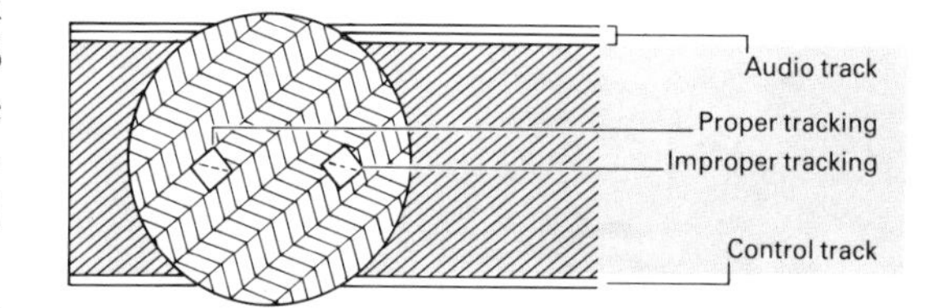

and replaced through one stylus to two separate amplifying systems
1930s Electric motors used to drive record turntables
1935 Magnetophone, manufactured in Germany, became first commercial tape recorder using new magnetized plastic tape; iron dioxide tape developed by AEG and I.G. Farben
1948 Long-playing 12-inch disk introduced by Columbia Records, made of vinylite and rotating at 33⅓ rpm
1949 7-inch, 45 rpm disk introduced by RCA
1950s Hi-Fi developed and became popular
1956 Video tape recording demonstrated by Ampex
1958 Standard system of two-channel stereo for LPs introduced
1959 First helical scan video recorder introduced by Toshiba
1960s Valves increasingly replaced by transistors in amplifiers
1963 Tape cassettes introduced by Philips
1967 Dolby system of sound recording introduced by R.M. Dolby in USA, eliminating background hiss and improving fidelity
1971 Quadrophonic sound systems introduced
1972 First home video-cassette recorder introduced by Philips
1978 Pure metal-particle tape introduced by 3M to improve high-frequency response
1980 Small portable tape recorder, known as Walkman, introduced by Sony of Japan
1983 Compact disk (digital audio disks) systems developed by Philips and Sony

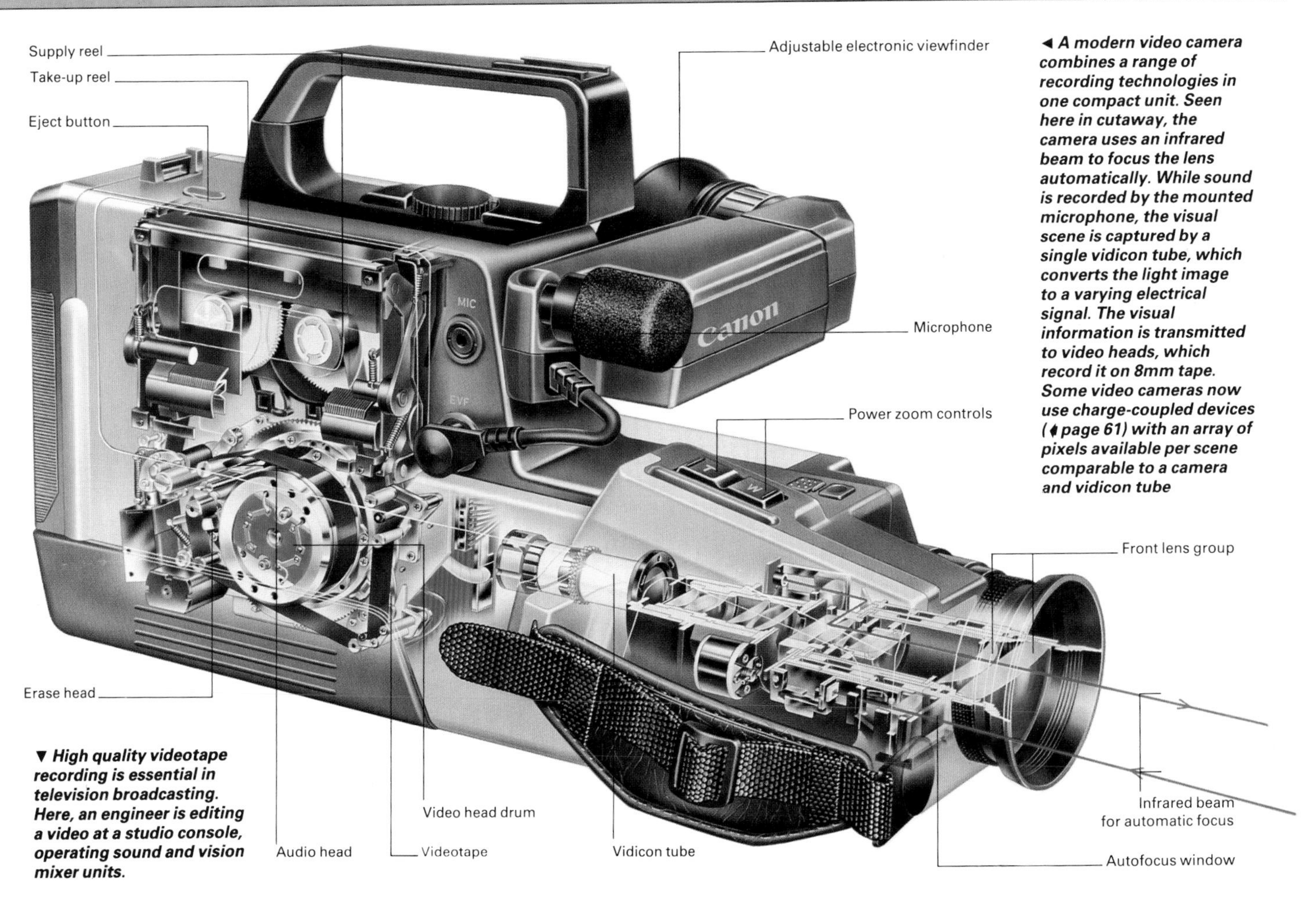

◀ A modern video camera combines a range of recording technologies in one compact unit. Seen here in cutaway, the camera uses an infrared beam to focus the lens automatically. While sound is recorded by the mounted microphone, the visual scene is captured by a single vidicon tube, which converts the light image to a varying electrical signal. The visual information is transmitted to video heads, which record it on 8mm tape. Some video cameras now use charge-coupled devices (♦ page 61) with an array of pixels available per scene comparable to a camera and vidicon tube

▼ High quality videotape recording is essential in television broadcasting. Here, an engineer is editing a video at a studio console, operating sound and vision mixer units.

The vidicon tube

As with printing, when a photograph is reproduced on a page by a mass of tiny individual dots, so in video technology a scene is recorded and played back in the form of thousands of pixels, each of which details the brightness of a single small area. One device for recording a scene in this way is the vidicon tube.

An optical image is formed on a target disk coated with a photoconductive material whose electrical resistance decreases as light falls on it. For each pixel, a separate measure of light intensity produces a tiny electrical signal; these are then amplified and scanned by an electron beam emitted from a heated cathode at the vidicon's back end and focused by magnetic coils. In a carefully synchronized process, the beam scans the pixel rows, one line after another, completing its coverage of each entire scene in about four-hundredths of a second. To record a picture in color, where color strength as well as brightness must be detected, video-cameras may comprise three vidicons, one for each primary color, as do television cameras. Alternatively they may incorporate a matrix target within a single tube.

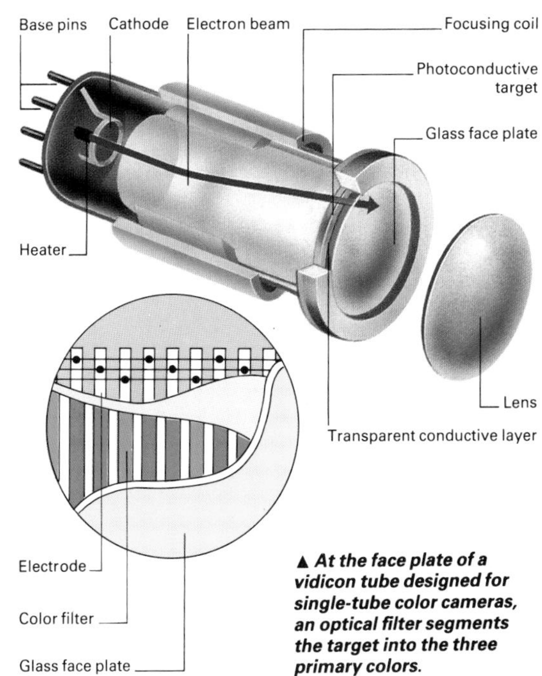

▲ At the face plate of a vidicon tube designed for single-tube color cameras, an optical filter segments the target into the three primary colors.

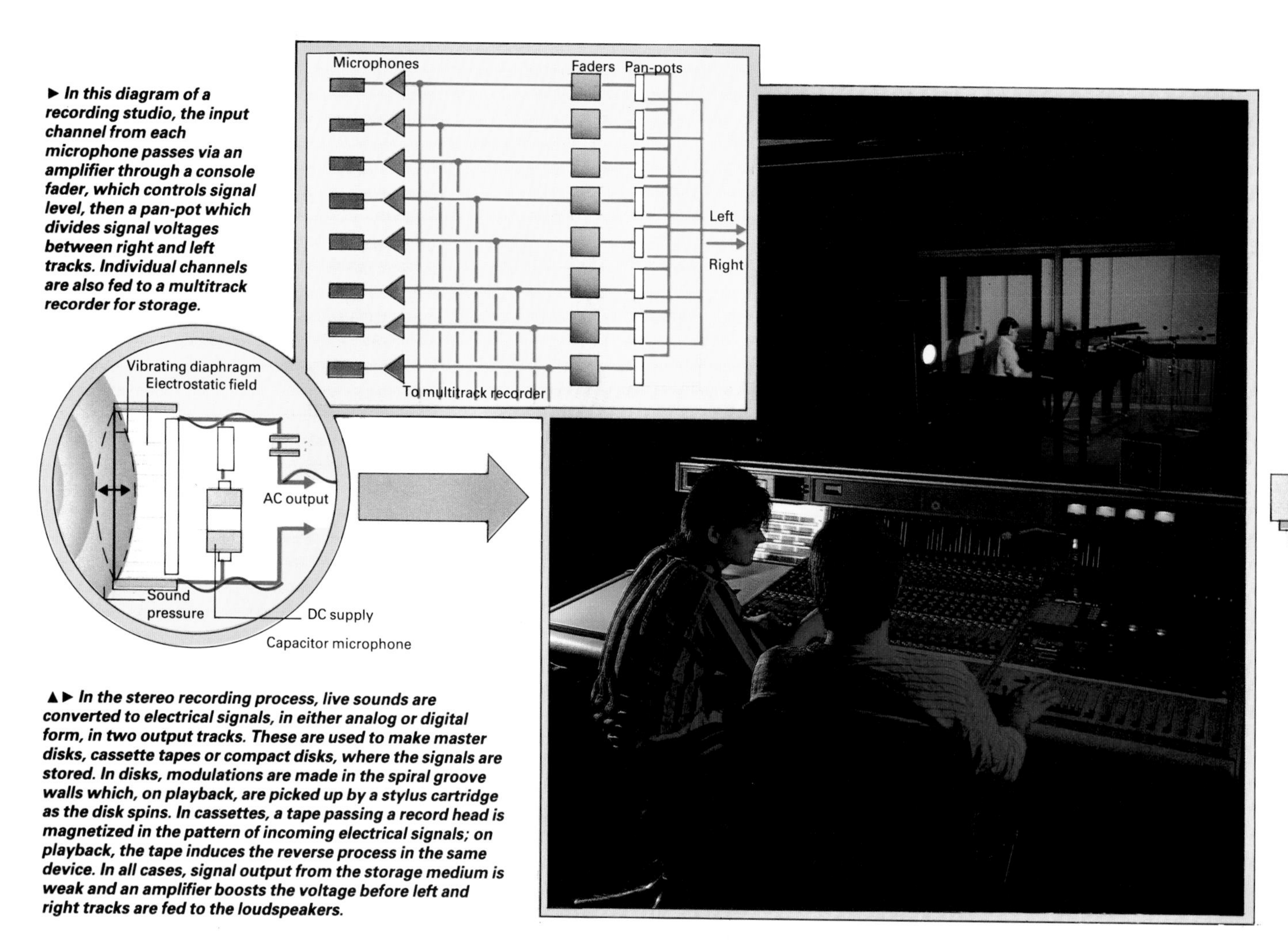

► In this diagram of a recording studio, the input channel from each microphone passes via an amplifier through a console fader, which controls signal level, then a pan-pot which divides signal voltages between right and left tracks. Individual channels are also fed to a multitrack recorder for storage.

▲► In the stereo recording process, live sounds are converted to electrical signals, in either analog or digital form, in two output tracks. These are used to make master disks, cassette tapes or compact disks, where the signals are stored. In disks, modulations are made in the spiral groove walls which, on playback, are picked up by a stylus cartridge as the disk spins. In cassettes, a tape passing a record head is magnetized in the pattern of incoming electrical signals; on playback, the tape induces the reverse process in the same device. In all cases, signal output from the storage medium is weak and an amplifier boosts the voltage before left and right tracks are fed to the loudspeakers.

The recording studio control room

Little by little the sound engineer in a recording studio has become as much of an artist as the performer, singer or instrumentalist. Recording has come to require practical skills and theoretical knowledge in both the artistic and technical fields. Microphones exhibit differing faithfulness with which they pick up sound, a property known as frequency response, and differing sensitivity to sounds arriving from one or more directions. Recording studios tend to choose a mix of microphones with best frequency response for the purpose in hand, and with the necessary directional qualities to cut down on the incidence of unwanted sounds or background noise during recording. Also to minimize noise, studios are provided with control rooms used by the producers and engineers, which are separated from the main studio by a sound-insulated observation window.

Recording large groups often calls for the use of several microphones – at least one for each performer. The sounds from these various microphones or sound channels are balanced and mixed at a special console. The studio acoustics are usually made a little "dead" and may be adjusted passively by the use of sound-deadening screens or electronically by the use of equalizing circuits that alter the frequency content of the individual channels to emphasize base or treble as required. Artifical echo may also be needed for a particular recording, especially in a television studio.

The microphone

Microphones are pressure transducers: they convert or transduce the pressure oscillations associated with a sound into an electrical signal (◆ page 43). An element common to all microphones is a component that moves in sympathy with the oscillating sound pressure. In the boom type of microphone this component is a metal coil that moves in the magnetic field of a permanent magnet, thus inducing a current which varies in accordance with the original sound. Such moving-coil microphones have been superseded for most purposes by condenser microphones. These are in common use in recording and broadcasting studios as stand-mounted microphones. In a condenser microphone the moving element is a small, thin, metallic diaphragm. This forms part of an electrical capacitance, the other part being formed by a fixed plate across which a high polarizing voltage is maintained. As the diaphragm vibrates, the capacitance of the system varies and an electrical signal is developed. This is amplified before further processing. Condenser microphones of this type can be made quite small but their sensitivity decreases with their size.

Lapel-type microphones, as used in many television interviews, are more often electret microphones. These are condenser microphones with a diaphragm made of a thin plastic sheet with a conductive metal coating on one side. The other side of the sheet rests on a perforated, metallic backplate with many small raised points that act as supports. As the plastic diaphragm moves, the capacitance between the conductive coating and the backplate varies, which generates an electrical signal since this arrangement can be contrived to be self-polarizing.

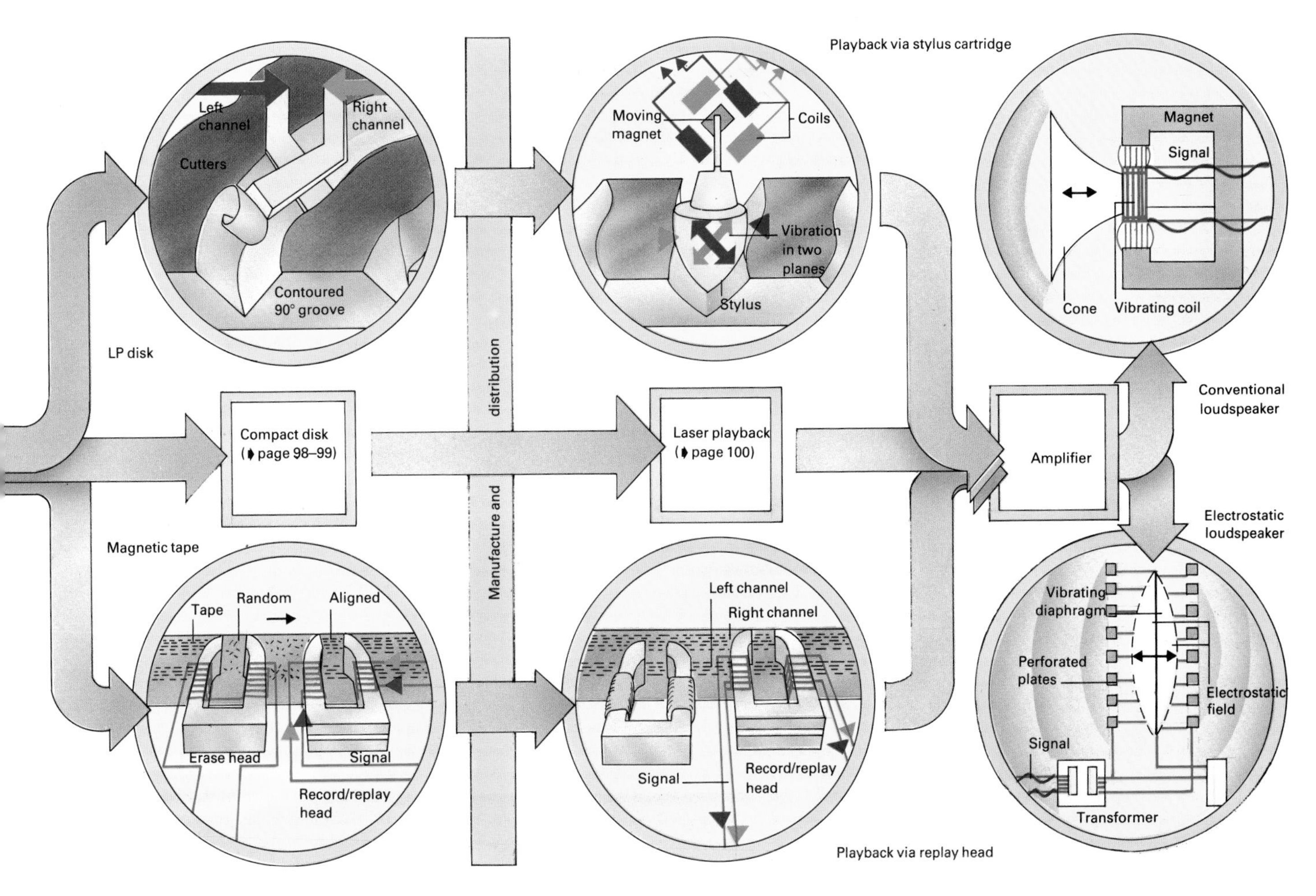

The loudspeaker

Loudspeakers reverse the transduction started by the microphone: they turn incoming electrical signals into sound. A typical cone loudspeaker operates more or less as the reverse of a moving coil microphone. The incoming varying voltage carrying the original sound information is fed to a magnet inducing a current in the coil (known as the voice coil) which moves and causes a fragile thin paper cone to vibrate and radiate sound. As with microphones, an important criterion of loudspeaker performance is frequency response. The larger the cone the better it is able to radiate low-frequency sounds; the vibration pattern on a large cone, however, tends to break up at high frequencies. Consequently a good quality or hi-fi loudspeaker consists of two or three individual systems containing cones of different sizes: a large or bass speaker sometimes called a woofer, a mid-range unit, and possibly a small unit designed for high frequencies (tweeter). The efficiency of the speaker system in radiating low-frequency sounds is much improved by mounting the units in a wooden baffle and joining it to an enclosure or cabinet.

A very efficient design of loudspeaker is the electrostatic type. It consists of a tightly-stetched conductive plastic diaphragm placed between two perforated conductive plates. An incoming varying voltage is applied to the central diaphragm; the constant attraction between it and the outer plates ensures that it remains central. Audio signals from the amplifier applied via a transformer to the outer plates then produce a push-pull motion in the diaphragm in sympathy with the audio waveform.

Mono, stereo and Dolby stereo

Until the 1950s most sound recordings were made from a single microphone, replayed through speakers that reproduced identical sounds. This single-channel, or monophonic, system could not match the quality of the original sound, as all directional information was lost.

By using two microphones carefully positioned to provide the same directional characteristics as the listener's ears, then mixing the output of each and directing the sound to left- and right-hand channels, an auditory perspective close to that of a live performance could be achieved. This is known as stereophonic recording.

A two-channel system is sufficient to achieve a quality of reproduction that comes close to the original. Multichannel systems, as developed by Dolby Laboratories, can offer further possibilities. In the Dolby stereo system, as used in some cinemas, several channels, each covering a different part of the original sound, are relayed through several speakers positioned around the auditorium, thus giving the listener the sense of being "immersed" in the sound field. In the recording studio and in some cassette tape recorders, the Dolby noise reduction system also utilizes multichannel recording techniques. Low-level, high-frequency sounds can be isolated and exaggerated to stand out from background noise. On playback, the recorded sound is reduced to its original state while the background noise recedes still further.

See also
Seeing with Sound 65–72
Seeing beyond the Visible 73–86
Photography 87–90
Digital Recording 97–100
Radio and Television 185–192

Assisted resonance

Designers of audio reproduction systems strive for perfection, and the expertise gained can also help to improve the quality of live sound. In a useful application of recording techniques, concert-hall acoustics that lack sufficient reverberation can be improved by the installation of an assisted resonance system. Hidden loudspeakers are placed in the ceiling, side and back walls of an auditorium and replay sound that is picked up by microphones elsewhere, amplified and delayed electronically.

When the Royal Festival Hall in London opened in 1951, it was thought by many to have too dry and dead an acoustic. In an attempt to improve this characteristic it received an assistance resonance system. More than 100 microphones are mounted in the roof, and a control room containing a bank of amplifiers and delay circuitry is situated at ceiling height at the back of the hall. The system has been so successful that several halls have since incorporated assisted resonance systems, among them the Neues Festspielhaus in Salzburg, built in 1960. A multi-purpose hall, plays, operas and concerts all take place here. Speech is best heard with a minimum of reverberation, so the hall was constructed to have a basic acoustic which is quite dead. The assisted resonance comes into its own during music performances.

Assisted resonance system

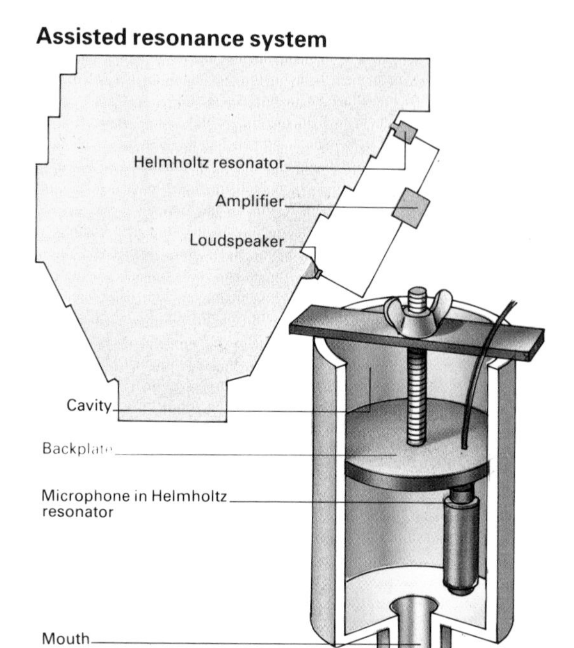

◄▲ The assisted resonance system of the Royal Festival Hall, London, consists of 168 audio reproduction units each comprising a microphone, amplifier and loudspeaker. Every unit covers a separate frequency band of a few Hz at the lower end of the audible spectrum between 58 and 700Hz. Microphones are positioned hidden from view around the ceiling at points where the sound at their pre-set modal frequency is at a maximum. They are encased in Helmholtz resonators, which further enhance the sound when adjusted to the selected frequency. Some new auditoria have installed computer control units to extend the operational flexibility of assisted resonance by "re-tuning" microphones automatically to cater for musical performances of all types.

Digital Recording

Analog and digital recording...Making a compact disk...Information playback...PERSPECTIVE...History of video-disk recording...The digitization process...Error correction

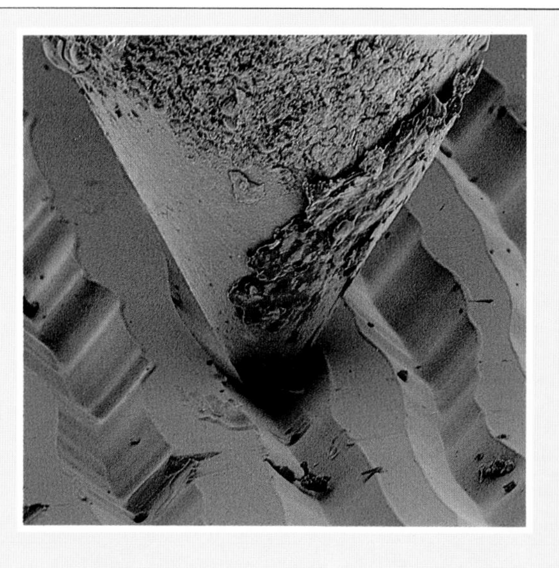

▲ **Conventional disks are played by a diamond stylus traveling along a spiral groove. The straightness of the groove varies with the amplitude of the recorded sound: the wavier the groove the louder the sound. The stylus transmits vibrations to the cartridge which turns them into an electric signal. Eliminating contact between soft disk and worn stylus is an advantage of the laser-read compact disk player.**

Sounds and images on disk or tape have traditionally been recorded by a system in which the depth of a groove or strength of magnetism in a tape has varied in a way closely following the variations in intensity of a musical sound or the variations in brightness of a picture. Known as an analog recording, it has the characteristic that reproduction improves the more closely the variations in the recording follow those of the original sound. However, any deviation results in distortion of the signal waveform. At recording, manufacture and playback, distortion and ambient noise may be imparted to the signal and create irretrievable imperfections.

The alternative, now widely used in recording audio and computer data as well as in telecommunications systems (➧ page 180), is to process the information as it enters the recording process and to express it in digital form. Signals are stored, transmitted and retrieved in the form of binary digits, and reconstituted in their original form only at the final playback stage, thus eliminating certain types of signal degradation. An important feature of this technique is that errors, even quite major ones, can be corrected in the course of reproduction. This calls for special treatment of the signal but approaches the aim of reproducing a signal precisely as it was recorded.

Compact disk players were first produced to give precise reproduction of recorded music but some versions are capable of reproducing video and computer data. These players are among the most sophisticated machines in common domestic use – their designers set out to provide error-free sound reproduction, not in the laboratory but in the home or in moving vehicles.

There are three key features to a compact disk system. First, the continuously varying sound is stored digitally: the signal read is precisely that recorded. Second, information about each piece of sound is held at more than one place on the disk, so that minor errors in reading can be detected and eliminated as the disk is played. Third, there is no mechanical contact between the disk and the reading head, which works not with a stylus but optically.

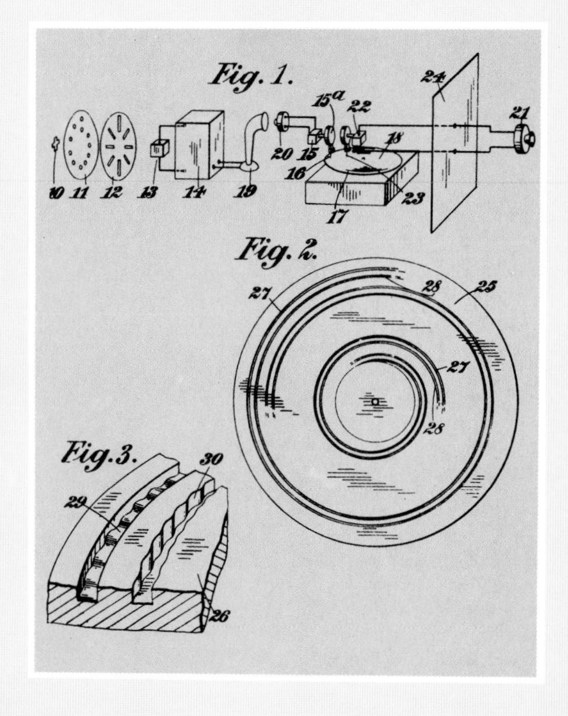

▲ **The Scottish inventor John Logie Baird filed a patent for his television recording system under the name of Phonovision in 1926. A crude image of the object was detected through a scanning disk by photocells, then converted to a sound signal recorded by conventional techniques of groove-cutting in a 78rpm shellac disk.**

Video disk recording

As early as 1926, the Scotsman John Logie Baird (1888-1946) had used 78rpm disks to record television signals for later replay, though, as with the television pictures of the day, the images were of poor quality. In the early 1970s Telefunken in West Germany and Decca in Britain combined to produce color video recordings on the long-playing disk with a system known as Teldec. But the disks needed to be played rapidly and wore badly; the system never won public acceptance. Shortly after this competing systems began to appear, including Laservision by Philips, from which the compact disk sound system was developed and video disk systems based on stylus tracking fell into disuse.

Unlike tapes, which allow the private user to record, compact disks have as yet made little headway in the video market. Future disks may be fabricated with a thin layer of magnetic material that enables home recording. For the present, developments in "data stream" reproduction, are currently being made in Japan and West Germany. Techniques whereby graphics, pictures and computer data can be all be produced on screen, along with sound through a connected hi-fi system, are leading the way in video disk recording.

On a compact disk, 30 tracks of information are drawn within the width of a human hair

Digitization

The first stage in preparing a disk is to make a tape with a digital representation of the sound. Sound is sampled at intervals of 25 microseconds, which corresponds to a sampling frequency of 40,000 times per second, or 40 kHz. This allows exact reconstitution of sounds of up to 20 kHz, or roughly the upper frequency limit of human hearing. At each sampling the intensity of the sound is assigned to one of some 65,000 levels and is described by 16 bits. This number of levels, between a very loud and very quiet sound, again approaches the limits of what can be distinguished by the human ear.

Digital records of the sound may be represented as a series of points on a grid rather than as a continuous graph showing actual amplitude variation. Such a record is limited in its precision. Nearest values are recorded when the amplitude has a value just above or below the set levels. If the amplitude fluctuates rapidly, perhaps over a number of levels, between successive samplings the fluctuation will not be recorded.

By using frequent samplings and many levels the detailed variations in the signal may be put into the record. Nothing in the record, however, implies that the reconstituted signal should be constructed in any particular way, for instance by straight lines or smooth curves. The record specifies only the points on the grid and the replaying device must provide a suitable means of connecting them.

On the disk itself the record is in binary rather than in decimal form. Each binary digit is able to switch the recording furrow on or off. In this form, and using just eight bits, the signal sequence "31 12 10 17" becomes "00011111 00001100 00001010 00010001". Encoded as pits using the convention that a pit stops or starts at each change from 0 to 1 or 1 to 0, this gives the track sequence shown right. The final electrical signal must be read against a time base so that the number of 0s and 1s in a sequence may be recognized.

Digitizing and recording

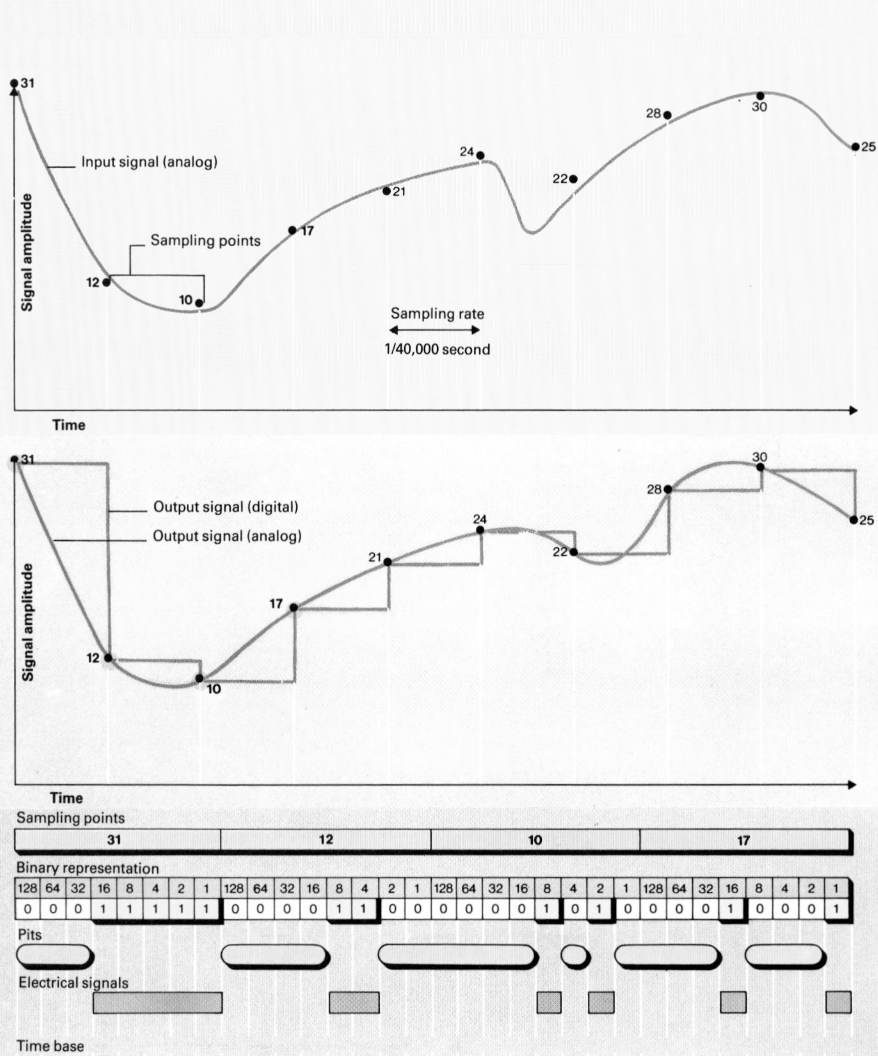

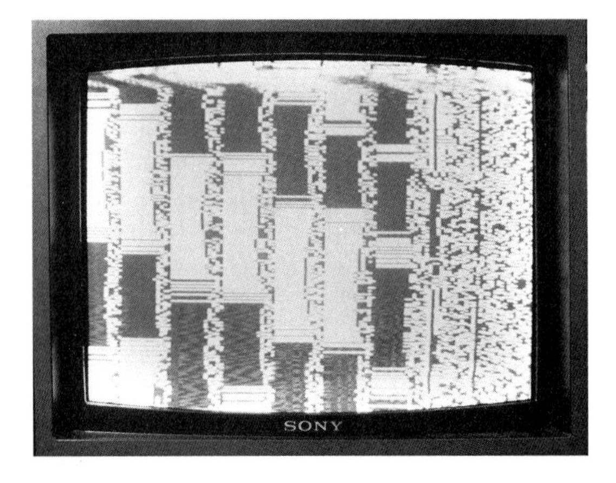

▲ *An alternative means to the compact disk of storing digital data is the videotape. Here, taped data is displayed on screen with black-and-white dots representing the binary digits. A videotape has a bandwidth necessarily large to accommodate visual information but commercial copying is expensive by comparison: with disks, entire programs can be stamped out at once.*

Making a compact disk

Disk manufacture begins with a glass blank, which becomes the master disk from which others are copied. Under conditions of strict cleanliness, the blank is ground and polished to an optical flatness. It is then carefully coated with a layer of photoresist, the thickness of which is controlled to a value of about 0·1 micrometers. Photoresist is a plastic material whose chemical properties are modified by light. Areas exposed to light may be dissolved away, leaving the unexposed areas intact.

Signals are cut into the photoresist by a continuously operating helium-neon laser, focused on the resist coating. As the disk revolves the laser is driven from the inner to the outer edge and the rotational speed reduced so that the speed of the illuminated spot along its track remains constant. The signal is written onto the blank by an acoustic switch which shifts the intensity of illumination between high and low values as required by the master tape.

When the resist is developed and etched, the result is a blank inscribed with an interrupted spiral furrow. The form of the track is unexceptional but its dimensions are impressive: the total track length

◄▼ The Compact Disk was developed jointly by the Philips and Sony Corporations. A standard was agreed in 1980: the disk is 120mm in diameter and 1·2mm thick, and can store 74 minutes of programming. A close-up of the surface reveals the pattern of pits which encodes the stored data.

is 5·7 kilometers and its pitch is just 1·6 micrometers; the width of the furrows is 0·6 micrometers and their depth 0·16 micrometers. Some 30 tracks are drawn within the width of a human hair.

Masters are too valuable for direct pressing of commercial disks, so copies are made. The master is coated with a thin layer of silver which makes the surface electrically conducting to allow a thicker, harder layer of nickel to be electroplated onto it. A reversed copy is made, which is in turn copied to make a submaster used to stamp out the copies that are sold to the public.

The base material of the compact disk is a polycarbonate plastics material which has stable optical and mechanical properties and is able to follow the fine detail of the master. The plastic disk is manufactured by injection molding from the center out, so that faults in the plastic are more likely near the edge. Precise monitoring of temperature is necessary at this stage to minimize unacceptable defects. After pressing, the disk is given a thin coating of aluminum to make it highly reflective, and then a protective lacquer coating. The disk is read through the base, which is 1·2mm thick, so the optical properties of the lacquer are relatively unimportant.

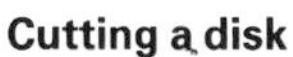

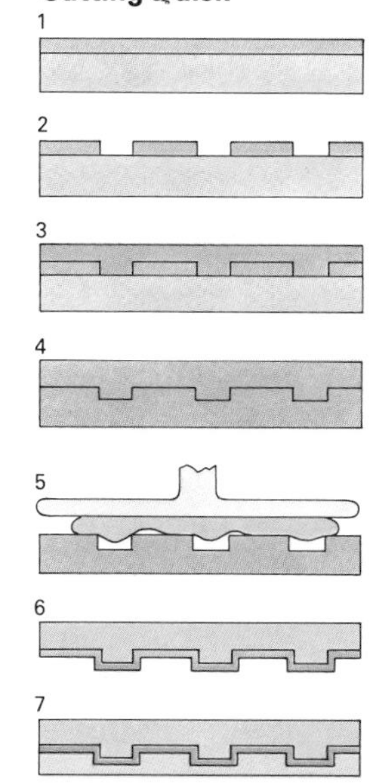

◄ The sequence shown in enlarged cross-section outlines the stages in the manufacture of a compact disk by injection molding. The base material is a glass blank coated with photoresist (1). Signals are cut into the photoresist by laser to create a master (2), which is strengthened with layers of silver and nickel (3) and used to make a reserve copy or submaster (4). Commercial disks are made from plastic material stamped by the submaster mold (5) to produce a "positive" disk (6) which is then given an aluminum coating to make it reflective, and a lacquer coating to protect it (7).

See also
Recording on Tape and Disk 91–96
Lasers and Holography 101–104
Telecommunications 177–184
Television and Radio 185–192
Road Transport 199–208

A disk is read by shining a fine beam of light onto its surface and detecting the amount of light reflected. The beam diameter at the reflecting surface is kept close to the width of a furrow, and the detector records not the level of light reflected but the sharp change in level caused by the transition from pit to blank and back again. Readings are taken through the thickness of the disk, and because the lens produces highy convergent light, the area through which the light enters and leaves the disk is many thousands of times that illuminated in the information layer. As a result dust particles on the surface have little effect on the light reaching the detector.

In the design of the compact disk player the severest design problem was that of tracking the information spiral, on too precise a course for a mechanical pickup to follow. Instead an electronic device detects the spiral and moves the lens mount so as to track it. A second control mechanism is needed to regulate the distance between the information spiral and the lens, which has a small depth of focus. This allows for warping or irregularities in the thickness of the disk.

Corrected signals are read from the information store in the player at precise intervals for conversion into a continuously varying voltage which drive amplifiers and loudspeakers. The digital approach confers some additional advantages here, too. The drive mechanisms for the disk may not provide signals at a uniform rate, but the store which holds signals for correction also acts as a buffer, so that outgoing signals are correctly timed. This eliminates the wow and flutter found in conventional playing systems, which arise from imperfections in the rate in which the signals are presented to the pickup.

Finally the digital signals sent out by the player are converted to continuous varying voltages which drive amplifiers and loudspeakers. Since it changes sharply from one digital level to another, an electronic filter is used to smooth off the edges of the digital signal, returning the output to the progressive variation of the original.

Optical path of tracking laser

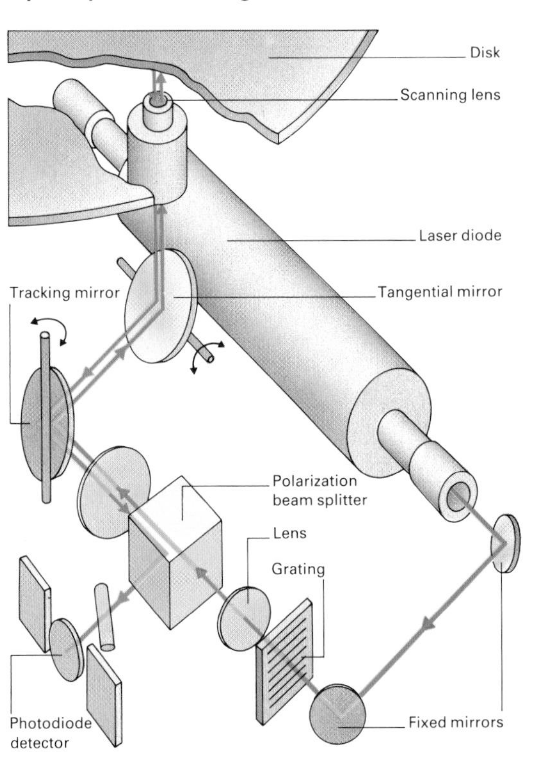

▲◄ An array of mirrors transmits the laser beam to a scanning lens which is focused on the underside of the rotating compact disk (seen in close-up, left) and electronically controlled to follow the spiral track. If a beam strikes a pit it is scattered; if it strikes a smooth spot the light is relayed back along the same path and detected by an optical sensing device.

Correcting errors

Any replay system makes occasional errors. The stylus in the groove of a conventional LP disk may not follow the contour exactly but it is unlikely to move far from its intended course so the sound produced will not be too far from the original. In a digital system, an error is as likely to occur in a binary digit representing a large number as a small one, so it is crucial to keep the error rate in reading very low. On the compact disk, each signal is copied several times, and the player reads all versions, compares them, and, in effect, takes a majority vote to select the correct one.

Errors are likely to be result of local blemishes, perhaps from scratches on the lacquer-covered side of the disk. Repeating the signal immediately on the information spiral is therefore likely to result merely in repeating the inaccuracy, so these repeated signals are spread out over a substantial part of the spiral. The player reads the disk into a small computer, which holds each signal until comparisons and corrections have been made. The corrected signal is then passed on to the next stage. If the signal is so disturbed that no correction can be made, the signal transmitted is either zero or the mean of adjacent clear signals. The error correction techniques are so powerful that correct output may be computed across a hole of 2mm cut in the information spiral.

Lasers and Holography

What is a laser?...Holograms and holography... Applications of lasers...PERSPECTIVE...Creating the laser beam...The origins of the laser

The laser is an extraordinary source of light. Unlike other sources of bright light (such as the sun or a camera flash) the laser concentrates all its power into a very narrow band of wavelengths. The output is monochromatic – of a single, pure color. The light from a laser at its output wavelength is incomparably more intense than from any other light source. Laser light is also highly directional: laser beams have reached the Moon having spread out to only 1·6 kilometers across.

The reason why laser light is so different lies in the way it is produced. When atoms change from high to low energy levels, they release surplus energy or photons of light in the form of finite parcels of radiation known as quanta. When the release is random, it is called spontaneous emission, as commonly experienced when the fluorescent tube in a lamp radiates light. However, if these photons encounter similarly energetic atoms, the atoms can be induced to emit photons also. This process is called stimulated emission and has the remarkable property that the second atom will emit photons at exactly the same wavelength and in exactly the same direction as the original light.

By shining a flashlamp onto a crystal or by passing an electric current through a gas, it is possible to create a situation in which there are more atoms able to emit light of a particular wavelength than are able to absorb it. This unusual condition is called a population inversion. Initially some atoms spontaneously emit photons and these then induce other atoms to emit. The light intensity grows in all directions. If a mirror is placed at either side of the crystal or glass tube, light traveling along the axis is reflected back into the medium and further amplified. If one of the mirrors is only partially reflecting, a beam of parallel light will pass through this mirror. This is a laser beam. The word laser is an acronym for Light Amplification by Stimulated Emission of Radiation.

The discovery of lasers

In 1953 the American physicist C.H. Townes successfully experimented with the principle of stimulated emission in amplifying microwave radiation. Seven years after this "maser" was devised, the world's first laser burst into life on 15 May 1960 in the Californian research laboratory of Hughes Aircraft Company. It was simply built from a cylindrical rod of ruby placed inside a spiral flashlamp. The ruby crystal absorbed the energy of the white light from the flashlamp and converted it into a giant pulse of red light – the first ever laser pulse.

Following this demonstration that visible laser action was possible in ruby, physicists started looking for other substances that could be made to "lase". In 1961 the first gas laser was constructed using a mixture of the inert gases helium and neon (He-Ne). Energy was put into the gases by passing an electric current through them. The red He-Ne is now common in school laboratories and in many applications where a high-power laser is not required. Recently a green-emitting helium-neon laser has been perfected (commercially marketed as the "Gre-Ne").

High-power lasers were developed in 1965 using the molecular gas carbon dioxide. These can now produce hundreds of thousands of watts of infrared radiation and are easily capable of cutting through metal.

Although hundreds of different materials have been found to lase at thousands of discrete wavelengths, it is the development of the tunable laser that has made the laser the ubiquitous and indispensable laboratory instrument that it is today. Using solutions of organic dyes as the lasing medium, laser light can be made in any color.

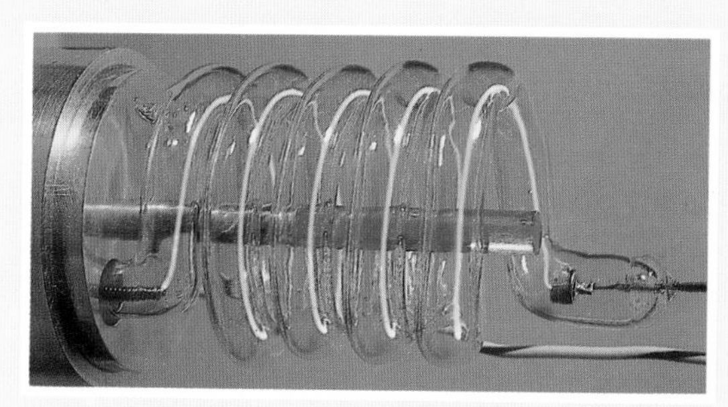

▲ The first working laser was produced by American physicist Theodore H. Maiman in 1960. Using a crystal of synthetic ruby, it momentarily produced light 10,000,000 times more powerful than sunlight.

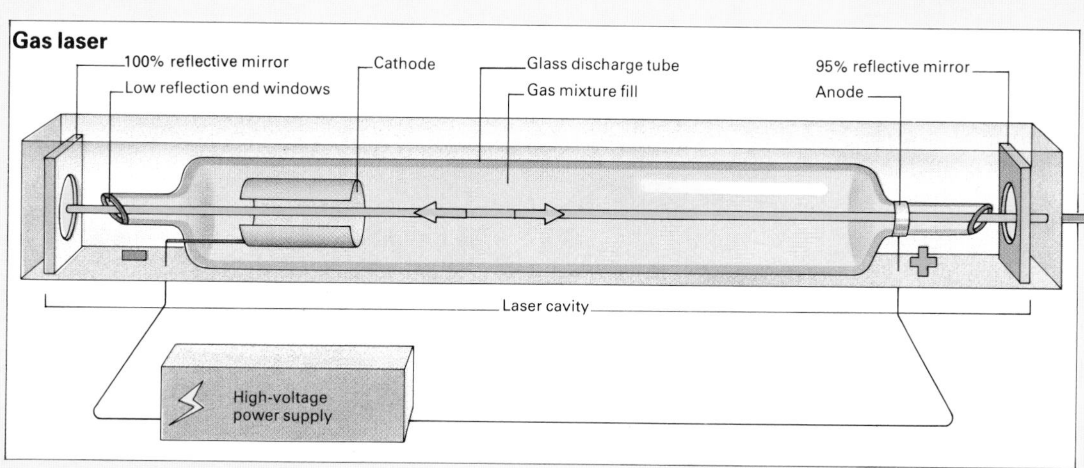

◀ In a typical gas laser, a high voltage excites the atoms. On descending to its original condition a gas atom releases a photon which collides with nearby atoms, inducing them to emit also. Photons reflect back and forth, and laser light is emitted.

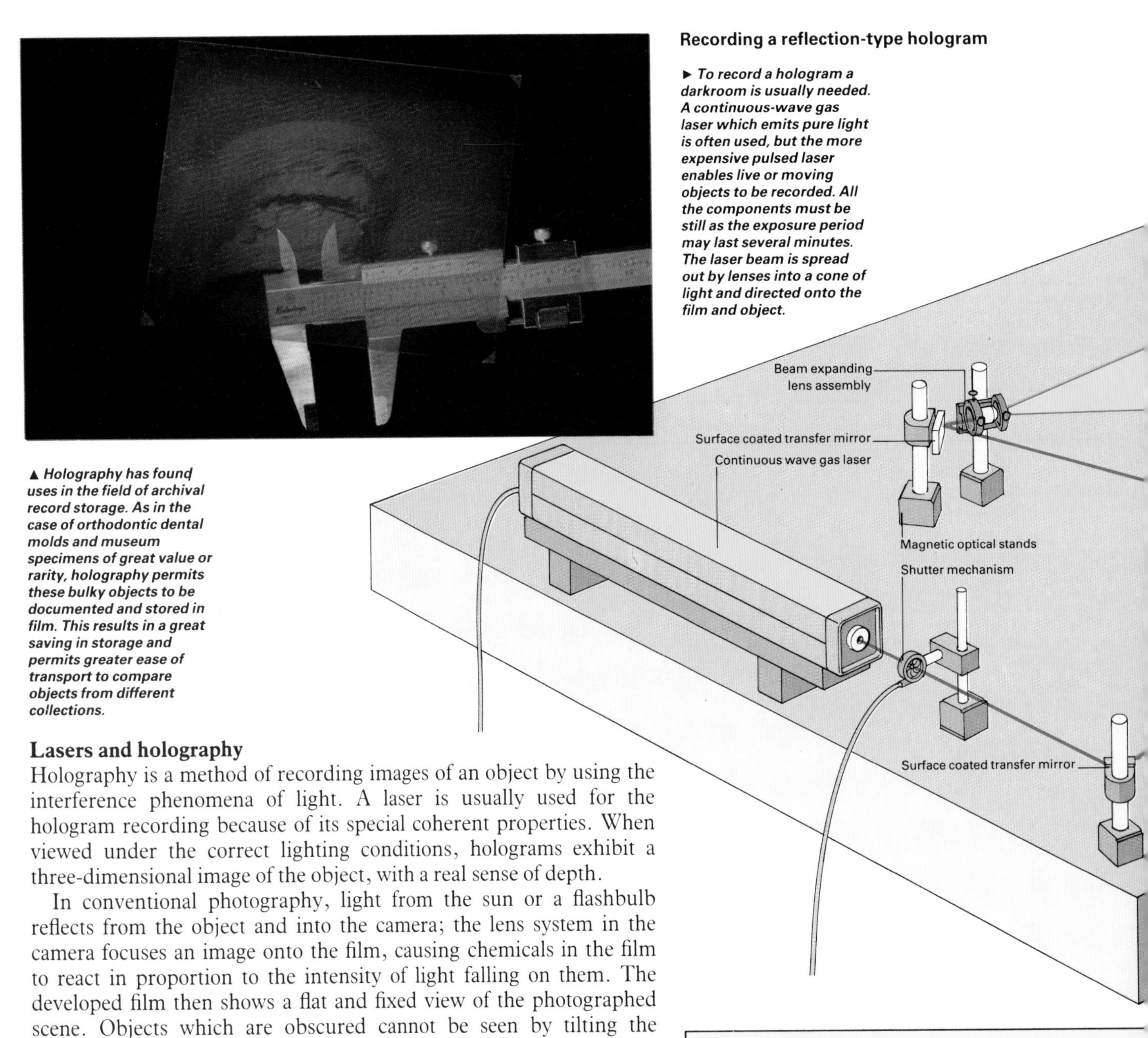

Recording a reflection-type hologram

► To record a hologram a darkroom is usually needed. A continuous-wave gas laser which emits pure light is often used, but the more expensive pulsed laser enables live or moving objects to be recorded. All the components must be still as the exposure period may last several minutes. The laser beam is spread out by lenses into a cone of light and directed onto the film and object.

▲ Holography has found uses in the field of archival record storage. As in the case of orthodontic dental molds and museum specimens of great value or rarity, holography permits these bulky objects to be documented and stored in film. This results in a great saving in storage and permits greater ease of transport to compare objects from different collections.

Lasers and holography

Holography is a method of recording images of an object by using the interference phenomena of light. A laser is usually used for the hologram recording because of its special coherent properties. When viewed under the correct lighting conditions, holograms exhibit a three-dimensional image of the object, with a real sense of depth.

In conventional photography, light from the sun or a flashbulb reflects from the object and into the camera; the lens system in the camera focuses an image onto the film, causing chemicals in the film to react in proportion to the intensity of light falling on them. The developed film then shows a flat and fixed view of the photographed scene. Objects which are obscured cannot be seen by tilting the photograph, and blurred objects cannot be made sharp by refocusing the eyes. These limitations are so familiar that they are subconsciously accepted; yet such pictures are not true representations of reality.

Holograms suffer from none of these limitations. A developed holographic plate resembles a window onto the scene which has been recorded. Different aspects of the scene can be observed by looking at the hologram from different angles. If the scene has a large depth of field, one must refocus one's eyes to see near and far objects.

Holography was invented in 1947 by the Hungarian-British physicist Dennis Gabor while working at the British Laboratory at Rugby. He was awarded the Nobel Prize for physics in 1971 for his "three-dimensional lensless method of photography". Holography is called a lensless technique as no image is formed of the scene to be recorded.

A hologram is formed by the recombination to two sets of matching (coherent) light waves from a single laser source. One set (the object beam) has reflected from the object under study onto the film; the

Holographic stereogram

► Some display holograms may appear animated when seen from different angles. Known as holographic stereograms, these originate from ciné pictures of the object, recorded by moving the camera around the object, or by turning the object while the ciné sequence is made (1). The ciné strip frames are used to make a set of compressed holographic images onto the film via a cylindrical lens system (2). The combined effect of these image strips results in a three-dimensional image (3).

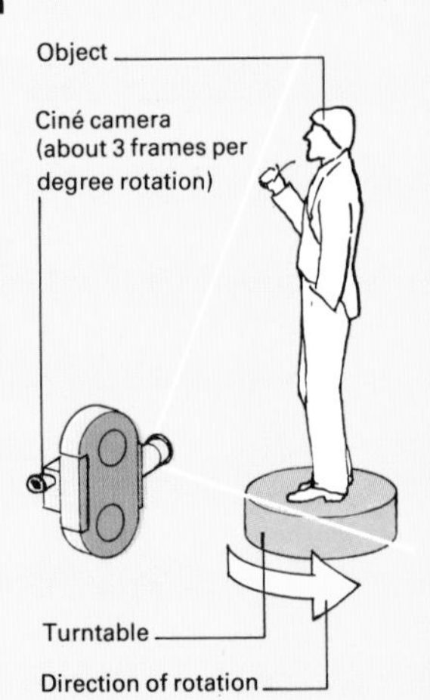

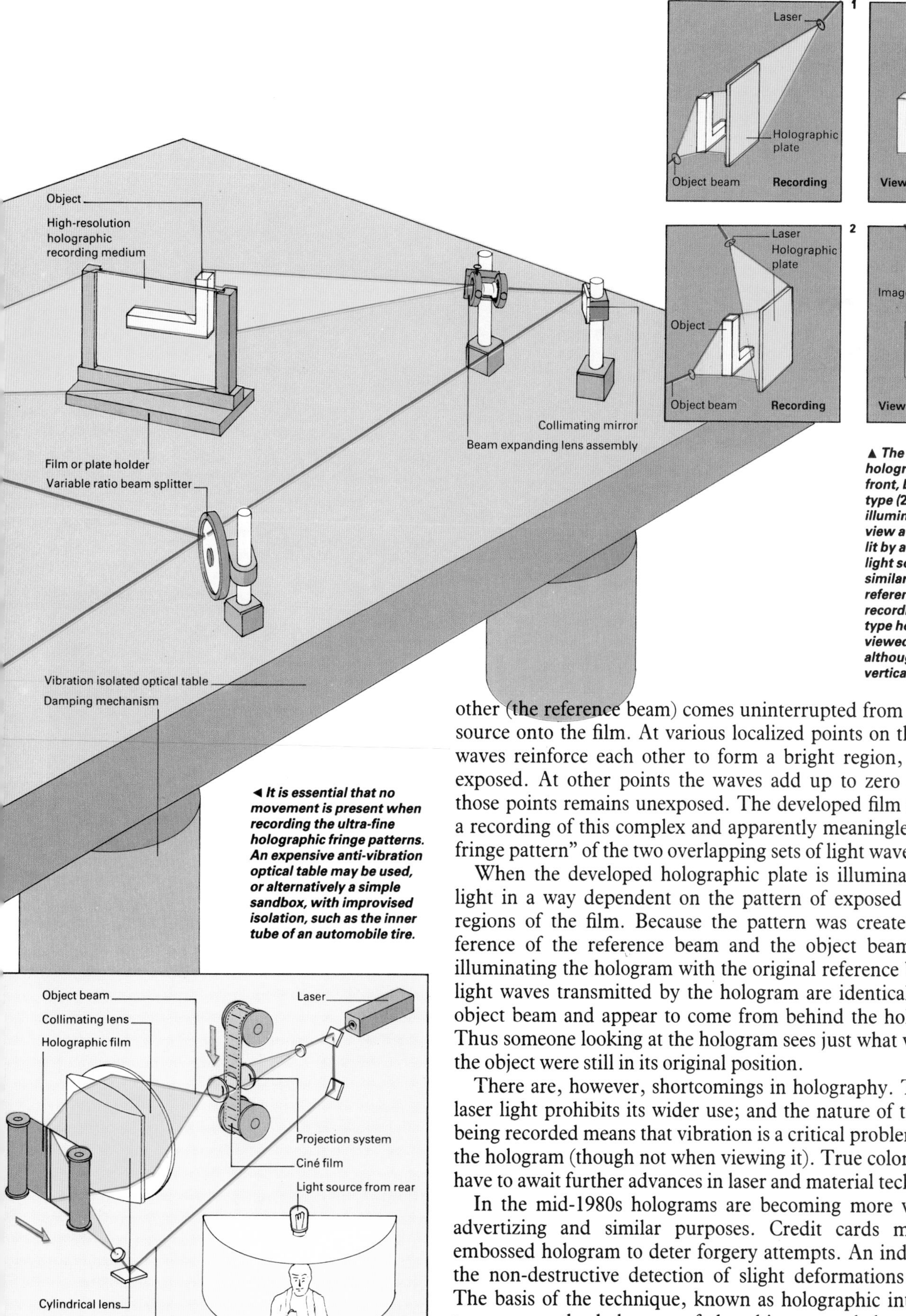

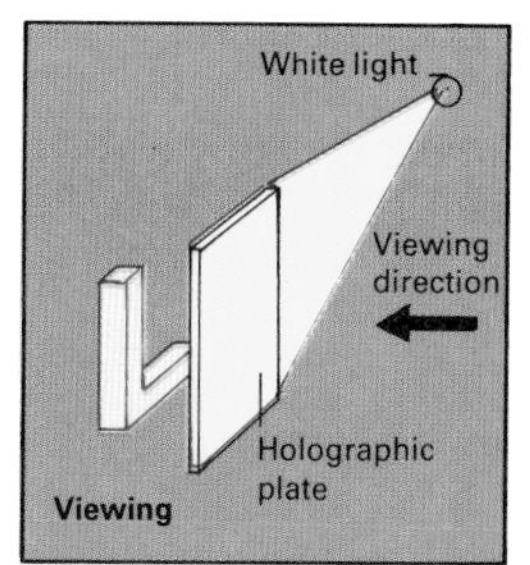

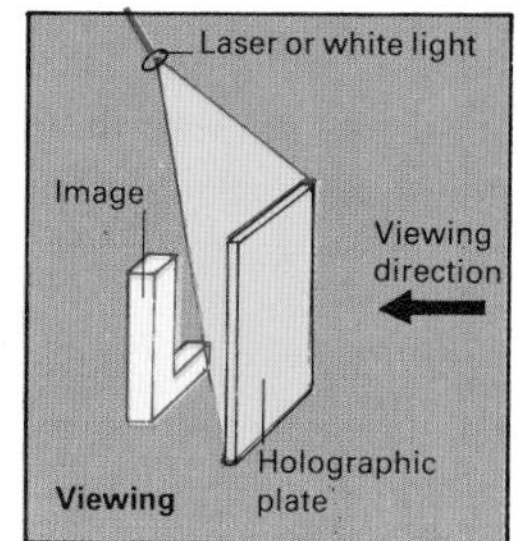

▲ The reflection type hologram (1) is lit from the front, but the transmission type (2) requires back illumination with a laser. To view a hologram, it must be lit by a strong directional light source from an angle similar to that of the reference beam used in recording. Transmission-type holograms may be viewed by white light, although with some loss of vertical parallax.

◄ It is essential that no movement is present when recording the ultra-fine holographic fringe patterns. An expensive anti-vibration optical table may be used, or alternatively a simple sandbox, with improvised isolation, such as the inner tube of an automobile tire.

other (the reference beam) comes uninterrupted from the spread light source onto the film. At various localized points on the film the light waves reinforce each other to form a bright region, and the film is exposed. At other points the waves add up to zero and the film at those points remains unexposed. The developed film simply contains a recording of this complex and apparently meaningless "interference fringe pattern" of the two overlapping sets of light waves.

When the developed holographic plate is illuminated it transmits light in a way dependent on the pattern of exposed and unexposed regions of the film. Because the pattern was created by the interference of the reference beam and the object beam, the result of illuminating the hologram with the original reference beam is that the light waves transmitted by the hologram are identical to the original object beam and appear to come from behind the holographic plate. Thus someone looking at the hologram sees just what would be seen if the object were still in its original position.

There are, however, shortcomings in holography. The need to use laser light prohibits its wider use; and the nature of the fine patterns being recorded means that vibration is a critical problem when making the hologram (though not when viewing it). True color holograms will have to await further advances in laser and material technologies.

In the mid-1980s holograms are becoming more widely used for advertizing and similar purposes. Credit cards may contain an embossed hologram to deter forgery attempts. An industrial use is in the non-destructive detection of slight deformations in equipment. The basis of the technique, known as holographic interferometry, is to compare the hologram of the object recorded under one set of conditions with the real object under another set of conditions.

See also
Introduction to Measuring 9–22
Surveying and Mapping 37–40
Digital Recording 97–100
Printing 171–176
Telecommunications 177–184

◄ Research is being carried out at the Lawrence Livermore Laboratory in California into nuclear fusion (➧ page 148), using a system of multiple laser beams to focus on a pellet of deuterium and tritium at a single point in order to reach the enormous temperatures required for fusion to take place.

▼ The heat and precise direction of the beam produced by an argon laser has useful application in a wide field of medical uses. The laser can be used to fuse a detached retina onto the eye, or to penetrate the inner ear in order to remove acoustic neuromas (tumors between ear and brain) or to drill into the ear bones. With the help of a microscope the surgeon directs the beam through a small funnel which gives the laser beam access past tissue restricting entrances to the ear.

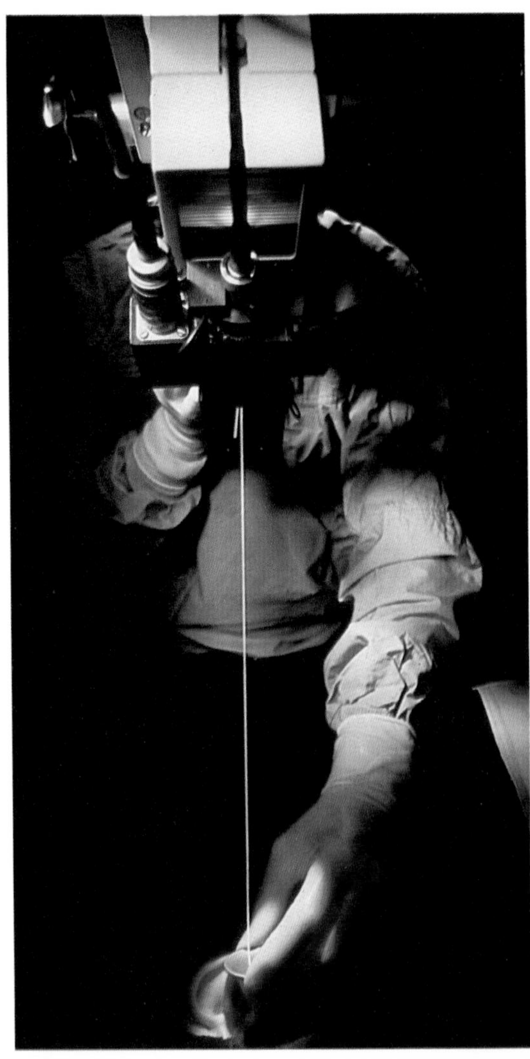

Applications of lasers

The invention of the laser aroused great interest within the scientific community, but had no immediate impact in industry or commerce. Now, over 25 years later, lasers have found a wide range of applications in areas as diverse as the battlefield and the operating theatre.

Powerful carbon-dioxide lasers are employed in heavy industry for the cutting, drilling and welding of sheet metal. Ultraviolet lasers are used in the semiconductor industry for the micromachining of components and circuit boards (➧ page 122).

With their higher frequencies capable of carrying a much higher level of information than radio waves, lasers are central to the development of optical fiber communications in the long-distance telephone networks (➧ page 182). The signals can be carried along the fiber by the infrared beam of a diode laser, a small and cheap semiconductor device.

The diode laser is by far the biggest selling of all types of laser because of its use in compact disk players (➧ page 97). The laser is used to read the recording on the disk instead of the conventional stylus. Other industries where lasers are common include retailing, for use in bar code scanners (➧ page 126), and printing, where desktop printers, typesetting machines and color scanners rely heavily on them (➧ page 173).

Lasers have a number of proven military applications, though none employ the laser directly as a weapon. Laser target designators provide bombs with an illuminated region to aim towards. Laser range-finders measure the return time of a reflected laser pulse to give the distance of the target (➧ page 166).

Many different types of laser are now being used in both the diagnostic and therapeutic medical fields. Cutting with lasers reduces bleeding because of the heat of the beam. Argon ion or copper vapor lasers can be used to remove tattoos and large birthmarks. An experimental cancer treatment involves the use of the gold vapor laser and a light sensitive chemical to search out and destroy cancerous cells.

Chemical Analysis

The uses of chemical analysis...Quantitative and qualitative analysis..."Wet chemistry"...Techniques of spectroscopy...Surface analysis...Catalysts... PERSPECTIVE...The discovery of chromatography...

Chemical analysis is important in many aspects of daily life. It is used to test the purity of foods and drugs, to maintain chemical manufacturing processes within safe limits, to ensure the quality of constructional materials, to detect and identify pollutants, to help authenticate antiques and identify forgeries, and to aid crime detection. Once based on laborious chemical procedures, chemical analysis is now primarily based on instruments that exploit physical phenomena, aided by electronic devices which have automated many procedures.

Analytic chemistry can be divided into two major branches, although the techniques often share a common basis. Qualitative analysis tries to answer the question: what is this? In recent years, for example, a number of countries have banned whale products; one of these, sperm oil, used to be used widely for softening leather and it is possible to analyze a sample of leather qualitatively to see whether it has seen treated with sperm oil or a substitute, such as jojoba.

Quantitative analysis answers questions about the amount of a particular substance. For example, it is essential to have precise percentages of trace metals such as molybdenum in special steels. The analytic chemist in a steelworks carries out analyses of steel to ensure that correct quantities of the alloying elements have been added.

Using color to analyze

Chromatography is a method of chemical analysis which depends on the different affinities between molecules in a mixture and an inert substance. The technique was discovered by the Russian botanist Mikhail Tswett at the beginning of the 20th century. He found that the color of grass could be separated into several different pigments by dissolving the colored material in a non-aqueous solvent and passing this down a glass column filled with powdered chalk. As the solution flowed down the column, the pigments were absorbed by the chalk. Passing pure solvent down the column desorbed them, but at different rates so that the different colored bands appeared as the column "developed".

Although column chromatography is still used, paper and thin-layer chromatography are more common today. These have the advantage that a chromatographic separation can be done in two directions on the same sample, increasing the resolving power of the technique. Paper chromatography uses special grades of paper developed for the purpose, while thin layer chromatography uses a layer of inert material such as silica spread onto a glass or plastic plate.

In gas chromatography, the sample is carried through a column by an inert gas. Such columns are usually heated and contain an inert material coated with a high boiling point liquid, chosen for its likely affinity with the substances to be separated. As the substances emerge from the bottom of the column, they can be detected by a variety of different devices. Gas chromatography has played an important part in detecting trace pollutants, such as fluorocarbons from aerosol sprays, in the atmosphere.

A more advanced form of chromatography, which is now widely used in laboratories dealing with very complex organic molecules, is high-pressure liquid chromatography. This can separate very small samples of complex mixtures by passing them through a chromatograph under high pressure.

◄ The moisture content of foodstuffs can be found by a technique using titration (adding precisely measured amounts of a liquid reagent), seen here in an automatic apparatus.

▼ The samples in continuous-flow automatic chemical analysis equipment, such as is used in studying blood samples, are kept separate merely by small bubbles of gas.

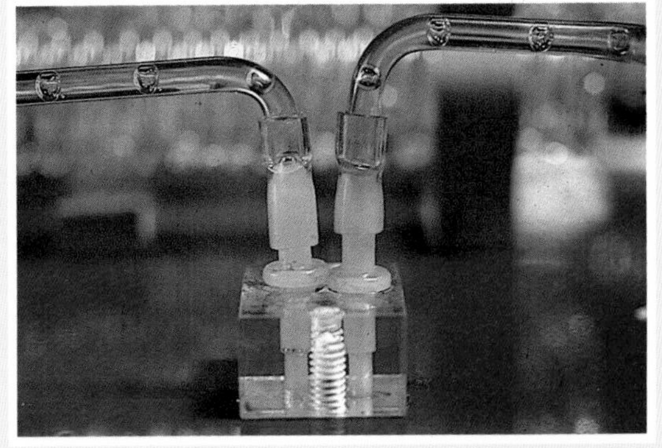

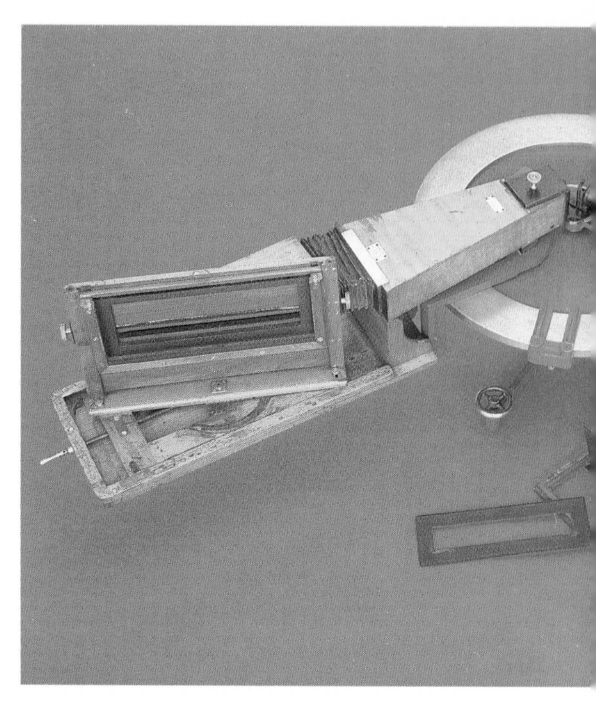

The older form of analysis, involving "wet chemistry", relied on the specific properties of different materials. Schemes of qualitative analysis were worked out by which an unknown mixture of inorganic ions could be identified by using a small number of special reactions. Often these schemes involved reactions which produced characteristic colors in solutions or precipitated solids under particular conditions. Quantitative analysis involved procedures such as the titration of acidic solutions of unknown concentration with alkaline solutions of known concentration. The change in color of an indicator showed when the procedure was complete. Alternatively, known derivatives of the element under examination were prepared and weighed.

Modern analysis frequently depends not on the chemical behavior of substances, but on the way in which they respond to different parts of the electromagnetic spectrum. Atoms and molecules can take up and give off energy, and consequently spectroscopy is divided into two main classes, emission and absorption.

Emission spectroscopy

If atoms are excited, by heating them strongly for example, they can take up energy which they then lose in "packets" of a particular size. In the 1750s it was noticed that when sodium-containing compounds,

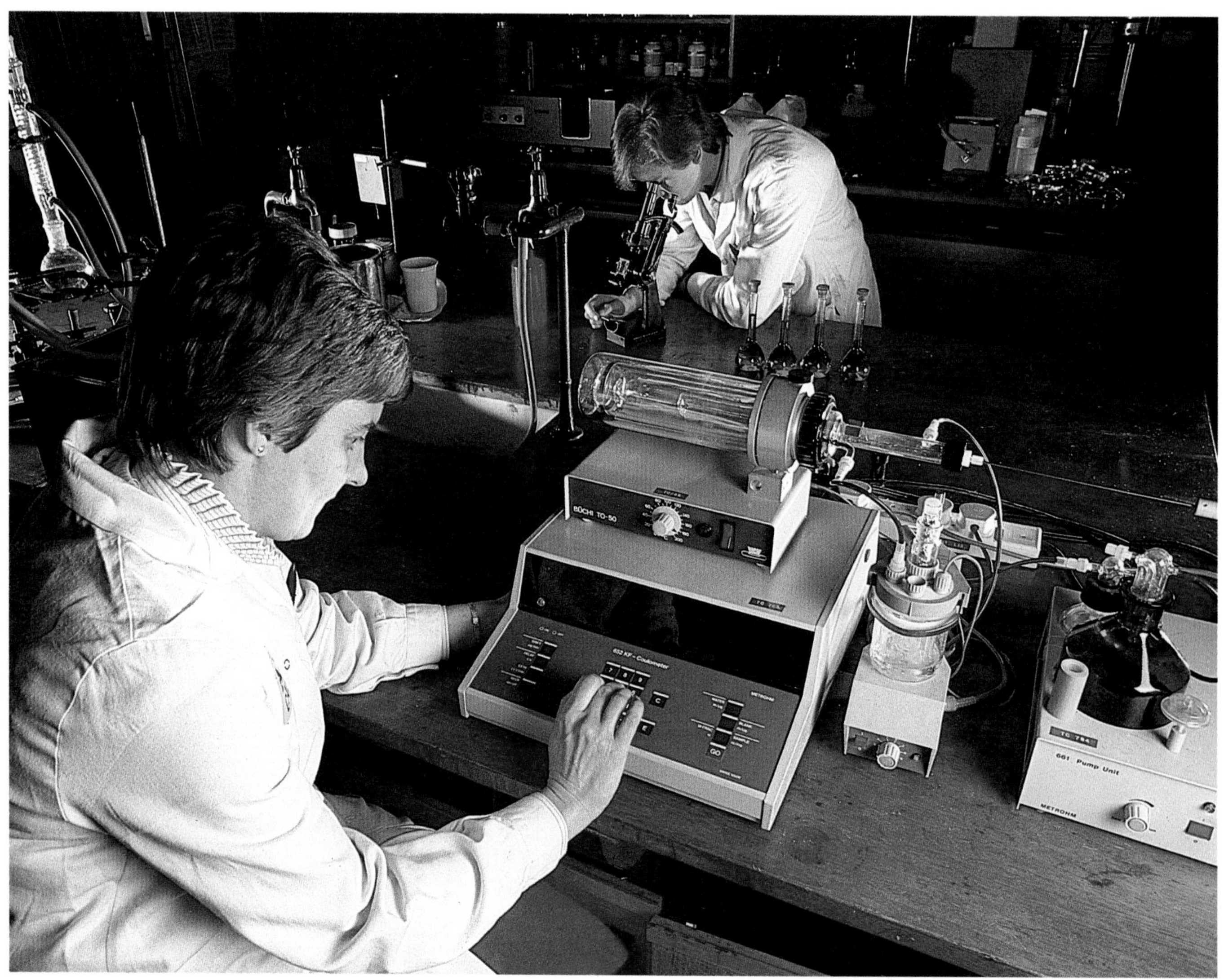

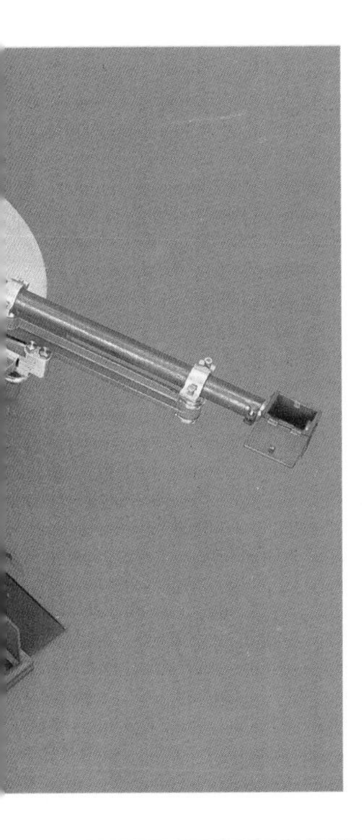

◄ The spectrograph of the English physicist W.N. Hartley (1846–1913), a pioneer of spectroscopy. He used it to study ultraviolet spark spectra. The sample was placed on the right, and light from it passed through a collimator (a device to obtain a parallel beam) and dispersed by a prism at the center. The spectrum was observed on the plate to the left of the picture.

such as common salt, were heated strongly an intense yellow light is produced, whereas potassium-containing salts color a flame lilac. Flame tests are a simple and economical method of testing unknown substances for the presence of particular elements.

Even when large amounts of energy are supplied to molecules, only a small percentage go into a high-energy or "excited" state. The advent of plasma torches and lasers has meant that there has been a revival of interest in emission spectroscopy. However, it is generally easier to measure the uptake of energy by a sample (absorption) than emission.

Absorption spectroscopy

If a sample of material is vaporized and exposed to an intense light source of a wavelength known to be absorbed by a particular element, then the amount of light absorbed is related to the concentration of that element in the sample. Atomic absorption spectroscopy is very widely used in analytical laboratories, for tasks such as checking food samples for unwanted contaminants like mercury.

Frequently it is the nature of the molecules in a sample rather than the individual elements that is of interest. The chemical bonds between atoms in a molecule can absorb energy from the infrared region of the electromagnetic spectrum (◆ page 73). Organic chemicals are those based on carbon atoms, which account for the vast majority of substances found in living organisms and in industrial products. In such chemicals, different types of molecular combination absorb different wavelengths of infrared.

By passing infrared radiation through a chemical sample and changing the wavelength over a range, it is possible to obtain an infrared spectrum which indicates the different groups of atoms that make up a molecule. Infrared spectroscopy can also be used to follow reactions while they are occurring. If the product of a reaction has a characteristic strong peak, a spectrometer can be set to that wavelength to measure the increase in concentration of the compound as the reaction proceeds.

Ultraviolet and visible spectroscopy depend on the more energetic part of the spectrum, with wavelengths of visible light and just shorter. They can similarly provide valuable information about the structure of molecules, as can spectroscopies which use microwaves, X-rays and gamma rays.

Mass spectroscopy (MS) exploits the behavior of molecular ions in a magnetic field as a means to identification. In mass spectroscopy, a substance is bombarded with high-energy electrons or laser light which cause it to form ionic fragments. These ions then pass through a magnetic field, which causes them to follow a particular geometric path dictated by the ratio of the mass of the particle to its electrical charge.

Spectroscopies developed more recently involve the interaction of either electrons or nuclear particles in atoms with strong magnetic fields. Magnetic resonance inspection (MRI), also known as nuclear magnetic resonance (NMR), spectroscopy is a powerful tool for analyzing the hydrogen content of materials and the way the hydrogen atoms are bound (◆ page 86).

Gas chromatography is a sensitive method for separating complex mixtures of molecules. Various forms of chromatography have been developed to provide an array of separative techniques which can be used not only for analysis but also for the purification of complex molecules on a production scale.

◄ A coulometer is an electronic device for measuring quantitative changes in a substance by measuring the quantity of electricity required to deplete a solution of the substance, usually by electrolysis. Here it is used for testing the moisture content of a sample of a plastic roofing material.

▲ Copper sulfate in an atomic absorption analyzer gives off a characteristic green flame. The sample is vaporized by the acetylene torch, and the flame can be tested for the presence of a particular element by shining the emission spectrum of that element through the flame, and measuring the amount of absorption that results.

See also
Optical Telescopes 57–64
Seeing beyond the Visible 73–86
Analyzing the Atom 109–114
Processing Raw Materials 237–244
New Materials 245–248

Detecting surface wear

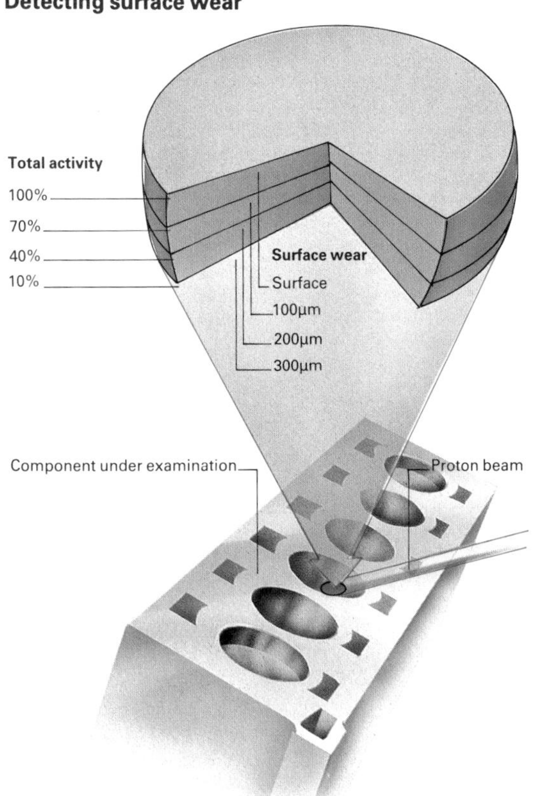

▼ Micrographs of the surface of a platinum alloy gauze catalyst before and after oxidation of ammonia to produce nitric acid. The process is widely used in fertilizer manufacture.

▲► Thin-layer activation techniques of checking wear in cylinder bores. A radioactive layer is placed on the material, and its radioactivity monitored to detect surface wear.

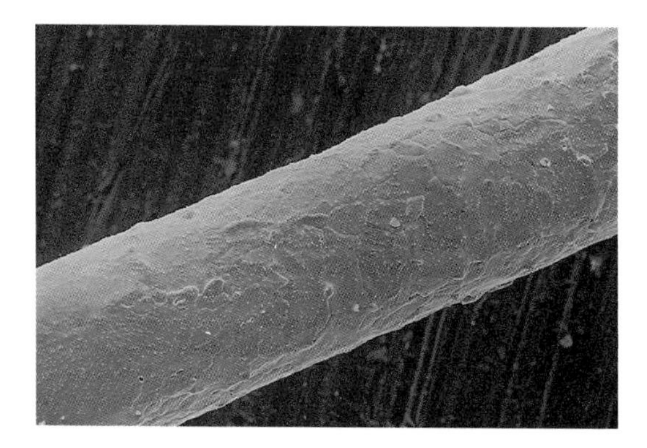

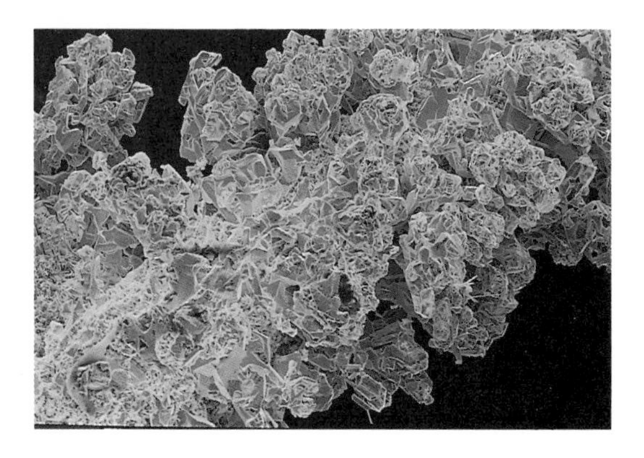

One of the most important areas of analysis to develop in recent years has been the study of surfaces. The surface of a material is often different from, and in technological terms more important than, its bulk composition. Heterogeneous catalysts operate through surface reactions, and are used to make many industrial processes work efficiently. Similarly, the resistance of a material to corrosion depends on the microstructure of its surface and how easily it can be attacked by atmospheric gases or complex solutions such as river- or rainwater.

The surface of a solid object can be bombarded with photons of different energies. The energy with which electrons are ejected as a consequence can provide important information about the structure and properties of the surface.

Whole objects can be bombarded with radiation to generate information about their structure. Neutron activation analysis exposes a small sample of a solid to high-energy neutrons from a nuclear reactor. These make some atoms in the material radioactive. By analyzing the gamma radiation emitted by the substance after neutron activation, it is possible to derive information about its composition. An area of growing interest in neutron activation analysis has been in studying core samples from geological surveys, notably those from under the sea, which may contain recoverable minerals. It can also be used for forensic purposes, such as determining arsenic in hair from a suspected murder victim.

Analyzing the Atom

15

The structure of the atom...Machines to study the sub-atomic world...Cyclotrons...Detecting particles: bubble and cloud chambers...Electronic detectors... PERSPECTIVE...The earliest techniques...Linear accelerators...The origins of the cloud chamber... Cerenkov radiation

Atoms are the building blocks of matter. Each element has its own unique atom, which characterizes the chemical and physical properties of the materials it forms. Thus hydrogen is a gas, copper a metal and so on. Atoms are minute, typically 10^{-10}m across. Even the largest can barely be discerned by the most powerful electron microscopes (◀ page 54). Yet atoms contain smaller objects still. Electrons surround a central nucleus, which consists of protons and neutrons, each about 10^{-15}m across. And the protons and neutrons are themselves built from quarks, which are a mere 10^{-17}m across or less. The electrons are likewise smaller than 10^{-17}m. But if atoms are at the limits of microscopy, how do scientists probe subatomic structure down to distances as small as 10^{-17}m?

The answer lies in "exciting" the atom or nucleus (giving it additional energy) and observing the outcome. A simple example is the flame test, in which the chemist burns a substance and observes the light emitted in the flame (◀ page 106). Some of the energy used in heating the material reappears as light, the color of which depends on the variety of atoms in the substance. By analyzing the emitted light the chemist can learn about the intrinsic energies of the atomic electrons and can assemble a picture of their configuration in the atom.

Physicists use similar techniques to learn about the structure of the nucleus and the particles it contains. But because the nucleus is much smaller than the atom, more energy is necessary to probe it, and nuclear physicists build huge, powerful machines to perform their experiments. The scientists describe the energies they use in terms of units called electronvolts. One electronvolt (eV) is the energy an electron gains when it passes between the terminals of a 1-volt battery. To study the nucleus, energies of several million electronvolts (MeV) are needed. And to probe the quarks requires energies as high as several hundred gigaelectronvolts (GeV). Giving an electron an energy of one gigaelectronvolt is equivalent to sending it across a battery delivering a billion volts.

The basic technique for investigating the nucleus is to accelerate subatomic particles, such as electrons and protons, to high energies and to fire them at a target. The accelerated particles energize the nuclei in the target, and the emissions (usually more particles) provide clues about the internal structure of the target nuclei. The machine that accelerates the particles uses electric fields to give them energy. There are several ways to do this and therefore several breeds of accelerator, each with its particular attributes. Physicists studying the structure of nuclei, who need energies of several million electronvolts, use devices known as Van de Graaff generators or others called cyclotrons (▶ page 111). But to accelerate beams to hundreds of gigaelectronvolts in order to study quarks, physicists rely on the machine known as the synchrotron.

Natural techniques

The first clues about the structure of the atomic nucleus came from early studies of radioactivity, which the French scientist Henri Becquerel (1852-1908) discovered in 1896. A radioactive material is one in which the nuclei are inherently unstable. In changing to a more stable form, the nuclei emit radiation. The radiation can take the form of helium nuclei (alpha particles), electrons (beta particles), or high-energy electromagnetic radiation (gamma rays).

Such transmutations betray the composite nature of the nucleus, as the New Zealand physicist Ernest Rutherford (1871-1937) realized in his work at Manchester and later at Cambridge. The first hints of the composite nature of protons and neutrons came to light in studies of cosmic rays. These are showers of particles produced when high-energy protons and nuclei from outer space collide with the Earth's atmosphere. In the late 1940s and early 1950s, cosmic-ray physicists found new particles similar to protons and neutrons, and these are now known to consist of configurations of the basic quarks slightly different from the familiar particles.

The atom

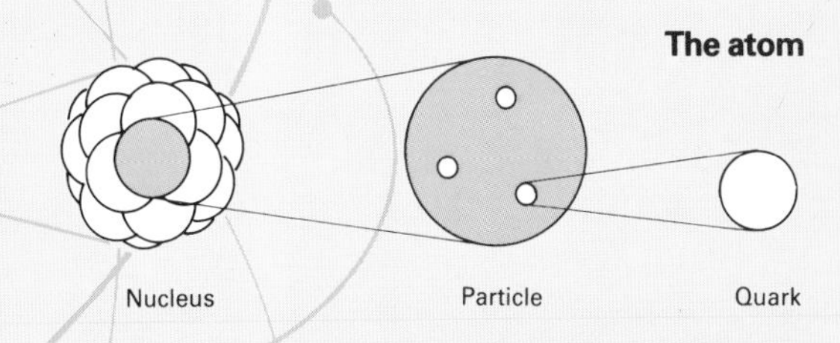

▲ A cloud chamber photograph taken in about 1930 in which nuclear disintegration was observed for the first time. The lines are tracks of alpha rays bombarding nitrogen; the transverse streak is a proton knocked off the nitrogen atom.

Stanford's long beam

Electrons pose problems in synchrotrons. Because they are very light, roughly one two-thousandth the mass of the proton, they radiate significant amounts of energy in the form of light and X-rays as they move round curves. This means that much of the energy pumped into accelerating electrons is immediately lost as so-called synchrotron radiation. To overcome this difficulty, physicists in Stanford, California, built a very long linear accelerator in which the electrons travel in a straight line. The machine at the Stanford Linear Accelerator Center, which first delivered an electron beam in 1960, is 3km long. It can now accelerate electrons to more than 30 GeV. The electrons start off from a thermionic gun (a heated cathode) at an energy of 40 MeV; they then travel through 960 accelerating stations, each 3·05m long and 10cm in diameter. The particles in a sense surf-ride on radio waves set up in these copper cavities, receiving a small push each time until they reach top energy at the end of the machine. Modifications begun in the mid-1980s will allow the accelerator to operate alternately on electrons and positrons (antielectrons with positive charge) at energies as high as 50 GeV. The plan is to direct the positrons and electrons in opposite directions around two arcs of a circle so that they meet head-on at a total energy of 100 GeV.

▲ The main accelerator of the synchrotron at Fermilab outside Chicago extends to the right of picture. Three beam lines extend tangentially from the accelerator and the triangular complex to the lower left is the antiproton source.

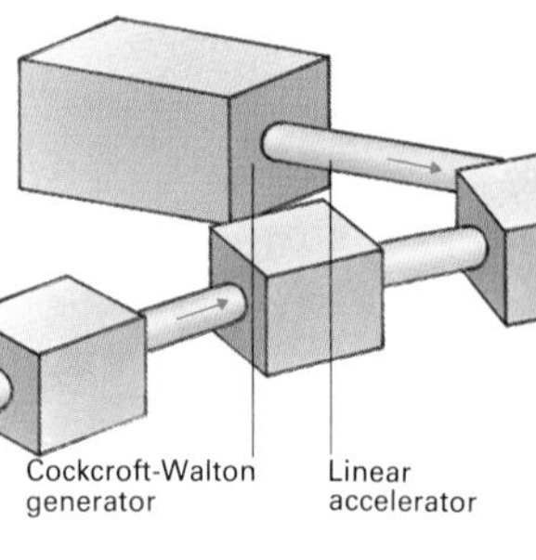

Cockcroft-Walton generator

Antiproton accumulator

► Before entering a synchrotron particles are energized in the Cockcroft-Walton generator; they pass along the linear accelerator and into the ring of the synchrotron. The ring is encased by magnets which focus and bend the particle beam. At one or more points around the ring, the particles absorb energy from radio waves set up in a copper cavity. The beam is directed onto a target in an experimental hall; the results are detected with bubble chambers and electronic detectors. In some instances antiprotons may also be accelerated in the opposite direction. They are first collected in the antiproton accumulator.

A synchrotron bends a beam of particles on a circular path so that it passes through the same accelerating field over and over again. By allowing the particles to circulate millions of times, the machine requires only modest electric fields (a few thousand volts) to reach energies of hundreds of gigaelectronvolts. The electric fields are set up in copper cavities by radio waves. Electromagnets provide magnetic fields that guide the particles on their circular path, and the beam travels through a pipe, several centimeters across, which is kept at ultrahigh vacuum to avoid collisions with air molecules.

The largest modern synchrotrons are at the Fermi National Accelerator Laboratory (Fermilab), near Chicago, and at CERN, the European center for nuclear research, near Geneva. Each machine lies in an underground tunnel forming a ring about 2,000 meters in diameter. The enormous size is dictated by the high energies these machines reach. CERN's Super Proton Synchrotron (SPS) accelerates particles to 500 GeV. With such high energies, the magnetic fields available from conventional electromagnets bend the beam only slightly, equal to the gentle curvature of a ring two kilometers in diameter. Fermilab's Tevatron, designed to accelerate protons to one teraelectronvolt (TeV; one thousand gigaelectronvolts), uses superconducting magnets to provide stronger fields than normal magnets.

The running of a large synchrotron involves the precise synchronization of many different complex operations. A machine such as CERN's Super Proton Synchrotron handles a beam, of some 10^{13} protons, injected into the ring every six seconds. As the protons increase in energy, the field in the guiding or bending magnets must increase accordingly to keep the particles on the same path. Moreover, additional focusing magnets must keep the beam focused to prevent it from hitting the walls of the vacuum pipe. In the trip to 400 GeV, which takes under four seconds, the protons travel over one million kilometers, more than twice the distance to the Moon.

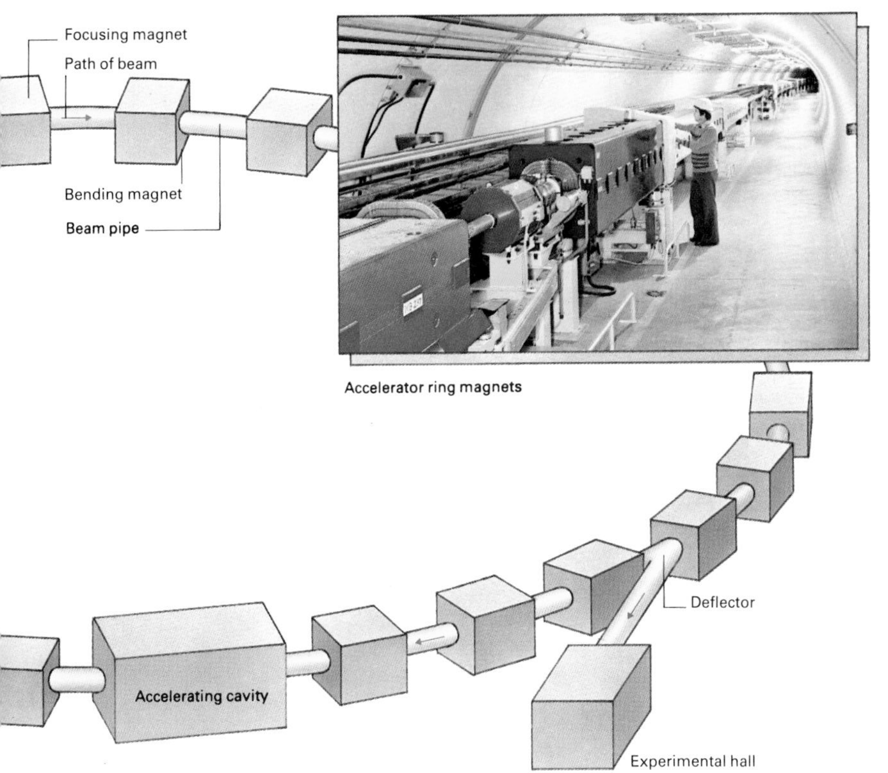

Heavy-ion machines

Research into nuclear structure usually refers to studies of the heavier nuclei, with many nucleons (protons and neutrons). Many experiments on nuclear structure take place at machines that accelerate heavy ions to energies of several million electronvolts. The highest-energy heavy-ion machine is the Bevalac at the Lawrence Berkeley Laboratory in California (where the first cyclotron was built). A linear accelerator accelerates the ions and then injects them into the Bevatron, a small synchrotron built in the 1950s. Together the two machines can take heavy ions to energies of over 2 GeV per nucleon. Another device is the electrostatic accelerator, invented in 1933 by the American physicist Robert Van de Graaff (1901-1967). This consists of a non-conducting (insulating) conveyor belt that carries electric charge from one end, and deposits it on a conducting (metallic) shield at the other end. The belt gradually charges up the shell, which is insulated from Earth, until it is at several million volts or more. In the so-called tandem Van de Graaff, a high positive voltage is set up at the center of an evacuated column. Negative ions are accelerated in the first half of the column. They are then stripped of some electrons so that they can be accelerated again as they move away from the central terminal in the second half of the machine.

Once the protons circulating in the cyclotron reach top energy, they are deflected out of the ring towards the experiment, and the cycle begins again. The SPS can also operate as a collider, in which beams of protons and antiprotons circulate in opposite directions before coming to meet head-on. Antiprotons are the antiparticles of protons: they have the same mass but the opposite electric charge. The negatively charged antiprotons travel round the same ring of magnets, while being accelerated by the same electric fields, provided they move in the opposite direction to the protons. CERN makes its antiprotons in the collisions of relatively low-energy particles with a metal target. Once an adequate number (around 10^{11}) have accumulated in a special ring of magnets, the antiproton accumulator, the antiprotons are injected into the big synchrotron to join protons there. The two beams are then accelerated and brought together with a total collision energy of 540 GeV or more.

The advantage in colliding beams is that in head-on collisions no energy is wasted in simply moving particles in the target. A head-on collision of protons and antiprotons at 270 GeV per beam (540 GeV total) is equivalent to a beam of 150,000-GeV antiprotons striking protons in a stationary target. This technique also figures in a number of electron synchrotrons. In this case the antiparticles are positrons (antielectrons) which carry positive charge, and again they can be accelerated in the same ring as the electrons, but in the opposite direction. Another alternative is to collide electrons with protons. In this case the particles must be accelerated in different rings, one above the other, because the protons are much heavier than the electrons. The rings must intercept at a number of points so that the two beams can collide once they reach maximum energy.

Observing the particles

Accelerators are the means for energizing the nuclei in collisions, but physicists must also analyze the results of the collisions if they are to learn anything about nuclear structure. To do this they need specialized detectors to track the debris that flies out from the high-energy interactions, and to provide some sort of permanent record of what happens. The particles themselves are minute and many are unstable and decay into other varieties of particle after fractions of a second. Fortunately, they are usually traveling close to the speed of light. A subtle effect of Einstein's theory of special relativity means that the longer-lived particles, those with lifetimes of around 10^{-8} seconds, travel several centimeters or more through a material.

All electrically-charged particles such as protons and electrons ionize the materials they travel through, losing energy by knocking electrons out of atoms that they pass. The process of ionization results in a temporary electrification of the material, and this in some instances can be made visible. This is the basic principle underlying most particle detectors.

One of the most widely used detectors is the so-called bubble chamber, invented in 1952 by the American physicist and Nobel Prizewinner Donald Glaser (b. 1926). The device consists of a vessel of liquid that is superheated. The energy lost when a particle ionizes the liquid is sufficient to induce boiling, and a trail of bubbles forms along the particle's path. To superheat the liquid, the pressure on it is released suddenly just before the particles arrive from the accelerator. To ensure that the whole volume does not boil, pressure is reapplied as soon as the trails have started to form.

Cosmic detectors

The first device to track the paths of subatomic particles was the cloud chamber, invented in 1895 by the Scottish physicist Charles Wilson (1869-1959). Wilson had been impressed by phenomena he observed on Ben Nevis (Britain's highest mountain) as sunlight struck the clouds and mist. Back at Cambridge he built a chamber to make his own mists to study in the laboratory. The device was a glass vessel filled with damp air, which Wilson could expand rapidly. The expansion cooled the air so that a mist condensed out. Wilson discovered that when X-rays ionized the air in the chamber, droplets of water condensed out on the charged ions produced. Later, in 1911, he used his chamber to observe the tracks of alpha particles (helium ions) emitted by a radioactive source. Other physicists developed Wilson's cloud chamber and used it to photograph the tracks of cosmic rays. In the 1930s and 1940s the cloud chamber revealed new types of subatomic particles in the cosmic rays. One of Wilson's students at Cambridge, Cecil Powell (1903-1969), used another technique to study the tracks of cosmic rays. He made the particles photograph themselves. Just as rays of light darken a film and give bright regions on the developed picture, so charged particles leave dark trails in film emulsion as they lose energy by ionizing atoms. Powell's team at Bristol University in the late 1940s took packs of special emulsion up mountains and attached them to high-flying balloons in their explorations of the cosmic rays. Their reward was the discovery of several new kinds of elementary particle; Powell, like Wilson before him, was awarded the Nobel Prize. Today, emulsions are still used at particle accelerators, especially when fine details are required.

▼ Wilson's original cloud chamber apparatus. If the globe is filled with moist air and the pressure reduced suddenly, drops condense on gas ions and make the particles visible.

Bubble chamber

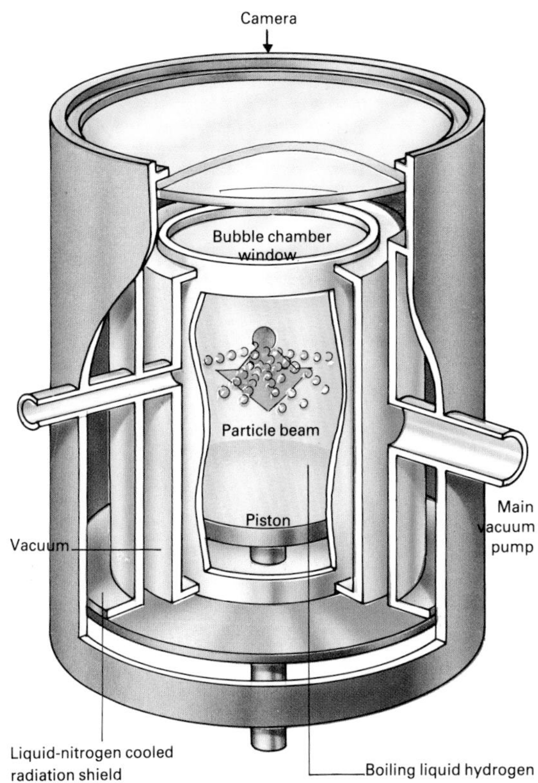

◄ The so-called Big European Bubble Chamber at CERN, which held 40,000 liters of liquid hydrogen cooled to −247°C, surrounded by a large superconducting magnet.

▲ As particles enter the bubble chamber via a hole in the wall, a piston causes the pressure to fall for a moment. Bubbles form in the liquid hydrogen and are photographed from above.

► A bubble-chamber image showing the tracks of particles after collision. The spirals are caused by the magnetic field. The faster or heavier the particle, the straighter its track.

See also
Introduction to Measuring 9–22
Measuring with Electricity 41–46
Chemical Analysis 105–108
Nuclear Power 141–148
Other Sources of Power 149–156

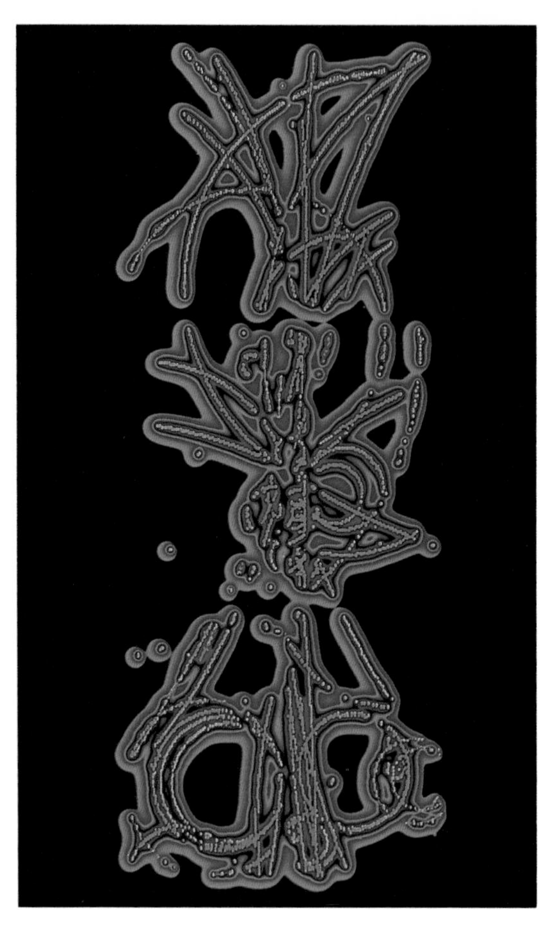

◄ ***The photomultiplier tubes that make up one of the vast particle detectors at CERN. A picture of the effects of a collison is built up from readings at many individual points.***

▲ ***An electronically produced and false-color image of a particle collision as detected by detector at CERN (shown left). This image led to discovery of the so-called Z particle.***

Another way to detect ionization is to use electrodes to collect the electrons or ions released, and to measure the resultant current or voltage. The technique works best with gases in a region of electric field. The field accelerates the electrons, which cause more ionization, and so on in a snowball effect until a detectable pulse of charge is produced. This principle underlies the Geiger counter, invented in the 1920s by the German physicist Hans Geiger (1882-1945). It also operates in the vast complex modern detectors used in association with synchrotrons. These incorporate thousands of wires to sense points along the paths of particles through large volumes of gas. The signals from the wires, and from other detectors, pass to computers which sift through the electronic information and record what is useful on magnetic tapes.

Whether a physicist has photographs of tracks in a bubble chamber to study or reels of magnetic tape recorded electronically, computers play an important part in the last stage of the experiment. This is when millions of images or electronic records of the complex patterns of debris from the collisions are analyzed in detail, and clues about the structure of nuclei and their particles begin to emerge. At this point the physicist finally has access to the subatomic world of 10^{-15}m and less.

Čerenkov radiation

Most methods of detecting charged particles rely on recording the ionization along a particle's path. Another phenomenon that reveals charged particles in particular circumstances is Čerenkov radiation. This is light radiated by a particle travelling so fast through a material that it is actually moving faster than light does in that material. The light is emitted at an angle to the particle's path, along the surface of an imaginary cone with its apex at the location of the particle. Particles travelling at different velocities produce Čerenkov radiation at different angles, so detection of the light not only reveals a particle's presence, but can also distinguish between particles which have the same electric charge and momentum but different mass and velocity. It is named for Russian physicist Pavel Čerenkov (b. 1904) who in the 1930s investigated the blue glow emitted when particles from radioactive radium pass through certain liquids. The effect is used in detecting high-energy photons in lead crystal. The lead in the glass encourages photons to produce charged electrons and positrons, which emit Čerenkov radiation in the glass.

Computer Hardware

Defining the computer...Computer memories...The microcomputer...Mainframes and supercomputers... Making the chips...PERSPECTIVE... Mechanical precursors of the computer...Electronic computers...Storing information on the chip...Uses of the supercomputer

A computer is a machine that can manipulate data according to a predetermined sequence of operations, and can retain data in a memory before, during or after manipulation. The set of instructions which determines its sequence is called a program, and may be changed at the command of the operator. The data may be in the form of numbers and in this case the sequence represents a calculation. Other possibilities are that the data represent words or even images, and such data are also open to manipulation. Devices which manipulate data according to a programmed sequence but cannot accept new operational commands are known as microprocessors rather than computers.

Data prepared for entry to a computer, whether a number, letter or punctuation symbol, are ascribed a value which is expressed in binary form, that is written using only the symbols 1 and 0. Such a symbol (or "bit", short for binary digit) can be stored electrically corresponding to the positions of a switch. Electronic switching allows the stored symbols to be treated according to arithmetic or logical rules. Such bits are processed in multiples known as bytes, which form the basic unit of computer information.

Computers operate by breaking down complex tasks into simple, but perhaps numerous, steps, which are performed sequentially at high speed. A vital element in computer design, therefore, is the ability rapidly to access any given piece of information in memory, and to treat that information in the sequence laid down by the program. The development of integrated circuitry in the 1960s and 1970s allowed computation speeds to become very fast as the computers themselves shrank in size.

The earliest genuine use of computers was in the decennial census of the United States in 1890. The data collected was so numerous that it was scarcely possible to complete the work on one census before the next was due. To speed the calculations the data was put on punched cards which were processed by means of electric tabulators and sorters. Punched tapes and cards are still used to input data, particularly for mainframe computers (➧ page 120).

The prehistory of the computer

Precursors of the modern computer have been in use for a surprisingly long time. As early as 1804 the French engineer Joseph Marie Jacquard (1752-1834), working in Lyons, developed a programmed loom for weaving patterns in silk. His machine used a punched card to control the rise and fall of a weft of up to 400 strands, and could produce patterns of great complexity.

The important feature of the Jacquard loom is that the punched card contained a permanent set of instructions, which could be taken from one machine to another. Conversely, any one machine could follow many different programs. Control of what the machine did was separate from the mechanics of machine operation.

Jacquard's machine was a process controller rather than a calculator. An Englishman Charles Babbage (1791-1871) developed the idea of the separated program to solve mathematical problems. In 1823 the British government gave him funds to make an "analytical engine" for use in solving a wide range of computational problems. It was to be controlled by a program written on a punched tape and driven by a steam engine.

Babbage, however, repeatedly improved his design, elaborated the concept and sought additional funds to carry on the work. He never finished the machine.

▲ The Analytical Engine designed by Charles Babbage and built in a simplified form according to his drawings by his son. It could perform four arithmetical functions.

▼ The Harvard Mark I computer (1943), the first fully automatic computer to be built. It remained in use for 15 years. It added or subtracted two numbers in 0·3 seconds.

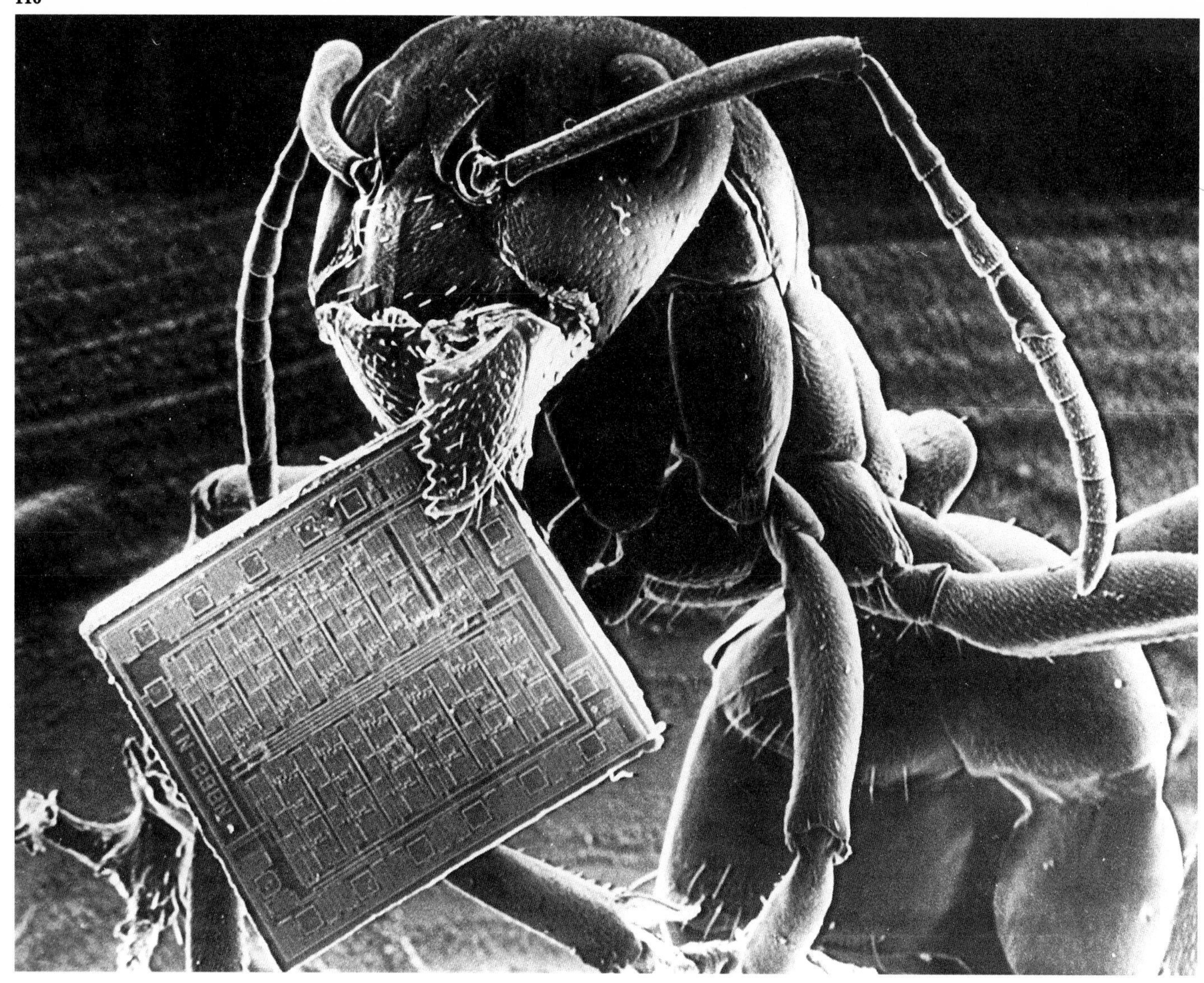

▲ *Miniaturization, speed and cheapness have been the goals of computer designers in the 1970s and 1980s. Even smaller integrated circuits than this are under development.*

The most widely used form of electronic memory uses solid-state components and takes the form of a single silicon chip, with an active surface of some 5 millimeters square. It may hold from 8,000 to 262,000 bits of data, though memories of larger capacity are also produced.

Each individual bit of data is held by a few circuit components, and these memory elements are arranged in a rectangular array. Thus each piece of data is associated with a particular place in the memory and may be described as "location addressable". This may be contrasted with the way in which information is stored in a dictionary, where, by specifying a part of the content – the first letters of a word – it is possible to find the rest. Such a memory is "content addressable".

Data is read from the memory, or written to it, by the computer's central processing unit. To read data from the memory, circuits are addressed which activate a chosen row and column of the array and so couple a single element to external circuits. Memories may be arranged to have a single input/output line for transfer of data, or they may have parallel lines so that several bits can be transferred simultaneously. Systems with 1, 8, or 12-16 lines are in regular use, corresponding to transmission by the bit, byte or word. When a memory is arranged in this way a piece of information can be written directly in any circuit element or read directly from it. It is necessary merely to designate the row and column of the array using the decoding circuits and then to put in or take out the signal. This sort of memory is described as being random access memory (RAM), to distinguish it from the sequential information patterns held on magnetic tapes or disks.

▼ *Section of a ferrite core memory of 1954. A ring changes its magnetic polarity when a current is passed through it. The direction of polarity at each "address" represents 0 or 1.*

Storing information on a chip

The actual storage of data within a chip is by the accumulation of electrical charge in a small capacitor. A transistor then acts as an on-off switch connecting the charge to the data line. To couple a particular capacitor to the line, the row decoder activates a complete row of transistors. At the same time the column decoder makes connection to the data line. With the connection made, a charged capacitor sends a current down the data line whereas an uncharged one does not. By selecting the correct row and column any capacitor in the line may be chosen. It may then be read, or it may be charged or discharged in response to new data coming in.

One limitation of the capacitor storage cell is that it rapidly loses its charge. This happens partly by leakage to other parts of the chip and partly by charge sharing with the data line, which itself has significant capacitance, each time the cell is read. Thus the memory chip has additional circuits to replenish the charge on cells. This is done every few milliseconds and after each reading operation.

The detailed design of memory cells may be varied to suit particular purposes. By adding more transistors to each storage cell, charged cells may be continuously refreshed. Such cells take more space and so are more expensive. In a different approach, calling for a different treatment of the silicon and for several transistors per cell, leakage can be cut to a very low level so that a small battery may sustain the memory for many days. Such memories are commonly used for data storage in hand calculators.

▶ ▼ *A chip carries many transistors etched in the silicon. A transistor is a semiconductor that has an n-type region (with a surplus of electrons) and a p-type (with a surplus of "holes"). Electrons and "holes" cross the junction and current flows if a polarity is applied in one direction, but not in the other. A memory cell (below) comprises a transistor "gate" or switch, and a capacitor which stores the charge.*

Construction of a RAM cell

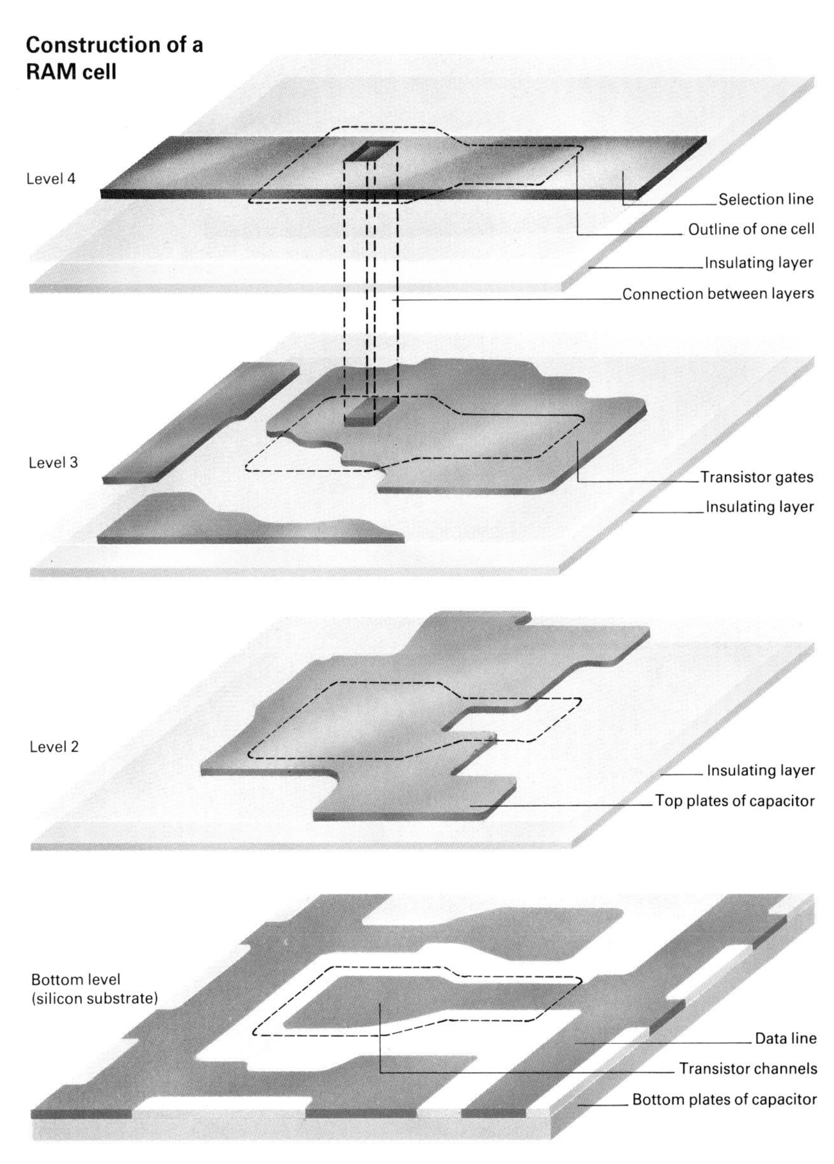

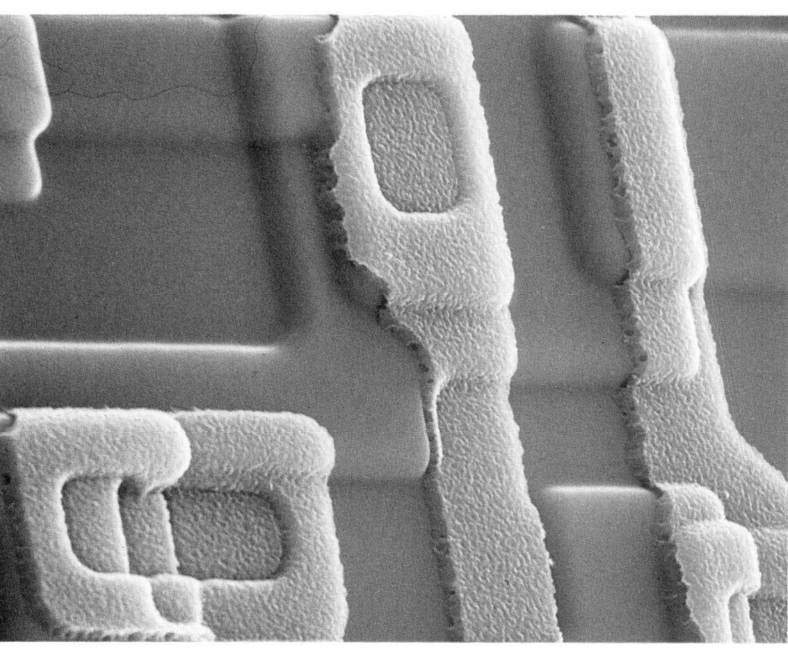

The development of the electronic computer

True electronic digital computing began in 1943 with a specialized code-breaking computer called Colossus. It was followed in 1946 by a general purpose computer known as ENIAC (Electronic Numerical Integrator And Calculator), developed at the University of Pennsylvania by J. Presper Eckert (b. 1919) and John Mauchly (1907-1980).

Electronic computing established the need for specialist memory devices. They are required, in particular, to hold figures evaluated at intermediate points in a calculation and so must be read quickly and permit frequent, rapid overwriting with fresh data. The first effective memories used tiny ferrite rings through which wires were threaded. Currents through these wires magnetized the rings and information was recorded by associating one direction of magnetization with the value 0 and the other with the value 1. Rings were read by currents induced in additional wires when interrogatory currents reversed the magnetization directions.

Ferrite memories were in production until about 1960 when integrated circuit methods began to capture the market.

The Microcomputer

Peripheral devices and the microcomputer

► The microcomputer can drive many peripheral devices which greatly extend its scope. Interfaces allow the CPU to receive input from keyboard or other devices such as joystick (for games) or mouse (for drawing), to store or read data or programs from memory, or to output data to screen, line printer or sound-generating devices. Most micro memory storage devices use magnetic storage on tape, floppy disk or (offering the most memory) hard disks, but read-only optical memory may also be available. The micro can be linked to a database or network via a telecommunications link; this can serve as an input or output device as required. Printer and modem cannot output data as fast as the micro can transmit it, so a buffer is needed to release the micro for other tasks.

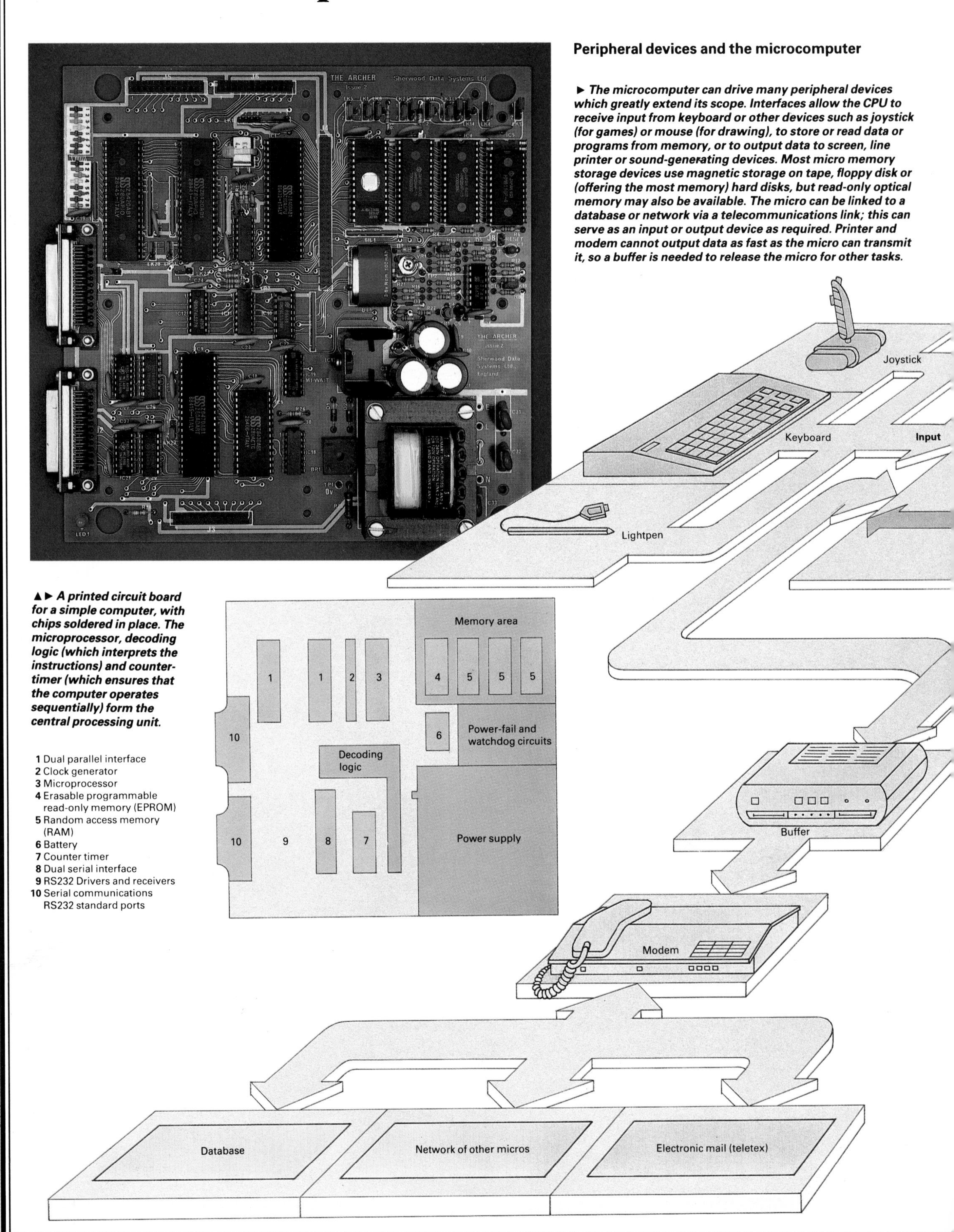

▲► A printed circuit board for a simple computer, with chips soldered in place. The microprocessor, decoding logic (which interprets the instructions) and counter-timer (which ensures that the computer operates sequentially) form the central processing unit.

1 Dual parallel interface
2 Clock generator
3 Microprocessor
4 Erasable programmable read-only memory (EPROM)
5 Random access memory (RAM)
6 Battery
7 Counter timer
8 Dual serial interface
9 RS232 Drivers and receivers
10 Serial communications RS232 standard ports

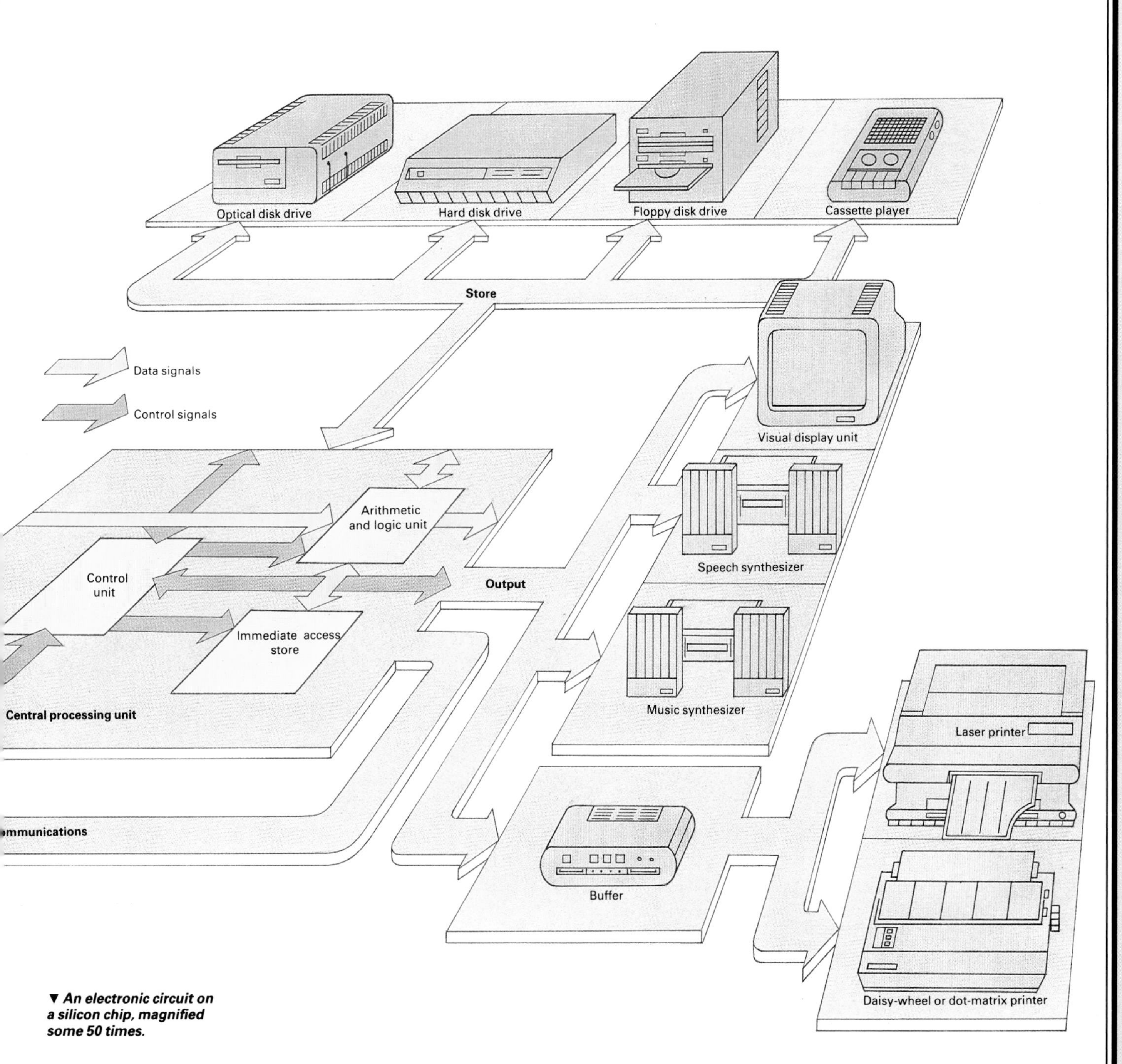

▼ ***An electronic circuit on a silicon chip, magnified some 50 times.***

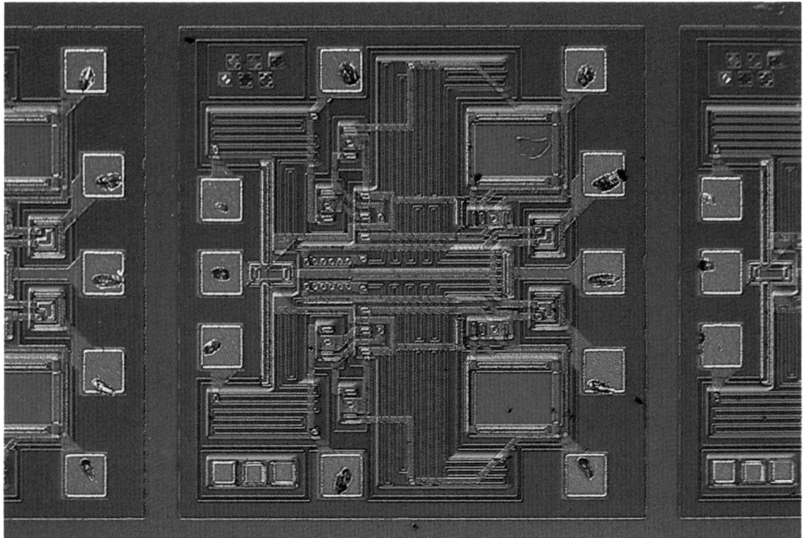

The central processing unit

A microcomputer is defined as having all the processing carried out on a single chip, unlike minicomputers which may have several functional devices mounted on a rack in a single unit. A micro can support only one application at any one time, whereas a mini can support several concurrently.

The "brain" of any computer is the central processing unit (CPU). In the CPU each byte of information is held at a particular register or location in the immediate access memory, and instructions from the program or keyboard are interpreted by the control unit, following a sequence established by the program counter and the clock generator. The control unit finds and fetches each byte of information in turn and passes it to the arithmetic and logic unit where the instruction is carried out; the value is stored in the accumulator (a small extra memory register) while further operations are carried out, then returned to the random access memory (RAM).

Supercomputers

The largest computers perform more than 100 million arithmetic operations each second. At this rate, a single computer could make one calculation for each citizen of the world in less than a minute, and might complete a tax assessment for a country in a single day. Yet manufacturers continue to seek ways of making computers that operate even faster.

One reason for this is that problems such as that of tax assessment are only superficially large ones. More extensive calculations are common in those branches of science and technology that call for the numerical simulation of continuous fields. Such calculations may be used to study the electromagnetic fields produced by broadcasting antennae, the propagation of seismic disturbances, or the distribution of temperature in nuclear reactor vessels (▸ page 144). Two well-known uses are in meteorological forecasting and the description of fluid flow around ships and aircraft. Large computers can work on such calculations at speeds that approach their theoretical maximum.

It is a major organizational task to keep such machines occupied. In any big computing center it is notable that the peripheral equipment – the terminals, disk and tape players and storage, communications, organization and printout equipment – dominates the scene; the central computer itself is relatively small.

Circuits suited to microcomputers or hand calculators may take about a microsecond to perform a switching operation, but in large computers the high priority given to speed means that a more complex fabrication process is used to make transistors that switch more rapidly. In these devices the switching time may be cut to a few nanoseconds, as in the computers built by Cray Incorporated in the United States. Unfortunately this speed has its price, for the circuits use more power and the number of elements that can be put on a single chip is consequently reduced. In the large Cray-2 computer this problem has been overcome by totally immersing the circuit boards in an inert cooling liquid.

▲ The computer room of a mainframe is dominated by input and output devices, storage and other elements relating to organization of the computer.

► The Cray-2 (shown below and diagrammatically at top of page) is one of the largest and fastest computers. It performs each calculation in 4·1 billionths of a second. It houses four processors which can operate separately or together. The computer is used for tasks, such as weather forecasting or calculating airflow over automobile or aircraft bodies, which involve very large numbers of related computations (center).

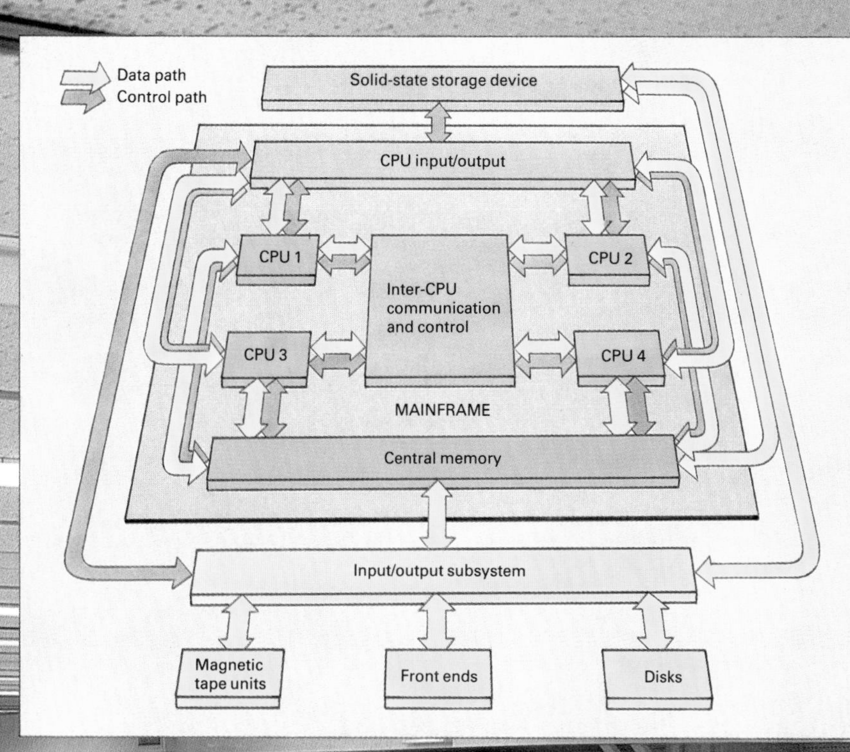

◄ ***The Cray-2 supercomputer has a foreground processor to coordinate the flow of data between the memory and the external devices across four I/O channels. The foreground processor, four background processors and external devices together give a high data throughput.***

Forecasting by computer

Meteorological problems call for a description of how physical variables such as pressure, temperature, humidity and wind velocity vary continuously with position and with time from a known initial distribution. To use a computer to forecast, the region of interest is divided into a large number of small elements, the value of all physical variables being uniform within each element. Adjacent elements are supposed to interact following accepted physical laws, and the change to be expected within a short interval of time is calculated for each variable and a new value derived.

The number of points and time increments treated may each number up to 1,000,000 and the number of physical variables to 30. For large calculations, the product of points, variables and increments may range to 100 billion.

See also
Introduction to Measuring 9–22
Digital Recording 97–100
Computer Software 123–128
Telecommunications 177–184
Radio and Television 185–192

Making the chips

Chips are made by a combination of processes, including photolithography, oxidation, diffusion and vacuum deposition; these create the required patterns in the silicon substrate and introduce transistors, resistors, capacitors and circuitry.

An initial pattern for each layer is produced by means of computer-aided design to a large size, then reduced photographically and copied to create a photolithographic mask, capable of producing up to a hundred identical chips from a single slice of silicon, 10cm in diameter and 4mm thick. Up to 12 such masks may be prepared, each used to mark out a different area in the chip. A dopant – often phosphorus or boron – is used to modify the electronic properties of particular layers. This diffuses through the mask windows to the required depth to create semi-conducting "n" and "p" regions in the silicon as required, and a heat treatment causes a silicon dioxide film to close over the doped regions. This protects them during etching of later layers of the circuit. New windows are etched through the oxide layer and a film of aluminum is placed on the chip; this is in turn engraved to leave the electrical interconnections. Finally the slice is given a protective layer of oxide.

▲ The layout of the electronic circuitry for a chip is designed, usually using computer-aided design, at a magnification of about 250 times, before being photographically reduced to final chip size and copied for use. A separate layout is produced for each layer of the chip.

► The individual chips are probe-tested before being cut off the silicon slice; then inspected visually under a microscope and installed in a package with fine wires providing connections between the chip and the pins. Finally they are retested ready for use. Fewer than 25 percent of the individual chips from a slice may survive the testing procedures.

Computer Software

User-friendly computers...Machine code and programming languages...Creating imaginary worlds... Databases...Expert systems...PERSPECTIVE...Writing a computer program...Computers and graphics... Spreadsheets and business applications...Data capture ...Games-playing computers

The journals that advertize computers may suggest that computer systems and the terminals by which we communicate with them possess human attributes. Despite the jargon, terminals are not actually smart or friendly, nor are computer systems intelligent or expert.

These attributes lie in that aspect of computer systems known as software. Visible parts of the computer, the keyboard, screen, disk and central processing unit, comprise the hardware, and these are controlled by instructions which detail what is done in response to each command. Such instructions constitute the system's software, and it is their versatility that determines the "feel" of the system.

Early computers were programmed in direct machine code. Instructions were given in a detailed form that required of the user a close understanding of how data, and the intermediate results of calculations, are held within the computer. In effect, only specialists could have access to computers, for the process of learning machine code was too involved for those whose ultimate interest was elsewhere.

Modern equipment is easier to use. One key feature of software is that it can build upon itself. If one person writes a program (a set of instructions) to perform a particular task, another person can incorporate the set as an element (a subroutine) within another program, without needing to know in detail what is contained in the original program. It is enough to know how to call the program into action and appreciate any circumstances under which the program will fail.

By building up the complexity of the languages, and thus arranging for more and more complex operations to be carried out in response to simple commands, computing became accessible to non-specialists. A number of computer languages (systems of computer instructions) have been developed. These languages are not, in a literary sense, extensive – each comprises perhaps one hundred separate instructions – but by compounding instructions many tasks can be performed.

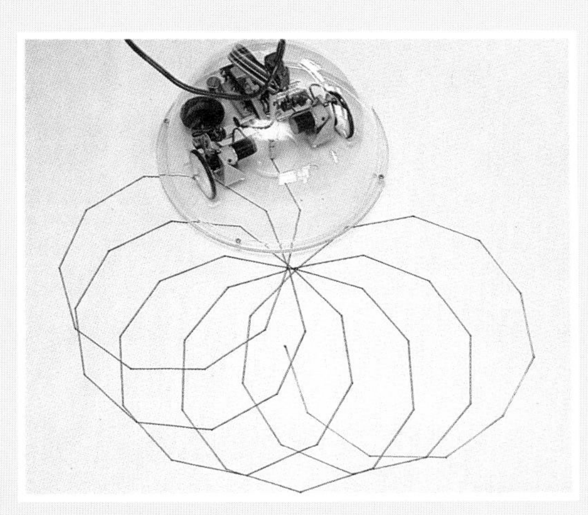

◀ **A "turtle" traces out a pattern on paper according to a computer programmed by schoolchildren using the computer language LOGO. This was developed as an educational language, teaching young children the elements of programming by concentrating on geometrical concepts, and although it is simple to use, it is a surprisingly powerful language, capable of generating images on screen in full color. The turtle is a simple robot, one of a number of input and output devices relating an image on paper to the one on screen.**

Computer languages

There are a number of different languages used by computer programmers – ALGOL, Lisp, LOGO, COBOL, BASIC, Pascal, Forth among them – each of which has a particular advantage in the way certain kinds of instruction can be put to the computer. COBOL, for example, was created for large-scale data processing in commercial or adminstrative contexts. By comparison FORTRAN, which was tailored to fit the requirements of the natural scientists who were the first non-specialist group to take up computing in large numbers, is optimized for mathematical calculation. BASIC is frequently used as an introductory language, especially popular in home computers.

```
10 PRINT "Enter highest integer"
20 INPUT N
30 PRINT "Enter maximum permitted value"
40 INPUT M
50 Z = 0
60 FOR I = 1 TO N
70 Y = I*I
80 IF Y>M THEN 110
85 PRINT I,Y
90 Z = Z+Y
100 GOTO 140
110 Y = M
120 PRINT I,Y
130 Z = Z+Y
140 NEXT I
150 PRINT "Total is ";Z
160 STOP
```

The program shown here is a trivial example of BASIC, designed to show the main features of that language. The task is for the computer to produce a cumulative total of the squares of a series of numbers, up to a limit determined by the user, and with a maximum figure, also determined by the user, above which no square is to be read. When preparing the problem, the programmer may first reduce it to a flowchart or "algorithm", in which the task is broken down into the sequence of discrete steps that the computer must follow; the points of input of data, comparison, choices and action to be followed are all laid out.

The next task is to convert the flowchart into computer language. In this case, each separate task is presented on a new line, identified by its line number. The computer follows these in order, unless given the command (as in lines 80, 100 and 140) to go to another line. Lines 20 and 40 allow the user to define the maximum values to be set in the calculation and refers to these as M and N. The sequence 60-140 is a loop, or set of calculations to be performed by the computer repeatedly; the instruction in line 80 tests if at any point the value of a squared number (Y, or I squared) exceeds the value M and, if so, directs the computer to line 110 which will substitute the value M for it. Z is the running total of the values of Y. When the loop has been repeated N number of times, the computer will go on to line 150, print out the cumulative value Z, and the program will finish. It may then be reused, or may be modified by the designer; it may be printed out, or stored on disk or tape and loaded back into the computer whenever needed.

Computer software allows imaginary worlds to be modelled, and permits us to check our knowledge of the real world against them

Formal computing languages are excellent vehicles for exploring the patterns underlying natural phenomena, since they enable complex problems to be treated in terms of mathematical models which can break the problem down into large numbers of simple individual steps. Conversely, computing languages can serve as a means to construct new worlds of the imagination. Since the languages use abstract logic, their programs need not be circumscribed by the laws of physics; they can reflect the theories or idiosyncrasies of the programmer, and their laws need not be too serious.

Exploration of imaginary worlds with computers is vital in computer simulation of events, such as the early moments of the universe, which cannot be examined physically. Aircraft flight and nuclear power station simulators similarly create a model world, designed to explore real events and provide the participant with a safe and cheap opportunity to develop decision-making and motor skills. But the imaginary worlds of computer software have moved beyond the confines of the academic or computer specialist.

Two things brought computer games into the hands of children. Small computers were produced at a price within the reach of many families, and fun packages, usually called software packages, made it simple to take part in the events of an imaginary world. In such games, the package determines the structure of the game, whether it be tic-tac-toe, war in space, or chases through a maze; the player works entirely within the rules that have been set by the software programmer.

In a number of business systems, on the other hand, the user, not the programmer, is responsible for creating the model. In what is called spreadsheet analysis (➧ page 127) the software is effectively a model-making kit; details of the business are inserted and the user may then query the model to assess the effects of change. A central feature of such systems is that the new user should be able to learn it quickly; good programs may be described as "user-friendly". The common feature of software packages at all levels is that they allow the user to concentrate on the application to which the program is directed, without needing to pay too much attention to the functioning of the computer itself.

▲▶ Images of the Mandlebrot set, a numerical sequence involving the "imaginary numbers" (derived from the square root of −1) which can be visualized on a computer to produce the most complex pattern known to mathematics. If a computer traces the sequence in two dimensions and colors each pixel relative to the value of that point, the pattern (right) results. If one portion is examined more and more closely, complex and unique whirls and loops can be seen. Finally, a tiny representation of the original pattern appears.

◀ Flight simulators generate an imaginary world that attempts to copy part of the real world, enabling the trainee pilot to learn without danger.

Computers and graphics

The "friendliness" of computers for problems of all sorts has been greatly aided by the development of graphics that can be manipulated by the keyboard or specialized input devices. Even small home computers can be programmed to produce charts and patterns.

*Computer-aided design (CAD) is used to analyze the properties of structures before manufacture; coordinates or dimensions are entered and plotted by the graphics software into a three-dimensional image that can be turned, amended or tested as well as colored and shaded (**♦ page 19**).*

For a preexisting image to be computer-enhanced, it must be entered into the computer memory either manually by an operator tracing out its features or automatically using an optical scanner. The image is divided into pixels with the tonal values of each recorded. By altering these values, the image may be modified in shape (so that it may appear to move) or color, details such as shadows may be added or dropped, or a multiple image may be built up from the original.

Databases allow researchers to check facts in many diverse fields without the need to visit a library

Databases

Computing techniques can put the user in touch with a vast store of information. By telephone or satellite, information kept at a central source where it is continually updated as necessary, is ready for access from a subscriber's terminal as required. As such it is the key to the store of instant and accurate information that is needed in many aspects of modern life.

Traditionally libraries have served the function of information stores or "databases", where concepts, ideas, experiences and information can be retained from one generation to the next. To make the store of information useful, it has to be arranged systematically so that the user can select material by subject, author or period, and follow up references from one text to another. A growing amount of material is available in a form suitable for direct access from a computer. Periodicals and books are often printed from magnetic tapes (♦ page 173), from which it is a simple matter to take abstracts for inclusion in a computer database. Much of the statistical data of government, business and research organizations is computerized. The result is a range of databases, already large and growing rapidly.

▲ A light pen is a laser beam that detects patterns of light and dark when drawn across a bar code; a quick method of entering information into a database much used in shops, libraries and warehouses.

◄ A visually controlled robot arc-welder; a camera sends pictures of its position to the computer, which detects deviations from expected positions and makes adjustments accordingly.

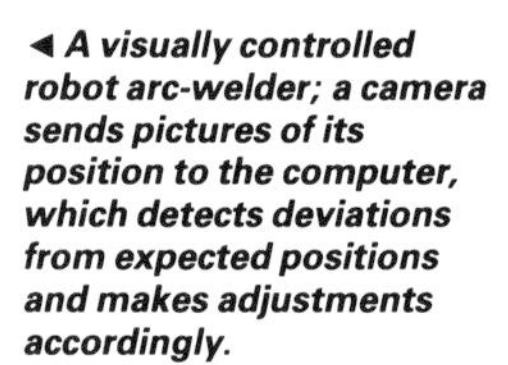

▼ A map of Manhattan with software that can show the quickest (blue) and shortest (pink) routes between two points.

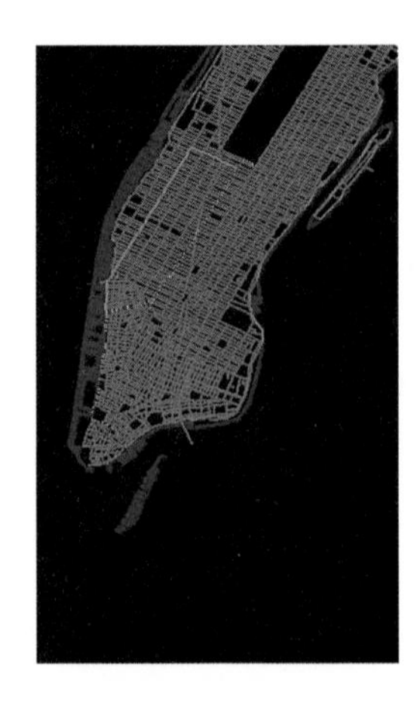

► A computer-generated image of the dimensions and reach of a robot. Software for engineering design (♦ page 19) has simplified the process of industrial design.

Access to these databases has extended greatly through the proliferation of the smart modem, or telephone connection between computers, which allows a user with a microcomputer in the office or laboratory to interrogate and receive answers from a database many kilometers away. Extensive databases cover the medical, legal and scientific professional areas; they also provide links to industry and commerce. This ability to cross-link interests is the really important feature of database interrogation. Without leaving the desk, a user can, for example, call up a research article from a US journal, relate the article to a patent held in Europe and check both against the industrial output of a company operating in, say, Taiwan.

Interrogation of the database is like making intelligent use of a library catalog. With every piece of data is associated a series of key words; these may be words of the titles of articles, names of authors, companies or important topic areas. The database will list all references it contains to a keyword, and scrutiny of these lists enables the user to find what is needed. Alternatively, the request may be for references that contain two or more keywords together.

Scientific databases normally provide the full reference to an article and an abstract of its contents. Such is the volume of material available that storage of complete articles is unusual, and once the reference has been obtained the user goes to a library to read the article itself. Business databases are a little different. The information sought is often more succinct: annual trading figures for a specific company, or the overlap of directors between two or more firms. For such queries the required information might be transmitted in full for display at the terminal.

A few databases handle complete texts. With these it is possible to search out new items or continuing stories reported over long periods of time. The search can be quicker than looking through newspapers, and careful use of keywords should give a full review of a story.

Databases may become available on laser-scanned optical disks (◀ page 98). It is not possible to overwrite information on these disks, so they would be used for data that change slowly, but very large amounts of data can be accessed quickly and cheaply. The optical disk database might be sited in a local library or information center.

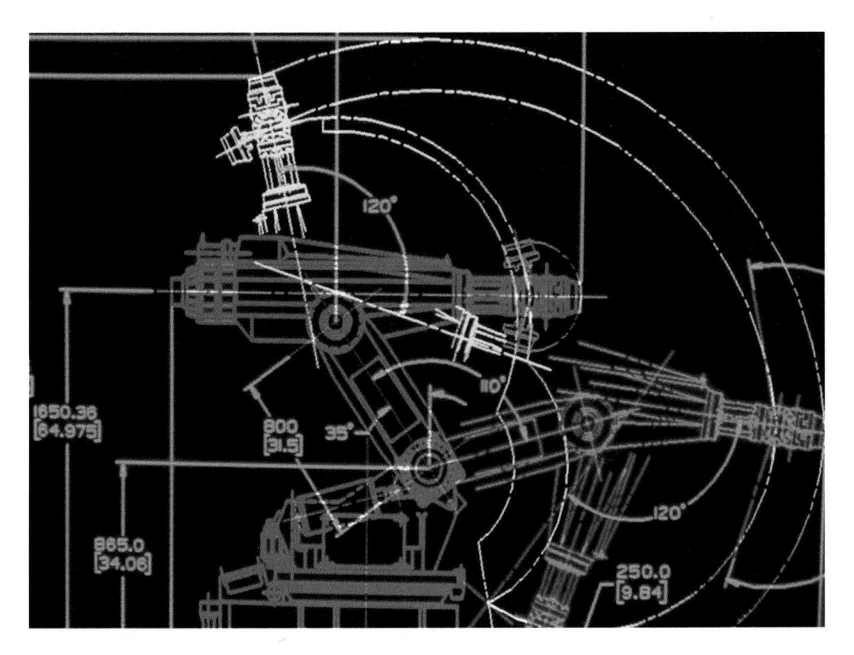

Software in businesses

Students in business schools, and managers of companies, continually face questions concerning the profitability of deals in which costs, quantities and currency exchange rates are in continuous fluctuation. The traditional method of dealing with such problems is to use a spreadsheet, in which the various elements in the calculation are laid out in a systematic manner, normally in cells on a grid, showing the logic of the analysis. Some cells might show unit costs, others might convert them to a single currency. Total costs and sales show in other cells and their difference, the profit, in yet another.

These paper calculations are obviously cumbersome; if one element changes, it may be necessary to recalculate the value of every cell. It is tedious and time-consuming but it must be done, since "what if..." questions are the stuff of management and must be answered.

Daniel Bricklin and Robert Frankston sought to relieve the tedium of answering business school questions by finding a way to place the spreadsheet on a computer. To use their program a rule is provided setting the value of each cell in the calculation. For some cells, as for component prices, the value is simply keyed in; for others, the cell value is the sum, difference, or more complicated combinations of the values in certain other cells. Any value rule can be used.

Using this system, derivative questions, such as what happens to revenue if there is a change in exchange rates, become trivial; the entry for a single cell is changed and all the others are modified as necessary by the program. Bricklin and Frankston sold this program under the name of Visicalc; its successors are in use around the world.

Data capture

A curious commercial competition has emerged in the supply of material to databases using full text. Many current books, and all older material, are available.only in printed form. To be put into a database they must be read from text and put into a computer-compatible form.

There are two ways to treat the problem. It is possible simply to retype all the text at a terminal. This is a laborious task, especially as newspapers are seeking to extend their record backwards in time. This work is mostly done in countries such as Taiwan or the Philippines, where the operators may not understand the language they are keying.

The alternative technique is to scan the text electronically, with a device capable of identifying the printed characters. This method has long-term potential, but costs are currently comparable to those for keyboarding text manually. Machines require sharply-defined characters to read unambiguously, and the text must be precisely aligned; they also may have to cope with a change of font within a page, if the text uses italic, bold or other typefaces. Machines are beginning to be able to deal with these problems, but a good deal of manual labor is needed in presenting the text to the machine in a manner that will keep the error rate to an acceptable level. The text often has to be cut into single-column strips, a practise that inhibits the copying of rare or valuable books by this method.

See also
Introduction to Measuring 9–22
Measuring with Electricity 41–46
Seeing beyond the Visible 73–86
Digital Recording 97–100
Computer Hardware 115–122

Expert systems

A new level of complexity in modern software is the treatment of problems that have no direct analytic solution. Such problems call for judgment drawn from experience, and the people to whom they have conventionally been referred are described as "experts". Typical of such problems is that of medical diagnosis. The medical practitioner asks a few questions, assesses the urgency of the patient's needs, makes an initial diagnosis and checks it with tests. If no firm diagnosis can be made, conservative action and more tests may be called for. The essential point is that throughout this process conclusions must be drawn from information that is frequently incomplete, and the practitioner has to draw on experience of what is likely in the circumstances.

Software is now available which can assess problems in this manner. To date these "expert systems" deal mainly with simpler issues than medicine – war-games for example – but within certain areas of diagnosis they have already shown their potential. Such systems are among the most advanced current projects in software writing, and require a quite different approach to the yes/no algorithms of simpler programs. They exploit their knowledge of a specialized subject area to reduce the number of options under consideration to those which experience shows are likely solutions. The earliest expert systems were produced in the late 1960s to deduce the structure of organic molecules. Games-playing programs can now be designed to learn from experience, rather than simply analyze the position a set number of moves ahead.

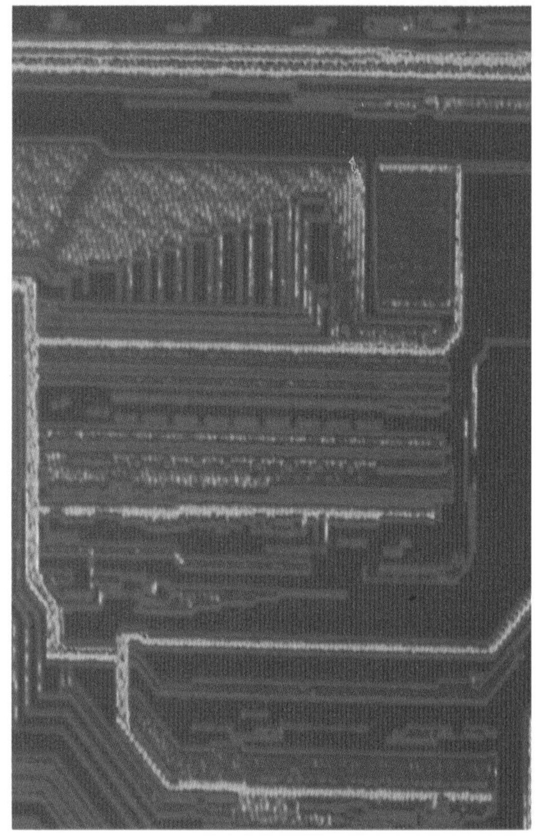

◄ Micros in a classroom may run their own educational software or may be connected to a mainframe which presents information to all but lets students work individually.

▲ Software is usually intangible, but can be made visible – this micrograph shows the flow of electricity through a microprocessor chip while in use; the colors key to the voltage strength.

Computers playing chess

Chess is a game of war. The combat is ritual, superficially archaic, and is conducted according to a rigid set of rules. Its attraction is that no strategy ensures that the first player may win or draw, and the number of available moves is huge; it has been estimated that there are in the region of 10^{120} possible positions of the pieces.

To software enthusiasts, chess has long offered a challenge: can a program be written to ensure a win against a top-class human opponent?

Early attempts to write chess-playing programs were somewhat unsuccessful, but more recent programs incorporate strategies derived from experience of play and have set high standards. At the World Chess Federation Congress in 1985 it was agreed to allow a team of chess computers to play in the Chess Olympiad.

A Computer Rating Agency has been started by the US Chess Federation. This seeks to measure the strength of a chess-playing program by having it play against humans of rated ability. Programs have been written with ratings equal to those of very strong club players. The programmers will not be satisfied until they can face up to a world champion; but in 1986 the world champion Gary Kasparov, playing blindfold, was able to beat ten computers simultaneously.

Engines

The origins of the steam engine...The steam turbine... The internal combustion engine...The gas turbine and jet engines...PERSPECTIVE...The earliest steam engines... James Watt and the classic steam engine...The Stirling engine...The diesel engine...Rockets

Engines are machines that convert heat energy into mechanical work. There are two main varieties: internal combustion engines, in which the fuel is burned within the working cylinder, and external combustion engines in which it is burned outside. The most important example of the external combustion engine is the steam engine, where the energy of hot steam is converted into work by moving a piston in a cylinder. Steam engines provided the driving force for the Industrial Revolution and in the 19th century they were used to power machinery of every sort, including factory tools, mine pumps and ships, trains and cars. The basic steam engine as it developed in the 18th century made use of the expansion of water as it turns into steam and its contraction as it condenses back to water. The alternation of expansion and contraction within a closed cylinder drove the piston up and down, and this motion could be used to power a pump, or converted to a rotary motion by means of a flywheel. Improvements to the basic design were made throughout the 19th century, allowing greater mechanical efficiency and flexibility of use.

The steam boiler is a closed vessel supplied with water that is boiled to provide steam. Steel-making technology in the later 19th century allowed boilers to be built that could withstand the pressures needed for superheating steam (above the temperature at which water normally boils); this was usually done by passing the steam through narrow tubes in the boiler casing. Steam at high temperatures and pressures can be used more than once, and "double-" and "triple-expansion" engines were developed in the 19th century. The superheated steam entered a small high-pressure cylinder, and the exhaust was passed to the larger second or third cylinder, its temperature and pressure falling as it passed through the system. The pistons were attached by cams to a single driveshaft. These engines proved light but powerful, well suited to marine purposes.

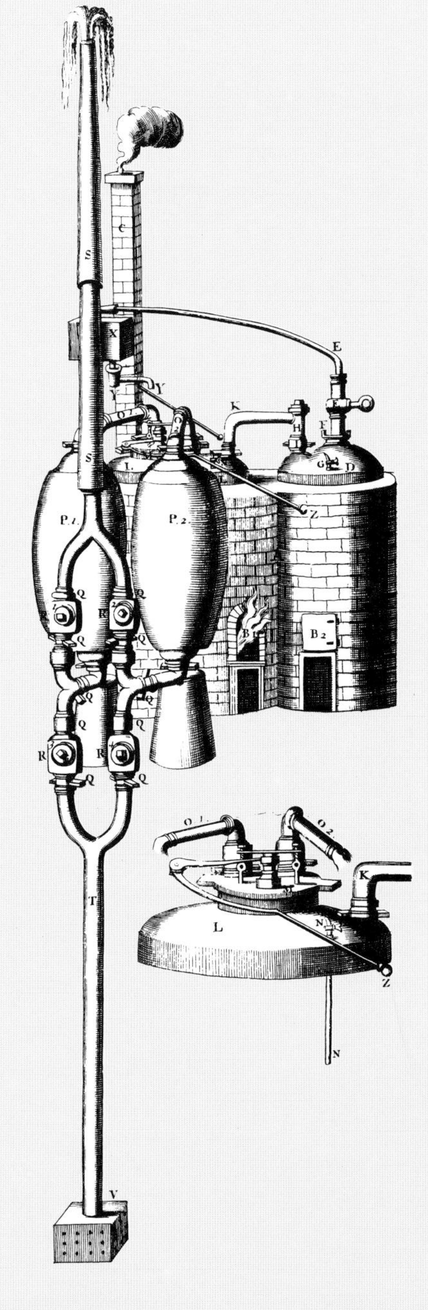

▲ Thomas Savery's steam pump of 1698 had no piston but can be considered the first effective external combustion engine. It had double furnaces with boilers, from which steam passed to the two cylinders. Water was then allowed to play onto the outside of the cylinders, cooling and condensing the steam. The ensuing partial vacuum caused water to be drawn up the pipe through a one-way valve, thus enabling the engine to be used as a mine pump. The engine was not fitted with a safety valve.

The earliest steam engines

The first practical steam engine was patented in 1698 by the English engineer Thomas Savery (c. 1651–1715), adopting principles originally advocated a few years earlier by the French inventor Denis Papin (1647–c.1712). Papin, a colleague of Christiaan Huygens and Robert Boyle and inventor of the pressure cooker and safety valve, had hoped to build a boat driven by an engine that employed the expansion and condensation of steam to move a piston within a cylinder. He built a prototype which he demonstrated in Marburg, Germany, in 1690, but found little support for his scheme. Savery's design used the vacuum created when steam was condensed in the cylinder to lift water in a pump shaft. The passage of steam from the boiler to the working chamber had to be controlled manually by a stopcock which required constant attention.

Savery's engines found their main use in the tin mines of Cornwall, and the design was improved for this purpose by the Devon ironmonger Thomas Newcomen (1663-1729), whose "fire engines" were installed in mines throughout Britain. Newcomen added a piston attached to a heavy overhead beam; the weight of the beam pulled the piston up, drawing steam into the cylinder. An automatic valve closed off the passage from the boiler and brought cold water to play in the cylinder, condensing the steam and creating a vacuum in the cylinder which pulled the piston down again to complete the cycle. Newcomen's engines were still costly in fuel as the cylinder had to be reheated for each stroke.

Chronology

1st century AD Hero of Alexandria designed device to turn steam into rotary motion
1690 Denis Papin of France described atmospheric steam engine using a cylinder piston
1698 Thomas Savery of England patented atmospheric steam engine design
1712 Thomas Newcomen's practical atmospheric engine entered service at Dudley Castle, Staffordshire, England
1718 Self-working valve gear mechanism invented by Henry Beighton
1769 Condensing steam engine patented by James Watt
1781 Epicycloid gear train patented by Watt
1782 Watt's double-acting steam engine patented
1787 Flyball governor invented by Watt
1791 Gas turbine engine built in England by John Barber
1799 Slide valve invented by William Murdock
1804 Iron-cased rockets with detachable sticks developed in England for use by the military
1811 Arthur Woolf's double-expansion or compound steam engine built
1816 Closed-cycle external combustion engine designed by Robert Stirling
1825 First practical horizontal steam engines built
1840 Spin-stabilized rocket developed by William Hale in Britain
1849 Valve gear for close control of steam engines patented by G.H. Corliss in USA
1860 Etienne Lenoir of France patented water-cooled gas engine and spark plug
1867 Nicolaus Otto and Eugene Langen build gas engine

The classic steam engine

The most important improvement in steam engine design was brought about by the Scottish inventor James Watt (1736-1819). In 1769 Watt introduced a separate condenser, a modification that separated the heated and cooled parts of the work cycle and allowed the cylinder to be kept at a constant temperature. Then, in 1782, he added a steam inlet at each end of the cylinder, developing the "double acting" engine, with the piston driven in both directions by steam power. His later engines, made in great numbers by the firm of Boulton & Watt, had two further modifications that allowed them to be used to drive factory machinery: like Newcomen, Watt used the piston to move a beam, but he converted the beam movement to a rotary motion by means of a flywheel driven by epicycloid or planet gears; and he developed the so-called fly-ball governor, one of the earliest automatic feedback mechanisms. As it revolved, centrifugal force threw the balls outwards, and a linkage caused a valve in the steam inlet pipe to close, thus slowing the engine. This device kept the engine operating at a steady speed.

Watt's fundamental design was used with many modifications through the 19th century. The most important development was the move to a compound engine using pressurized steam, a development that Watt himself had resisted as he had not believed a sufficiently strong boiler could be built.

▲ Watt's double-acting engine had a vertical cylinder (left) and piston driven by steam alternately from top and bottom. The cylinder was kept at a constant temperature as the exhaust steam passed to the separate condenser (below). This machine was about four times as efficient as Newcomen's engine of 70 years previously.

The steam turbine

Although the conventional steam engine became more and more complicated with refinements introduced to adapt it to particular needs, it was never able to convert more than 12 percent of the latent energy of the fuel into mechanical power. In the 1880s Charles Parsons (1854-1931) designed a form of steam-driven engine that was much simpler in principle – as simple, in fact, as the windmill – and more efficient. In the steam turbine, the movement of the steam is used to rotate vanes attached to a central shaft; the drive is therefore a directly rotary one. The first turbine demonstrated by Parsons worked at 18,000 revolutions per minute.

The modern steam turbine normally runs at much lower speeds (usually 1,000–3,000 revolutions per minute). It has alternating rows of rotating and stationary blades, with sliding seals between the tips of the rotating blades and the outer casing, and between the tips of the stationary blades and the rotating shaft, so that very little of the steam's energy is wasted. Steam enters at the center and flows outwards in opposite directions through the turbine blades. The cylinder is emptied again by a condenser, which produces a partial vacuum and sucks the steam out. Large turbines may have three cylinders, with the steam entering the first at high temperature and pressure (perhaps at 565°C and $1{\cdot}6\times10^7Nm^{-2}$); it is reheated before entering the second cylinder, then gradually cools and loses pressure as it passes through the second (intermediate pressure) and third (low pressure) cylinders.

1867 Water-tube boiler for raising steam patented in USA by G.H. Babcock and S. Wilcox
1871 Triple-expansion steam engine introduced
1876 Otto's four-stroke gas engine demonstrated
1878 Two-stroke engine built by Dugald Clerk of Scotland
1884 Internal combustion engine capable of 800rpm built by Gottlieb Daimler
1884 C.A. Parsons introduced steam turbine
1887 Float-feed carburettor invented by Edward Butler
1888 Parsons' turbine used to drive electricity generators
1891 Dugald Clerk introduced supercharging
1892 Rudolf Diesel patented compression-ignition engine in Germany
1895 Light high-speed gasoline engine, capable of 1,500rpm, invented by Albert de Dion and Georges Bouton in France
1900 Gearboxes with three forward and one reverse gear introduced by Packard in USA
1903 Russian teacher K. Tsiolkovskii published seminal ideas on rocketry
1926 Liquid propellant rocket designed by R.H. Goddard flown for first time in USA
1930 Liquid-oxygen- and gasoline-fueled rocket tested by Hermann Oberth in Germany
1933 Solid/liquid propellant rocket, GIRD 09, flown in Russia built by M. Tikhonranov
1939 He-176 rocket-powered aircraft flown in Germany
1942 A-4 rocket, designed by Wernher von Braun, flown at Peenemünde, Germany
1947 Two-stage rocket, "Bumper", flown by Holgar Toftoy in USA

The Stirling engine cycle

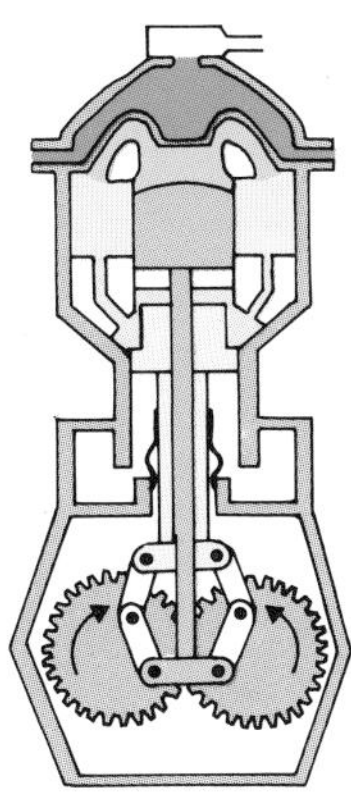

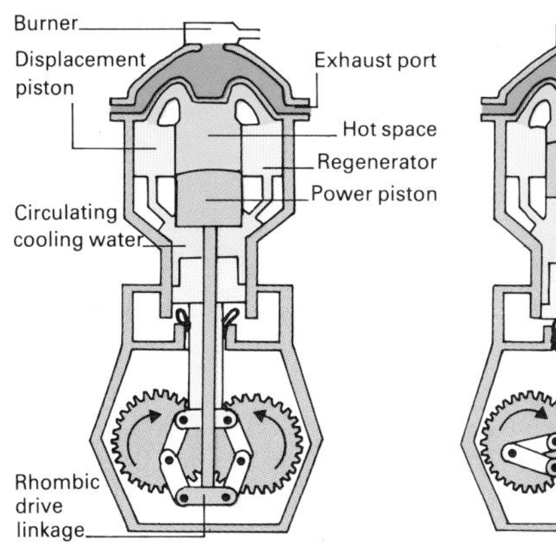

The Stirling engine

In 1816 the Scottish clergyman Robert Stirling invented a free-piston or hot-air engine. This depended for its power on the expansion and displacement of air (or other gases) inside a cylinder heated by external combustion. Stirling devised a cycle of heat transfer that was ingenious and economical; indeed the "Stirling cycle" comes closer to being thermodynamically ideal than any other practical engine. The design was ignored for many years because of constructional problems, but it has recently aroused renewed interest for its quietness, cleanliness and high efficiency. The working gases (hydrogen, helium or nitrogen) are sealed within the system and are heated and cooled by being passed through an external heat exchanger. Heat is stored in the regenerator as the gases move towards the cold space, and is subsequently released as it returns to the hot space. The engine consists of two enclosed pistons oscillating up and down, the displacer piston leading the power piston by 60-90° in the 360° cycle.

The gases trapped beneath the power piston move it upwards, compressing the gases above. As the pressure equalizes above and below the displacer piston (which has a larger surface area on its upper face), it starts to move down under the influence of the greater force from above. As the displacer piston moves down, externally heated gases enter above it, maintaining nearly constant temperature during expansion. In doing so the gases below are compressed and the power piston is forced down. The displacer piston then moves up under the influence of the gases trapped above the fixed stem, which act as a gas spring (as do those beneath the power piston). The gases above the displacer piston are forced back through the regenerator and cooler, completing the cycle. The net power produced results from the difference in work done between expanding high temperature gases and compressing lower temperature ones.

◄ C.A. Parsons' boat "Turbinia" brought turbine power to general attention by reaching 35 knots at the Queen Victoria's Jubilee naval review of 1897.

▲ The blades of a modern power station turbine. Steam enters the turbine where the blades are smallest, and is forced to the larger end blades.

The internal combustion (IC) engine emerged during the late 19th century as an alternative to steam power. In an IC engine the fuel is burned inside the engine, as the name implies. The cannon is in a sense a "single stroke" IC engine, and several early investigators, including the Dutch physicist Christiaan Huygens (1629-1695), experimented with gunpowder as a fuel for driving a moving piston. In designing a practical IC engine, there were two critical problems: to find a suitable fuel and to devise a reliable means of igniting it on a repeatable cycle. The early availability of town gas provided an answer to the first problem. The second proved more intractable.

The first successful gas engine was built by Etienne Lenoir (1822-1900) in Paris in 1860. It was an engineering success but a commercial failure, and was soon replaced by the popular "Otto" engines of the German engineer Nikolaus Otto (1832-1891), introduced in 1876. Gas engines soon became an ecomonic alternative to steam engines for the power needed to supply smaller factories, but the future of the IC

▲ *Otto's horizontal gas engine of 1876. A mixture of gas and air was drawn into the water-cooled cylinder via a slide valve and ignited on the return stroke by a flame brought in, through another slide valve, from a continuous gas jet.*

The internal combustion gasoline engine

The four-stroke cycle

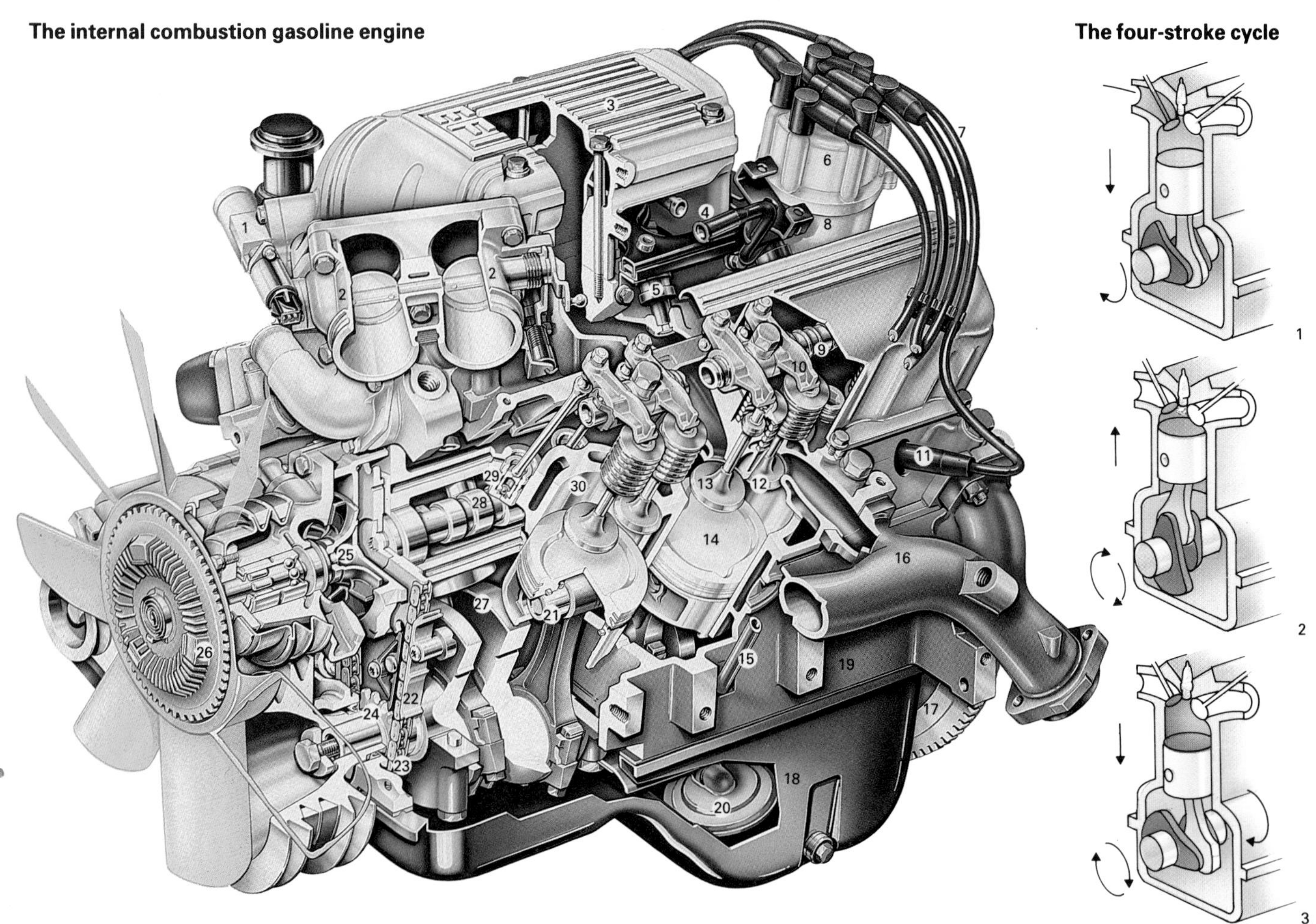

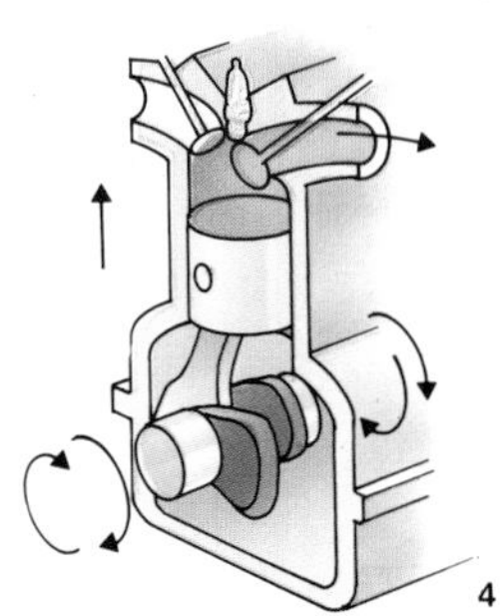

▲► *A cutaway of a modern V-6 six-cylinder engine, manufactured by Ford. Each cylinder fires in turn. The petrol pump draws fuel into the carburettor, where it is mixed with air and passed into a cylinder via inlet valves, which open as the camshaft (to which all the pistons are attached) pulls the piston down (1). On the compression stroke (2), the valves close and the mixture is compressed. The distributor, driven from the main camshaft, passes an electric charge to each cylinder in order, timed to coincide with the end of the compression stroke.A spark is generated at the plug, and this ignites the mixture, forcing the piston down in the power stroke (3); as the piston returns, the outlet valve opens and the exhaust escapes (4).*

1 Idle speed control valve
2 Throttle body
3 Air intake chamber
4 Fuel rail
5 Fuel injector
6 Distributor cap
7 High tension lead
8 Distributor
9 Rocker shaft
10 Rocker
11 Spark plug connector
12 Exhaust valve
13 Inlet valve
14 Piston
15 Exhaust recirculation valve
16 Exhaust manifold
17 Flywheel
18 Sump
19 Cylinder block
20 Oil intake pipe strainer
21 Piston pin
22 Timing chain tensioner
23 Timing chain
24 Crankshaft gear
25 Water pump
26 Thermo-viscous fan
27 Crankshaft
28 Camshaft
29 Hydraulic tappet
30 Cylinder bore

The diesel engine

In 1893 the German Rudolf Diesel (1858-1913) described his concept of an engine that ignites the fuel by compression and uses a heavy oil as fuel. As air is compressed to a high pressure, it becomes hot enough to ignite the fuel, which is sprayed into the cylinder at the end of the compression stroke. Combustion actually takes place as the piston moves down on its power stroke, avoiding the sharp increase in pressure produced by the spark ignition of the Otto cycle.

The diesel engine has to be heavier and bulkier than the gasoline engine to accommodate the high pressures, but it is reliable and economical and has a higher thermal efficiency than the gasoline engine. It has been a great commercial success for large transport vehicles on road and rail. It is also used for private cars in which economy is of a higher priority than performance.

engine was to be not with town gas but with the new fuel, gasoline.

Of all the new forms of transport devised during this period, only the railways and ships have the sheer size to accommodate the massive boilers required by steam engines. Although some steam cars were produced, individual forms of transport needed a lighter but powerful engine that would carry its own fuel in a safe, practical manner.

Oil-based bituminous products had been available from antiquity, but it was the expansion of the United States westwards towards the Pacific that led to the exploitation of the now readily-available sources of crude oil for the production of kerosene (paraffin). However, the most volatile fraction of this oil, petroleum, was regarded as a waste product until it was found to be an ideal fuel for the IC engine (▶ page 240). The invention of the carburettor (one of technology's most undervalued inventions) effectively matched a light, powerful and robust engine to a suitable fuel. The results have changed the world more than any other single technological development.

The diesel cycle

The diesel engine

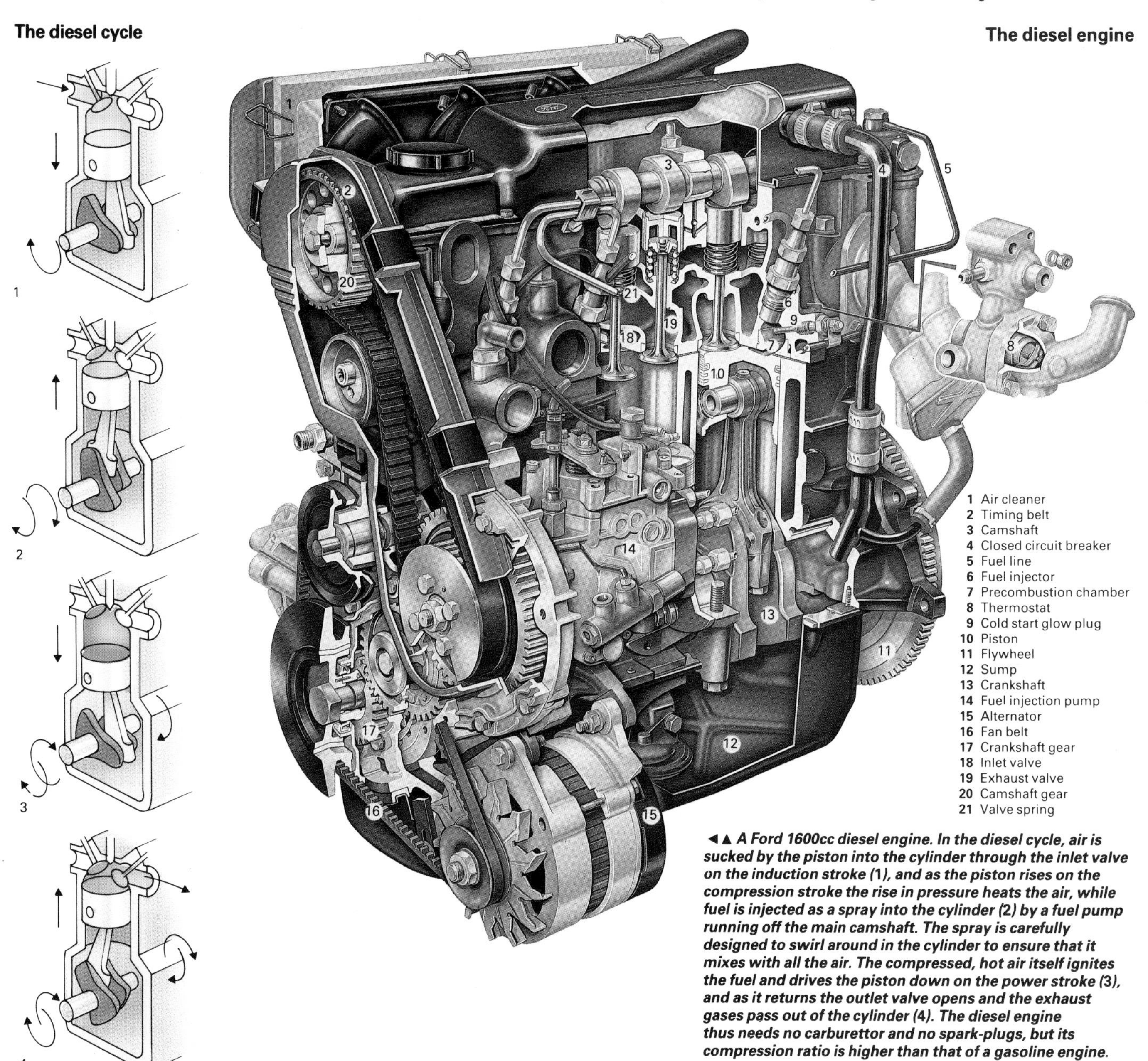

◀▲ A Ford 1600cc diesel engine. In the diesel cycle, air is sucked by the piston into the cylinder through the inlet valve on the induction stroke (1), and as the piston rises on the compression stroke the rise in pressure heats the air, while fuel is injected as a spray into the cylinder (2) by a fuel pump running off the main camshaft. The spray is carefully designed to swirl around in the cylinder to ensure that it mixes with all the air. The compressed, hot air itself ignites the fuel and drives the piston down on the power stroke (3), and as it returns the outlet valve opens and the exhaust gases pass out of the cylinder (4). The diesel engine thus needs no carburettor and no spark-plugs, but its compression ratio is higher than that of a gasoline engine.

See also
Electricity 135–140
Rail Transport 193–198
Road Transport 199–208
Air Transport 217–224
Space Transport 225–228

The jet engine

The gas turbine or "jet" engine is to the petrol engine what Parsons' steam turbine was to the traditional steam engine: it transformed a mechanically complex reciprocating engine into a much simpler, more powerful rotary engine. The jet was developed during the Second World War and by 1945 both the Germans and the British were flying jet-powered fighter planes. Jet engines entered civil aviation with the de Havilland Comet in 1952 (➧ page 220) and have dominated it ever since, whether in pure jet form, as "turbofans" or as "turboprops" (in which part of the thrust is used to drive a propeller).

The jet engine, like the steam turbine, consists of alternate rows of stationary and rotating blades through which a working fluid passes. The rotating blades are attached to a central drive shaft. In the simplest form of jet engine, the working fluid is air which is taken in and compressed by the front-end compressor turbine. It is then mixed with kerosene and ignited in the combustion chambers surrounding the central shaft. The hot gases pass through another set of turbine blades to exit from the engine at high velocity. It is the thrust generated by these exhaust gases that provides the motive power of the jet engine, with a small fraction of the thrust being used to power the front-end compressor via the central shaft. In the "turbo-prop", additional turbine blades extract extra energy from the exhaust gases and convert it into rotational shaft energy to drive a conventional propeller. The turbofan, which is used to power the largest aircraft, has a large fan in the air intake which channels air around the outside of the compression chamber; this provides about 75 percent of the total thrust of the engine. A variation on this design, tested in 1986, is known as the propfan. This has the fan blades *outside* the engine casing. Whatever its form, the gas turbine is at the same time one of the simplest and yet most sophisticated sources of motive power.

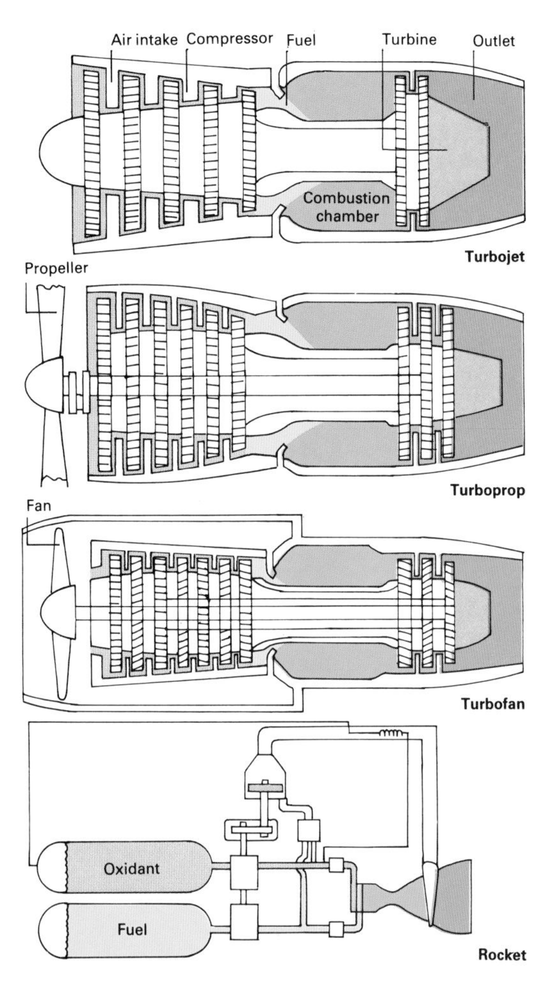

▲ The turbojet takes in air, which is compressed before passing to the ignition chamber where it mixes with fuel and ignites. The turboprop also has a propeller driven by the exhaust, whereas the turbofan uses a fan to take in extra air and force it into the tailpipe for more thrust. A liquid-propellant rocket uses turbopumps in the preburners to inject fuel and oxidant into the combustion chamber.

Rocket motors

The most powerful propulsive motors of all are rockets. They burn a fuel to produce a hot gas which leaves the combustion chamber at high speed through a carefully directed orifice so producing a thrust in the opposite direction. The possibility of using rocket power for transport was first proposed by the Russian K.E. Tsiolkovskii (1857-1935). The pioneering work was done by the American Robert H. Goddard (1882-1945) who launched the first liquid propellant rocket in 1926.

The operation of today's vast multistage rockets remains in principle similar to that used by Goddard. Two liquified gases are usually used, an oxidant and a fuel (often hydrogen). These are pumped by a turbine to the combustion chamber where they are ignited and the gases expelled from the exhaust to create thrust. The thrust is controlled by opening or closing a valve from the fuel tank to the combustion chamber. Some rockets (including the booster rockets used to lift the Space Shuttle ➧ page 226) are powered by a solid fuel, in which the oxidant and propellant are already mixed.

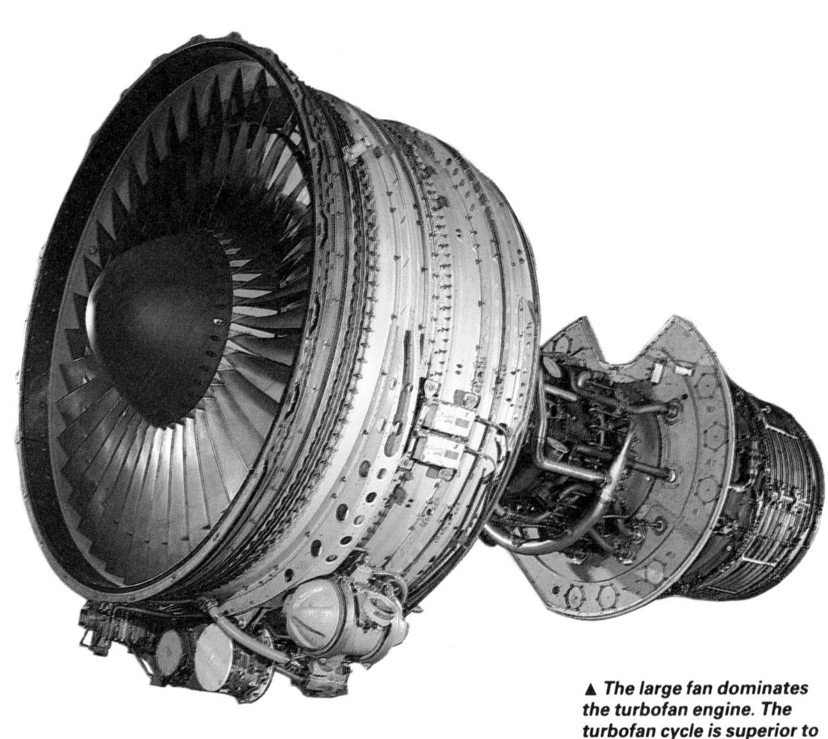

▲ The large fan dominates the turbofan engine. The turbofan cycle is superior to other designs at subsonic speeds and provides good thrust at takeoff and while climbing.

Electricity

Electric power...Storing electricity: batteries and cells...The motor and the generator...Commercial generators...Transmission...PERSPECTIVE...Fuel cells... Faraday and the discovery of electric power...Early uses of electric power...The development of electric light... The social costs of electricity

Electricity is the most convenient and flexible medium used to transport power from its source to where it is needed. Whereas in previous centuries it was necessary to set up inefficient mechanical links between, say, a waterwheel and a millstone, today the power of the wheel can be converted into electricity, transported instantly and with minimal losses to where it is needed, and turned back again into power. There are four major elements in this system: the power source, the generator which it drives, the transmission line, and the motor which converts the electricity back into power.

Batteries

Small, portable or remote electrical devices may not be suited to the permanent connections available with mains electricity, and for these an electrical cell (or battery, if two or more cells are wired together) is used. This relies on the flow of electrons between two different elements (known as electrodes, the positive one being the anode, the negative one the cathode) set in an electrolyte, or electron-freeing substance. It thus provides electricity by exploiting a chemical reaction (◀ page 41). The cell was demonstrated by Alessandro Volta (1745-1827) in 1800, and it became a viable commercial device after 1868, when the French physicist Georges Leclanché patented a cell using a carbon anode, zinc cathode and sulfuric acid electrolyte. A dry version of the Leclanché cell substitutes ammonium chloride paste for the acid. These cells form the basis of the batteries used in many devices today, although a car battery has electrodes of lead and lead oxide, and unlike a Leclanché cell it can be recharged. For operations where continuous use is required, zinc-mercuric oxide batteries are available, and ultra-light batteries can be made from lithium, with a lithium iodide electrolyte. These batteries are used for such purposes as powering heart pacemakers.

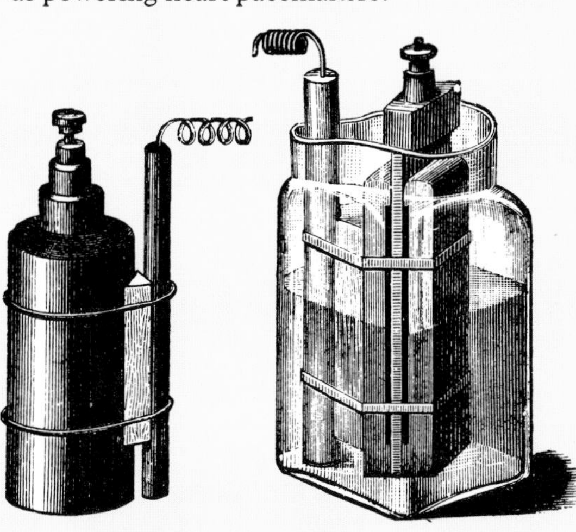

► The Italian physicist Alessandro Volta showing his electric pile or battery to Napoleon. This consisted of disks of copper and zinc separated by layers of felt soaked in acid, and was the first device to give out a continuous flow of electricity.

◄ Leclanché cells of the 19th century. If the terminals are connected, the zinc cathode dissolves and gives up electrons which flow via the wire to the carbon cathode to form ammonia and hydrogen ions with the ammonium chloride electrolyte.

Fuel cells

A fuel cell is a device that converts the chemical energy of a fuel to electricity in a highly efficient and environmentally acceptable manner.The principles of operation are similar to those of batteries, except that fuel cells do not need to be charged and do not discharge like batteries. They use the chemical potentials of two substances, an oxidant and a fuel, that will react with each other to create electricity. Many fuels and oxidants have been tried in fuel cells since they were first proposed in 1839, but the most practical fuel is hydrogen, and the most practical oxidant oxygen. Many types of fuel cell have been developed, such as alkaline, molten carbonate, solid oxide, phosphoric acid and even solid polymer electrolyte (SPE). However, only phosphoric acid electrolyte fuel cells are available for commercial use.

The conversion efficiency of fuel cells is very high, whatever the size of the plant and at loads from 25 percent to full load. Electrical efficiencies as high as 90 percent are theoretically possible; in practise, today's fuel cells are capable of 50 percent efficiency with phosphoric acid fuel cells at atmospheric pressure, and up to 60 percent with pressurized systems.

A fuel cell can reach full power in 30 milliseconds, a response that cannot be matched by any other form of power plant.

Fuel cells are reliable because they are essentially simple. The fuel cell itself has no moving parts and therefore the stack requires little maintenance. They have a projected life of 40,000 hours or more of operation at full power. Fuel cells have been used as the power source in spacecraft, and power stations have been built using fuel cells in the United States and Japan (▶ page 155).

Chronology

1800 Alessandro Volta invents battery, first source of continuous electric current
1820 Hans Oersted discovers principle of electromagnetism
1821 Michael Faraday develops principle of DC motor
1836 Daniell cell, first efficient battery, introduced
1858 Arc lighting introduced on lightship at North Foreland, Kent, England
1868 Leclanché cell, prototype of dry battery, developed
1873 Zénobe Théophile Gramme develops first viable electric motor
1878 First practical filament light demonstrated by Joseph Swan in England
1879 Thomas Edison shows his filament light design
1881 Lead-acid accumulator introduced
1882 First public power stations open in New York and London
1886 George Westinghouse and William Stanley transmit electricity for more than 100 meters in Massachusetts, using Stanley's AC transformer
1888 Nikola Tesla designs AC motor
1891 First 3-phase transmission of AC power takes place between Lauffen and Frankfurt, Germany
1892 First pumped-storage power station opens near Zürich
1895 Electric hand-drill invented by Wilhelm Fein in Germany
1901 Electric typewriter developed
1903 First alternator with rotating field magnet installed at Newcastle upon Tyne, England
1905 Tungsten filament lamp introduced
1908 Commercial vacuum cleaners introduced in USA by W. Hoover

Generators and motors

If a loop of wire spins about an axis which lies across the loop and across a magnetic field, a current is induced in the loop which alternates with the changing direction of the field traversing the loop. If a current passes around a loop which lies across a magnetic field, forces are exacted on the loop which tend to make it spin about an axis which lies across the loop and across the field. These two physical laws provide the basis for a whole series of machines in which electrical power is converted to mechanical power by a motor and mechanical power is converted to electrical power by a generator.

Many geometries are possible. Simple motors and generators, such as are found in battery-driven toys or bicycle dynamos, use permanent magnets to produce the magnetic fields, and reversible connectors called commutators to make motors rotate continuously and generators to produce a continuing current in a single direction (direct current, or DC).

Larger machines use electromagnets to produce their magnetic fields. They generate alternating currents and, in turn, use these to drive alternating current (AC) motors. A key idea in AC machinery is that, by using alternating currents to energize electromagnets disposed in the correct way, it is possible to create a volume in which the magnetic field rotates just as if it lay between the poles of a rotating permanent magnet. When a coil fixed to the shaft of a motor is put into this volume, currents are induced which react with the magnetic field to produce forces that tend to rotate the coil so as to follow the field. Motors like this are called induction motors.

A linear induction motor resembles the rotary motor except that its magnets have been "flattened out" so that the motion produced is linear; one major modern application for the linear induction motor is in "maglev trains" (▶ page 198).

AC Generator

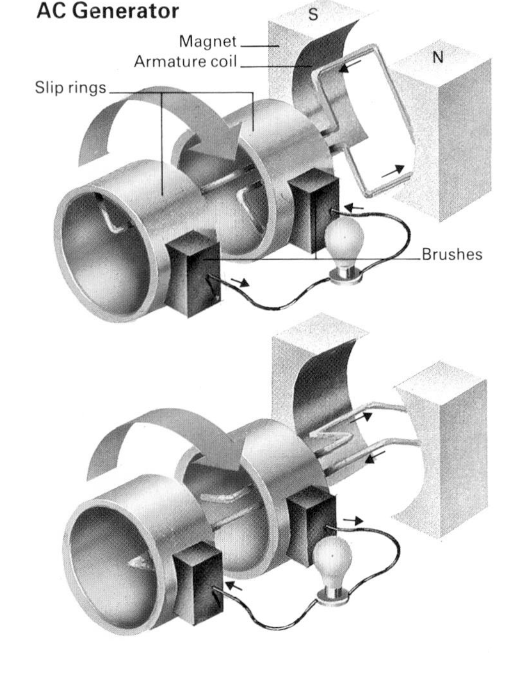

DC Generator

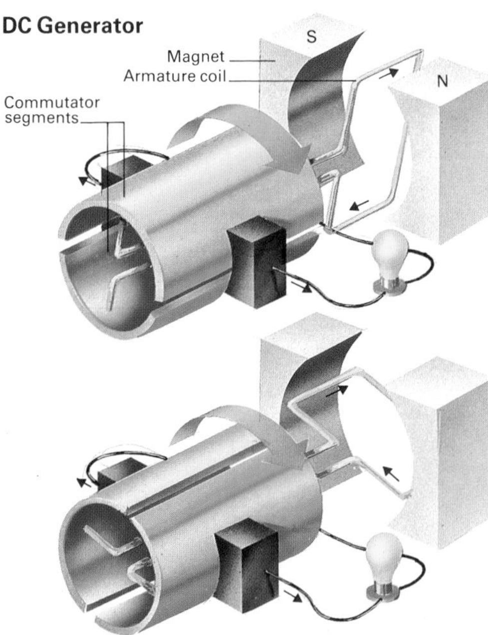

◀ *In the schematic alternating current (AC) generator a single wire is turned through the field of a stationary magnet, and is connected to a circuit by means of slip rings and brushes. A current is generated which reverses each time the wire passes from one pole of the magnet to the other.*

◀ *In the DC generator and motor, the coil is connected to the circuit by means of a commutator. This is split in such a manner that as the current reverses in the coil, the opposite half of the commutator connects with each terminal of the circuit, thus generating a single-flow current or a smooth rotary motion.*

1908 Electric washing machines available in USA
1913 Geothermal power station opened at Lardarello, Italy
1913 Domestic refrigerators available in USA
1919 Domestic electric cookers available
1920 Three-core power cables installed by Martin Hochstadter in USA, and oil-filled cables by Luigi Emanueli, Italy
1920 Immersion heaters become practical
1923 Photoelectric cell developed
1939 Gas turbine first used in electricity generation, in Switzerland
1941 Electric power generated from wind power by 1·25Mw windmill in Vermont, USA
1947 Microwave ovens available
1954 Photovoltaic cell developed at Bell Telephone Co, USA
1957 First nuclear power plant opens at Calder Hall, England
1957 Mercury cell battery developed
1959 Fuel cells developed in Britain by Francis T. Bacon
1964 Magnetohydrodynamic (MHD) electricity generation achieved in Britain – current produced by passing hot gas through strong magnetic field
1966 Superconducting magnets developed in Britain
1967 World's first large-scale tidal power station opens, at La Rance, St Malo, France
1968 Zinc-air battery developed commercially in Britain
1972 Sodium-sulfur battery developed in Britain
1977 Gallium arsenide solar cell developed by IBM in New York
1986 Aluminum-air battery introduced

▲ *The city of Los Angeles, emblematic of the modern dependence on public electricity supply for power, heating and lighting. Refrigeration and air conditioning are so common that the hottest days can produce an even higher demand for electricity in such cities than cold nights.*

► *Thomas Edison's first successful incandescent lamp (1879) used an element made with carbonized sewing thread; it burned for 40 hours. By the following year he had set up a method of mass production. He advertised the lamps by lighting his laboratories and showing people around in the evenings.*

The discovery of electrical power

It is curious that, although the electric motor is so commonplace, it uses one of the most mysterious forces of nature – electromagnetism. Scientists still do not fully understand what the electromagnetic force actually is, but engineers are extremely adept at making use of what it does. The first demonstration of these abilities was given by the British physicist Michael Faraday (1791–1867). In 1821 he exhibited the first electric motor, consisting of a copper disk rotated between the poles of a permanent magnet. Ten years later he was to demonstrate the principle of induction, on which the generator depends – a principle discovered independently at the same time by the American physicist Joseph Henry (1797–1878).

Early applications

In the years following Faraday's demonstration of the electric motor innumerable attempts were made to apply the device to practical uses. In 1842 a Scottish inventor, Robert Davidson, tested an electric locomotive on the Edinburgh to Glasgow line. It weighed 5 tonnes and reached a speed of 6 km/h. By the mid-19th century battery-powered electric lathes, drills, printing presses, trains and boats had been developed, but none was a lasting success owing to the high costs and low power available. In the 1840s, however, the British chemist John Woolrich developed a huge electric generator which he installed in a silverplating and gilding plant. This device made possible the mass-production of electroplated goods.

Electric light

The earliest electric lights were of the carbon-arc type. In this design the light derives from a continuous spark generated by passing current across a gap between two carbon electrodes. An early example was installed in the South Foreland lighthouse, Kent, England, in 1858, powered by a specially designed generator. By 1880 Billingsgate fish market in London and part of the Thames Embankment were illuminated by arc lights. Similar systems were installed in the United States and in Paris, but by this time the first incandescent lights were already appearing. These have a filament, originally made of vegetable material but today more often made of metal, through which the current is passed. The filament is placed in a vacuum or an inert gas to prevent oxidization.

The simultaneous development of the incandescent carbon filament lamp in the 1870s by the British chemist Joseph Swan (1828-1914) and the American inventor Thomas Edison (1847-1931) was to revolutionize the electricity industry. Edison planned lighting on a grand scale to compete with gas lighting; the Edison Electric Lighting Company was founded in 1878 and their Pearl Street generating station in New York was opened in 1882, one of the first central power stations.

Large-scale electricity generation

Whether it is coal-, oil- or nuclear-powered, a conventional power station's electricity is generated in the same way: the fuel is used to create steam which drives a turbine (◀ page 130). A hydroelectric system, on the other hand, uses a water-driven turbine directly (▶ page 153). This produces a rotary movement which in turn powers the electricity generators, which are essentially large electromagnets that turn within a hollow cylinder of conducting coils, and convert the rotational movement of the turbines into electrical power in the stator (the fixed coils); this is then fed into the national grid system (▶ page 140) as alternating current.

Power stations are notoriously wasteful in their conversion of fuel to useful energy; often some 60 percent of the fuel's energy value is lost, mostly in the form of waste heat within the turbines. The generators themselves are highly efficient machines, losing no more than about 2 percent of the energy. The nature of the conversion process means that economies of scale may be made and power stations have become ever larger, and are now often built with a capacity of over 2,500 megawatts.

Efforts have been made to reduce the wastefulness of the process. Combined heat and power (CHP) schemes have been introduced in some towns to collect the waste heat and supply it direct to industry and homes in the locality in the form of hot water. Meanwhile studies have been made of different forms of large-scale generation avoiding the turbine altogether. Some use fuel cells, others seek to create an electrical current by passing a high-temperature ionized gas through a magnetic field. This procedure, known as magneto-hydrodynamic generation (MHD), has been known in principle for many years and might cut the waste heat by 50 percent, but it has yet to be demonstrated as a commercial means of generating all or part of the electrical power available from a furnace or nuclear reactor.

The modern power station

▲ The turbine generators at a modern 660 MW power station in Britain. Superheated steam passes from external boilers to the high-pressure turbine, then is taken through heat exchangers to be reheated before driving the medium-pressure and the two largest, low-pressure, turbines. The generator is on the right at the end of the turbine shaft. In front is the condenser which takes exhaust steam for cooling and returns it to the boilers.

◀ The generator installed by Thomas Edison at the Paris Opera in 1887. Edison designed the first large-scale generators to supply the lighting industry that he had himself created. Reciprocating steam engines drove the generators via belts.

▶ Generators at the hydro station at Snowy Mountain, Australia. Water turns turbines fitted with variable vanes to control the speed of rotation; the turbines drive vertical generators.

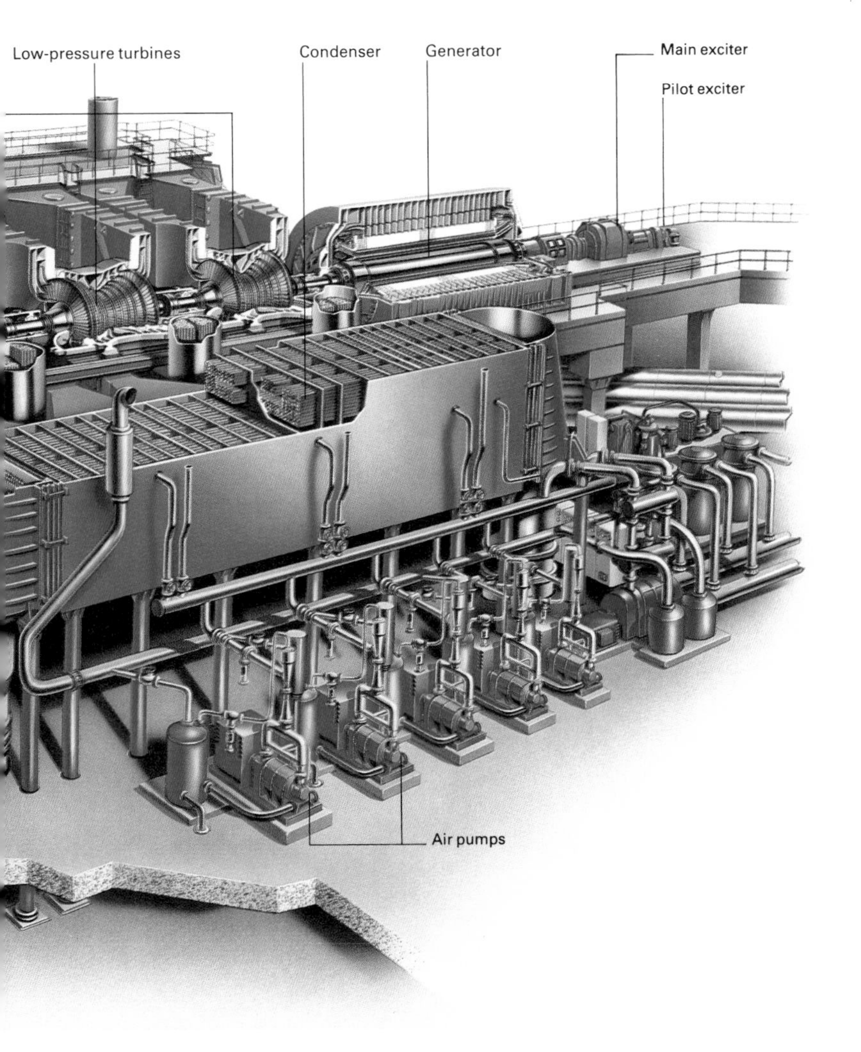

The social costs of electricity

Power is produced at a cost, in environmental degradation, in resource depletion and in human health and safety. The advantages of nationwide electric power are judged to be greater by far than the disadvantages, but great efforts still have to be made to minimize any adverse effects.

The principal byproducts of power stations are waste heat, gaseous effluents and radioactive wastes. Of these, waste heat is the most manageable and can, if properly harnessed, be positively advantageous with the establishment of combined heat and power schemes, serving, for example, fish farms sited near power stations.

Flue gas effluents can be treated in several ways. The principal effluents are smoke, ash and sulfur dioxide (SO_2). The emission of smoke can be reduced to practically nothing by improvements in boiler combustion conditions. Emissions of ash can be effectively controlled by the use of electrostatic precipitators. Much of this ash can be recycled into building materials.

Emissions of sulfur dioxide, however, are a much more difficult problem. Although some sulfur can be removed from the fuel before it is used and some can also be removed from the flue gases, sulfur removal is an expensive business. Most modern power stations, therefore, minimize local effects of SO_2 emission by building very tall chimneys. The plumes from such chimneys can punch their way through the lower atmosphere, but this simply passes the problem to someone else. SO_2 emissions from British chimneys are thought to be polluting lakes and rivers as far away as Scandinavia and Germany with acid rain – rain which is literally dilute sulfuric acid (H_2SO_4). This problem grows as the demand for electricity rises.

See also
Measuring with Electricity 41–46
Analyzing the Atom 109–114
Engines 129–134
Nuclear Power 141–146
Other Sources of Power 149–156

◄ ***Circuit-breakers are fitted in the electrical grid between the transformers and the high-voltage transmission lines. These are designed to act as main on/off switches for the system, and prevent arcing between the terminals. Circuit-breakers operate automatically when there is a significant difference between the current entering and leaving a circuit, or when a preset level of current is exceeded. A transformer consists of two coils of wire around an iron core, one comprising many more turns than the other. An alternating current passing through one induces a current in the other; the voltage of the secondary current is proportional to the first, reflecting the number of turns on the two coils.***

Transmitting electricity

The currents produced by a modern power station are measured in terms of thousands of ampères. Such large currents are unsuitable for transmission over long distances since there would be significant losses in the form of heat in the cables. There are two ways of approaching this problem: by reducing the resistance of the cable, and by reducing the amount of current that is passed along it. The fact that the cable diameters are limited by cost and weight means that their resistance cannot be reduced indefinitely, however high their conductivity. As a result, the output of the power station is transformed from large currents at relatively low voltages to smaller currents at very high voltages (often between 300,000 and 400,000 volts). At these voltages electricity can be transmitted around the country on overhead cables suspended from transmission towers. In urban areas, underground cables may be used, though these are very costly to install and maintain. Overhead cables are usually made of strands of copper or aluminum, and copper-clad steel is used over very long spans such as river crossings. At various grid supply points, the supply is transformed down to about 33,000 volts, and transmitted to intermediate substations where is is brought down to about 11,000 volts. Finally, it is taken to distribution substations where the voltage is reduced to 100 or 240 volts, the potential at which it is supplied to homes and offices. Factories and hospitals may take their electricity at a higher voltage, and the railway system may have its own substations beside the tracks, drawing power from the grid supply points.

A complex network of supply in many countries, with the output of many power stations interlinked and controlled from a central point, allows power to be switched around the country as it is needed. This ensures that switching on or off a major power station does not have a noticeable effect on the supply that reaches consumers, and allows the peaks and troughs in demand to be spread out over many stations.

Alternating current suffers more voltage drop in transmission than direct current, and in some countries the current is converted to DC before transmission, and reconverted back to AC near the point of use; this is done, for example, in the undersea cable link between the British and French electricity networks.

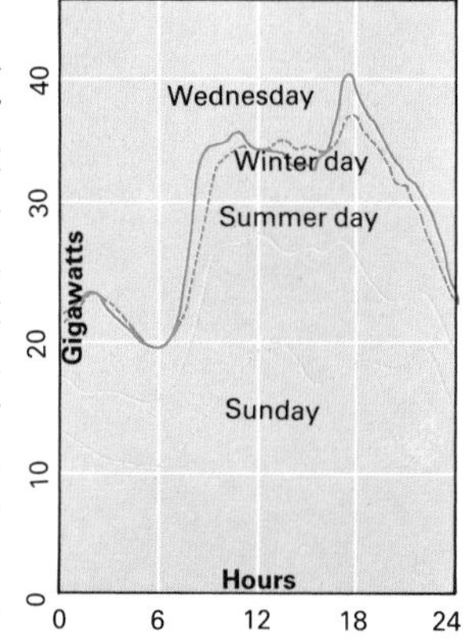

◄ ***Electricity demand varies during the day and through the year; a national supply system can match supply to changing demand at once. The largest national grid system is Britain's, with over 140 power stations.***

▼ ***Steel pylons carrying high-voltage power lines in Illinois, USA. The conductors may be made of a steel core covered in aluminum, with insulation made up of strings of glass or porcelain.***

Nuclear Power

The argument for nuclear power...Splitting the atom... The nuclear fuel cycle...The nuclear reactor...Reactor designs...PERSPECTIVE...The Chernobyl accident...The problems of waste disposal...The pressurized water reactor...Other designs...The CANDU reactor...Nuclear fusion

The world's supply of fossil fuels is limited. In the early 1970s some alarming projections were made according to which, if current rates of economic growth continued, there would be a devastating shortage of fossil fuels in the world by the year 2000. Some of these projections have been shown to be too gloomy. The rise in oil prices in 1974 caused a cutback in demand and economic growth in many industrialized countries while at the same time it encouraged oil companies to seek out and develop what had previously been reserves of marginal viability.

All projections of the world's major energy sources into the 21st century nevertheless emphasize the growing importance of nuclear power. Its expansion has not been as rapid as expected in the 1960s, owing to fears about safety and waste disposal, persistent environmental problems and the massive capital investment required. Despite these objections, which to many people were confirmed by the consequences of the disaster at the Chernobyl nuclear power station in the Soviet Union in 1986, the supporters of nuclear power still claim that it is the cheapest, safest and most economic means available at present of generating electricity on a large scale.

The earliest work on nuclear power was done in the United States in the early 1940s by the Italian-American physicist Enrico Fermi (1901-1954), and nuclear power was first used to generate electricity in the early 1950s in Britain. Since then nuclear power stations have been built in most industrialized countries including Third World nations such as Brazil.

Harnessing the power of the atom

At the heart of each atom is a nucleus which contains a number of positively charged protons and uncharged neutrons. The energy required to bind the nucleus varies with their number. When a nuclear reaction occurs, the number of neutrons and protons in the nucleus alters, thus changing the atom from one element into another. In this process some of the energy binding the nucleus together is released. One such reaction occurs when the heavy element uranium 235 is split apart by bombardment with neutrons. The total mass of the remnants of this rupture is slightly less than that of the original U-235, and the difference is released as energy (following the equation of mass with energy, derived by Einstein). Other neutrons are also released, which in turn bombard nearby atoms and set up a chain reaction. If the process continues unchecked and the uranium is packed densely enough, a nuclear explosion results; in a nuclear reactor, however, the density of the uranium fuel and the rate of reaction are carefully calculated to ensure that it "burns" evenly. The energy given off in the fission process is used to heat steam, which drives the generator turbines (◀ page 138).

▲ In the seven days after the accident at Chernobyl large areas of Europe received significant contamination as rain washed radioactive elements out of the radioactive cloud. Children in Eastern Europe were particularly at risk.

The Chernobyl accident

Just after 1 am on Saturday 26 April 1986 an explosion and fire destroyed the core of reactor number four at the Chernobyl nuclear power plant, north of Kiev in the Ukraine. Radioactivity from the stricken plant was first detected in Scandinavia, over 1,500km away, on the following Monday.

The Chernobyl reactor is of a design unique to the Soviet Union (▶ page 145). Graphite is used as the moderator, and the nuclear fuel is housed in more than 1,600 pressure tubes, with the steam led off directly to turn the turbine generator.

The nuclear chain reaction was shut down in the initial seconds of the accident at Chernobyl, but the fuel rods continued to generate heat through the decay of radioactive atoms within them. The resulting explosion destroyed the cooling system, and the rods set fire to the graphite moderator; the heat of the fire melted the fuel rods. The reactor had no outer containment building and radioactivity was released directly into the environment.

It was a situation without precedent. If water were poured on the flames, further explosions might have resulted, and the water would in any case have contaminated drinking supplies. Finally the Soviet authorities dropped 4,000 tonnes of sand, clay and lead on the core by helicopter. However, this solution gave rise to the possibility of a "China syndrome" type of meltdown, a threat averted by pumping nitrogen gas into the core.

The immediate danger ended two weeks after the accident had begun. Some 30 men died from injuries and acute radiation poisoning in the immediate aftermath of the explosion. The long-term consequences of the accident are more doubtful – many may die prematurely from cancer induced by the extensive radiation contamination.

The nuclear fuel cycle produces dangerous wastes that must be stored for thousands of years

Uranium and the nuclear fuel cycle

Uranium exists naturally as a mineral ore which is extracted by surface or underground mining. The crude ore is then crushed and chemical solvents are added to dissolve out the uranium in the form of "yellowcake" – a mixture of uranium oxides. Only a small part of this (seven parts in a thousand) is uranium-235, the fissile substance needed to produce a chain reaction in most designs of reactor; the rest is the isotope uranium-238. For reactor use it is necessary to increase the proportion of U-235 in a process known as uranium enrichment. This is done by gaseous diffusion after the yellowcake has been converted to uranium hexafluoride ("hex") – a highly corrosive gas. Reactors such as the pressurized water reactor (PWR; ▶ page 144) that use water as the moderator require only a 2–4 percent enrichment in the form of uranium dioxide (UO_2). To produce this, the hex is first converted to a uranyl nitrate solution from which the oxide powder can be made. This powder is then processed into small, dense, cylindrical pellets and packed into rods for use in the reactor core.

After the fuel has been used in the reactor, it contains some valuable materials (which can be recycled) and many dangerous waste products. Some of these "fission products" break down into safe substances fairly quickly but others, such as plutonium-239, remain highly radioactive for thousands of years. These wastes cannot be processed further and must be stored; much effort has been spent on trying to find a safe method of disposal with no danger of leaks or spillage for many centuries. Until such a method has been developed, much of this "high-level waste" remains stored in tanks, while other, less toxic "low-level" wastes may be buried or discharged at sea. The safety of this practise remains highly controversial.

Plutonium may be recovered from the used fuel by reprocessing, for use in breeder reactors (▶ page 145) or in nuclear weapons. Reprocessing is a highly complex, technologically difficult and dangerous operation, carried out by remote control behind heavy shielding. The process produces its own highly radioactive wastes which require disposal.

In addition to the formidable engineering and security problems imposed by this fuel cycle, there is a further worry of fuel transportation. Fresh fuel elements can be shipped in ordinary cases, but once they have been used they are highly radioactive and require heavy shielding: two tonnes of irradiated fuel requires a 50-tonne containment vessel. Fresh reactor fuels are shipped by air, sea road and rail. Irradiated fuel needs remote handling, massive containers and specially protected transport systems.

▲ ***The fuel rods are loaded into the reactor core under controlled conditions. Some reactors have to be shut down for refueling. The layout of the core varies from design to design; here the compact core of a fast breeder reactor (▶ page 144) is under construction.***

▲ ***The ore is refined, and the proportion of U-238 may be enriched. The uranium is powdered and baked into pellet form to be loaded into thin metal tubes – zircaloy or stainless steel – for use as fuel. The slightly radioactive fuel rods are transported to the reactor.***

▶ ***Uranium ore, sometimes 1 percent uranium or less, is mined by underground or surface techniques. Mining is mainly done in North America, southern Africa, Australia and the Soviet Union. Deposits are found by detecting gamma rays with airborne instruments.***

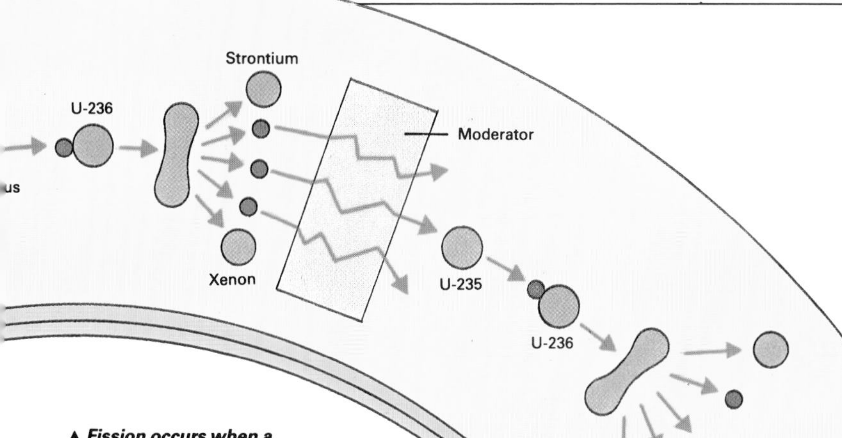

▲ *Fission occurs when a neutron bombards a uranium atom so that its nucleus divides. Spare neutrons fly off and create a chain reaction. The neutrons can be slowed by substances known as moderators, and absorbed by others (such as boron).*

▲ *The spent fuel has to be transported away for reprocessing or disposal, and many fission products, including plutonium-239 and strontium-90, are highly radioactive. The fuel is transported in heavily screened containers by rail or ship.*

◄ *Low-level wastes may include reactor coolant and utensils or clothing. These may be contaminated only slightly or the substances involved may be only mildly radioactive. Quantities of these wastes are stored in tanks, discharged at sea or put in trenches.*

■ *At the reprocessing plant the spent fuel may be taken from its rods, and the uranium and plutonium are dissolved away from the wastes. The plutonium is then separated out and stored, as is the uranium in its oxide form. All this is done by remote control.*

► *The high-level wastes are reduced by evaporation, then stored in special cooling tanks. In some countries they may then be vitrified or fused into pillars of dense glass for final disposal. A French vitrification machine is shown here.*

■ *Reprocessing began in the 1950s through a need to conserve uranium supplies and preserve plutonium. Not all reactor types take fuel in a form suitable for reprocessing. And even reprocessed fuel has large quantities of toxic wastes that need disposal.*

The problems of nuclear wastes

Nuclear power remains highly controversial. Apart from fears about the safety of the reactor itself, there is an overriding problem of what to do with the waste products. Throughout the nuclear cycle, the materials involved are all to some extent radioactive. Neutrons emitted from reactor cores tend to make everything in their vicinity radioactive. During refueling, used reactor fuel must be removed and processed, and some of this spent fuel is highly contaminated with products such as strontium-90 and cesium-137, which remain dangerous for hundreds of years, and plutonium-239 and americum-241, which are not safe for thousands of years. The actual quantities involved are small but, to date, no scientist has come up with satisfactory methods of disposing or containing them for this timescale.

Some countries such as Sweden have built large-scale underground stores for their nuclear waste in geologically-stable areas of granite; others have investigated the possibility of disposal on or under the sea-bed, either by dropping the wastes straight into deep ocean trenches, by dropping them in containers designed to penetrate the ocean floor in less deep areas, or by drilling deep beneath the sea-bed.

Low-level wastes (which may range from reactor coolant to paper towels) may be discharged into the sea or buried in containers in shallow clay trenches. Such practises assume that radioactivity will not seep out to contaminate the environment, and give rise to the moral question of whether we have the right to leave our technological ordure as as insoluble burden for future generations.

A typical pressurized water reactor uses three kilograms of fuel per day

The reactor

A nuclear reactor itself is essentially a pressure vessel contained within a "biological shield" to contain the radiation release. The fuel rods are placed in a vessel full of a coolant, which prevents the reactor core from overheating by conducting the heat away. The vessel also contains a moderator, a substance that slows down the neutrons set free by fission to the optimum speed to continue the reaction. The fuel rods are arranged so that control rods can be raised and lowered between them; these comprise materials such as boron or cadmium which absorb neutrons and therefore slow down or stop the reaction. Various substances are used as coolants and moderators.

The pressurized water reactor (PWR) is the most common design of commercial nuclear reactor. It uses enriched uranium oxide as its fuel, with ordinary water circulating under high pressure as both moderator and coolant for the reactor.

The basic structure of a PWR is a large pressure vessel of welded steel with a "lid" fixed by heavy bolts. Inside the pressure vessel is the reactor core with its control rods, the whole core being surrounded by ordinary water at high pressure (150 atmospheres). The core is made up of fuel elements in the form of tubes of zircaloy, four meters long and about one centimeter in diameter, filled with small cylindrical pellets of enriched uranium dioxide. These are inserted and removed from above.

Coolant water leaves the core at the top through heavy pipes and enters the steam generator or boiler, where it passes through thousands of tubes immersed in more water (at atmospheric pressure) at a much lower temperature. Heat is transferred to this water, which vaporizes and is used to drive the generator turbines (◆ page 138), while the coolant is pumped back into the base of the reactor core. This arrangement ensures that the highly radioactive coolant water does not enter the turbines. The coolant loop also contains a pressurizer which, with the help of immersion heaters, maintains coolant pressure under varying reactor loads and prevents the water from boiling. Control rod assemblies are usually suspended above the core, with drive mechanisms functioning through the lid from above.

The PWR must be switched off in order to refuel. When the reactor is cool, a pool-shaped chamber above the reactor, called the reactor well, is flooded with water to provide cooling and shielding. The lid is then removed and the fuel elements changed. This procedure usually takes place about once a year. The whole reactor is enclosed in heavy shielding of concrete, two or more meters thick.

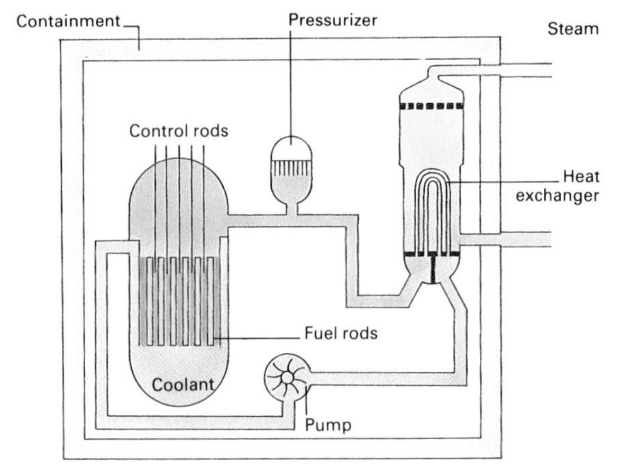

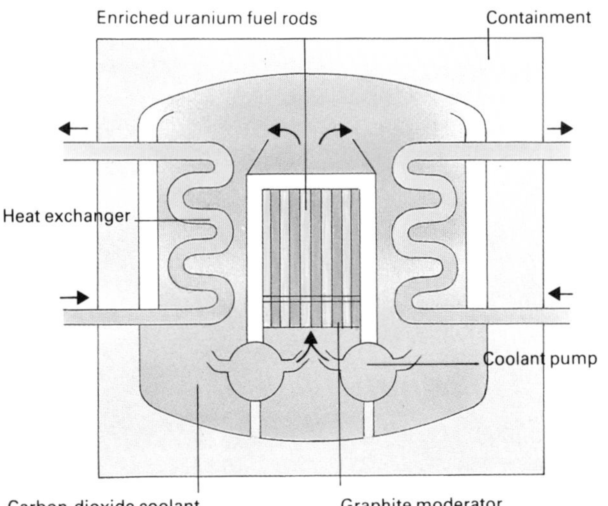

◄ The advanced gas-cooled reactor uses 2 percent enriched uranium fuel. The core is machined graphite, and the heat exchangers are within the pressure vessel. It produces steam at 650°C whereas the high-temperature gas-cooled design operates at 1,000°C.

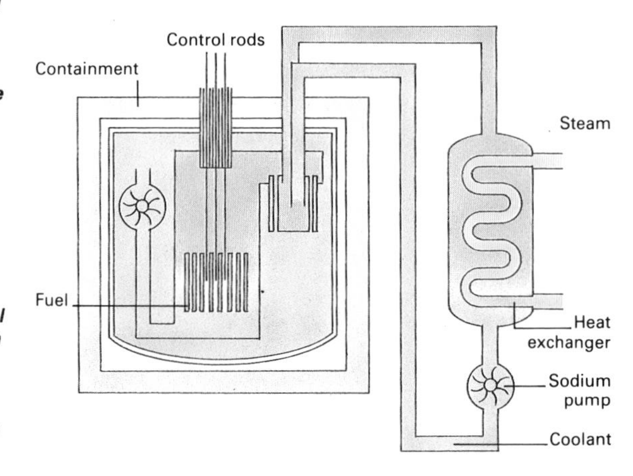

► The fast breeder reactor has a sodium coolant and no moderator. The uranium-plutonium mix fuel is in a compact core, set in a larger tank of sodium. Heat exchangers inside the shielding pass heat to a second sodium flow to take it to steam generators.

Reactor designs

In addition to the PWR, a number of other reactor designs are in use. Each one has its own configuration, and different designs produce different reactions and therefore produce varying quantities of the radioactive waste elements. Some reactors are run purely for experimental purposes; others, usually run by the military, are used to generate plutonium as a "waste product" for use in nuclear bombs.

*Among commercial designs, gas-cooled reactors have a better safety record than the PWR, and burn the uranium efficiently. The Advanced Gas-cooled Reactor (AGR), a British design, uses a carbon dioxide coolant, although a more recent version, known as the High Temperature Gas-cooled Reactor (HTGR), replaces this with helium. The Canadian Deuterium Uranium design (CANDU) uses heavy water (D_2O) as both coolant and moderator (**▶ page 147**). In addition to the PWR, ordinary or "light" water is used as coolant and moderator in the Boiling-Water Reactor (BWR) design; here the pressure is kept down, and a less massive pressure vessel is required. Each of these designs has its advantages, although the proponderance of the PWR in the United States tends to make this design economically more attractive than its alternatives in many countries.*

The so-called "breeder" reactor works somewhat differently. The fuel is packed more densely than in other reactors; there is no moderator. A liquid metal coolant is used, to cope with the very high temperatures generated. The advantage of the breeder reactor is that it uses as its fuel uranium-238, the isotope found naturally, mixed with a little plutonium; its name derives from the fact that, in the course of the reaction, part of the U-238 is converted to plutonium, so that the spent fuel contains more plutonium than it started with ("breeds" plutonium).

Critics of the breeder (sometimes known as the fast breeder, since no moderator is used to slow down the neutrons released during fission) point to the dangers of creating ever more plutonium, and argue that the compact design of the reactor core means that in the event of a malfunction there is more likelihood of reaching a meltdown resulting in a catastrophic accident than is the case in a conventional design.

▲ Three Mile Island, Pennsylvania, where a serious incident occurred in 1979. Coolant and emergency pumps failed; a valve on the pressure vessel also jammed open although instruments indicated it was closed. The reactor pressure fell, steam formed and a release of radiation and meltdown were feared. No injuries or leak occurred in the two-day long crisis.

► A reactor control room.

◄ The pressurized water reactor layout.

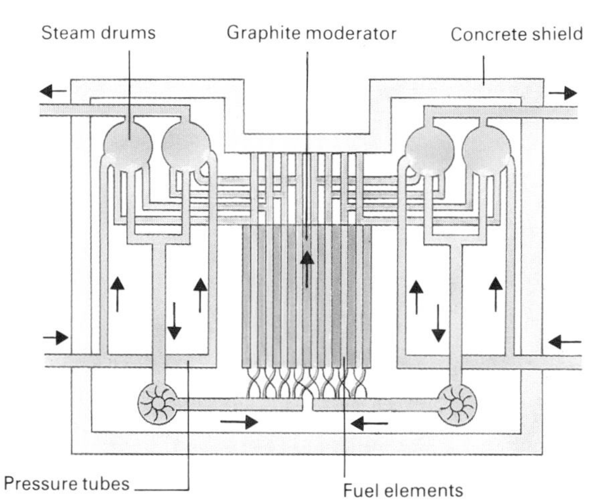

◄ The RMBK reactor, as used at Chernobyl, has a graphite moderator, with the fuel in more than 1,600 vertical pressure tubes filled with light water. The water is allowed to boil under controlled conditions and the steam is led off directly to drive the turbines.

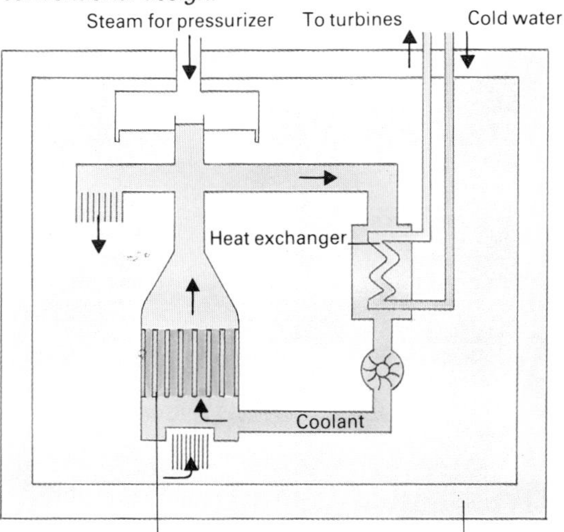

► The prototype PIUS design of reactor is intended to be inherently safe. It incorporates a tank of cold water and boric acid, which absorbs neutrons. Any disruption to the reactor's activity would cause this to flood the core, shutting down the reactor.

The CANDU reactor

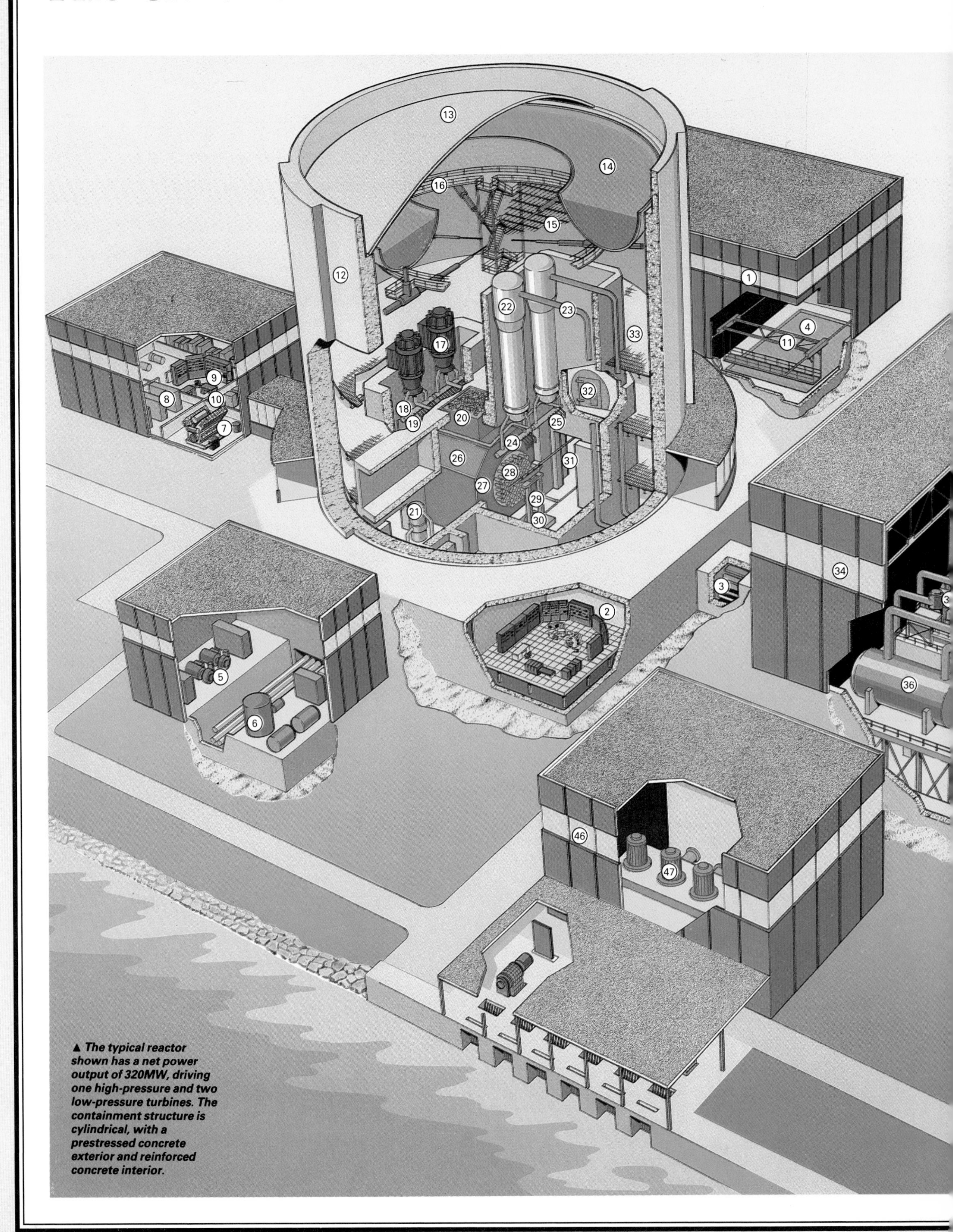

▲ *The typical reactor shown has a net power output of 320MW, driving one high-pressure and two low-pressure turbines. The containment structure is cylindrical, with a prestressed concrete exterior and reinforced concrete interior.*

The CANDU reactor

The Canadian heavy-water reactor (HWR), known as CANDU (CANadian Deuterium Uranium) reactor, is known as one of the safest and most reliable designs in commercial service.

As early as 1945 Canada was experimenting with a reactor at Chalk River, Ontario; this took unenriched uranium fuel (of which Canada has plentiful supplies) and used deuterium (heavy water) as the moderator. A second successful design was in operation in 1947. Interest in this form of reactor was maintained, and in 1971 the first of a series of CANDU reactors opened at Douglas Point, Ontario. Since then, the design has been exported to several other countries, including Argentina, India and Pakistan.

As a reactor moderator deuterium has the advantage over normal water in that its neutron absorption rate is slower; this means that fewer neutrons are wasted so that natural uranium oxide, rather than enriched uranium, can be used as the fuel. Another advantage is that the fuel elements can be more widely spaced. The disadvantage is that deuterium is a highly expensive component, so the reactor core is designed to minimize the amount consumed.

The fuel elements, which can be changed for refueling without shutting down the entire reactor, are contained in a lattice of zircaloy pressure tubes. Deuterium, pressurized to about 90 atmospheres, is pumped through these to carry the heat of the reaction away to a heat exchanger, turning normal water to steam for the turbines. The pressure tubes are themselves housed within a tank, or calandria, which contains the deuterium moderator. About half the energy output of the reactor derives from the fission of plutonium, which is itself created during the fission of the U_{235}.

The good safety record of the CANDU design results from the fact that the reactor is fairly slow to respond; there is the further advantage that the fuel needs neither enrichment before use nor reprocessing afterwards.

1 Administration and maintenance
2 Main control room
3 Containment bulkhead
4 Irradiated fuel bay
5 Recirculated cooling water heat exchangers and pumps
6 Demineralized water storage tanks and pumps
7 Standby diesel generator
8 Control equipment room
9 Secondary control area
10 Emergency core cooling system
11 Irradiated fuel bay crane
12 Reactor building
13 Dome
14 Dousing tank
15 Dousing system spray headers
16 Access platform
17 Heat transport system pumps
18 Reactor inlet header
19 Reactor inlet feeders
20 Reactor mechanisms deck
21 Moderator system heat exchanger
22 Steam generators
23 Main steam lines
24 Reactor outlet header
25 Reactor outlet feeders
26 Shield tank
27 Calandria
28 Reactor channels
29 Fueling machine
30 Fueling machine trolley
31 Fueling machine maintenance lock
32 Airlock
33 Steam generator room floor
34 Turbine hall
35 Low-pressure heater
36 Reheaters
37 Turbine stop valves
38 High-pressure turbine
39 Low-pressure turbines
40 Generator
41 Low-pressure steam piping
42 Turbine building crane
43 Deaerator storage tank
44 Reserve feedwater tank
45 Condenser cooling water inlet
46 Pumphouse
47 Condenser cooling water pumps

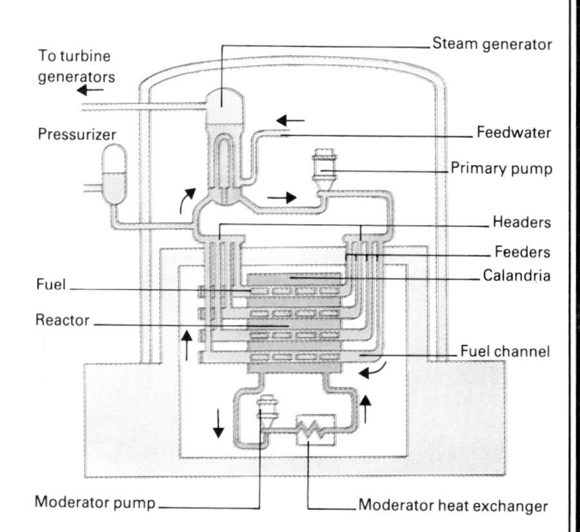

▲ The CANDU reactor consists of a lattice of 208 fuel channels contained within a calandria made of stainless steel and 780cm in length. There are two steam generators, supplying steam at 260°C, but only a single coolant loop.

See also
Introduction to Measuring 9–22
Analyzing the Atom 109–114
Electricity 135–140
Other Sources of Power 149–156
Water Transport 209–216

The tokamak

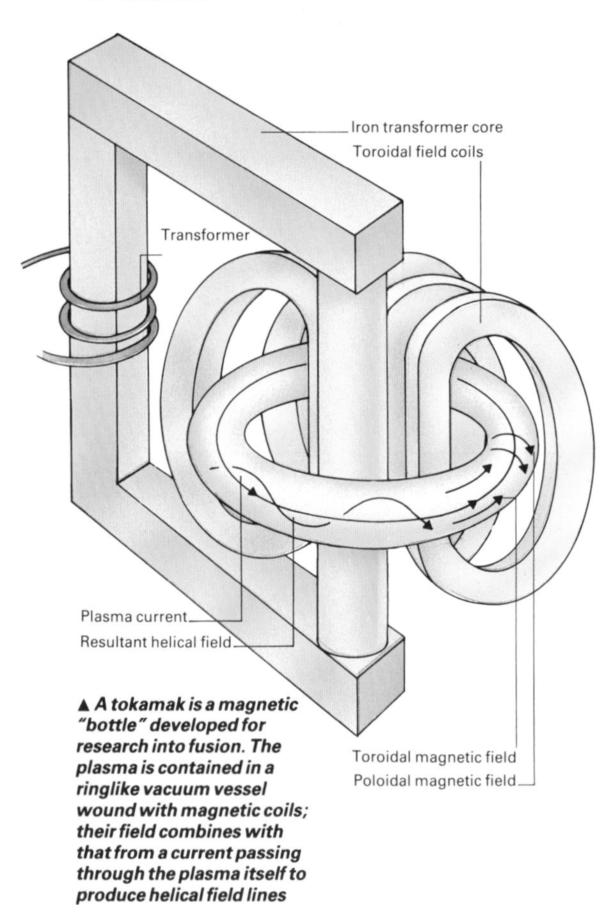

▲ A tokamak is a magnetic "bottle" developed for research into fusion. The plasma is contained in a ringlike vacuum vessel wound with magnetic coils; their field combines with that from a current passing through the plasma itself to produce helical field lines around the plasma.

Controlled thermonuclear reactions

Nuclear fusion – deriving energy from fusing the nuclei of two atoms – is a long-term hope for a safe and virtually infinite source of energy. Despite intense research, however, it is unlikely that it will become a practical option until after the year 2000.

A dramatic demonstration on Earth of nuclear fusion was the explosion of a hydrogen bomb in 1952. By this date projects were already under way for producing a controllable fusion reaction as a source of energy. Its main attraction is the prospect of an inexhaustible source of fuel, the sea. Fusion reactors will use the heavy isotopes of hydrogen known as deuterium and tritium. Deuterium exists naturally in water, and although tritium does not exist naturally, it can be produced from lithium (a relatively common element) in the normal operation of the fusion reactor. Fusion reactors will also produce less severe problems of radioactive wastes than fission reactors.

Current plans for fusion energy involve two alternative approaches: magnetic confinement and laser implosion. In the first, the hydrogen isotopes exist in a plasma (a gas of such high energy levels that the electrons separate from their nuclei). At sufficiently high temperatures the hydrogen isotopes will collide and fuse to form helium, giving off energy in the process. The temperature required is at least 100,000,000°C, at which all metals would be vaporized, so the plasma must be contained within a magnetic field or "bottle". The second concept uses powerful beams of laser light to compress and heat a pellet of deuterium and tritium to cause fusion. The first approach has received most support, and the race is on to contain tiny portions of matter for several seconds at the necessary temperatures and pressures. By the mid-1980s scientists had succeeded in generating temperatures well over 100,000,000°C, but not simultaneously with other conditions that are required.

Nuclear fusion

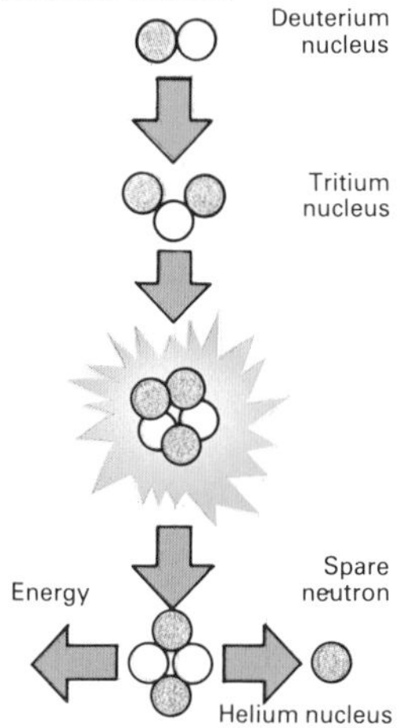

▲ A fusion reactor will exploit the energy released by the fusion to two heavy isotopes of hydrogen.

► The Joint European Torus (JET) tokamak, built in Culham, Britain, came into operation in 1983. It is scheduled to reach full power by 1990, when the first experiments with deuterium and tritium will be carried out.

Other Sources of Power

21

Renewable and non-renewable sources of energy... Energy from the Sun: solar collectors and photovoltaic cells... Wind, wave, tidal and hydroelectric power... Geothermal energy... PERSPECTIVE... The many forms of solar energy... Wind farms... Heat from the oceans

Energy is the fundamental currency of modern society, yet Western society, in particular, is profligate in its use of existing energy stocks, predominantly fossil fuels of limited supply. It has become necessary to plan for a future in which the energy used is appropriate to needs and obtained from renewable sources that are convenient to our condition. Fossil fuels cannot be renewed. They were formed over many millions of years from living organisms which obtained *their* energy, via photosynthesis, from the Sun. The process continues in the formation of peat, but far too slowly to replenish the stocks that are used. In searching for alternatives different societies have responded in different ways. Brazil makes fuels from sugar cane (➧ page 154) while China is investing in digesters to convert farm wastes into fuel. Iceland harnesses geothermal heat to warm homes (➧ page 156), while the United States is investing in the development of photovoltaic cells (➧ page 150), a spinoff from space technology.

Research into these forms of energy has often not been consistently carried through, and has reflected the price and availability of fossil fuels. When oil was cheap, and when nuclear power appeared to offer virtually limitless potential for expansion, large-scale installations of wind- or wave-powered electricity generators seemed like futuristic dreams. The rise in oil prices in the mid-1970s encouraged research in these fields, and now huge wind-farms are found (➧ page 152) and wave machines are used to generate electricity commercially. But recession and low oil prices in the early 1980s made the often high development costs for such schemes less attractive again.

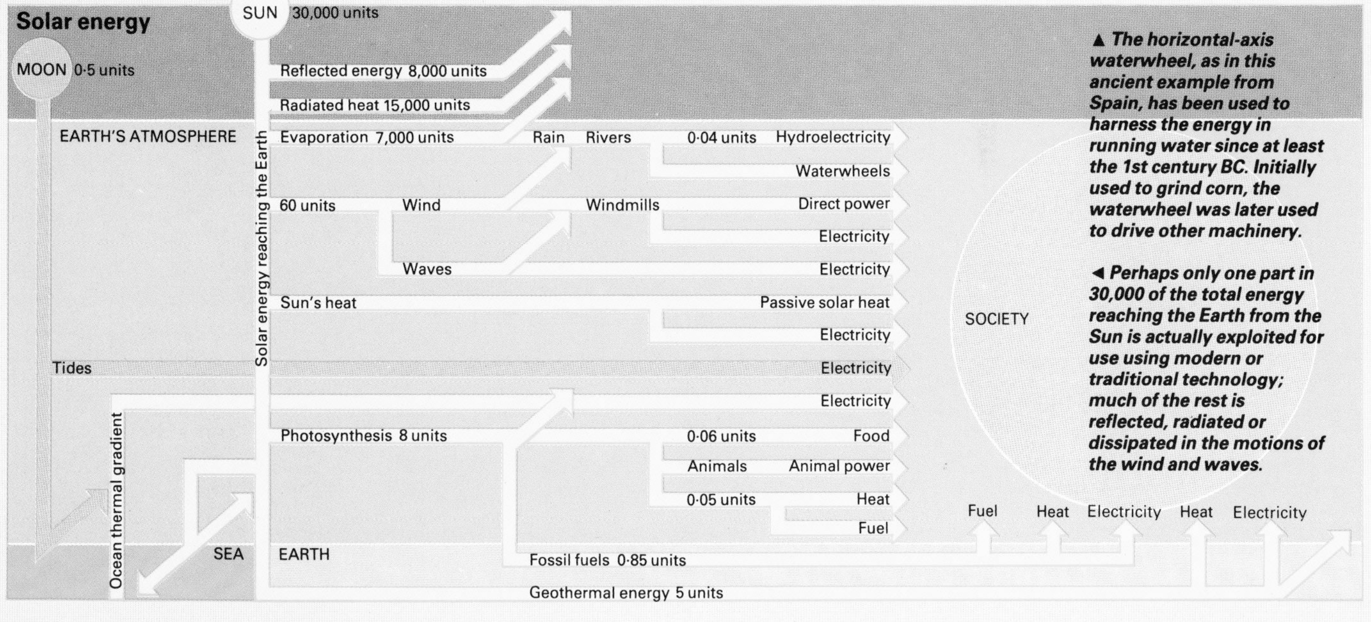

▲ ***The horizontal-axis waterwheel, as in this ancient example from Spain, has been used to harness the energy in running water since at least the 1st century BC. Initially used to grind corn, the waterwheel was later used to drive other machinery.***

◄ ***Perhaps only one part in 30,000 of the total energy reaching the Earth from the Sun is actually exploited for use using modern or traditional technology; much of the rest is reflected, radiated or dissipated in the motions of the wind and waves.***

Photovoltaic cells can be used to harness the power of the Sun to run pocket calculators or huge power stations

▲ A photovoltaic cell consists of a semiconductor with a thin layer of n-type above a thicker p-type layer (◀ page 116). When photons strike the surface, electrons diffuse through the junction between the layers in both directions, giving rise to a current, which is collected by metal conductors on the upper and lower surfaces. The photograph (top) shows a massive photovoltaic array.

Photovoltaic cells convert sunlight directly into electrical energy. They are electronic, need little maintenance, and can be used at any scale from hand-held electronic devices to power stations.

The first practical silicon solar cell was developed in 1954 by Bell Laboratories in the United States as a by-product of transistor research. First used on an American satellite in 1958, photovoltaic cells have become a major source for power in space. They have a layer of silicon, from which electrons are knocked free when struck by photons. The electrons are drawn off by a grid of metal conductors, setting up a flow of direct current. The cells require virtually no maintenance, and can be set together in small groups or huge arrays. Early photovoltaic cells had an efficiency of only 8-11 percent; by the mid-1980s efficiency had risen to almost 20 percent. As the demands of the space program grew, the cost began to fall; from over $600 per

▲ *"Solar One", California, is the largest solar thermo-electric power station. 1,800 individually-programmed sun-tracking mirrors reflect heat to a central collector. Peak output is 10MW.*

◀ *This experiment tests the use of a huge array of photovoltaic cells as a means of collecting solar energy in space, before it is dissipated by the Earth's atmosphere. The plan is for an orbiting collector 8km across; this would be linked to a control station, which would beam the power down to Earth as a microwave radio signal.*

peak watt in the early 1960s, the cost was down to less than $10 per peak watt 20 years later. Between 1977 and 1982, sales increased tenfold, but solar cells still generate power at more than ten times the cost of electricity from conventional sources.

Another way of using solar energy is with collectors that absorb the heat directly and pass it to water or some other storage medium. Domestic heating schemes using passive solar collecting panels can reduce heating bills dramatically in a suitable climate, especially if linked with some form of heat pump (➧ page 162). Larger-scale solar-thermal devices use arrays of mirrors to focus the Sun's rays onto a central collector, where steam is created to drive turbines and generate electricity. A solar-powered furnace at Odeillo, in southern France, has generated temperatures of 4,000°C. Such devices, however, require cloudless conditions to work effectively.

Wind power

The wind and the waves are produced by the effect of the Sun's rays on the atmosphere, and power from them is solar energy received second-hand. Windmills have been used since antiquity for grinding corn and pumping water, and wind generators with blades spanning up to 100 meters are used today to generate electricity. The 1970s and 1980s have seen much research into wind power, which is particularly suitable for coastal regions, including much of western Europe and California, and commercial exploitation has begun. Horizontal-axis windmills have to be turned to the wind, and the inevitable fluctuations in the rotation speed of the arms – particularly in the three-arm design – make it difficult to match their movement with the requirements of the electricity grid. The speed of the blades is determined by the angle they present to the oncoming air, so this is controlled by computer. Several designs have also been tested with a horizontal axis; these do not need to be turned to face the wind, and the generator need not be placed at the top of a tower.

Power from water

Water wheels, like windmills, have been used for many centuries to harness the power of fast-flowing streams; their modern equivalents are the turbines of hydroelectric schemes, which drive electricity generators. Some such schemes, known as pumped storage systems, use periods of off-peak demand to pump the water back to the higher level and recycle the water through the turbines when needed.

Wave power has often been claimed as a promising source of alternative energy for coastal regions, and many designs have been produced to convert the up-and-down movement of the water into electric power. The many problems that remain to be overcome include the difficulty of producing a reliable design in materials that would not need constant attention, the remoteness of most suitable sites, and the vast scale that would be needed for commercial exploitation. There are also environmental dangers.

The tides are another possible source of energy (lunar rather than solar energy). A tidal power plant operating at La Rance, in northern France, since 1966 works by allowing water to enter a river estuary as the tide rises and passing it through turbines as the tide falls. A similar scheme for the Severn estuary in Britain is designed to provide 7,200 megawatts or 6 percent of the nation's electricity for a capital outlay equivalent to that of a single nuclear power station. As always there are environmental problems which may prevent planning approval.

► *The Hoover Dam, Nevada, USA, one of the world's largest dams (221m high) and suppliers of hydroelectric power.*

◄ *A wind-farm at Altamont, California, where 86 mills were installed in 1985 to generate 26 Mw of electricity. The windmills used here are mostly three-blade machines with a rotor diameter of 45m; the blades are made of wood and epoxy laminate. The rotors are turned to the wind by hydraulic motors, and the mills shut off automatically in high winds.*

◄ *A vertical-axis wind generator. Windmills are usually installed with another generator for use if the wind fails; methods that have been tried to store wind-generated power include batteries, pumped-water storage, flywheels, and a hydraulic accumulator that stores power in the form of high-pressure oil.*

▼ *Vertical-axis windmills in Palm Springs, Florida, in the design patented by the French inventor G.J.M. Darreius in 1931. The blades are in a "troposkein" curved to reduce bending stresses, but they are expensive to manufacture and only part of the blade extracts power from the wind efficiently.*

Power from plants

Rotting organic matter can, under certain circumstances, produce inflammable gases, notably methane. Rural areas have a great bulk of crop residues and animal wastes suitable for the production of this form of energy. Anaerobic digestion, the process of fermentation in the absence of air, converts complex organic material into methane and other gases. It is a simple process, an effective treatment for wastes that could otherwise be a health hazard, and leaves a residue that is useful as a fertilizer. The foremost exponent of this "biogas" is China, where hundreds of thousands of methane generators are in operation. By the mid-1980s they produced the energy equivalent of 22 million tonnes of coal.

In the most fertile regions, plants capture and store the equivalent of one and a half times the total world human energy demands each year. Attempts to make use of this vast reservoir of stored energy began in the 1970s, when processes aimed at the production of ethanol (ethyl alcohol) for blending with gasoline to make "gasohol" were developed. Ethanol is produced directly from sugar by fermentation or from starches that are first converted to sugar and then fermented. It is derived from crops such as sugar cane, beet and sweet sorghum, from root crops such as cassava, and from all major cereals.

The use of ethanol as a motor fuel is not new. In the United States the industrialist Henry Ford (1863-1947) promoted the production of gasohol, and it was widely used in the Midwest in the 1930s. After the Second World War the abundant supply of gasoline virtually destroyed the alcohol fuels industry.

Of the countries experimenting with ethanol as a fuel, Brazil is the leader: by 1980 production of alcohol had reached some four billion liters. The United States has also embarked on an ambitious alcohol fuel program, with the goal of 10 billion liters of ethanol by 1990.

▲ The aim to make Brazil self-sufficient in oil by AD 2000, by encouraging the use of alcohol as a fuel, is widely promoted.

► A fuel-cell electricity generator in Japan; fuel cells (♦ page 135) run on hydrogen and oxygen, which can be derived cheaply from abundant plants and feedstocks.

▼ A methane generator in use in a village in India. Animal dung is mixed with water and put in the slurry hopper, which passes it to the generator. Here it ferments anaerobically, and the resulting methane is stored in a small gas-holder. The waste slurry has a high accessible-nitrogen content and is a useful fertilizer.

► Biomass is the great quantity of organic matter, mainly plants, which absorb some 2 percent of the Sun's energy that reaches the Earth. It can be converted to useful energy in a number of ways. In temperate and industrial regions, biomass can never provide more than a small proportion of

BIOMASS

energy requirements, but in tropical regions it has great potential, and marine and water plants also offer possibilities. Useful sources of biomass include natural vegetation and trees; algae; crops such as sugar and starch; reeds, rushes and water hyacinths; and waste products.

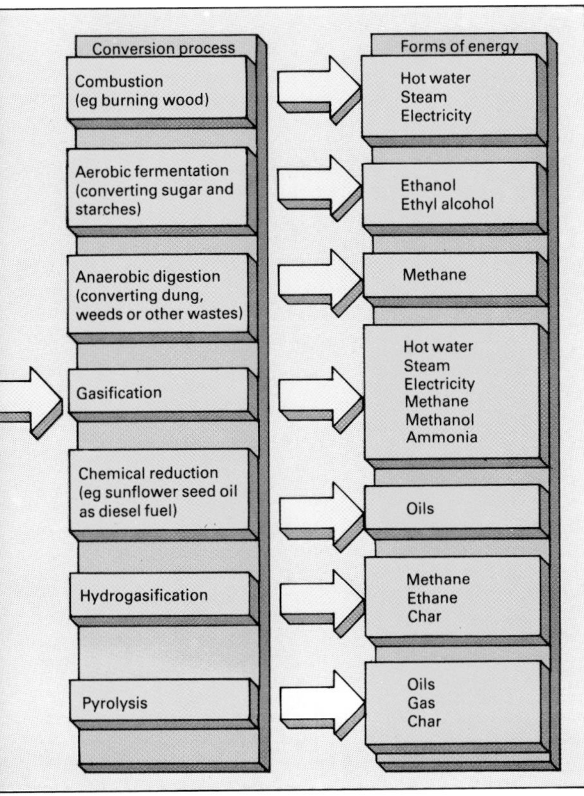

Diesel fuel substitutes

A number of plants produce an oil that can be substituted directly for diesel oil. Brazil has planted some 1·6 million hectares of dende palm, the oil of which can either be mixed with diesel fuel or used directly in slightly modified engines. The hope is that this will eventually supply some ten percent of the country's fuel needs. Sunflowers grow on poor quality land and with very little water, and sunflower seed oil is being tried in the United States and South Africa as a fuel for farm equipment. In the Philippines coconut oil is used as a diesel fuel additive ("cocodiesel"), and scientists at the University of Arizona are hoping to extract at least 22 barrels of oil a year per hectare of gopherweed, a variety of milkweed that grows prolifically in the arid lands of the southwestern United States.

Urban wastes

Vast quantities of potentially useful energy are dumped daily by town dwellers. One tonne of urban refuse contains on average the energy equivalent to 200 kilograms of coal; and every year the average American throws away 650 kilograms of garbage. Munich, in West Germany, derives more than 10 percent of its electricity from garbage, and three plants in Paris burn 1·7 million tonnes of waste per year to produce the energy equivalent of 480,000 barrels of oil. Luxemburg and Denmark convert more than half their urban wastes to heat or electricity, and Japan is close behind, with more than 85 plants for waste utilization.

See also
Measuring with Electricity 41–46
Analyzing the Atom 109–114
Engines 129–134
Electricity 135–140
Nuclear Power 141–148

Geothermal energy

The Earth's internal store of heat can be tapped as an energy source. This heat is the product of gravitational collisions, atomic reactions and radioactive decay in the magma, the molten rock that lies under the Earth's crust. On average temperatures increase around 25°C with each kilometer of depth, but in many parts of the world there are "hot spots" where temperatures as high as 360°C can be found within two kilometers of the surface. In such areas (which include Iceland, New Zealand, parts of Italy and the United States), geothermal power is a practical possibility. Electricity from geothermal steam was first produced in 1904 in Larderello, Italy; continuous electricity generation began there in 1913 and today output is 360 megawatts.

There are three types of thermal reservoirs: hydrothermal, geopressured and dry hot rock. The first type is the most suitable for energy production. It consists of magma topped by an impermeable layer which prevents heat loss. Water and steam can be tapped for the direct production of power. Geopressured reservoirs are produced by the clays in a rapidly subsiding basin area trapping heat in underlying water-bearing formations. They are of little practical use as an energy source. In dry hot rock systems there is no permeable aquifer and therefore no steam. Several ways of extracting heat from such systems have been proposed, based on fracturing the rocks and then pumping in water to be converted to steam by the residual heat. One method (hydraulic fracturing) proposes pumping high-pressure water into a well, causing the rock to crack around the borehole. Another (nuclear fracturing) suggests that the fracturing can be done by an array of multiple nuclear explosions. Hydraulic fracturing has been shown to work at an experimental site at Fenton Hill, New Mexico.

Dry rock geothermal power

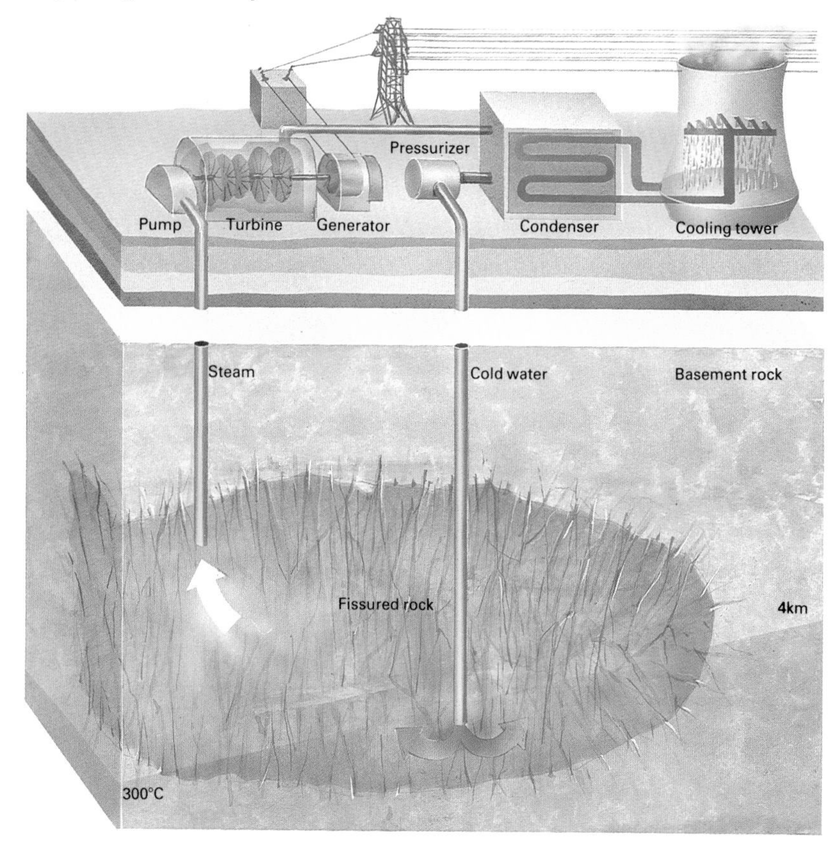

Ocean thermal gradient plant

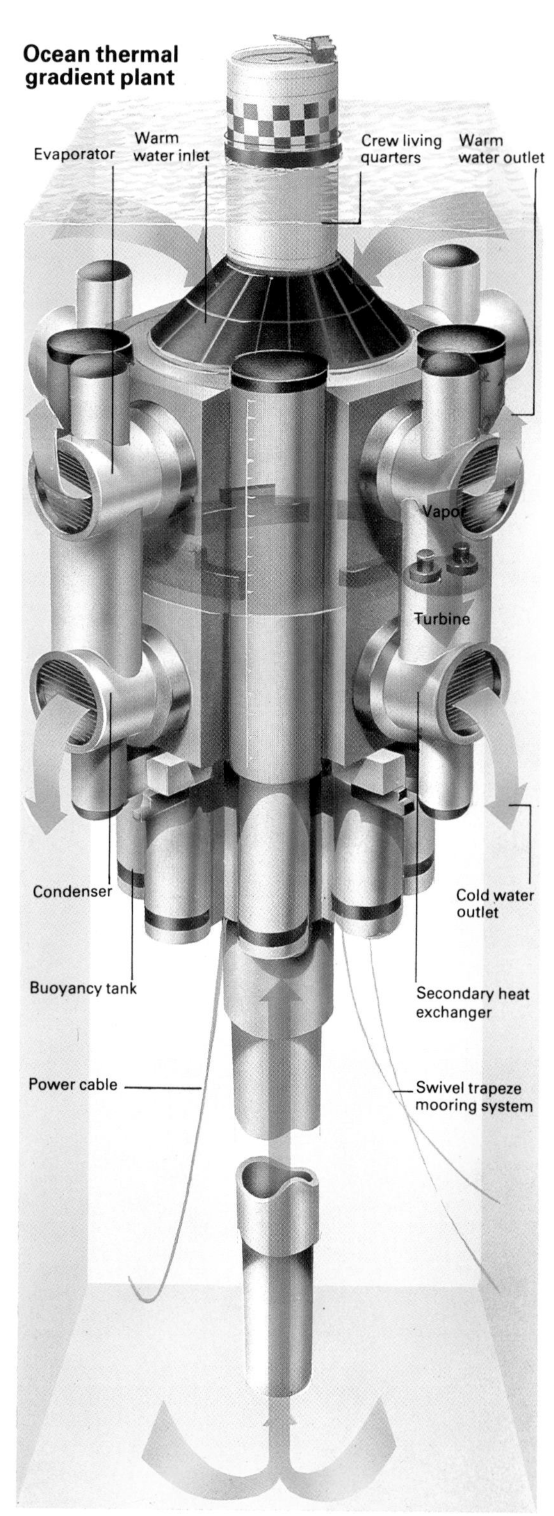

◀ A geothermal power station extracts heat from the Earth by fracturing the hot, dry rocks, at a depth of 3,000m or more, pumping water down and extracting the heat in the form of steam. A similar technology uses an opposite approach – fractured rock is employed as a store for surplus solar energy or waste industrial heat.

▲ Power may one day be extracted by ocean thermal electric conversion. This exploits the fact that the water from 1,000m or below may be 25°C colder than at the surface. The cold water is pumped up to a heat exchanger where it is used to evaporate and condense ammonia as part of a cycle to drive turbines to generate electricity.

Shelter

The need for shelter...Influences on building styles... Climate, materials and society...Steel and glass... Reinforced concrete... Plastics...Building for efficiency: conserving energy...PERSPECTIVE...The shape of buildings...Timber-framing...Domes over towns... Plastic igloos...The autonomous house

Shelter is a basic human need. Its provision ranges from the simplest form of animal furskin, wrapped around the body to keep out the cold and damp, to complex modern buildings. Earliest people found what shelter they could in caves; today people go even beyond the Earth, taking their special shelters with them.

One kind of sheltered environment that everyone needs is a house, and almost every society has built and continues to build its own kind of housing. To meet this need, a technology of shelter has developed that takes raw materials and transforms them into houses that protect their occupants from the extremes of the natural environment, and provides comfortable spaces for the personal and social activities typical of that society. Their form is influenced by three interacting factors: climate, materials available and social requirements.

For most of history this technology has been relatively simple. The basic need for a house has been so important that almost everyone knew how to design and build their own. In modern societies this knowledge has become more specialized, the preserve of architects and building technologists. At the same time, housing has ceased to be a simply-satisfied need. Instead it has become a problem, needing ever more resources, specialized knowledge and equipment.

Climate and the shape of houses

To provide a comfortable indoor climate efficiently, with the minimum of additional heating or air-conditioning, the basic form of a house should be adapted to the natural exterior climate. In cold regions the need to conserve heat is important; the surface area of the house should therefore be kept to a minimum. The hemispherical form of an igloo (➧ page 161) is ideal in this respect, as well as being structurally suitable for building with blocks of snow. For more conventional materials and construction, a cube shape may be appropriate. In temperate regions, a rectangular form is desirable, with the long side turned towards the sun to catch the available solar heat. In hot-arid desert regions, a more compact shape is preferred, perhaps with features such as an internal courtyard cooled by a pool or fountain to help create a more comfortable internal "microclimate". In hot-humid regions, such as tropical forests and swamps, the houses tend to be light and open, to catch any cooling breezes.

These forms remain as applicable to modern housing design as to traditional buildings, and are one factor taken into account by designers of energy-conscious homes.

▲ The Bedouin tent of the Middle East is made of sheets of woven goat's hair or wool, supported on timber poles: a form of shelter well adapted to a nomadic and pastoral way of life. It offers warmth at night and protection from strong winds, but allows breezes to cool the interior.

◄ In contrast to the nomads' tents, the construction of modern high-rise buildings, such as these in the business district of Washington, DC, require deep foundations and the importing of a multitude of materials, equipment and skills. Concrete, steel and glass are the most widely-used materials in modern building; steel is often employed for the basic skeleton, which, in the case of "stressed-skin" buildings, may be on the outside.

The dome and the external frame are two of the oldest, as well as two of the most exciting modern, structural forms

One purpose of a house is to provide shelter from the climate. Indeed, this is often described as the single most important factor influencing house design, though it is only one of many. The type of climate affects the principal sheltering functions of the house. If rain is the most important element, then roofs are usually pitched and jut well clear of the walls to shed the rainwater away from the house. In hot, dry climates, on the other hand, the roof can be flat and conveniently serve as an extra outdoor "room".

Different building forms are also influenced by the kinds of materials that are available for use. For instance, it is difficult to build a flat roof in materials such as brick or unreinforced concrete. These materials are weak in tension: that is, they do not withstand the stretching that results, for example, from the force of gravity on an unsupported beam. To span an opening or to roof a building, architects have had to devise methods, such as the arch, to direct the load down into the supporting walls. Alternatively they have sought to introduce secondary materials into the building to span the opening. A lintel over the door of a brick house is an example of this approach.

In the past, all but the most lavish buildings have tended to be made of materials available locally, and the difficulty of working them has affected the choice of whether the post-and-lintel, the arch, the frame or the dome became the predominant structural style. Today, a much greater range of materials is available, and these offer the builder the opportunity to explore these ancient architectural forms in new ways and contexts. Thus the dome was used from prehistoric times to span enclosures in stone, originally by placing concentric rings of stones one upon another, slightly narrowing the gap with each course (the corbel dome). The lightness and stability of the dome has long been appreciated, and its possibilities were extended in revolutionary fashion by the American architect R. Buckminster Fuller (1895–1985), who constructed geodesic domes with frames made up of triangles or other simple shapes of metal or plastic. These are said to offer the cheapest and most efficient method of enclosing space, and can be built on any scale.

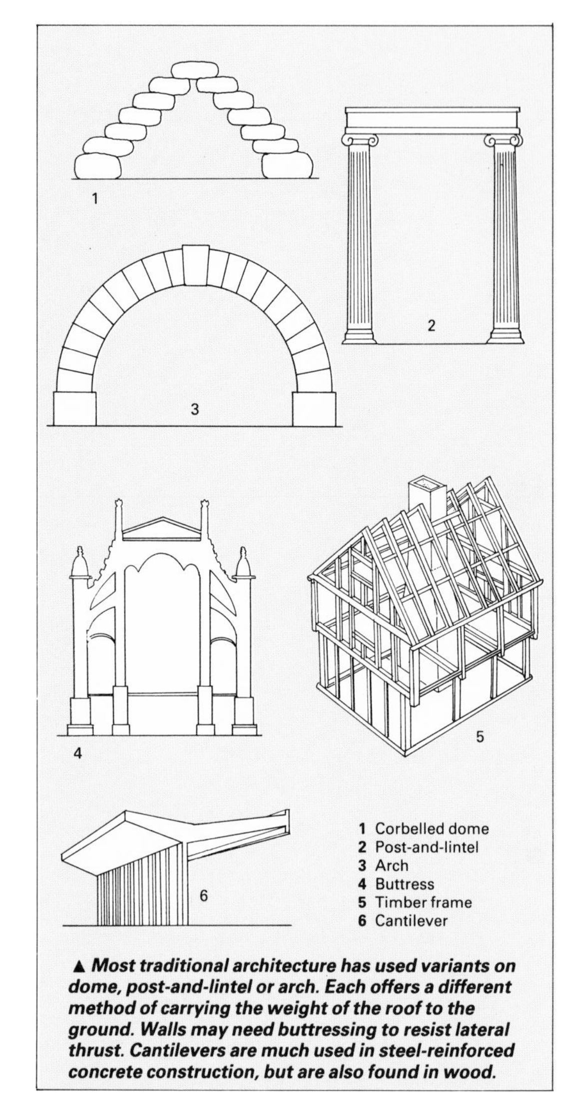

▲ Most traditional architecture has used variants on dome, post-and-lintel or arch. Each offers a different method of carrying the weight of the roof to the ground. Walls may need buttressing to resist lateral thrust. Cantilevers are much used in steel-reinforced concrete construction, but are also found in wood.

Building with timber

Timber has been used from earliest times, both as a structural component and for cladding and enclosing. Wood has particular properties that account for its suitability as a building material. It resists both tensile and compressive structural forces; it is relatively light and easily manipulated; it can be worked with simple tools, to shape it and to construct the joints between separate members; it is warm to the touch and has fairly good thermal insulating properties; and it has been readily available in most regions of the world.

Some of the earliest timber structures were little more than thatched, tentlike enclosures. The interior space was limited, and could only be increased by using very long timbers. When iron tools became available, it was possible to cut and fit together large frames, using hardwoods such as oak. These box-frames could be infilled with brick or, more usually, wattle and daub.

The lightness and tensile strength of timber makes it especially suitable for building roofs, and many ingenious designs have been produced since Roman times to cover spans greater than the longest timbers.

▲ *The Hongkong Bank headquarters (1986), built around a steel frame.*

◄ *A grass hut in Thailand; a frame of thin wooden poles supports a thatched roof.*

▼ *A large geodesic dome in Florida, by Buckminster Fuller. The geometry of this design means that the structure can be quite rigid even if built from a material such as cardboard.*

Steel, glass and concrete

Cement and concrete have been used since Roman times, but modern reinforced concrete uses steel rods embedded in the concrete to strengthen it. Concrete itself is strong only in resistance to compressive forces, whereas steel is particularly strong in resistance to tensile forces. The steel rods are placed within the concrete at points where tensile forces will be experienced. The rods are laid within a framework into which the liquid concrete is poured. When the concrete is set hard, a very strong composite structure is formed. This enables forms to be built that would not have been possible in unreinforced concrete, including cantilevered structures of all kinds. Reinforced concrete and steel structural frames enable the building of multistoried blocks, while the production of glass in sheets led to the introduction of large windows. The technology of glass continues to develop, with heat-reflective and thermally-insulating types. Steel and glass were used effectively together by the German-born American architect Mies van der Rohe (1886-1969) in his office and apartment blocks, and some private houses. These have an open, light and airy structure, following a simple, traditional post-and-lintel form; the glass provides an all-over "curtain" wall, with no structural purpose. His style has been copied all over the world.

Plastic domes built as disaster relief in central America have become part of indigenous architectural style

Reinforced concrete is often cast at the particular site of construction, but it can also be precast and delivered to the site for erection. This has the advantage of permitting the use of indoor, factory methods for making the concrete panels. One implication of this is that the panels are made to standard, modular sizes so that buildings can be constructed from them like a kit of parts. This technique has been used particularly in the design and construction of high-rise apartments in attempts to improve both the quantity and quality of housing by "mass-production" building methods.

Building with plastics

Some of the most modern materials are the various kinds of plastics. These are generally strong in relation to their weight, resistant to chemical action, and flexible; until recently they have also been combustible and adversely affected by sunlight. In buildings, plastics have been used particularly to form exterior cladding panels and interior linings for floors, walls and ceilings, especially as insulation boards. These uses are normally in the form of flat sheets, but plastics can also be poured, sprayed and molded into virtually any shape. Plastic roofing panels are particularly suitable for surfaces where natural light is required but glass cannot be used. Other important uses of plastics in buildings include epoxy resins for adhesives and surface coatings, mastic waterproofing materials for sealing joints, and the adoption of plastic (often polyvinyl chloride, or PVC) guttering has brought a great saving in expense in house building and maintenance.

Packaging towns

For many years planners have had a vision of enclosing entire towns under a dome or ceiling which would keep off the worst of the weather; the American designer Buckminster Fuller had plans for a geodesic dome which would cover most of Manhattan. Possibly a more practical alternative has been developed with a scheme to construct a dome "tent" of tough fluorocarbons, some 65m high and covering 14 hectares, with a layer of fiberglass insulation. The intention is to provide sufficient protection to moderate the effects of the Arctic winter enough to permit permanent settlements for oil-drilling or mining communities. A coating of Teflon or silicone would be placed on the outer surface to ensure that it remained clear and clean enough to give the inhabitants the impression of living in open country.

The material is already used for sports stadia, and has a projected lifespan of 40 years; the tent would be held up by keeping the air pressure within the dome very slightly higher than that outside. Sportsdomes held up in this manner need little extra support and have a flat profile that is not affected by wind. This design is not suitable for the Arctic, where a steeper pitch is needed to prevent snow from settling; hot air can also be pumped into the insulating layer to promote melting.

The weight of a fabric dome is about one-thirtieth of that of a steel roof covering the same span, and the fabric dome takes a quarter of the time to erect.

Plastic igloos

Emergency housing is often needed after natural disasters, and such crises have sometimes provided an opportunity to explore some new uses of modern building materials. Some 50,000 people were killed and 250,000 made homeless in an earthquake in northern Peru in May 1970, and an international relief operation was quickly under way. One of the most interesting developments was an igloo-type domed house made from plastic, and used as temporary housing that served for disaster relief. The igloos are made on site by inflating a hemispherical rubber balloon and then spraying it with plastic materials which form a foam skin about 10cm thick. The skin sets hard in a few seconds, the balloon is deflated, a hole is cut for the door and windows, and the lightweight dome, weighing little more than 100kg, is ready for habitation. These domes remained in use for several years, and in some cases were extended and adapted by their inhabitants.

▼ *The Louisiana Superdome, topped with an inflated fabric roof. Other sports stadia have been built with the fabric stretched tightly in "inverted pleats" resembling a horse's saddle, and hanging from supports. Tears are repaired by sewing in a new panel.*

▲ *A Greenland Inuit (Eskimo) with a newly-built igloo, a temporary shelter for hunting trips. The blocks of snow are laid in a continuous spiral, gradually closing in to form a dome. A tunneled entry passage offers protection from the wind, and the interior will be hung with furs.*

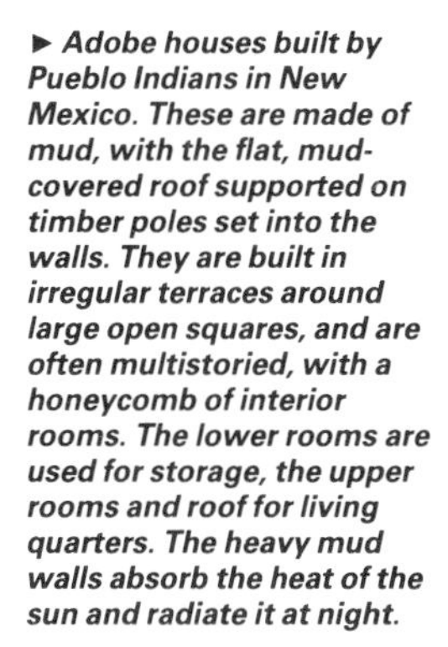

▲ *A modern apartment block in Peking, built in the "international" style of precast concrete sections assembled on site.*

► *Adobe houses built by Pueblo Indians in New Mexico. These are made of mud, with the flat, mud-covered roof supported on timber poles set into the walls. They are built in irregular terraces around large open squares, and are often multistoried, with a honeycomb of interior rooms. The lower rooms are used for storage, the upper rooms and roof for living quarters. The heavy mud walls absorb the heat of the sun and radiate it at night.*

An autonomous house

The energy crisis of the early 1970s led to a variety of projects in the design of "low-energy" houses. Some of these were just like conventional houses, but very heavily insulated. Others used relatively new devices such as solar collectors to absorb the "free" heat of the Sun. A few were radical experiments to make the house completely "autonomous" of all mains services and reliant solely on the energy sources that were available on site. These sought to use wind or water for electricity, and often generated gas by methane generators using household wastes (◀ page 154), while the water system might be planned to recycle as much as possible and collect whatever was needed from rain or dew.

One such experiment was a dome house designed and built by Dr Robert Reines near Albuquerque, New Mexico. The dome is about 10m in diameter and is made of pressed steel segmental sections. It is lined internally with an 8cm sprayed coating of flame-resistant polyurethane foam. Like an igloo, it has an airlock entrance tunnel to prevent heat loss. It is provided with a 200cm diameter window and air-vent at its apex, and a number of smaller porthole windows.

Heating is provided by a hot-water radiator ring around the base of the dome. The water is heated by means of solar collector panels, and stored in a heavily-insulated tank. Electricity for the pumps, lights and domestic appliances is provided from windmills, and stored in batteries. This house is entirely autonomous in energy terms. Its internal air temperature stays fairly constant at around 20°C, despite the extreme winter and summer external temperatures which can range from −20°C to 40°C.

◀▲ This house in New Mexico relies on wind-power to pump water from a well. The water is stored in drums which cover the south-facing walls, which can be covered at night or exposed to allow the Sun to heat the water.

► One form of solar space heating is a wall built of a dark, heavy material and covered with glass. The wall is heated by the Sun, and air between glass and wall rises and passes out of a vent at the top, drawing cool air in below.

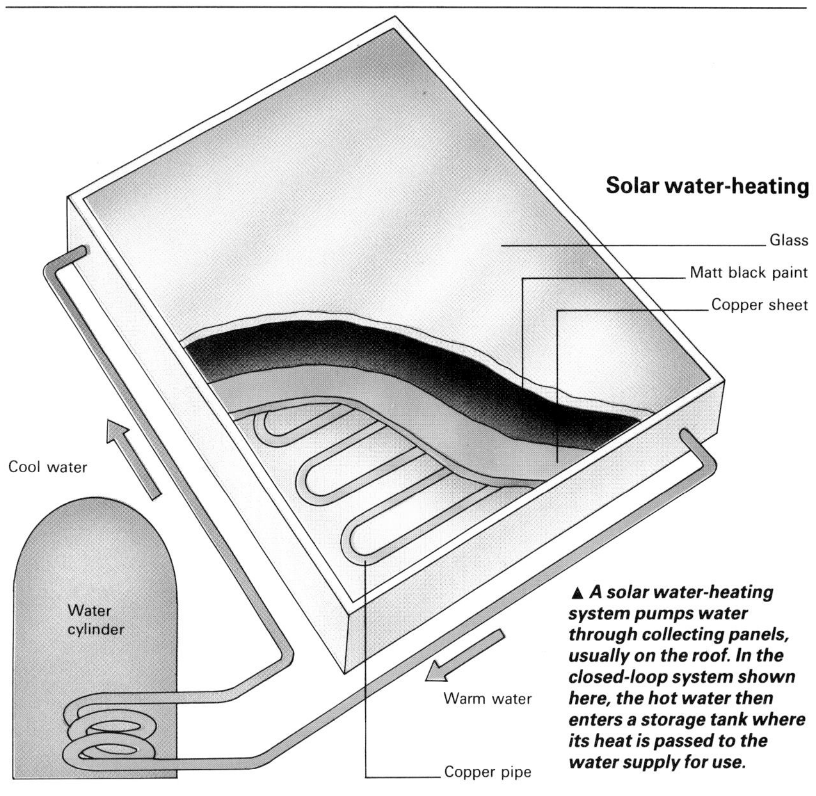

▲ A solar water-heating system pumps water through collecting panels, usually on the roof. In the closed-loop system shown here, the hot water then enters a storage tank where its heat is passed to the water supply for use.

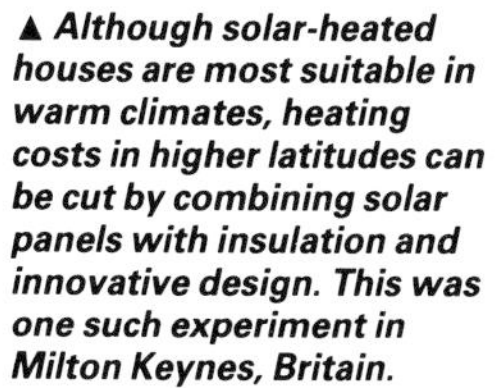

▲ Although solar-heated houses are most suitable in warm climates, heating costs in higher latitudes can be cut by combining solar panels with insulation and innovative design. This was one such experiment in Milton Keynes, Britain.

► One problem in solar houses is to balance energy efficiency with ventilation. The skylights of this house in the United States open and close according to the amount of sunlight. The mechanism is triggered by a photoelectric cell.

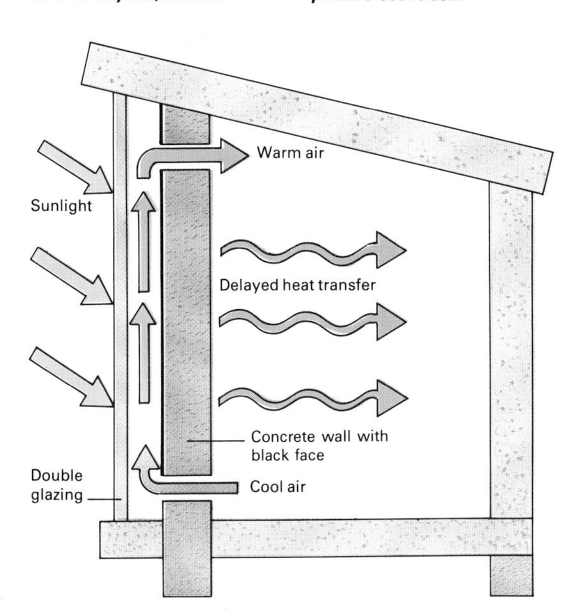

See also
Electricity 135–140
Other Sources of Power 149–156
Defense and Attack 165–170
Processing Raw Materials 237–244
New Materials 245–248

▲ This house is built with a heavy layer of polyurethane insulation on its roof. This material can be used easily and quickly by foaming it around an inflated balloon mold, and then fireproofing it. Other ways of improving insulation include filling cavities in walls with plastic foam or aluminum foil; using slabs with vacuums in the cavity between insulating materials; using foam-filled glass-fiber reinforced plastic; and applying a translucent layer of insulation to the external surface of solid walls to reduce heat loss and increase solar gain.

Insulation and energy-saving

In industrialized societies, about 20 percent of all energy consumption is in the domestic sector, and half of this is for providing space heating. The need to conserve fuel has therefore led to a concern with energy-saving in the home, particularly with savings in space heating. One way of making such savings is to improve the thermal insulation of houses. Some countries – such as the Scandinavian countries – have traditionally had better insulated houses than some others – such as the United States and Britain.

The material of which a house is made can profoundly affect how much heat it needs to maintain a comfortable internal air temperature. No material completely prevents heat from passing through it, so whenever there is a temperature difference between the inside and the outside of a house, heat flows from one to the other through the building fabric. The rate of heat flow is different for different materials: aluminum has a very high rate of flow; glass, stone and brick have much lower rates, and wood and expanded polystyrene much lower still.

The standard measure for the rate of heat flow through building materials is known as the U-value. This is measured by the amount of heat flow, in watts, through one square meter of surface area of the material, with a temperature difference of one degree Celsius between the two sides of the material. Thus a higher U-value means that more heat flows, and insulation is poor. A typical house wall, with a cavity between two leaves of brickwork, has a U-value of about 1·4. If the air cavity is filled with a plastic foam, the U-value is halved to 0·7, and the wall becomes twice as good as an insulator against heat loss. A solid brick wall, on the other hand, 105mm thick and with a plaster covering, may have a U-value of 3·0.

A single glazed window has a typical U-value of 5·6, but this may be reduced to 3·2 if double-glazing is used. A pitched tile roof with felt, roof space and a plasterboard ceiling may rate 1·9, but if 25mm glass fiber insulation is added the U-value will fall to 0·7, whereas if the insulation is 75mm thick, the rating may be better than 0·4.

In choosing materials for insulation, it is necessary to consider economic factors – such as how long it will take for the savings in running costs to offset the installation expenses – and also to consider the particular climatic conditions of the house. Thick, heavy forms of construction tend to absorb and release heat slowly. During the hot day, houses of materials such as adobe feel relatively cool inside, because they only warm up slowly. Then, during the cool night, they slowly release the stored heat again and so feel relatively warm inside. Conversely, in hot-humid climates very light forms of construction are preferred. This avoids any temperature build-up inside the house, which would bring higher humidity, and provides instead a shady, well-ventilated, relatively cool interior.

Defense and Attack

Technology and weaponry...Anti-tank guided missiles ...Surface-to-air missiles...Portable SAMs...Radar defense against airborne attack...PERSPECTIVE...Missile development after 1945...Missile warheads...Swingfire... Air-defense gun systems...Radar and warfare

For almost as long as warfare has been known, technology and military operations have been inexorably linked. History echoes with occasions when the side having the technical advantage has the capability to develop a tactical advantage as well. In the First World War, the first appearance of the tank on the battlefield for the Allied forces was just such a case. The invention and application of atomic weapons in the Second World War illustrated just how overwhelming the technical advantage could be, and how drastic the implications for the whole of humanity. If the German nuclear scientists had reached the same development stage as their rocket designers, the history of the 20th century would have followed a very different course.

In the last quarter of the century the link between technology and military operations has never been stronger. The NATO countries rely heavily on technical superiority: the greater the technical lead, the more confident NATO feel about containing the much more numerous Warsaw Pact forces. As a consequence, the military in the West demand ever more sophisticated weapons which necessarily draw on technology at the leading edge of research. This is nowhere better exemplified than in the development of battlefield missiles and the successive innovations made to improve the accuracy and versatility of guidance systems.

Missile development since 1945

The origins of modern missile technology are found in the German rocket program of the Second World War. Both the V-1 and V-2 rockets were the forerunners of today's surface-to-air and long-range surface-to-surface missiles. At the same time the Germans were also pursuing research into anti-tank guided weapons. Their X-7 missile, which was trialed before the end of the war, was a genuine first-generation missile incorporating a trailing wire for reception of command signals.

At the end of the war, some of the German scientists involved went to the United States and others to the Soviet Union, but as both countries were actively developing nuclear weapons, all their effort was concentrated on continuing the V-2 program in order to provide long-range vehicles for nuclear warheads. Consequently work on the X-7 ceased.

The development of missiles to carry nuclear weapons and the advances in radar together provided the expertise for the design of surface-to-air missiles, such as the Bloodhound and Nike-Hercules, in the late 1950s. So effective were they that aircraft were forced to adopt much lower attack levels, and this has resulted in a steady progression of low-level air-defense systems since the 1960s. Other technological advances have led to the introduction of miniaturization, allowing the development of man-portable shoulder-launched weapons.

▼ The Nike-Hercules missile, seen here at Homestead, Florida, succeeded the Nike-Ajax in 1960, since when it has been the United States system for high-altitude air defense. A solid-propellant rocket, Hercules had a larger warhead and greater range and speed than the Ajax. Its effectiveness was demonstrated when it intercepted another Nike-Hercules at an altitude close to 30,000m and at a speed of Mach 7.

Modern shaped-charge warheads generate great pressure, piercing the target and flowing through the hole in a long, thin jet

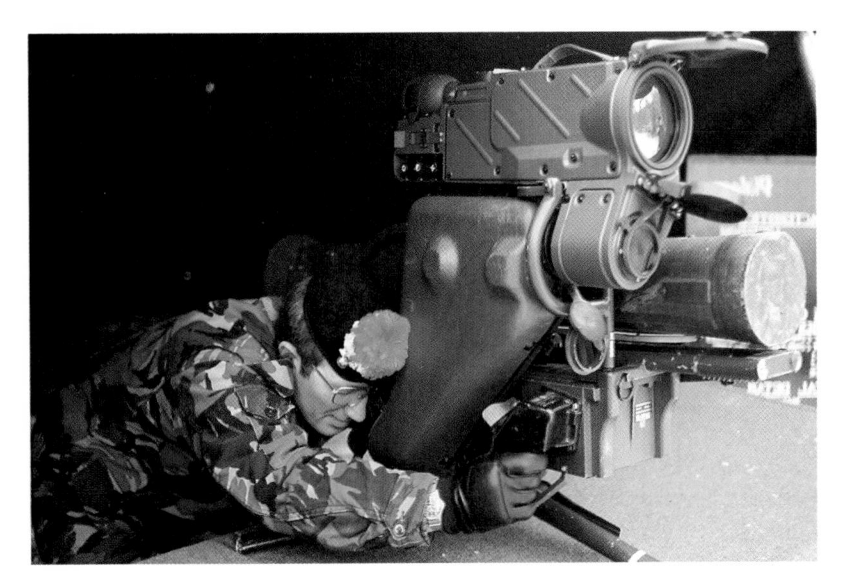

► The Apache attack helicopter can launch its Hellfire missiles at very low levels, enabling it to take advantage of any available cover. When operating in conjunction with a ground-based laser target designator, the Apache presents only a fleeting target at the moment of launching.

◄ The man-portable MILAN AGTW is carried to the firing position in two separate units, the missile in its container and the launcher. On firing, the container acts as the launch tube. The operator keeps his sight trained on the target while the goniometer automatically fixes the missile on course.

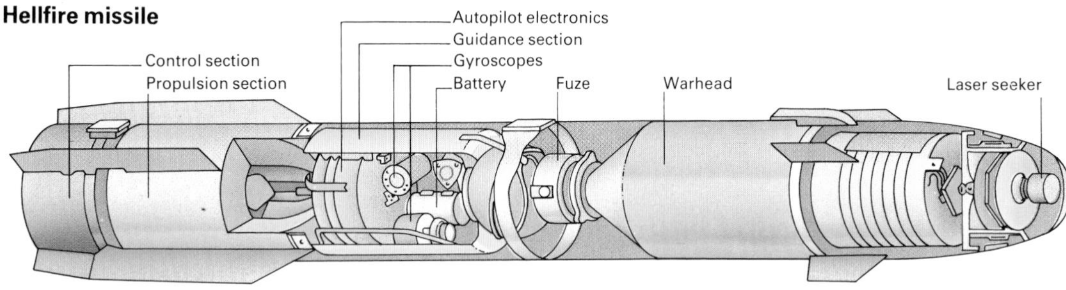

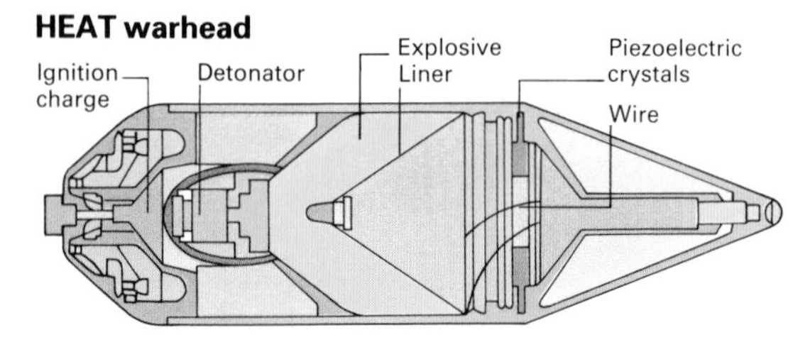

◄ Modern technology is incorporated in the detonation device for this modern HEAT warhead, which otherwise retains a standard design based on the shaped charge principle.

▲ The Hellfire missile carries a laser light detector and a shaped charge warhead in its front section, with control and guidance systems and the rocket motor at the rear.

Missile warheads

All anti-tank missiles now in service employ High-Explosive Anti-Tank (HEAT) warheads based on the shaped charge principle, also known for its inventor as the Munroe effect.

The shaped charge gets its name from the conical recess formed in the explosive at the front of the warhead. To enhance penetration, a thin metal cone liner is inserted into the cavity. The explosive is detonated at the rear so that the detonation wave passes through the explosive and over the copper cone, causing the cone to collapse inwards and forwards. The pressures involved are so great that although the copper may remain in the solid state, it is forced to flow like a liquid. By standing the warhead off from the target a free space is provided in which the liner can be focused into a long, thin jet. Such is the pressure of the jet that it forces the steel armor aside, producing a narrow hole as it penetrates. Any remaining jet not used up in penetration is then available to strike vital parts of the vehicle such as the fuel or ammunition. Typically a warhead with a cone diameter of 125mm penetrates well in excess of 600mm of steel armor.

Anti-tank missiles

Most anti-tank missiles have command signals from the tracker transmitted to them down a trailing wire. Radio communications across a battlefield can be unreliable owing to the nature of the terrain or to interference from other radio transmissions. Early anti-tank guided weapons (ATGW) required the operator to steer the missile by a joystick while keeping the target in the center of his sights. Any ATGW using this technique, known as manual command to the line of sight (MCLOS), is designated first-generation. MCLOS is largely superseded by the semi-automatic command to the line of sight (SACLOS).

The United States SACLOS weapon, TOW, has a range of 3,750 meters and weighs 19 kilograms, which makes it more appropriate for vehicle-mounting than for use by infantry. A Franco-German missile, MILAN, on the other hand, weighs a mere 7 kilograms, but its range is only 2,000 meters and its warhead is smaller than TOW's. The advantage of the SACLOS system over MCLOS is that the operator has only to track the target, while the missile is controlled automatically. TOW employs television to track the target by recognizing its tail flare. MILAN, a more recent device, contains a goniometer, an instrument which detects the infrared heat component of the tail flare, and determines the bearing of the course the missile should follow

to its target. As well as relieving the operator of the task of flying the missile, SACLOS also improves the flying characteristics of the missile since adjustments are made immediately and continuously, thus reducing any delay and eliminating the more severe changes in flight demanded by a human operator.

The evolution of the ATGW from MCLOS through SACLOS has now led to automatic command to the line of sight (ACLOS) with missiles such as the Hellfire. The operator has merely to launch the missile in the general direction of the target. Working in conjunction with the operator is an observer forward of the launch position, who illuminates the target with laser light. On-board detectors enable the missile to home in on the reflected beam by tracking the light back to the source of the reflection.

So far as the operator is concerned, the attack of armor is now "fire-and-forget", and the next step will be to dispense with the services of the forward observer as well. Missiles of this type are under development in the mid-1980s. Two methods in particular are being considered for missile guidance: one is to detect the infrared signature of the vehicle since it is at a different temperature to its surroundings; the other is to site a transmitting and receiving radar in the nose of the missile itself.

The Swingfire

Although a first-generation anti-tank missile, the British Swingfire will continue in widespread use into the 1990s. The attraction of this design is its unique facility for changing its flight-path by up to 45° immediately after launch. It has, as a result, a wide range of coverage which makes it ideal for vehicle mounting. Another unique feature is that the control and sighting unit can be separated from the launcher by up to 100m; this means that the launcher vehicle can be kept completely hidden behind rising ground with only the tracking equipment exposed.

This versatility at launch, which gives Swingfire its name, is achieved by a device known as a jetavator, a swiveling nozzle mounted at the rear of the missile and through which the propulsion jet passes. Immediately after the missile has been fired, the control unit transmits directional commands to it along the trailing wire. These are received by actuators inside the missile which then angle the jetavator to produce the prescribed change in course. The range of the Swingfire missile is 4,000m; it weighs 27kg and has a 125mm warhead.

The Rapier missile system

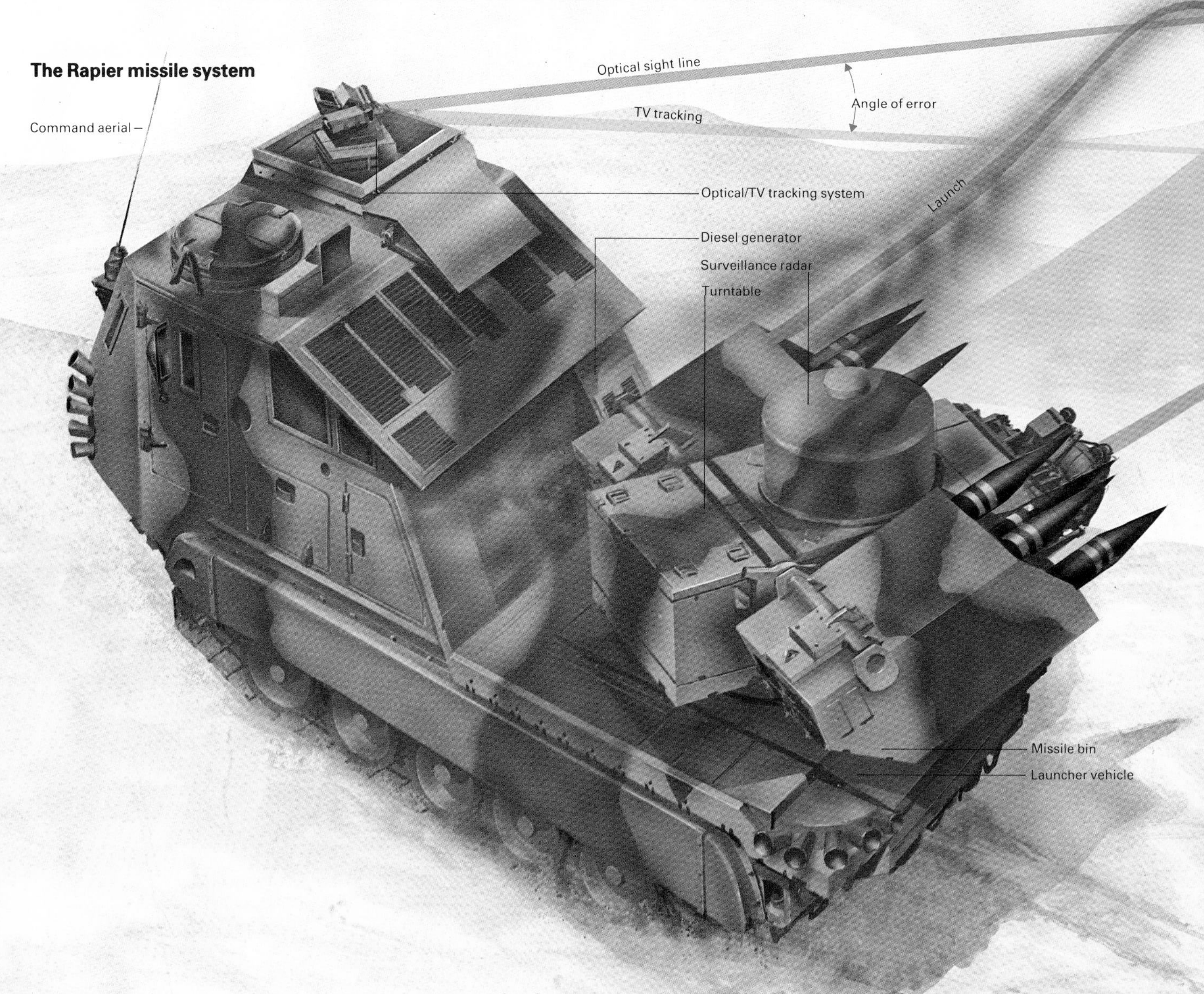

▲ The Rapier system incorporates a surveillance radar rotating through 360°. When the IFF system indicates a contact is hostile, the tracker head is automatically turned onto the target bearing. The operator searches to acquire the target, then tracks it until it comes within range. After firing, the missile is sent course corrections. It flies along the line of sight from tracker to target aircraft until impact occurs.

Surface-to-air missiles

The success of the early surface-to-air missiles (SAM) is measured by the complete change in aircraft attack profiles. The strike rate of these missiles was so discouraging for aircraft that they are now forced to operate at very low levels. So a whole new breed of SAM has evolved. A leading missile system operating at low-level altitude is the British Rapier. Rapier has its own radar to detect incoming aircraft. This makes the system ready, positions the missiles for launch and activates the Interrogation Friend or Foe (IFF) unit. The IFF unit determines whether the aircraft is hostile by transmitting a coded radio signal to which a friendly aircraft would react by transmitting back an identifying response. If the appropriate signal is not received and the aircraft is within range, a missile is launched. The operator then tracks the target visually through the optical system in the tracker head while a television camera follows the tail flare of the missile. Flight data are fed into a computer which determines the necessary flight-path corrections and these are transmitted to the missile via a radio link. All this occurs very quickly: the time from detection to destruction is approximately ten seconds.

Since its introduction, Rapier has undergone continual modification and improvement. A second radar, known as Blindfire, has

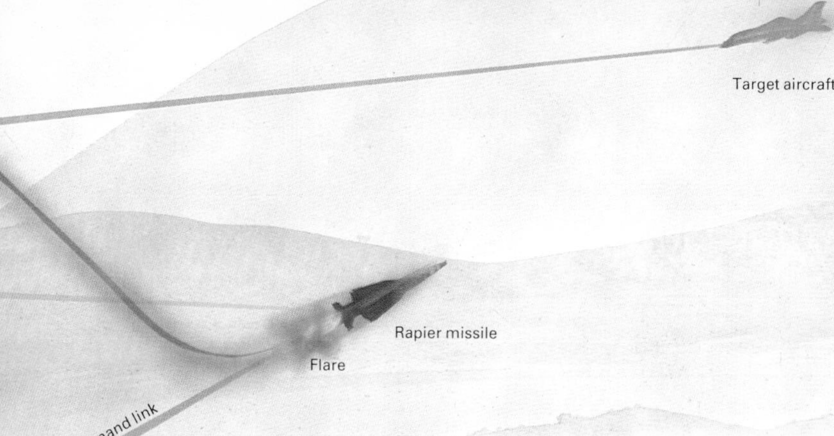

The modern air-defense gun

Whereas many countries have developed surface-to-air missiles capable of bringing down all types of aircraft including helicopters, none has yet produced a missile system that can provide adequate defense against helicopters attacking in massed formation. The problem is simply one of being able to engage a large number of targets in the short time available.

As yet the only viable solution is the rapid-firing air-defense gun. More effective still would be a missile-gun mix, but this option is very expensive, since suitable gun systems may cost more than a main battle tank, and large numbers would be needed to provide adequate defense. However, countries that can afford air-defense guns gain a system that would be effective against all types and formations of aircraft and helicopters. By combining a radar and a computer to lay the gun on a predicted interception point, high hit rates can be achieved. Indeed, so advanced are some of these guns that it is possible to determine the fall of shot, and with this information compensate for any error by re-laying the gun before the next salvo is fired.

◄ Blindfire is a radar system with a narrow pencil beam, allowing simultaneous tracking of missile and target. The radar is mounted on a chassis; the base housing of the radar contains the electrical, electronic and hydraulic power assemblies. The upper housing carries the main reflector, subreflector assembly and television gathering unit. Blindfire extends the application of Rapier to a 24-hour all-weather capability.

been incorporated to overcome the limitations of the optical system when operating in bad weather or at night. Tracked Rapier is also now in service, with the weapon system mounted on the chassis of an armored personnel carrier.

Portable SAMs

Such systems provide a very effective defense for a wide variety of potential targets, but for highly mobile troops, for those who are heliborne or those engaged in covert operations a man-portable system is essential. To meet this requirement the United States have developed the Stinger missile. Based on passive infrared homing, this guides itself to the target by detecting and tracking the heat emitted by the aircraft, particularly the engine exhaust. While Stinger has the advantage of being a "fire-and-forget" missile, it is susceptible to countermeasures such as heat flares ejected from the aircraft which act as decoy infrared sources. The alternative is to give control of the missile to the operator or to control it automatically. In the former case the operator is required to track the target optically while steering the missile using a thumb joystick; in the latter a television tracking system may be used instead of the joystick, thus making the missile semi-automatic in operation.

▲ The Sgt. York is a high-mobility air-defense gun system which employs two radars to pinpoint enemy aircraft. Both operate continuously, one rotating in the horizontal plane, the other in the vertical. Using an on-board computer, the twin 40mm Bofors guns then deliver shells to a predicted intercept point at a combined rate of ten rounds per second. It can engage aircraft attacking in massed formation.

See also
Measuring with Electricity 41–46
Seeing with Sound 65–72
Seeing beyond the Visible 73–86
Air Transport 217–224
Space Transport 225–228

▲ ***The outstanding feature of this Boeing 707, modified for AWACS, is its 9·2m rotodome, rotating at 6rpm, which houses surveillance radar and radio link antennae.***

At the onset of any battle, the advantage always lies with the aggressor, since it is he who selects the point of attack and has the ability to concentrate his forces. The defender must deploy his resources more widely in order to protect all his key positions. This is nowhere illustrated more clearly than in the provision of land-based defense of ground targets against attack from the air. Aircraft, because of their speed, have the ultimate capability for surprise attack and are difficult to hit and bring down. Modern jet aircraft cover one kilometer in three seconds, which means that if the aircraft is flying low (at a height of 50 meters) the engagement time will be extremely short, almost certainly less than ten seconds. However, the aircraft have only the same few seconds in which to identify and engage their targets, so if defensive positions can put up sufficient fire to distract or deter the pilot, the engagement will be incomplete and the aircraft payload will remain undelivered.

Besides engaging in ground attack, aircraft have an important surveillance role especially behind enemy lines where there are few alternative methods. If enemy reserve echelons can be detected and consequently engaged deep behind the front line, they can be destroyed long before they reach the combat zone. The problem is that aircraft in the conventional combat role remain just as vulnerable to SAMs both at high and low level.

However, this limitation has been overcome by mounting the radar on an aircraft and then flying at about 10,000 meters. An excellent example of this is the American Airborne Warning and Control System (AWACS), a converted Boeing 707 with a radar mounted on top which has a range of 400 kilometers to the distant horizon. Thus the aircraft can fly in the relative safety of friendly airspace while identifying potential targets for engagement by indirect fire. This is the strategy that will be employed by NATO to negate the large numerical advantages in equipment held by the Warsaw Pact countries.

▼ ***The Exocet missile has a range of 40km, so defensive measures must concentrate on detecting, destroying or avoiding the missile while in flight, rather than attacking its launch vehicle.***

Radar at war

The farther away a radar is required to search, the weaker the echo signal will be against surrounding "noise". Methods of enhancing the signal-to-noise ratio are important in radar, particularly in its military uses, either by making the aircraft or ships very "quiet" to radar or by making a lot of "noise" (jamming). Thus modern warships are designed without vertical surfaces to reflect as few radar waves as possible, and they may scatter clouds of metallic "chaff", to confuse the enemy radar. The fact that a radar is operating can be detected by an object within its field. Exocet missiles use this principle to home in on radar-carrying targets. An aircraft pilot enters the radar range of its target before launching the Exocet, to "program" it with the location of the target. The missile then flies below radar range, and switches off its own radar when approaching the target to avoid detection.

Printing

The origins of printing...Typesetting, by hand, linotype machine, photography and laser...Color origination... Printing presses...PERSPECTIVE...Letterpress, lithography and gravure...The history of typesetting...Electronics and newspaper production... Xerography and laser printing...Printing inks

From ancient times writing has proved to be a very effective means of human communication, and it provides the principal record of what happened in the past. Until the Middle Ages, however, making a written record was a laborious task that had to be carried out by hand. In the 15th century the technique of printing was developed, and for the first time it became possible to distribute information identically and cheaply to a great many people. Since that time, printing has developed as a fundamental technology underpinning virtually all human societies. Provided that adequate education is given in reading, printing provides a means by which information and understanding can be available to all. Modern printers have adopted highly complex and advanced technologies to improve the efficiency and effectiveness of the medium.

Printing is a complex process, and the press itself constitutes only one link in the chain of printing a book or paper. Before press, the words have to be set into type and arranged on the page; and the illustrations sized and prepared so that they too can be reproduced cleanly on the press. Each of these tasks has spawned its own industry, known respectively as typesetting and reproduction. The paper on which the job is printed must also be carefully selected for its ability to take ink without smearing, to maintain its brightness, its resistance to warp and so on as well as its cheapness; and the inks must match standard tones, maintain an even consistency and dry quickly.

◄ **The original flatbed printing press used by Johannes Gutenberg in the 1450s. The type was placed in a sliding tray and inked; the paper was placed on top and the press was screwed down to print the page. This technique remained virtually unchanged until the late 18th century.**

Letterpress, gravure and lithography

The basic principle of printing is to transfer a well-defined pattern of ink, representing letters or illustrations, from a printing plate to paper. There are three main ways of localizing the ink on the printing plate. First, if the pattern stands out in relief from the rest of the surface, these high points can be inked by a roller and placed against the paper in a press. This is the traditional method of printing, often called letterpress.

Another way of localizing the ink is to form patterns of pits and depressions in the plate. If these are full of ink and the rest of the surface is wiped clean, ink will only reach the paper where it is required. This is the gravure process, and is used for jobs (such as printing postage-stamps) that require fine quality but long print-runs.

The third technique uses a flat printing plate. The pattern is defined with a greasy ink and the rest of the surface covered with a thin film of water; the ink and water repel each other and the pattern remains distinct. This technique was invented by the German printer Aloys Senefelder in Munich around 1800; he termed it lithography (from the Greek word for stone) because it required a fine stone surface. Modern lithography uses metal or plastic plates, and has been growing in popularity steadily throughout the century. It is now the most common of the three techniques, used worldwide for newspapers, magazines, books, leaflets and other forms of general printing (▶ page 176).

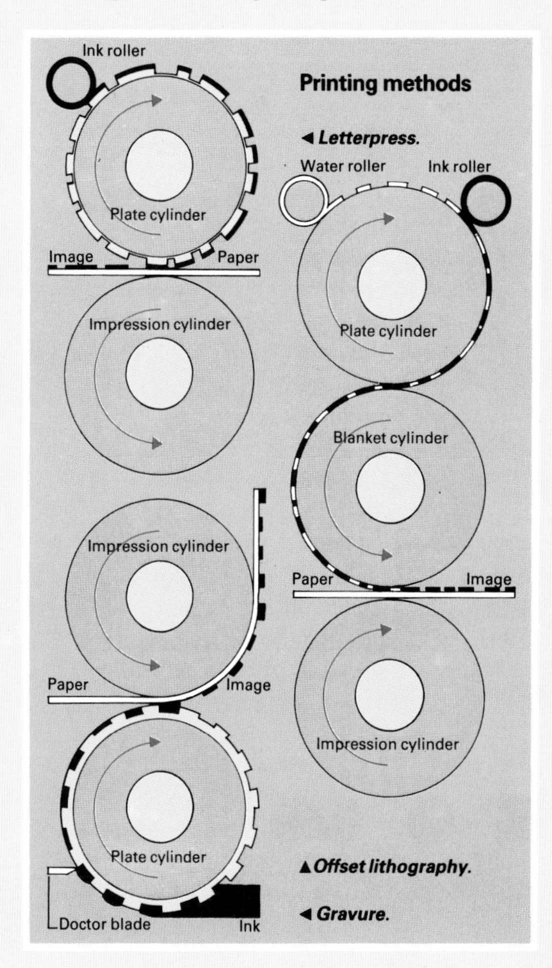

Printing methods

◄ ***Letterpress.***

▲ ***Offset lithography.***

◄ ***Gravure.***

Chronology

c1450 Printing press and movable type invented by Johannes Gutenberg in Gemany
1642 Mezzotint process of printing in graduated tones invented by Ludwig van Siegen
1719 First three-color engraving process invented by Jakob Le Blon
1768 Aquatinting method of engraving invented by Jean Baptiste le Prince
1775 Standard units for measuring type sizes introduced by Francois-Ambroise Didot
1790 Rotary movement incorporated into printing press by William Nicholson
1790 Stereotypy – copying type in papier-mâché from which to cast printing plate – introduced in Paris
1796 Lithography introduced by Aloys Senefelder in Germany
1800 First all-metal press built by the Earl of Stanhope, in England
1803 Fourdrinier machine for making continuous rolls of paper and cutting specific sizes installed in London
1810 Steam-powered printing-press invented by Friedrich König and Andreas Bauer; it was installed by *The Times* in London in 1814
1822 Typesetting machine patented by William Church
1837 Chromolithography process of color printing patented by Godefroy Engelmann
1844 First rotary printing press invented by William Hoe in United States
c1850 First mechanically operated press introduced for photolitho process
1855 Collotype printing process patented in France

Evolution of typesetting

For over 450 years metal type had to be laboriously yet skilfully assembled by hand to form the printing block. Often the printer cast his own type in suitable molds, and had to hold a very large stock of the individual letters and symbols in a range of sizes. Then, in the 1880s, machines appeared which made this process of casting and setting partly automatic. Linotype machines cast a line at a time, Monotype and Intertype letter by letter, as the copy was keyed in at a keyboard resembling that of a typewriter. However, the printer still had to assemble the pages of metal type and the separate blocks which held the illustrations. Changes to the text or page design could only be carried out by casting new type and reassembling the page. In the 1950s photosetting machines were developed which held a range of letters as photographic negatives. As the keys were pressed these negatives were assembled and exposed onto a sheet of photographic film. A complete page was then made up by hand and converted to transparent film from which the printing plate was made. This system is still widely used.

▼ The Linotype machine, invented in the 1880s, is still used on newspapers. As the operator keys the copy, a matrix for each letter falls into place. The assembled line is filled automatically with spaces to justify it, then closed with a crucible containing hot metal. The line of type is thus cast and placed in the galley, while the matrices are returned to their magazines at the top of the machine.

1865 First roll-fed rotary press built by William Bullock
1880 Half-tone process for reproducing photographs introduced
1884 Linotype hot-metal setting machine invented by Ottmar Mergenthaler in United States
1887 Monotype system of typesetting patented by Tolbert Lanston in United States
1890 Flexography patented in Britain
1895 Rotogravure printing process invented by Karl Klíč
1904 Printing press for offset lithography built by Ira Rubel
1911 Typecasting machine for large type perfected by Washington Ludlow
1937 First practical dry photocopier demonstrated by Chester Carlson in United States
1939 Photocomposition machine invented by William C. Huebner
1947 Mechanical phototypesetting introduced
1949 Functional typesetting introduced, with Lumitype
1954 Programmed composition using perforated tape introduced
1960s Cathode ray tubes introduced to phototypesetting; magnetic tape replaced perforated tape for phototypesetting
1960s Web offset printing machines introduced for newspapers
1965 Alphanumerical phototypesetters introduced
1970 Computer-controlled composing systems introduced
1973 Color duplication photocopiers introduced
1976 Laser typesetters introduced by Monotype
1984 Image-processing color scanners introduced

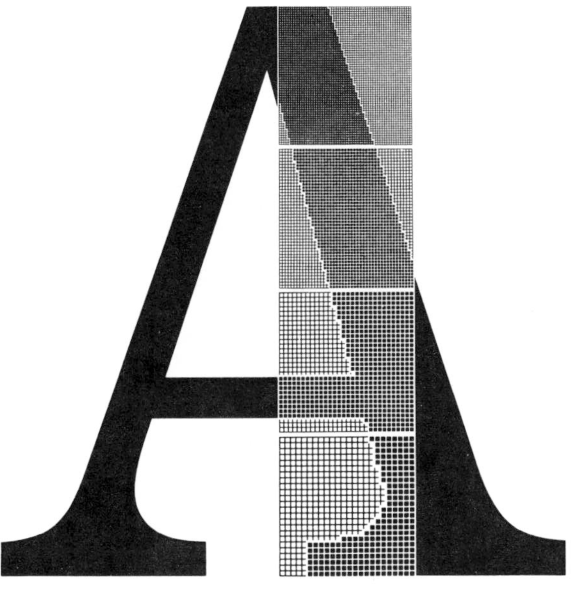

◄ Traditional typesetting involved assembling lines of metal characters by hand on a composing stick, then locking them in place in a "forme". Each character was made of an alloy of lead, antimony and tin, and was cast from a copper mold.

▲ In digital typesetting, the character is stored as tiny dots and reproduced by a cathode ray tube; the characters are ragged in outline. A laser typesetter creates the character as a series of horizontal line-scans by deflecting a laser beam off a spinning mirror.

Setting the type

The assembly of the right combination of letter shapes to form a printing plate is called typesetting, after the traditional method of setting the individual pieces of metal type together in a frame. The modern method of typesetting relies on computers, lasers and photography to do the job. The text is typed into a computer keyboard, with the result displayed immediately on a VDU (visual display unit); this task is increasingly done by the authors themselves, who may pass word-processed disks straight to the typesetters. Typesetting commands, instructions to the computer to change typeface or size, to add justification, columns and so on, are then added. Most modern systems allow the operator to see the effect of these commands to the layout of the page, so that corrections and amendments can be made to uneven lines or overlength copy before setting. The copy is then stored on magnetic disks or punched paper tape.

When the text is satisfactory, the next step is to convert it into a typeset version, which can be used in combination with the illustrations to make up the printing plate. The computerized text, together with its typesetting commands, is passed to a typesetting machine. The characters that go to make up each font (or set of typeface designs) are stored digitally in the memory of the typesetting machine, and are generated by a cathode ray tube or laser beam; the character is focused and resolved at the required size on photographic paper or film. Lines are justified automatically, and are spaced by moving the paper or film. Complete lines are set at a time, arranged in position so that the output can be transferred directly to a printing plate. The fastest machines can set 1,000 narrow-measure lines (as used in newspapers) a minute.

The image on the film is transferred photographically to the printing plate, which has been coated with a light-sensitive chemical. The film is placed on top, the plate is exposed and, on developing, the parts that have not been exposed to light are washed away. The remaining pattern of coating acts as a protective mask for a chemical etch which forms changes in relief for letterpress and gravure plates or in surface properties for lithographic plates.

◄ Visual display terminals such as the Linotron Typeview allow the compositor to see the effect of makeup instructions added to the text. Such machines have changed typesetting by permitting pages to be assembled on screen rather than by hand.

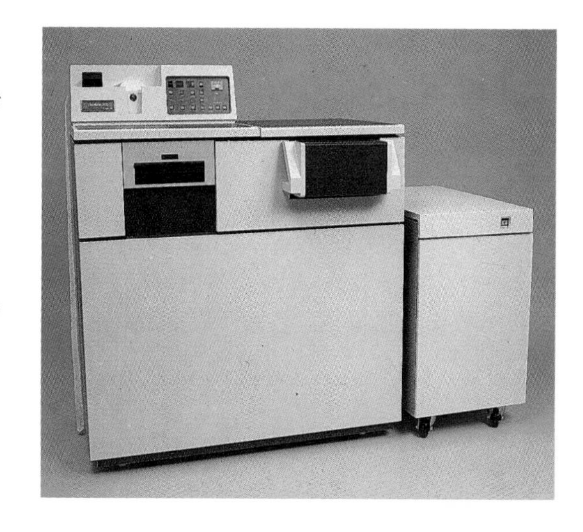

► The Linotron 202 is a widely used digital phototypesetter for general typesetting. It can access details of 60 fonts instantaneously. These are held on magnetic disks and are converted via the optical fibers of the faceplate as a continuous series of fine strokes.

Text and photographs can both be handled electronically in a modern newspaper office

Color reproduction

The retina of the human eye contains three groups of cells which are sensitive to the three primary colors – red, green and blue – respectively. We perceive all other colors by the relative responses of these three groups. Thus light which makes the red and green sensors give a roughly equal response is seen as yellow; an equal red and blue response as magenta (a purple color); and green and blue as cyan (a light blue). White light contains equal proportions of the three primary colors, and white paper reflects these colors equally. Ink absorbs certain colors and reflects others, and to reflect the primaries inks must be used that absorb other wavelengths. Thus the secondary colors (cyan, magenta and yellow) are used as the inks in color printing. Magenta ink absorbs only green light and reflects red and blue; yellow absorbs blue and reflects red and green; by overlapping yellow and magenta inks, green and blue are absorbed and red is shown. Virtually any color can be printed from a combination of three inks, cyan, magenta and yellow, and by varying the proportions of the three colored inks most other colors can be synthesized.

In letterpress and lithography the amount of ink deposited cannot be varied continously across the printing plate: either an area is printed or it is not. So areas of tone in illustrations are divided into a fine, regular pattern of dots by a process called screening. The size of the dots determines the density of ink at that point: a pattern of small dots gives the effect of a light tone, while a similarly spaced array of larger ones gives a deeper tone. To print a color picture, each of the three colors is printed separately using a different screen pattern to control the final hue in all parts of the picture. Black is also printed as a fourth "color" to give a sharper tone and more accurate shading.

The four screened images that will form the printing plates are derived from the original illustration by means of a scanner. A light beam or laser scans over the surface of the original picture and the relative amounts of the primary colors are measured, point by point, by photocells. From this information a computer determines the dot pattern, and outputs four pieces of transparent film, or separations, on which the dot patterns of the illustration are produced photographically, scaled to the size required for the printed image.

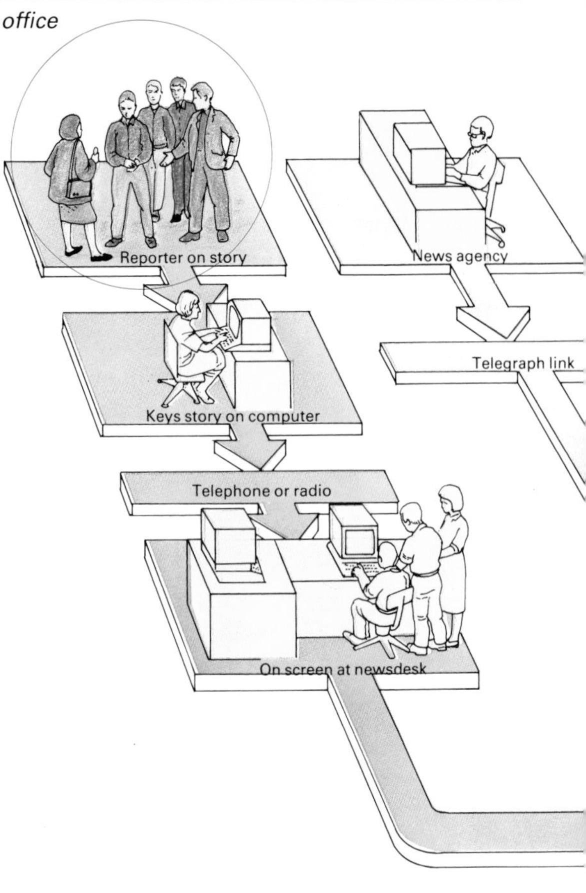

Electronic newspaper production

► ***A modern newspaper can take advantage of the new technology to speed communications. By keeping stories, artwork and photographs stored on a single memory bank, they can be accessed as required by journalists, editors, designers and printers. The central element is the page layout terminal at which text and pictures can be assembled on screen and juggled to produce an attractive page. It allows the designer to make amendments instantly.***

▼ A scanner converts an image, such as a transparency or piece of artwork, into film for printing. The image is placed on the drum; a laser spins inside the drum, traveling across the image to record the color density at each spot.

◄ This series of color separations show how the image circled is broken down by the scanner into the four printing colors, with each color further broken down by screening into a pattern of standardly spaced dots. The image is exposed with filters permitting only one color to show through at a time, while a screen of dots covers the photographic emulsion. These are graduated in density such that a strong exposure to light produces a large dot and a lesser exposure creates a smaller dot. When the four screened images are put together the original image is reproduced. The screens are arranged so that the dots do not overlap exactly – densely colored areas show a roseate pattern.

Color scanner

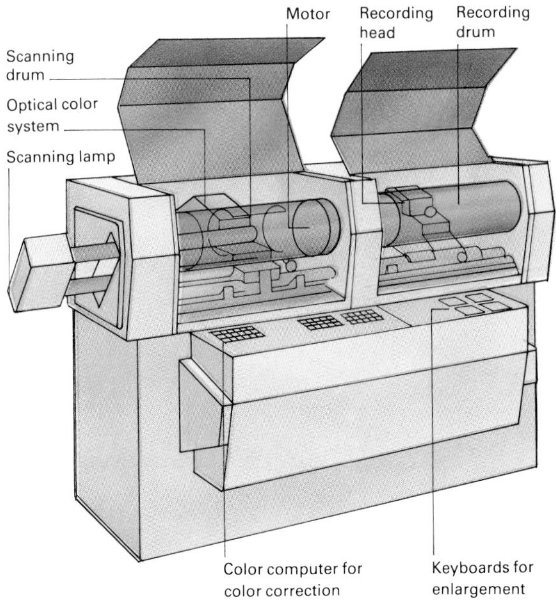

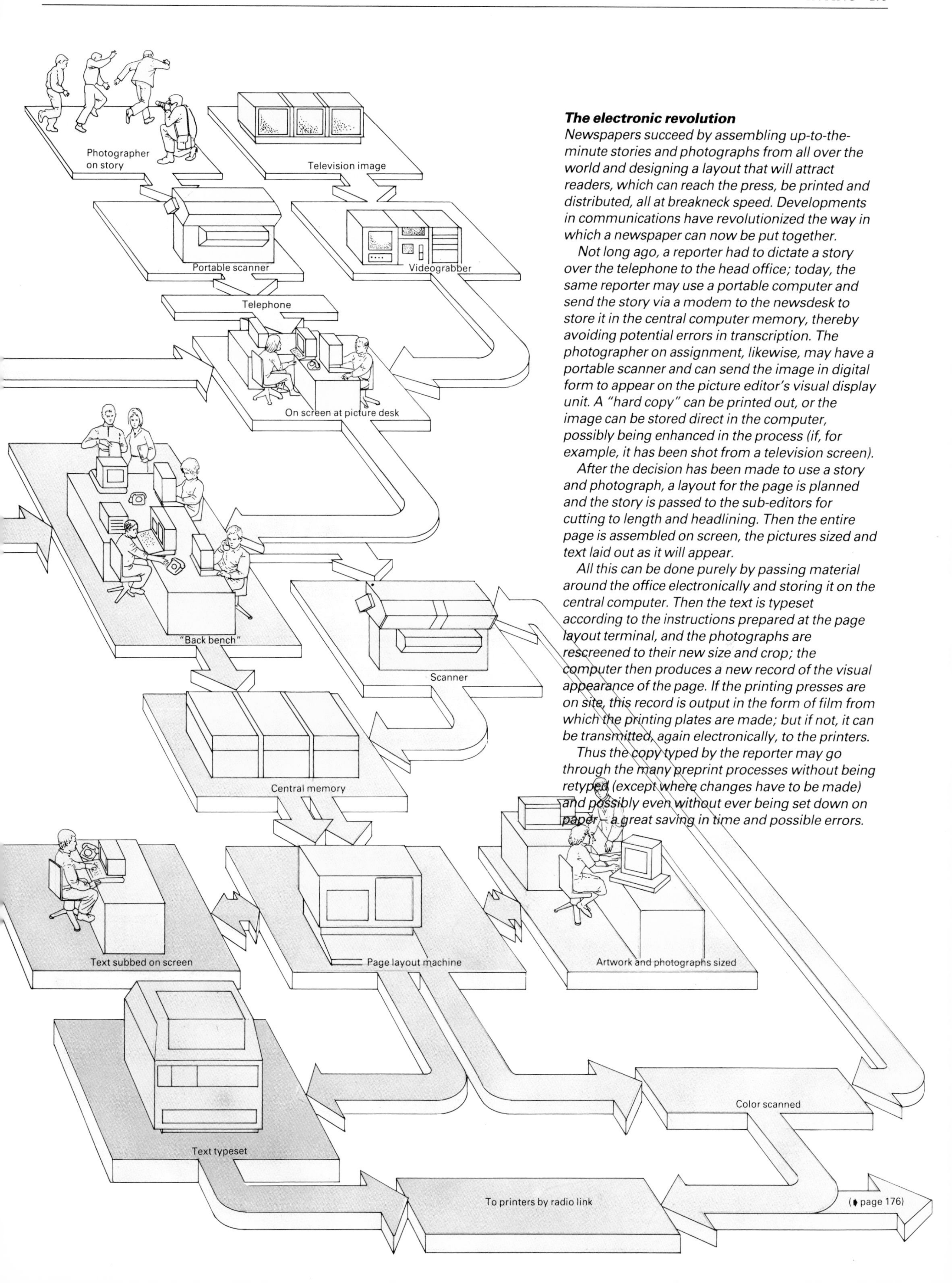

(▸ page 176)

The electronic revolution

Newspapers succeed by assembling up-to-the-minute stories and photographs from all over the world and designing a layout that will attract readers, which can reach the press, be printed and distributed, all at breakneck speed. Developments in communications have revolutionized the way in which a newspaper can now be put together.

Not long ago, a reporter had to dictate a story over the telephone to the head office; today, the same reporter may use a portable computer and send the story via a modem to the newsdesk to store it in the central computer memory, thereby avoiding potential errors in transcription. The photographer on assignment, likewise, may have a portable scanner and can send the image in digital form to appear on the picture editor's visual display unit. A "hard copy" can be printed out, or the image can be stored direct in the computer, possibly being enhanced in the process (if, for example, it has been shot from a television screen).

After the decision has been made to use a story and photograph, a layout for the page is planned and the story is passed to the sub-editors for cutting to length and headlining. Then the entire page is assembled on screen, the pictures sized and text laid out as it will appear.

All this can be done purely by passing material around the office electronically and storing it on the central computer. Then the text is typeset according to the instructions prepared at the page layout terminal, and the photographs are rescreened to their new size and crop; the computer then produces a new record of the visual appearance of the page. If the printing presses are on site, this record is output in the form of film from which the printing plates are made; but if not, it can be transmitted, again electronically, to the printers.

Thus the copy typed by the reporter may go through the many preprint processes without being retyped (except where changes have to be made) and possibly even without ever being set down on paper – a great saving in time and possible errors.

See also
Photography 87–90
Recording on Tape and Disk 91–96
Lasers and Holography 101–104
Computer Software 123–128
Telecommunications 177–184

Printing presses

Once the text has been set, pictures scanned and pages made up into final film, the job is ready to go to press. The film for the pages is imposed, or assembled such that up to 48 book pages may be printed on a single printing plate; their order and orientation is planned so that they can be folded neatly into a single section of a book or magazine. The photographically prepared printing plate, usually made of aluminum but sometimes plastics or paper, then receives the image of the page so that some areas of the plate hold ink while others are clean.

Many different designs of press are in use, even within the most common category of lithographic printing. The main types of large press include rotary or web presses, for high-speed and large print-run jobs, and sheet-fed presses which are more frequently used for books. On an offset litho press, the plate is attached to a cylinder which is inked as it rotates. The ink adheres only to the image areas, and is passed onto a second cylinder from which it is transferred to the paper. Color presses have four plates and four sets of cylinders, which have to be lined up or registered exactly to achieve a crisp image. Some presses print both sides of the paper; on others the sheet must be fed through twice. It is crucial to achieve an even flow of ink across the paper and throughout the job; ink flow can now be controlled by a computer.

Ink-drying can be a problem with high-speed presses such as webs which may run at speeds exceeding 60,000 copies an hour. In these cases drying is accelerated by means of high-energy dryers or by ultraviolet or infrared radiation. After printing, the paper must be folded, cut or perforated, and perhaps sewn or glued. Add-on units make it possible to carry out these operations as part of the printing process, and the finished product might, in the case of a magazine, include such variations as tear-off strips, glued-in cards and pop-ups or cut-outs.

Xerography and laser printing

To produce a small number of copies of a document, the conventional forms of printing are uneconomic. Various forms of low-cost duplication have been developed, and one of the most successful has been xerography. This relies upon the properties of photoconductive materials, which are electrical insulators in the dark, but then become conductors where light shines upon them. The heart of most xerographic copiers is a rotating drum coated with a photoconductive material such as selenium. First of all, in the dark, the surface of the drum is given an electrical charge from a high-voltage electrode. Then an image of the document to be copied is projected onto this charged surface. The light parts of the image cause the coating to become conductive, and so the charge leaks away from these areas. When a very fine black powder (called the toner) is sprinkled over the drum it adheres by electrostatic attraction only to the parts which are still charged – that is the areas of the projected image which were dark. This pattern of toner is transferred from the drum to a sheet of paper, where it is made permanent by heating to fuse the toner particles together. Color copying is achieved by repeating the process with three different toners.

The xerographic copier can be turned into a printing device by scanning a laser beam across the drum to draw out the pattern of letters or illustrations. Laser printers of this sort can be very fast, producing up to one hundred copies a minute. A particular advantage with this technique is that because each page is "written" separately by the laser, each page can be different from the last. A single copy of a complete book can be printed out if required, instead of a batch of a single page.

Printing inks

Ink consists of pigments, which form the color, and additives, which control properties such as the stickiness or drying speed. Both are held in a base called the vehicle. The first printing ink, based on a gum, was developed for woodblock printing in China in the 3rd century AD. Later, oil-based inks were found to be more suitable for metal type, and nowadays a very wide range of oil and solvent-based inks are available.

A letterpress ink should be tacky and not flow very easily. Lithography involves distinguishing between areas of the plate that receive ink and others that repel it, so for litho printing a balance must be struck between the thickness of the oil-based ink and the water which sits on the non-ink areas. If too much water is used, the ink will thin; if too little, the ink smudges. A gravure ink should be thin enough to fill the tiny cavities of the plate, with good adhesion to paper so that the ink will be drawn out during the impression. On all high-speed presses, fast drying is desirable so that the ink does not smear or transfer to another sheet. There are many special inks, such as magnetic inks for security printing and electronic reading of cheques. Metallic effects are not usually achieved with inks, but by blocking thin foils onto the paper.

◀ A web offset press prints a continous roll of paper.

Telecommunications

25

The first electronic communications...Telephones and exchanges...Carrying the signal...Integrated circuits... Optical fibers...Communication between machines... PERSPECTIVE...Early days of telegraphy...Chronology... Modern digital exchanges...Transatlantic communication...Semiconductor lasers...Manufacturing optical fibers...Electronic Funds Transfer

In 1842, Sarah Hart was found dead in her home in Slough, England. A man identified as John Tawell had been seen leaving her house, and later he was spotted boarding a train for London. A message was relayed along the recently installed telegraph to Paddington Station, London, and Tawell was arrested when he arrived. Eventually he was tried and executed for the crime.

The events behind this story made a large contribution towards establishing the electric telegraph in the eyes of the public as an invention with a significant future. Various experimental devices had been built since the late 18th century, but the telegraph between London and Slough, completed in 1841, was the first commercially successful system. It used two wires to transmit electric currents, which deflected two magnetic needles. The number and direction of deflections indicated a certain letter, and skilled operators could spell out messages at rates of up to 20 words per minute. For the first time messages could be transmitted over long distances virtually instantaneously.

Most of the pioneering systems were installed by railroad companies to improve the speed and efficiency of their signaling, but events like the arrest of Tawell quickly led to the establishment of public telegram services and press networks. Most of these used single wires transmitting to a sounder which produced an audible output of Morse code dots and dashes. It was more than 50 years before voice telephony superseded this type of coded telegraphy as the principal means of communication at a distance.

Two productive partnerships

William Fothergill Cooke (1806-1879) was a medical officer in the Indian Army when he became interested in telegraphy. He formed a partnership with Sir Charles Wheatstone (1802–1875), a professor of physics at King's College London, to develop and commercialize a telegraph system. Their first trials in 1837 were with a six-wire system which used combinations of five needles to indicate a letter. They soon realized, however, that practiced operators could rapidly interpret more complex codes, so they simplified the system to two needles and then finally to one.

Meanwhile, the American painter Samuel Morse (1791-1872) was learning about the new subjects of electricity and telegraphy. By 1835 he had made a crude prototype of a system that would produce a record of a message directly onto paper. Most of the subsequent technical development was due to others, notably the mechanic Alfred Vail (1807-1859) with whom Morse went into partnership in 1838. Their system, patented in 1840, included an electromagnetic relay which, by regenerating the signal every few miles, allowed indefinitely long links to be built. At the receiving end, an electromagnet moved a pen to record automatically the coded message on a strip of paper. Vail was probably responsible for what was the most important contribution in this work, the combination of short and long pen marks (dots and dashes) known as Morse code. When this system began to be used, it was found that operators could interpret the sounds coming from the pen much faster than they could read from the paper tape. So the pen was replaced by a "sounder" which became the standard method of receiving telegraphy throughout the world.

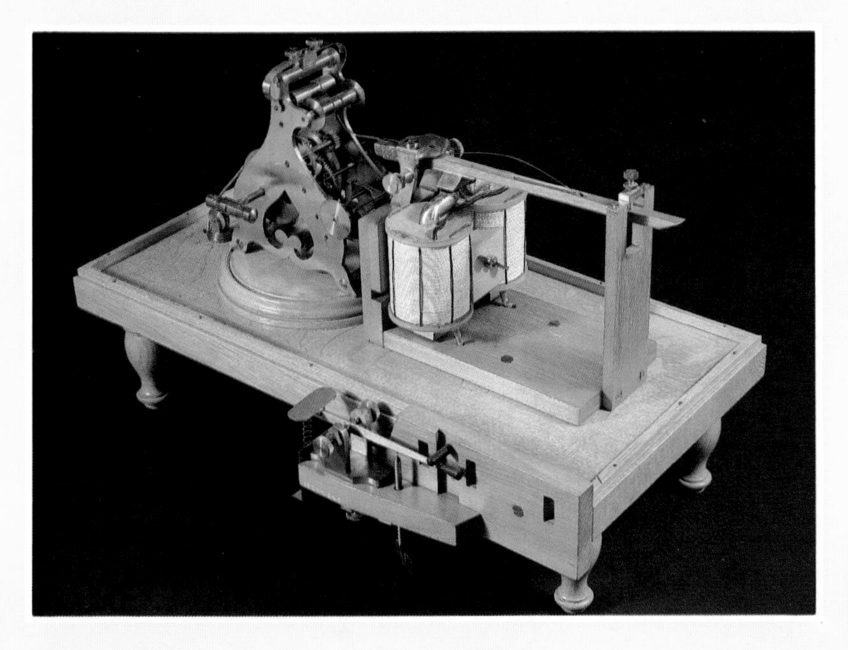

◄ The Morse receiver was a single-needle telegraphic system: electromagnetic signals activated a pen which recorded a coded message in the form of long and short dashes.

▲ Cooke and Wheatstone's two-needle telegraph. The handles were turned to connect the battery to a circuit. The resulting current deflected the magnetic needles.

Chronology

1774 Telegraph using electrostatic machines built by Georges Lesage of Geneva
1792 Optical-relay semaphore system pioneered by Claude Chappe in France
1837 Six-wire, five-needle electric telegraph patented by Charles Wheatstone and William Cooke
1840 Electromagnetic message relay system patented by Alfred Vail and Samuel Morse
1844 First morse telegraph line opened between Baltimore and Washington DC
1846 Printing telegraph invented by Royal House
1858 High-speed automatic morse system developed by Charles Wheatstone
1866 First successful transatlantic cables laid
1872 Time-division multiplexer printing telegraph system invented by Jean Baudot in France
1874 Quadruplex telegraph system designed by Thomas Edison
1876 Telephone invented by Alexander Bell
1878 First commercial exchange opened at New Haven, Connecticut
1881 Two-wire all-metal circuit telephones introduced
1887 Typewriter adapted for telegraph use
1889 First practical system of automatic telephony invented by Almon Strowger in USA
1892 First public automated exchange installed, at La Porte, Indiana
1893 Centralized battery for telephone speaking and signalling first used commercially at Lexington, Massachusetts

Telephones

To make a telephone call, it is first necessary to convert the sound wave of the voice into an electrical signal for transmission along a wire. This is done by the microphone, which consists of a circular diaphragm pressing against a chamber full of carbon granules. Carbon is not a particularly good conductor of electricity, so this loose arrangement of granules acts as an electrical resistor. The amount of resistance to current flow that it presents depends on how firmly the granules are squashed together. When a sound wave reaches the microphone, the diaphragm vibrates in sympathy; the granules are compressed and released, and the electrical resistance changes accordingly. If a voltage is applied to the microphone, the current varies and an electrical waveform is produced which closely follows the original sound waveform.

This electric current then passes through wires and switches until it reaches the receiver of the other telephone. The receiver has the opposite job – that of converting the electrical waveform back into sound again. This is done with a coil wound around a magnet, placed close to a steel armature. When a current passes through the coil, it changes the magnetic field that is acting on the armature. The armature in the receiver moves in response to the electrical signal that was generated by the microphone, and the speaker's voice is re-created as a sound vibration in the air.

Exchanges

A major feature of the telephone system is the ability to be connected to any one of a large number of other subscribers, which means that some form of switching is needed to route a call to the correct destination. Initially callers could only be connected to others who were attached to the same exchange. As the service grew, connections between exchanges became necessary and exchanges themselves became larger. A long-distance call had to be routed by three or four operators, resulting in a slow, cumbersome and expensive service. So there was a need for automatic switching methods, operated by dials or push-buttons on the telephone. These controls cause a series of pulses to be transmitted to the exchange; these direct the switching apparatus at the exchange to route the call to the correct person. Early automatic systems used various types of motorized switches. Each dialed pulse causes the switch to move to the next position along, thus connecting to different set of wires. However, most modern exchanges use solid-state electronic devices to carry out the switching process, and computers to control the operation. To do this efficiently the voice signal itself must be converted to a digital form, so that the electrical waveform no longer looks like the soundwave, but itself is a series of pulses. The powerful computer circuitry used in these electronic exchanges allows the switching to be carried out more cheaply and reliably than with electromechanical systems. It also provides the opportunity for new facilities, such as the automatic redirection of calls to another number, to be incorporated comparatively simply. As more and more exchanges become electronic, more of this type of facility can be added.

▲ ***Before the arrival of Strowger's automatic exchange, subscribers were connected to a small local exchange where the operator would make connections by means of a plug-and-socket arrangement.***

The birth of the telephone

The first practical telephone was built in 1876 by the Scottish-American inventor Alexander Graham Bell (1847-1922), who started his career as an expert on the physiology of speech. In 1874 he patented a method for transmitting several telegraph messages simultaneously along a single wire by sending each one as a different tone. This lead him into ways of converting speech into electrical signals. His first successful telephone system, developed in 1876, used an electromagnet and a diaphragm, rather like a modern receiver, for both microphone and receiver. This invention won very rapid acceptance, and in 1877 the Bell Telephone Company was selling pairs of instruments for social purposes. The first commercial exchange was built at New Haven, Connecticut, in 1878.

Avoiding human error

The first practical automatic exchange was invented as early as 1889 by an undertaker from Kansas City, Almon B. Strowger. His telephone operator was the wife of a rival undertaker, and he suspected that she was diverting his calls to benefit her husband's business. Strowger devised a switching system using electromagnetic stepping switches that would give him the "girl-less, cuss-less" telephone. Although his prototype consisted of a hat-box with a hat-pin through it, the system was so successful that many "Strowger" exchanges based on this principle are still in use.

1900 Load circuits introduced to reduce telephone signal distortion
1910 Successful multiplexing of telephone systems achieved
1915 Transcontinental telephone circuit opened in USA using triode valve signal repeaters
1926 Short-wave radio used to transmit transatlantic telephone signals
1927 H.S. Black devised negative feedback to improve repeater performance and prevent interference between channels
1943 Submerged vacuum-tube amplifiers incorporated to telegraph systems
1946 High-frequency microwaves used to transmit telephone signals
1956 Telex introduced in Canada
1956 Transatlantic telephone cables with submerged repeaters laid
1962 Telstar demonstrated satellite potential in relaying telephone signals
1963 Regular commercial satellite relays for telephone
1964 Code teleprinters introduced
1965 First electronic telephone exchange installed in New Jersey
1967 Battery-operated cordless telephones developed in USA
1970 First long optical-fiber for telecommunications produced by Corning Glass, USA
1972 Video-telephones introduced by Bell Systems
1976 First optical-fiber cable trunk for telecommunications installed
1981 Teletex communication system launched

► In a digital exchange, multiplexed calls have to change their time-slots as well as channels, in an operation lasting no more than a few microseconds. Thus digital exchanges have time switches as the call enters and leaves the exchange, as well as the usual space switches. "Bits" enter the time switch and are placed in a buffer in rotation, then removed during a later time-slot. Their path through the space switch is provided by means of electronic crosspoints which route them to an outgoing time switch. The outputs of time switches in the exchange can be multiplexed together for best use of the space switch crosspoints.

► A trunk exchange serves one or more numbering plan areas into which a country is divided. A call from one area to another is routed via the exchanges; the number dialed informs each control unit of the direction required.

▼ A digital exchange.

Telephone signals

To make switching and transmission more efficient, the continuous voice waveform is usually converted to a digital signal. The amplitude of the original waveform is measured at regular intervals (usually 8,000 times a second), and then the value of these samples is converted to a binary code. This code is transmitted as a sequence of pulses which represent the binary digits 0 and 1.

The next question is how to provide an adequate number of channels on a long distance route. Between two cities there might be a regular demand for 10,000 telephone calls simultaneously. One solution is to lay 10,000 cables alongside each other, each carrying a separate call. However, this is normally very expensive, both in the cost of the wires or optical fibers and in maintenance. A much cheaper solution is to combine calls in such a way that they can be transmitted simultaneously along a single wire and separated again at the other end. The combination of signals in this way is called multiplexing. With digital signals it can be done simply by increasing the rate at which the binary pulses are sent out and interleaving many signals into a sequence. It is quite practicable to send binary information over long distances at rates of several hundred million pulses per second, and at this rate many thousands of voice signals can be combined. Using multiplexing, thousands of simultaneous phone calls are normally transmitted over a single wire, optical fiber or other link.

Digital telecommunications will give subscribers access to cable TV, home shopping and banking facilities

Carrying the signal

The connection between a telephone and its local exchange is usually a fairly cheap piece of wire, suspended between poles or buried underground. But for the much longer routes between exchanges – maybe halfway round the world – it is much more difficult to transmit a signal without significant loss of quality. One problem is that even the best metallic conductors have a significant electrical resistance, so that the signal gradually weakens as it travels along. Hence, with wires and cables, amplifiers are generally needed every few kilometers to make up for this loss of power, and these add considerably to the cost of the installation and maintenance. The new technology of optical fibers (➧ page 182) is rapidly replacing ordinary metal conductors for many long-distance applications because they can span distances of up to 100km without needing amplifiers. Microwave radio links (➧ page 187) are often the most economical for medium-range "line of sight" routes, and satellites (➧ page 192) provide an increasingly high-performance and low-cost alternative to long overland or trans-oceanic cables.

Integrated networks

The digital signals that are transmitted in the telephone system need not only be the result of converting audio signals. The same binary pulses could just as well be encoded video signals, or simply digital data being sent between computers. So there is the opportunity to integrate audio, video and computer data systems onto one network, giving subscribers access to an immense range of facilities, such as cable TV channels, databanks and home shopping, using electronic funds transfer. Such systems, called integrated services digital networks (ISDNs) are only beginning to develop, but the home of the 21st century will have access to enormously powerful electronic communications with the rest of the world.

▲ Before the advent of multiplexing and radio transmission, individual telephone links had to be made by wire. By the turn of the century, the demand for telephones had increased rapidly, especially in cities. Where underground cables could not be installed, the result was aerial chaos.

◀ Integrated networks were once a futuristic dream. This 1930's magazine anticipates a time when audio and video telecommunication technology would become widely available for domestic use.

▶ The new technology of optical fibers, a cheaper, more efficient means of transmission, is itself generating another new technology: that of cable-laying at sea. Optical fibers are smaller and lighter than conventional coaxial cables. Here, a submarine cable is laid during trials, prior to completing the first transatlantic fiber-optic link in 1987.

▲ A large information-carrying capacity makes microwaves highly suitable for trunk transmission. The radio tower (left) consists of an array of dishes, each 2-3m in diameter, which are highly efficient in focusing microwaves into a narrow beam. Signals are received on a traveling wave tube, a wire helix containing an electron beam. The interaction between radio waves and electron beam serves to amplify the signal.

Transatlantic telegraph...

Very soon after the regular introduction of the telegraph, various short and shallow underwater links were made. But the real challenge came when it was decided to lay a cable across 2,700 kilometres of the North Atlantic from Newfoundland to Ireland. This cable was much longer and deeper than anything attempted before, but backed by the finance and enthusiasm of the American newspaper tycoon Cyrus Field, work went ahead. There were two unsuccessful attempts in 1857 and 1858 when the cable broke while being laid, before a link was completed in August 1858. In view of the technology available at the time, this success must rank high against any subsequent achievement in telecommunications. It was, however, short-lived, as the cable failed after a few weeks when the operating voltage was increased too much in an attempt to improve the operating speed. But the value of the link had been established – in one of the 732 messages which had been transmitted, the British Government had saved £50,000 by directing troop movements.

With better cable design and handling techniques, the first commercially successful telegraph links across the Atlantic were completed, after an earlier abortive attempt, by the steamship "Great Eastern" in 1866. Over 20 more telegraph cables were laid, the last being abandoned in 1966.

...and telephone

It took almost a century of submarine cable technology before a fixed telephone link could be established across the Atlantic. The main problem was to maintain adequate sensitivity and bandwidth, amplifiers being needed every few kilometers along a copper cable. Because of the difficulties of repair, these amplifiers had to last at least 20 years to make the cable an economic proposition. It took many years for vacuum tube technology to reach this degree of reliability, and so the first transatlantic cable, TAT-1, was not laid until 1956. The first telephone link across the Atlantic was established in 1927 when a long-wave telephony service was established between England and the United States. Telephone calls were soon being relayed by short-wave radio links, but these were still often unreliable and of poor quality. TAT-1, with a capacity of 36 telephone channels, was a great improvement.

*Later cables using transistor and integrated circuit amplifiers had much greater capacity. The latest, TAT-8, due to be laid in 1987, uses optical fibers to provide 6,000 telephone channels. Also, since Intelsat 1 (Early Bird) in 1965 (**▶ page 192**), there is a growing number of geostationary satellite links, for video and high-speed data transmission as well as a huge additional number of telephone calls.*

Telecommunications links with light

Throughout recorded history, light has provided a simple means for long-distance communication in the form of bonfires, signaling lamps, heliographs and so on. But sending light through the atmosphere is not very reliable – rain, snow and fog can all have a severe effect on its propagation, and make it unsuitable for a link which must work all the time. To make a useful telecommunications link, the light must be sent through a medium that does not vary with the weather, and the most successful systems have been built around thin fibers of glass. In the 1960s scientists realized that this might be a cheaper and better means of telephone transmission than electric current in a copper wire, but it took a decade to prove its performance in the laboratory, and another to make it effective in real applications. However, it is now so successful, with immense potential for further improvement, that optical communications technology is likely to be dominant for years to come.

The optical fiber

When a light ray traveling in glass meets a boundary with another material of lower refractive index, such as air or another type of glass, then one of two things can happen. If the ray impinges upon the interface steeply enough, then it will pass into the second medium, undergoing refraction (bending) as it does so. However, if the angle of impingement is near enough to glancing incidence, the ray will be completely reflected back into the glass. This phenomenon is called total internal reflection. A single, unprotected optical fiber consists of a glass core surrounded by a glass cladding of lower refractive index contained within a plastic sheath. Total internal reflection keeps low-angle rays propagating within the core, with virtually no power loss at each reflection. The speed characteristics of the fiber are optimized when the core is so narrow that there is only one angle with which a ray can propagate. This is called single mode operation, and the core of a single mode fiber is usually less than ten micrometers in diameter.

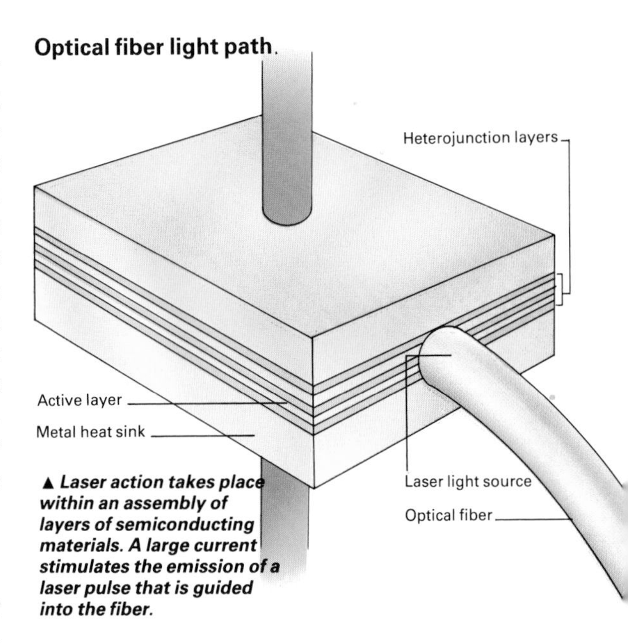

▲ *Laser action takes place within an assembly of layers of semiconducting materials. A large current stimulates the emission of a laser pulse that is guided into the fiber.*

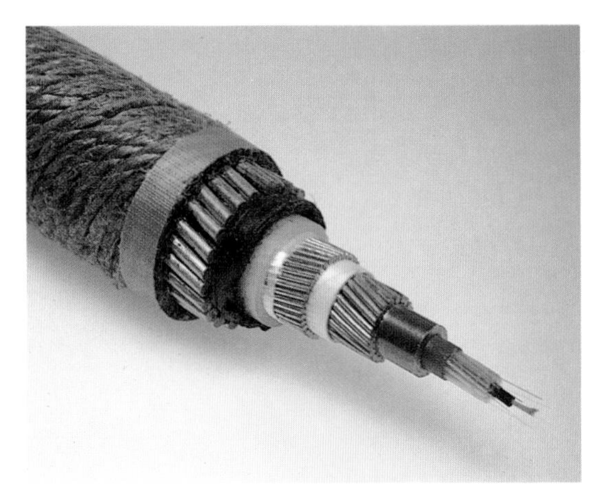

▲ *Submarine optical fibers cable links consist of bundles of "stretched-out" glass wrapped in protective and reinforcement layers. Six polyethylene tubes, each containing a fiber, are enclosed in a paper heat shield, itself wrapped in a lead moisture barrier and successive layers of protective twine and plastic. Finally, the cable is contained within a polyethylene sheath reinforced with steel wires.*

◀ *Alexander Graham Bell's photophone of 1880 was an early experiment in optical communication. Light was focused onto a reflective surface that vibrated in response to sound from a mouthpiece (right). The modulated pulse of light received at a selenium cell (left) was converted to a current and amplified to drive a speaker.*

► Optical fibers are manufactured with a glass outer cladding which makes the transmitted light bounce back into the very pure central core. Ordinary window glass is too light-absorbent for this purpose, so fibers are made from extremely high-grade quartz glass. A tube of this material coated on the inside with even purer glass is heated and collapsed into a solid rod called a preform. Finally, as shown here, it is drawn into a hair-thin fiber a few kilometers long.

▼ In single mode transmission, the thin core, only eight times wider than the infrared wavelength used, imposes a regular wave pattern that allows only one angle with which light rays can propagate.

Plastic sheath
Cladding
Core
Light signals
Metal contacts
Optical fiber

▲ At the photodetector, photons are absorbed in a silicon semiconductor "sandwich". In a uniform electric field the photons give rise to a modulated electron flow.

Fiber manufacture

During the 1960s and 1970s scientists developed techniques for producing fibers of such high purity that a light beam of the right wavelength in the infrared part of the spectrum only lost half its power over a distance of 15km. In these fibers, contamination by other materials had to be held at less than one in ten thousand million – a figure even higher than for many semiconductor materials used for integrated circuits. It is now normal for optical systems to transmit several thousand simultaneous telephone calls down the tiny core of a single fiber for distances up to 100km without amplification along the route. Not surprisingly, optical fibers are now being widely installed throughout the world, replacing less efficient and more expensive copper cables.

Semiconductor lasers

*Optical fibers can only achieve their astonishing potential if there are suitable devices to transmit and receive the light signals in them. A great deal of effort has been put into developing semiconductor lasers to produce very short pulses of light at very stable wavelengths, and these are used on most telecommunications links. Typically, they emit digitally encoded pulses of light at a wavelength of 1.3µm – in the infrared region of the electromagnetic spectrum (**◀ page 73**). Semiconductor lasers use the advanced technology of heterostructures, which involves forming many layers of different materials, including gallium, arsenic, phosphorus and indium, with the atoms arranged in a single continuous structure throughout the device. But the most dramatic feature of semiconductor lasers is their size: a laser capable of transmitting thousands of simultaneous telephone calls down hundreds of kilometers of fiber is fabricated on a tiny chip of semiconductor less than 0·2mm across. The electrical power put into these tiny devices is such that they are often mounted on solid state refrigeration elements to prevent them from overheating.*

Coherent light

The way that light is used at present in optical fibers closely parallels the way that Marconi used radio waves in the early days of wireless telegraphy (▶ page 185). The source of radiation spreads over a range of frequencies that is much larger than the bandwidth of the signal, and is simply switched on and off. Radio transmission improved immensely in information capacity and range when pure frequencies could be generated and modulated with the required signal. Exactly the same principles are now being applied to optical transmission along fibers. A large number of lasers, each emitting at a particular frequency and each modulated by a different signal, can all transmit down a single fiber at the same time. At the other end, tuned receivers can separate the signals again. Using this technique, called coherent optical transmission, the information capacity of a single fiber rises from the present figure of a few thousand simultaneous voice channels to at least a few million. Also, since a coherent receiver can be designed to be more sensitive than the present untuned devices, signals can be sent through thousands of kilometers of fiber without amplification. These ideas are still at the research stage, but the results are so successful that very high performance coherent optical links are likely to be in operation by the end of the century.

See also
Introduction to Measuring 9–22
Computer Hardware 115–122
Computer Software 123–128
Printing 171–176
Radio and Television 185–192

Data communications

Modern telegraphy – communication between machines – is a growing area at the forefront of technology: the power of computers and other "intelligent" machines is effectively increased if they can communicate with one another. Thus home computer terminals can access distant data sources to provide a vast range of information displayed on the user's visual display units (◆ page 126). Salesmen in the field can send back records of their transactions stored on portable computers to a central system.

Many of these computers now exchange data along the public telephone network. Since computers work with high-speed binary signals, this form of communication usually involves the use of a modem (modulator-demodulator), which converts the binary signal into a code of audio tones that is compatible with telephone circuits originally designed only for voice communications. Message speed may thus be quite slow compared with the capabilities of the computer, and the growth of much faster digital networks is an important aspect of future data transmission services. Facsimile transmission (fax) is another means of conveying information. A fax machine electronically scans a document – which may contain diagrams and pictures as well as text – then transmits the visual information point-by-point at high speed down a telephone line.

One of the main areas of machine communication that is beginning to affect daily life is electronic funds transfer (EFT). At present most personal money transactions use pieces of paper – notes, checks, drafts and so on. Cash-dispensing machines maintained by banks using debit cards were the first form of EFT, but now such services are being extended to other forms of transaction. If there is a link between a point of sale terminal in a shop and a bank's computer, then the cost of an item purchased can be transferred immediately from the buyer's bank account to the seller's.

Smart cards

An EFT card is simply a plastic card that holds a personal account number, recorded as computer-readable characters, as a bar code and as a magnetic strip. However, no system can be guaranteed to be entirely fault-free, and a good procedure for error-checking and correction must be built into any EFT system. This can also protect against the possibility of fraud, particularly if the system is connected through the public telephone network.

The key to the latest EFT systems are "smart cards" – plastic cards containing an integrated circuit memory chip and possibly a microprocessor as well. Security is aided by the complexity and unique codes associated with each card. Some protection against errors is given by the on-card memory which stores an independent record of each transaction.

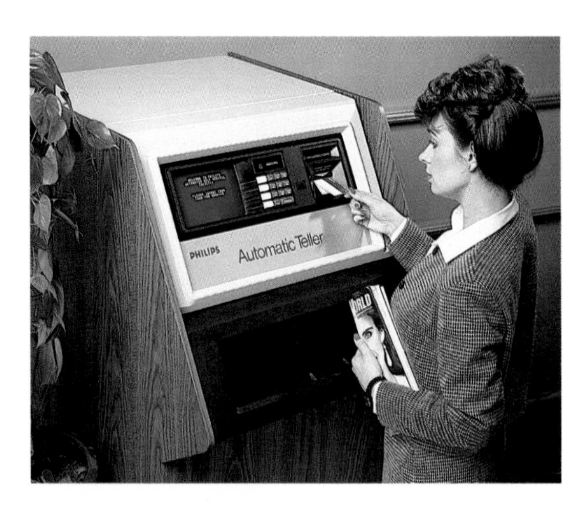

▲ EFT cards have a number of applications in the field of everyday financial transactions. Once simply a means of accessing automatic cash dispensers, they can now be used to call up the bank's main computer directly, either to notify details of a sale or to take instructions on transferring money from one account to another.

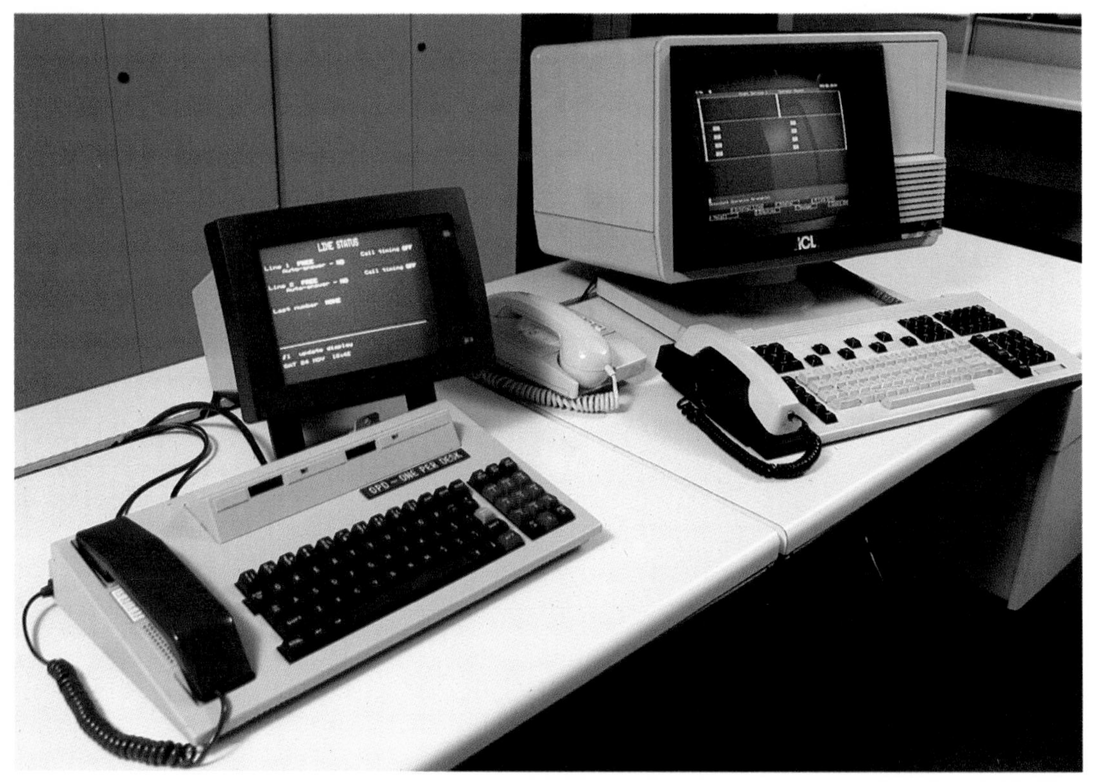

◄ With a variety of makes of computer hard- and software on the market, data communication systems are now being designed to connect a terminal with as wide a range of other systems as possible. The apparatus shown here incorporates a variety of interfaces and modems to translate its own software language into that assimilable by another machine. This allows cost-effective communication from one terminal to another, whether it be complex data or a simple telex message.

Radio and Television

Communication without wires...Radio transmission... Picture transmission...Developments in television... PERSPECTIVE...Marconi's experiments...Chronology... Radio signals...Cellular radio...How television broadcasting works...High definition television... Satellite communications

In 1886 the German physicist Heinrich Hertz (1857-1894) performed a classical physics experiment which paved the way for another communications revolution. He arranged for a high-voltage spark to be generated across two electrodes, while some distance away a similar pair of electrodes was connected to a loop of wire. When a spark passed between the first two electrodes, another was generated between the second two, even though there was no direct electrical connection between them. Somehow energy from the first spark was being propagated through the air and was concentrated by the receiving circuit to yield a second spark. Hertz even established that the energy was being transmitted at almost 300,000 km/s, the speed of light.

The medium of this energy transference is what is termed electromagnetic radiation (♦ page 73). Whenever an electric current passes through a wire, electromagnetic waves are emitted from it in all directions. In a simple form they consist of electric and magnetic fields that vibrate at the same frequency at right-angles to each other. When these vibrating fields strike another conductor, they set up new currents at the same frequency. Thus energy can be transmitted from one circuit to another even through empty space.

Hertz did not think that this ability to transmit information without wires would be a practical means of communication. But others did, including Guglielmo Marconi, (1874-1937) the Italian who developed the first successful "wire-less telegraph". In 1896 he sent messages over a distance of a few kilometers, using a spark gap transmitter and a simple receiver which indicated the presence of a signal by clicks in a telephone earpiece. In 1901 he even used a system like this to receive Morse signals across the Atlantic, although the spark gap had to be driven by a 25,000 watt generator connected to 50 aerials on masts 70 meters high.

▼ Gugliemo Marconi was the first to explore the potential of radio. Shortly after his arrival in England from Italy in 1896, Marconi established his Wireless Telegraph Company. His simple apparatus, consisting of a spark transmitter, powered by a generator, and a kite aerial receiver, achieved the first "telegraphy without wires".

▲ Marconi's success with long-distance wireless transmission resulted in the appearance of hundreds of radio stations all over the world. Early radio-telephony used long waves, and receivers, such as that seen in this 1920s photograph, comprised arrays of beam aerials, positioned to receive waves from specific directions.

It cannot possibly work

Radio signals travel in straight lines, yet Marconi managed to transmit signals one-eighth of the way around the globe. The contemporary opinion was that the curvature of the Earth would limit practical communications by electric waves to a maximum distance of 300km. Marconi apparently achieved the impossible because the radio waves he transmitted from Cornwall, England, bounced off a layer of charged atoms called the ionosphere. The long waves that Marconi used reflected at a height of about 100km, while shorter wavelengths, as he later demonstrated, reached higher levels. The structure of the atmosphere was not understood in 1901, and many scientists thought that Marconi's experiment had no chance of success. However, his gamble paid off and the principle of long-distance radio transmission was established.

Chronology

Radio
1886 Presence of radio waves demonstrated by Heinrich Hertz
1890 Coherer invented by Edouard Branly
1901 Wireless telegraphy used by Guglielmo Marconi to transmit a morse message across Atlantic
1904 Diode vacuum tube invented by John Ambrose Fleming
1906 Carborundum crystal radio detector patented by H.C. Dunwoody
1907 Triode invented by Lee de Forest
1912 Heterodyne radio system invented by R.A. Fessenden
1917 Superheterodyne electronic circuit introduced
1917 Neutrodyne circuit developed by L.A. Hezeltine
1921 First medium-wave broadcasts
1921 Short-wave transmissions made across Atlantic
1921 Stabilizing properties of quartz crystals discovered by Walter Cray
1928 Pentode developed by Philips of Holland
1933 Frequency modulation developed by Edwin Armstrong
1940 Pulse modulation developed
1948 Printed radio circuits introduced by John Sargrove
1948 Transistors invented at Bell Laboratories, United States
1954 First pocket transistor radios introduced by Texas Instruments, USA
1958 First practical integrated circuits made by J. Kilby
1958 Stereophonic radio transmissions made by BBC in Britain
1977 Quadrophonic radio transmissions made by BBC

Wireless transmission was made more efficient by the invention of the diode vacuum tube or valve by J. Ambrose Fleming (1849-1945) in England and the triode by Lee de Forest (1873-1961) in the United States. Radio waves could now be concentrated into a single frequency, and receivers tuned to the transmission frequency could be built. Frequency is measured in units called hertz (Hz), where one hertz is one vibration per second. Radio transmissions now extend over a range from less than 10,000 hertz to over 10 billion hertz.

Modulation

A uniform radio wave carries no information, so to be used for communication it must be varied (modulated) in some way by a signal. The two main quantities that can be changed are the amplitude and the frequency, hence the techniques called amplitude modulation (AM) and frequency modulation (FM). Variations in these parameters on the propagating wave are detected by the receiver, and so the signal is reconstructed. Amplitude modulation is the simpler technique, but frequency modulation is less affected by interference and so tends to be used when the quality or accuracy of the link are important. Both techniques can be used with analog or digital signals.

Bandwidth

The amount of information in a signal determines the maximum frequencies that it contains. This is called the bandwidth of the signal. Morse code sent by hand has a comparatively low bandwidth of only a few hertz. To give an acceptable reproduction of speech, telephone signals need a bandwidth of about 4,000 hertz, while hi-fi music needs about 20,000 hertz. Video signals, however, which contain information on the shape, color and movement in a continuously changing scene, often occupy a bandwidth of more than five megahertz. In radio transmission, stations are allocated different frequencies so that a receiver can tune into a particular station and exclude all others. To avoid interference, transmission frequencies must be spaced apart at least by the bandwidth of the modulating signal, and often by much more. So, for example, around 16 TV channels can be fitted between 400 and 500 megahertz, each spaced by six megahertz. As the pressure for more and more transmissions increases and as technology improves, higher frequency bands are being opened. The extreme example is coherent optical communication, where the frequency range available is vastly more than the whole of the radio spectrum (◀ page 182).

Microwaves

At extremely high frequencies, above about one billion hertz, it becomes easy to focus the radio waves into a beam, like the beam of a torch. This makes transmission along a single direction highly efficient and less susceptible to interference from other radio sources. These so-called microwaves have thus been developed as an effective means for line-of-sight telecommunications links, particularly for relays between broadcast TV transmitters, and as an alternative to cables for the telephone network.

▲ The invention of the triode vacuum valve in 1907, which could generate, detect and amplify radio waves, provided the first technology for radio transmission of the human voice. Early radio receivers, or crystal sets, picked up single-wavelength signals from nearby transmitters, and were listened to through headphones.

► Transmission by radio at short, medium and long wavelengths developed rapidly in the 1920s. To avoid congestion of the airwaves, international agreement on frequency allocation was needed. One result was the appearance of radio receivers which could be tuned to a variety of wavelengths broadcast from over long distances.

The transistor

The first transistor was invented accidentally, in 1948, by John Bardeen and W. Brattain at the Bell Telephone Laboratories in the United States. It was similar in function to a triode valve, but the currents flowed through a single crystal of semiconductor instead of through a vacuum. Compared with vacuum tubes, transistors were much smaller, lighter, cooler, more reliable and used less power. They have lead directly to smaller, low- power-consumption amplifiers for telephone lines, portable radios and TVs, and more reliable electronic switching in telephone exchanges. But, most significantly of all, they are capable of being made cheaply and in huge quantities in the form of the very large-scale integrated circuits that today are transforming telecommunications.

Television
1884 Television system using mechanical sequential scanning patented by Paul Nipkow
1897 Cathode ray tube with fluorescent tube introduced by K.F. Braun
1907 Crude television pictures transmitted by Boris Rosing
1926 First practical demonstration of television by John Baird in Britain
1928 First demonstration of color television by Baird
1931 Iconoscope camera tube patented by V.K. Zworykin
1936 First high-definition electronic television system launched in London
1937 Outside broadcasts began in London, using portable transmitter
late 1940s Cable television launched in United States
1949 First closed-circuit television system installed in Guy's Hospital, London
1950 Vidicon television camera introduced by RCA
1954 First regular color television broadcasts began in United States
1957 Plumbicon color television camera introduced by Philips
1962 Television pictures relayed by Telstar satellite
1964 Geostationary Syncom 3 satellite relayed first continuous television transmissions
1965 Miniature transistorized television introduced in Japan
1972 Digital television introduced by BBC
1974 Direct-broadcast television begun using the ATS-6 satellite
1977 Optical fibers used to transmit cable television signals
1981 First pocket television launched by Sinclair Radionics in Britain

Electromagnetic waves

▼ ***Electromagnetic waves consist of electric and magnetic fields vibrating at the same frequency at right angles to each other.***

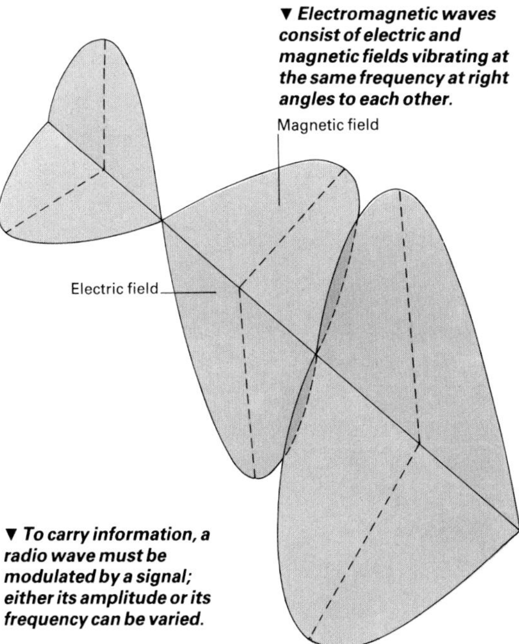

▼ ***To carry information, a radio wave must be modulated by a signal; either its amplitude or its frequency can be varied.***

Amplitude modulation (AM)

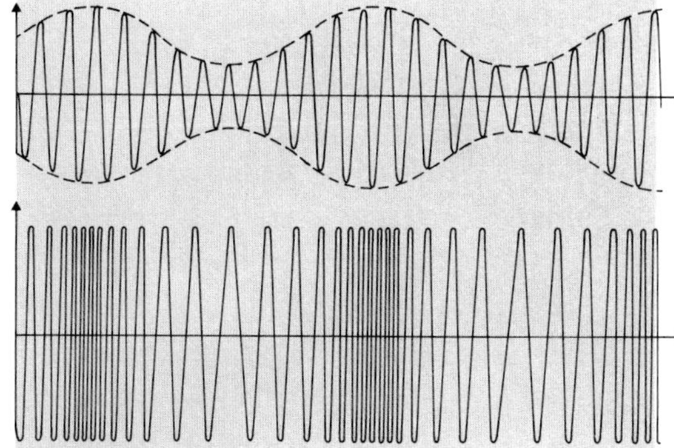

Frequency modulation (FM)

Cellular radio

One of the most demanding applications of the radio spectrum is that of person-to-person communication, such as a large-scale mobile telephone network for use in cars. To accommodate this while using a limited allocation of radio channels, the country is divided into a series of cells, each with a low-power transmitter that serves only the callers that are within the cell at that time.

A cellular telephone system like this relies heavily on powerful computer technology to set up calls between cells, keep track of where the callers are and reroute the calls if necessary. Groups of channels can be reused for cells that are not adjacent. In a practical system a network of 100 cells with 200 audio channels will allow over 1,000 simultaneous calls to be made.

Cellular radio

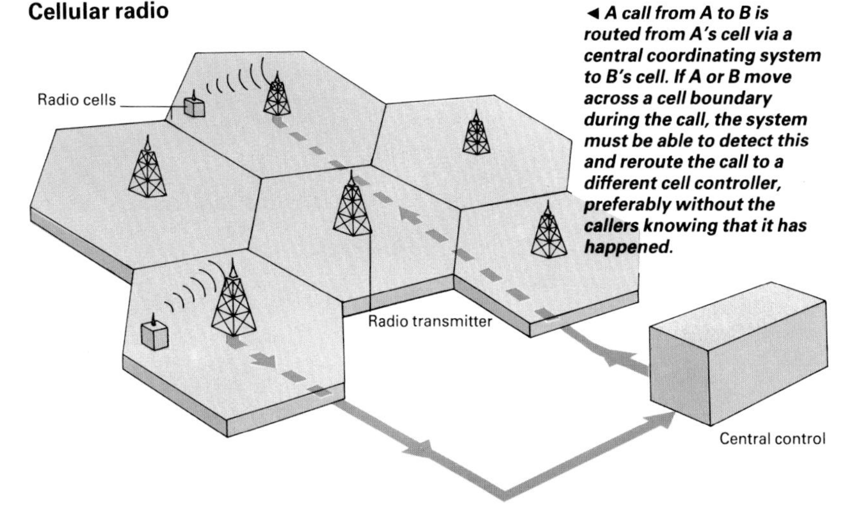

◄ ***A call from A to B is routed from A's cell via a central coordinating system to B's cell. If A or B move across a cell boundary during the call, the system must be able to detect this and reroute the call to a different cell controller, preferably without the callers knowing that it has happened.***

The Televised Image

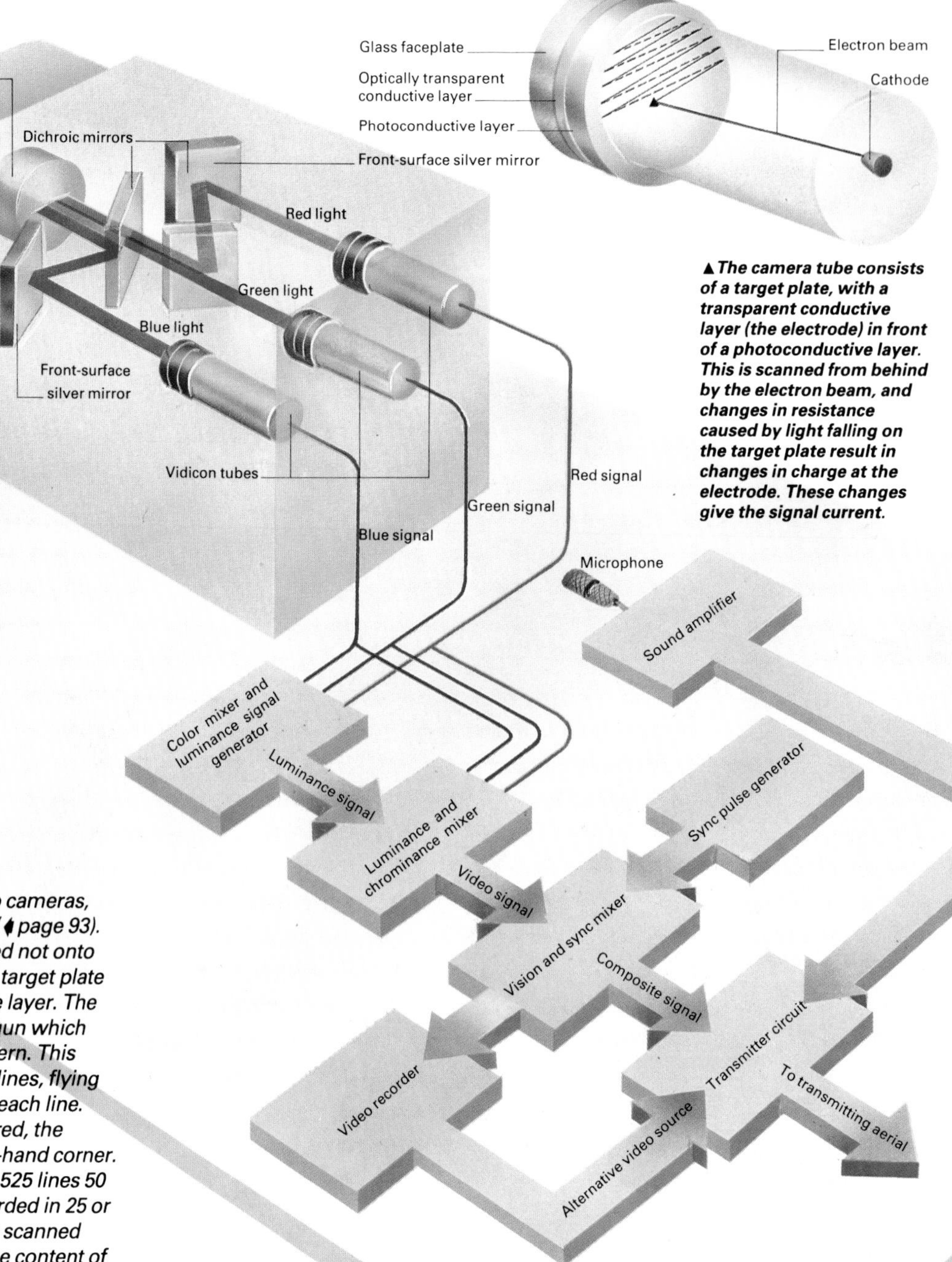

► *The light that enters the camera is made up of various wavelengths, corresponding to the colors reflected from the object to be transmitted. This is focused by the lens and then split, in a color camera, into its constituent primaries, red, blue and green. This splitting is done by means of dichroic mirrors, which reflect light of a certain wavelength and allow other wavelengths to pass through. Ordinary mirrors then reflect the split light beams onto the faceplate of two of the three camera tubes (one wavelength passes straight through to a tube). Each tube is then scanned by an electron gun 50 or 60 times a second; the brightness of each of the primary colors falling on the faces of the tubes is translated into equivalent red, green and blue signal currents from the three tubes.*

▲ ***The camera tube consists of a target plate, with a transparent conductive layer (the electrode) in front of a photoconductive layer. This is scanned from behind by the electron beam, and changes in resistance caused by light falling on the target plate result in changes in charge at the electrode. These changes give the signal current.***

The television image

Television cameras, like amateur video cameras, use vidicon tubes to record the image (◆ page 93). Light passing into the camera is focused not onto film, as in a normal camera, but onto a target plate that carries a charged photoconductive layer. The target plate is scanned by an electron gun which covers its entire surface in a raster pattern. This sweeps down the screen in a series of lines, flying back to the left-hand side at the end of each line. When the entire screen has been covered, the beam returns diagonally to the top left-hand corner. The normal TV camera scans in 625 or 525 lines 50 or 60 times a second (the scene is recorded in 25 or 30 frames per second, but the image is scanned twice each frame to prevent flicker). The content of the picture is scanned point by point and the charge on the photoconductive layer is translated into a varying signal current.

In a color TV camera the light is broken down into the three primary colors (red, blue and green), by means of dichroic (color-separating) mirrors. Each primary color is detected by a separate picture tube, and produces its own signal. A black-and-white luminance signal is created by combining information from the three color tubes, and this is mixed with the composite color signal to form a video signal for transmission. Sync pulses are incorporated in the signal before transmission to ensure that the receiver operates in synchronism with the transmitted signal. These pulses are controlled by a master timing device, and when more than one camera is used to record a scene, the same sync pulses are fed to all cameras.

The camera tube

If no light falls on the target plate of the tube, the electron beam scans the photoconductive area and charges it up like a series of minute capacitors. The inner surface of the photoconductive plate thus acquires a negative charge with respect to the outer, which is in contact with the transparent signal electrode.

If light now enters the tube, it falls on the photoconductive layer and causes a drop in resistivity at that point; the charge drops in proportion to the amount of light. When the beam scans the point again, it replenishes the charge, causing a current to flow from the electrode in proportion to the charge that has been lost. The signal current therefore varies in accordance with the pattern of light that has fallen on the tube.

▲ ***The signals from the three tubes are combined in the adding unit to form a luminance or brightness signal, while a color encoder merges the outputs of the three tubes into the chrominance (hue) signal. These are then encoded into one signal such that the receiver can separate them again. Different systems for combining the two signals are in use in different countries. The sound signal is also added to the signal before transmission, as is a sync pulse regulating the sweep of the beam across the screen.***

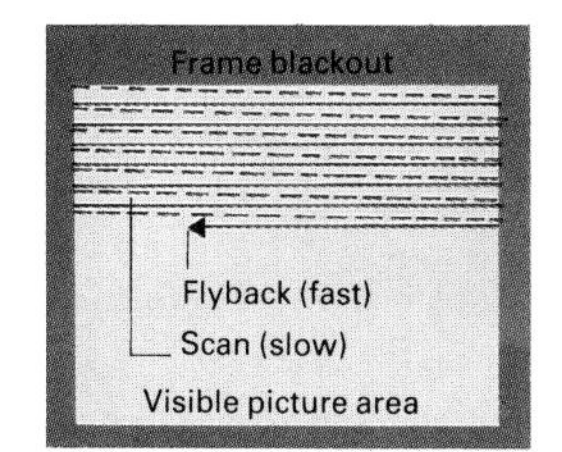

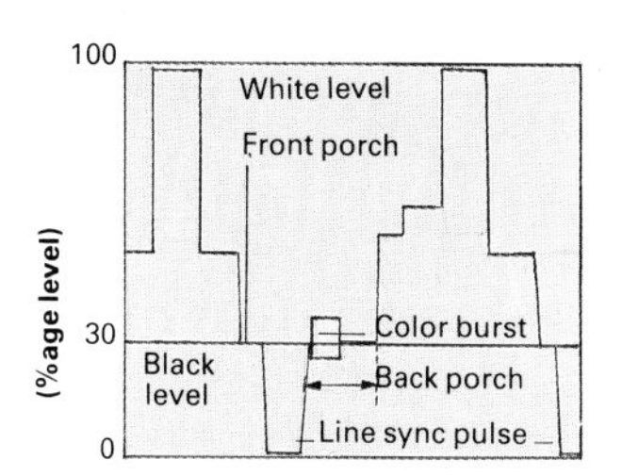

▲ ***The transmitted signal not only carries information relating to image area but also defines the picture frame. Teletext services are carried in the frame blackout located above the visible area (top). The color burst which controls the relationship between the chominance and luminance information is contained in the line back porch. The color burst, like the Teletext information, is located outside the picture area and is down the left-hand side.***

► ***The signal received by the aerial is split up into its luminance, chrominance, sound and sync pulse elements. The luminance signal controls the overall output of the electron guns in the cathode ray tube, and the chrominance signal controls the relative strength of the three beams. The sync pulses are separated into line and frame sync pulses to control the sweep of the beams across, and down, the screen.***

Television transmission

The video signals from the three camera tubes carry huge amounts of information. By taking into account the color characteristics of the human eye, it is possible to compress the bandwidth requirement so that a color signal can be transmitted in the same bandwidth as that required for a black-and-white picture (5·5MHz, or 5·5 million cycles per second) without any apparent loss of definition in the color picture. This bandwidth compression is achieved by dividing the video signal into two parts, a luminance signal of full bandwidth, plus a modulated wave of very much lower bandwidth carrying the chrominance information. The complete signal also includes synchronizing pulses, a color reference burst and a black level which acts as a reference against which the signal can be measured from black to white.

The television receiver

The signal from the receiver is decoded and the chrominance signal split into the three color signals. Each one is passed to an electron gun in the cathode ray tube. The electron beams scan across the screen in the same way and at the same speed as those in the camera tubes, but in varying intensity following variations in the signal. Thus each picture is made up of lines of varying brightness and color.

The electron beams themselves are not colored. The screen is coated with phosphor stripes that glow red, blue or green when hit by electrons. The beams pass through a grille and diverge before hitting the screen. The "green" beam produces a dot of green light from the green phosphor, and so on. The image consists of dots of the primary colors which merge in the eye to form the range of hues.

Color television receiver

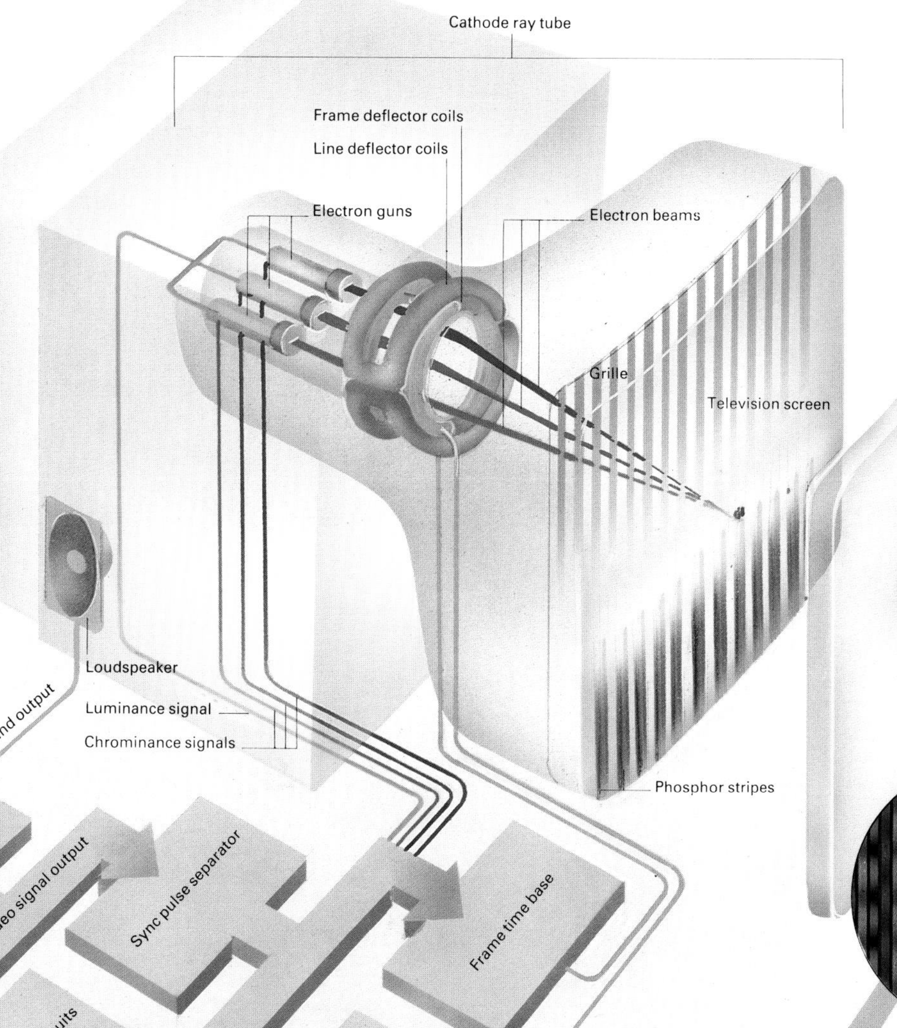

▲ ***The three electron guns each fire an electron beam varying according to the red, blue or green signal fed to them. Deflector coils controlled by the line and frame time bases sweep the beams across and down the screen through the grille.***

▲ ***A Trinitron screen has a composite bundle of phosphor stripes. Each of the three electron beams can only energize its particular color phosphor stripe, owing to the positioning of the electron guns and the grille.***

Developments in television

Although the basic signal format has not changed since color TV was introduced, various additional functions are now provided by the clever use of electronics. The trick is to add some digital information to those parts of the analog signal waveform which are used for synchronization and which do not directly affect the picture. Circuitry in the receiver selects this data, decodes it and puts it into a form suitable for display on the screen. This technique is the basis for various Teletext information services broadcast in many countries along with the ordinary signal.

In the studio high-speed digital converters can also convert analog signals into a form suitable for computer processing. Using digital picture stores, effects not possible with analog techniques are created – zooms, tumbles, swoops, skews, flips and many others. Computers can also generate pictures, not only simple captions and diagrams but even long sequences of lifelike simulation. At present, all these facilities are isolated digital links in the otherwise totally analog chain from camera through to display. Digital signal formats will eventually replace the present analog one, and picture quality and the range of effects and facilities will markedly increase.

Although most TV is broadcast by radio waves, cables are also used as a means of distribution, largely for the commercial reason that it restricts the service to those who have paid. The new technology of optical fibers (◆ page 182) is rekindling interest in this area. Because fibers are cheap and have extremely high bandwidths, they are an economically attractive method for distributing not only a great many TV channels, but also other data services linking to home or business computers. If such an integrated digital network included a local broadcast satellite receiving dish, then the environmental problem of many domestic aerials and the commercial problem of restricting access could be solved.

Modern high bandwidth transmission channels of satellites and optical fibers are also opening up another possibility – that of providing higher definition TV pictures. The proposed standards use well over 1,000 lines, a wider screen than at present and better resolution of detail. This bigger, sharper picture needs a bandwidth of more than 20MHz, almost four times that required by current systems, though still well within the capabilities of the technology.

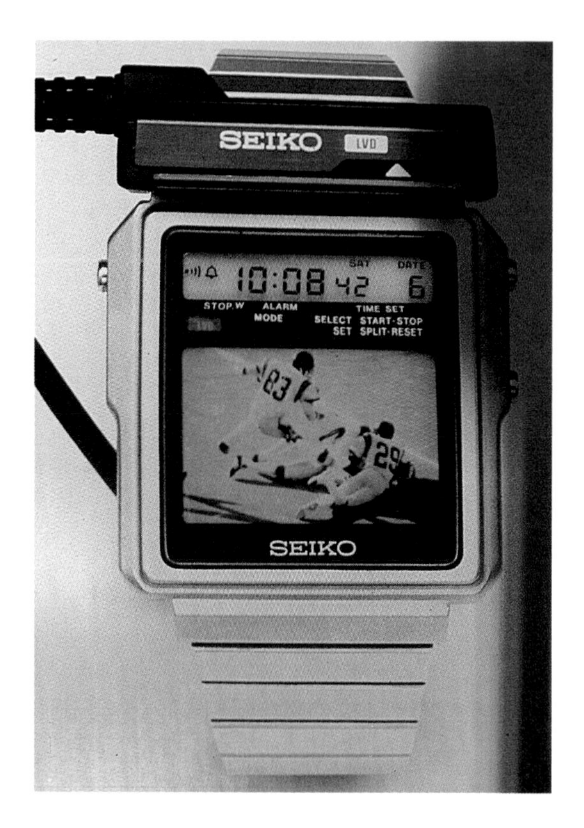

▲ Watch-sized TV receivers use a liquid-crystal display. The screen contains liquid-crystal cells that change their reflectivity when an electric field is applied to them.

▼ Portable TV sets incorporating a small cathode ray tube. Other miniature televisions may have the CRT at 90° to the screen; the electron beam is bent by deflector coils.

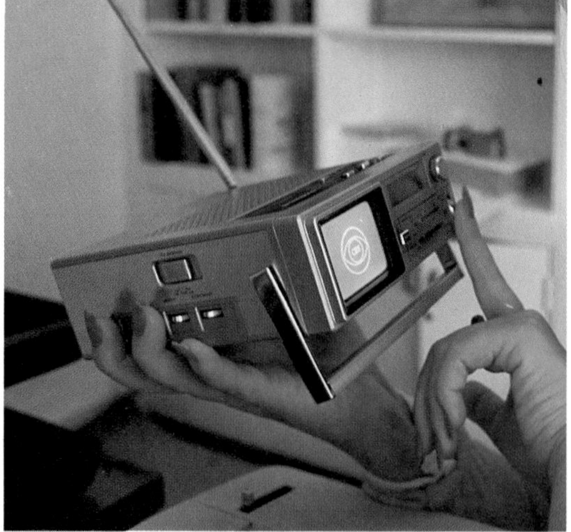

◄ Quantel's Paintbox is a digital electronic TV design system. Working with an electronic palette and stylus, results are displayed on a screen.

► High definition pictures are the next generation of TV technology. Current TV pictures are defined by 525 or 625 lines per frame (left). The Japanese propose a 1,125-line picture at 60 fields per second, interlaced on screen to give 30 frames (right). The larger carrying bandwidth can be reduced by transmitting only the moving parts of the picture and equipping the receiver with a memory. Digitially assisted TV uses a digital control signal which tells receiver circuits how to recreate the picture.

See also
Seeing beyond the Visible 73–86
Photography 87–90
Recording on Tape and Disk 91–96
Telecommunications 177–184
Space Transport 225–228

◀ **Despite ever-increasing sophistication in satellite technology, problems of malfunction or failure can persist. A valuable part of the Shuttle program (▶ page 226) is the retrieval and repair of damaged satellites. Here, two astronauts on a 1984 mission prepare to begin work on Westar VI.**

▼ **An Earth station antenna consists of an aerial bowl reflector with an adjustable mounting, so that it always points at the satellite. The parabolic dish directs received radio waves to a sub-reflector suspended above it, which reflects them in turn down through a hole in the dish to the receiver station.**

Communications satellites

Satellites are a vital link in modern telecommunications. Because microwaves do not reflect from the ionosphere they cannot be used directly for long-range communications, but they can be transmitted up to an orbiting satellite, which then relays them back to a distant part of the Earth. Experiments in satellite communications date from the early days of the space program, in the late 1950s. The earliest successful telecommunications satellite was Telstar, which in July 1962 provided the first transatlantic television link. It orbited the Earth once every 2½ hours, but was only usable for about 20 minutes during each orbit, when it was in the right position to be followed by the Earth stations. Then it could transmit 60 telephone calls or one television channel. Eliminating such gaps in transmission can be achieved by launching the satellite into a higher orbit (at a height of about 36,000km) around the Equator, so that it encircles the Earth once every 24 hours. Then, because it exactly follows the rotation of the Earth, it appears to be fixed over a particular place. This is called a geostationary orbit, and geostationary satellites can be used either as permanent links between two Earth stations, or as a means of broadcasting over a wide area. Countries too far from the Equator to use geostationary systems are served by very elliptical orbits in which the satellite is usable for a large proportion of the time.

Most international satellite telecommunications are controlled by a consortium of more than 100 countries, called Intelsat. Their first geostationary satellite, Intelsat I (originally called Early Bird), was launched in 1965 and was able to relay 240 telephone circuits or one television channel across the Atlantic. Intelsat VI began service in 1986. Each satellite can link 33,000 telephone calls and up to 60 television channels, using very narrow beam aerials that are accurately directed towards a few Earth stations. The higher effective bandwidth is achieved by multiplexing the antenna beams. Worldwide there are now more than 170 communications satellites in service for point to point links, navigation systems and broadcast television.

▼ ***Radio waves of different frequencies can defy the Earth's curvature in different ways. Long and medium wave transmitters propagate waves in all directions, but only those close to the Earth's surface are received, at a range of hundreds of kilometers. Short waves are reflected by the ionosphere and made to travel around the Earth by multiple reflection. The higher the frequency the higher a radio beam will penetrate the ionosphere. Those with very high frequency will pass directly through, and must be reflected from orbiting satellites. Alternatively, high frequency waves can be transmitted and received from high land-sited aerials located in a line-of-sight series.***

Long-distance radio transmission

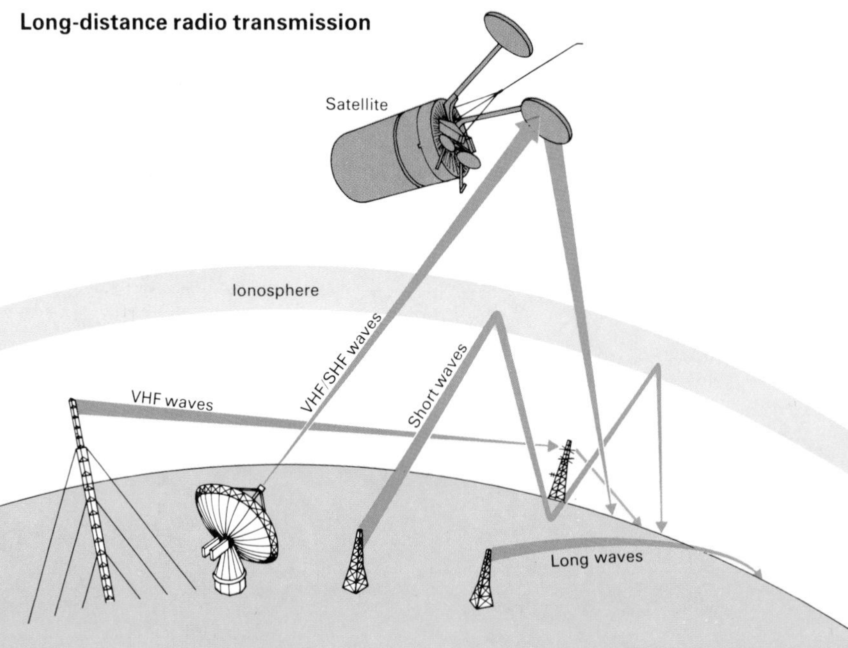

Rail Transport

Early history...Steam, diesel and electric trains...Maglev trains...PERSPECTIVE...The Stephensons...Chronology ...How locomotives work...Train à Grande Vitesse... How maglev trains work

The development of rail transport in the early 19th century provided a vital contribution to the progress of the Industrial Revolution. It remains a reliable, economical and high-volume system of conveying goods and passengers by land. Two important innovations, both achieved in Britain, made this possible: the successful introduction of reliable wrought-iron rails in Coalbrookdale, Shropshire in 1767, and the design of powerful steam-driven engines such as the *Rocket*, built by the George Stephenson in 1829.

Between 1775 and the 1790s the great British engineer James Watt (1736-1819) developed the steam engine into a reliable source of power, and the famous Boulting and Watt engine was being produced at the rate of 20 a year by the end of the 18th century (◆ page 130). On the expiry of Watt's patent in 1800, Richard Trevithick (1771–1833) built a high-pressure engine and used it to power a steam carriage, which duly became the first vehicle to convey passengers by steam power in 1801. Steam traction became a commercial proposition when a private industrial line was relaid from Middleton to Leeds in Yorkshire in 1812, in order to take steam locomotives working as coal-hauling vehicles. Rivalry between collieries prompted rapid development of the steam locomotive and improvements to the track, culminating in the achievements of the Stephensons during the 1820s.

In 1830 the world's first passenger railway opened, between Liverpool and Manchester, and only 14 years later there were almost 8,000 kilometers of railway line completed in Great Britain. Over the next few decades railroads carrying passengers and goods extended throughout Europe, crossed North America and spread to most countries. For the first time humanity had the means of traveling on land at speeds greater than that of a galloping horse.

▲ George Stephenson.

▲ Robert Stephenson.

The Stephenson family

The engineering team of George Stephenson (1781-1848) and his son Robert (1803-1859) were the foremost pioneers of early rail transport technology. George Stephenson, originally a colliery fireman, rose steadily to a position of responsibility in charge of machinery in a large group of north England collieries, building 39 large stationary steam engines for hauling coal by cable-drawn wagons on rails. Rapid developments in steam engine design following Richard Trevithick's success led Stephenson to experiment with steam-powered locomotives. In 1814 he built his first, the "Blücher". He was chief engineer of the first real cargo-carrying railway, linking Stockton and Darlington in north-east England, and together with his son Robert he designed the "Locomotion", which hauled the first train on that line in 1825. Its success led to Stephenson's appointment as engineer of the Liverpool-Manchester line. At trials in 1829 to decide the best locomotive available, the "Rocket", also designed by the Stephensons, decisively outperformed its rivals. Its novel multi-tubular boiler became the standard design for all subsequent steam locomotives.

Robert Stephenson became a well-known mechanical and civil engineer in his own right. In 1833 he became engineer-in-chief of the London to Birmingham line, and later won international repute for his bridge designs, which included work in northern England, Wales and Canada.

▼ ***The Liverpool and Manchester Railway, 1831.***

Chronology

1804 The first steam rail locomotive, the "New Castle", built by Richard Trevithick and run in South Wales
1812 Modern bogie design patented by William Chapman
1812 Steam locomotives used commercially at Middleton Colliery, Leeds, England
1813 First adhesion locomotive (with smooth wheels on smooth tracks) "Puffing Billy" introduced
1814 George Stephenson's first locomotive, "Blücher", built
1825 First steam locomotive in North America built by John Stevens
1825 First public railway, from Stockton to Darlington in England, opened
1829 George Stephenson's "Rocket" built, equipped with multitubular boiler
1830 First line relying exclusively on freight and passenger traffic built, between Liverpool and Manchester, England
1830 First scheduled service opened in North America – the South Carolina Railroad
1855 Interlocking points and signals installed in France
1859 First sleeping car built, designed by George Pullman
1863 First underground railway opened in London
1867 Airbrake invented by George Westinghouse in USA
1879 Invention of direct-current electric motor by Werner von Siemens made electric locomotives possible
1881 First public electric railway opened near Berlin
1889 First articulated steam locomotive developed by Anatole Mallet in Switzerland
1890 London's first electric underground line opened

Steam locomotive driving gear

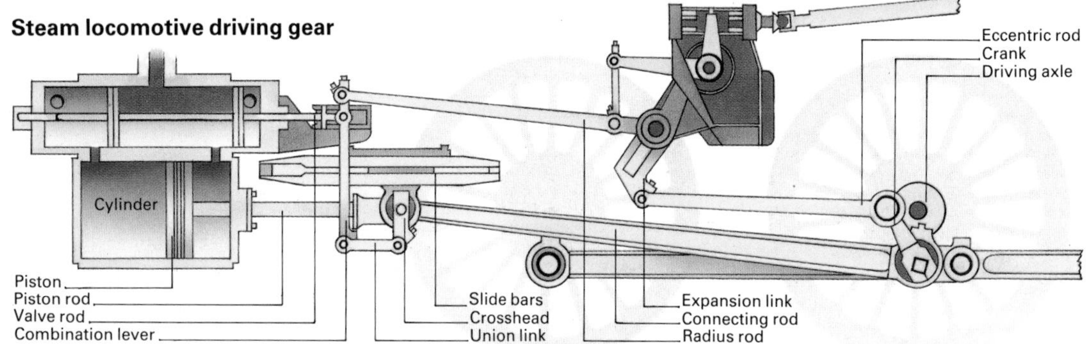

▲► The Union Pacific railroad was inaugurated in 1863. From the earliest days the Union Pacific has established a reputation for the size and power of its steam locomotives. In 1941 the largest of all, the "Big Boy" freight locomotive (above) made its first appearance, designed for service on mountain routes. Articulated underframes allowed the huge engine to negotiate turns at speed. Diesel-electric locomotives superseded steam engines during the 1950s (the last "Big Boy" ran in 1959) but the Union Pacific has set records with this form of traction as well. The turbocharged engines (right) are the world's most powerful, rated at 6,600 horsepower each. Typically, in the United States and Canada freight trains use combinations of units which develop sufficient power to pull trains extending over several kilometers in length.

How a steam locomotive works

Drive is transmitted from engine to wheels through the action of valves and pistons, housed in cylinders, and an assembly of interconnecting rods and linkages. The most widely adopted type of valve-gear is the Walschaert system, an interacting mechanism in which the valve controls the piston, which drives the crosshead and eccentric that together activate the valve.

A connecting rod is joined to the piston rod at the crosshead and, guided by slide bars, passes the motion of the piston to a crank attached to the wheel axle. Also linked to the axle is the eccentric, which transmits a reciprocal motion to the radius rod via the eccentric rod and expansion link, a device that enables the driver to engage forward and reverse gears and to control the speed of the train. The combined movement of the radius rod and crosshead, which are themselves connected by the combination lever, moves the two-headed valve backwards and forwards to control the admission and exhaustion of superheated steam to the cylinder. Steam locomotives usually have two or more cylinders, which drive one wheel-set directly and others of the same diameter indirectly by means of coupling rods.

1902 Steam superheaters invented by Wilhelm Schmidt of Germany
1913 First diesel locomotive built in Germany
1913 First diesel-electric passenger railcar ran in Sweden
1915 First electric locomotive using high-frequency current built by Hungarian Kalman Kando: locomotives could be run using public electricity supply
1916 Articulated railway coaches introduced on Great Northern Railway in Britain
1924 First diesel trains introduced in United States and Tunisia
1927 Centralized Traffic Control introduced to railroads of USA
1933 First high-speed railcar, between Hamburg and Berlin
1933 Continuous welded rails introduced in USA
1938 Locomotive "Mallard" in Britain achieves all-time record for steam traction, 203 km/h
1941 First gas-turbine locomotives introduced by Swiss Federal Railway
1960 First electronic speed and braking control system introduced on Leningrad subway
1960 First computer-controlled trains introduced on iron-ore freight routes in northern Sweden
1964 First monorail train between Tokyo and Haneda airport
1965 Japanese Shinkansen high-speed train entered service
1972 British Rail's Advanced Passenger Train tested experimentally
1979 Maglev passenger service introduced at Hamburg International Exhibition
1981 French Train Grande à Vitesse (TGV) introduced

The development of the train

Today, trains are powered by three types of power unit – steam, diesel and electric. Steam reigned supreme until the 1950s and is still the major source of power for the rail systems of China and parts of Asia, Africa and South America. The electric engine was developed in the late 19th century and met a growing need for cleaner trains and operating convenience, particularly suitable in running underground routes in cities. The diesel engine became popular in the 1950s in countries where fuel supplies were plentiful, especially in the United States, as well as in those unwilling to meet the high capital installation costs of electrification. The rising price of oil in the 1970s, however, moved some advanced countries towards electrification. Both sources of power were applied to highly-efficient locomotive designs that opened the way for a new era of high-speed travel, begun by the Japanese Shinkansen in 1965 and culminating in the most sophisticated system to date: the French *Train à Grande Vitesse* (TGV).

Electric and diesel locomotives

A similar drive mechanism is used in both diesel and electric locomotives. Both are mounted on wheel-base trucks or bogies – pivoted supports which enable the train to negotiate curves more easily. Coupled by springs to each axle are electric traction motors which receive power in the form of direct current (DC). In the case of locomotives driven exclusively by electricity, the power is usually collected in the form of an alternating current (AC) and fed to the motors via an air-cooled transformer, and rectifier, which converts AC to DC. The diesel-powered locomotive – which is more accurately described as diesel-electric – is essentially an electric locomotive that carries its own power unit. An internal-combustion engine fueled by diesel oil (◀ page 133) drives an electric generator which then feeds a current to power the traction motors.

The French TGV is part of a fully integrated high-speed system, in which a control center instructs the train driver on speed

◀ ***Assisted by computers, the operations center in Paris controls train movements, points, signaling and safety installations for the length of the TGV's journey.***

▲ ***The electronic display in the driving cab indicates the maximum permitted speed in km/h, according to the information relayed to the train from the operations center.***

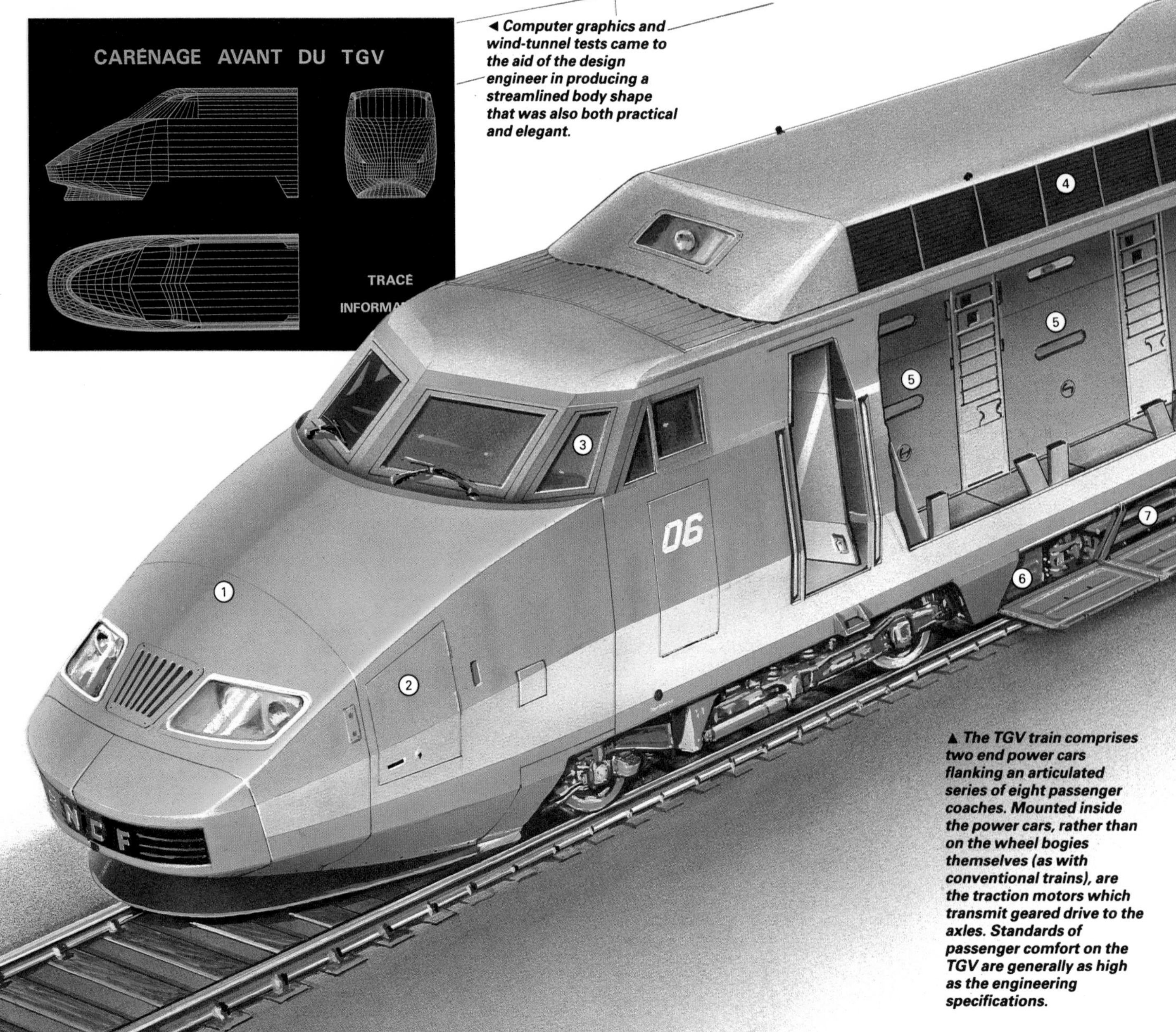

◀ ***Computer graphics and wind-tunnel tests came to the aid of the design engineer in producing a streamlined body shape that was also both practical and elegant.***

▲ ***The TGV train comprises two end power cars flanking an articulated series of eight passenger coaches. Mounted inside the power cars, rather than on the wheel bogies themselves (as with conventional trains), are the traction motors which transmit geared drive to the axles. Standards of passenger comfort on the TGV are generally as high as the engineering specifications.***

▼ The TGV's pantograph, the link between the overhead electric power line and the motors on board the locomotive, is designed to collect the current at high speeds.

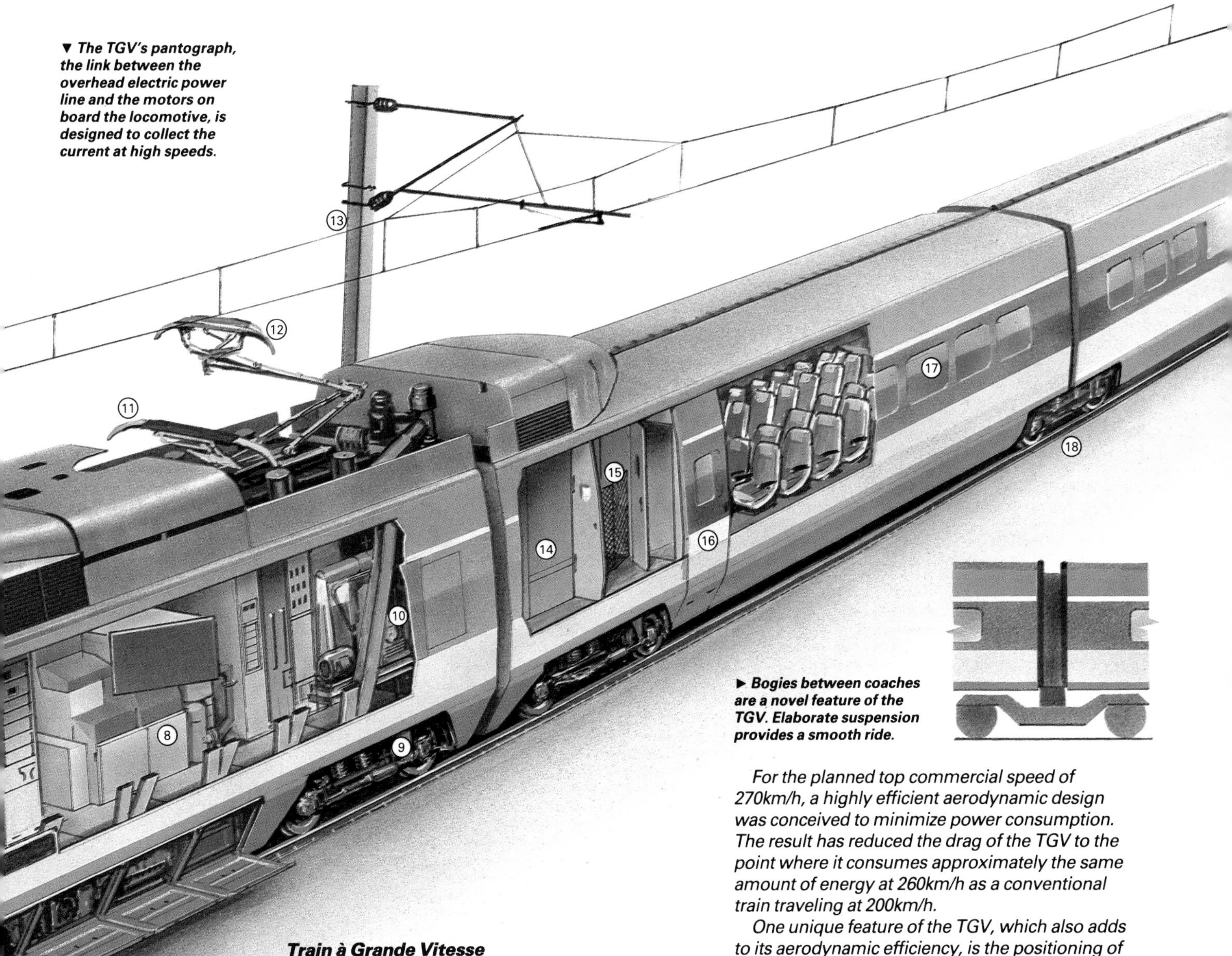

► Bogies between coaches are a novel feature of the TGV. Elaborate suspension provides a smooth ride.

1 Collision protection
2 Brake gear
3 Driver's cab
4 Cooling air vents
5 Traction motors
6 Driver's cab air-conditioning
7 Battery compartments
8 Main transformer
9 Suspension
10 Main compressor
11 Pantograph (1,500V DC)
12 Pantograph (25,000V AC)
13 Overhead wires
14 Rectifiers
15 Baggage compartment
16 Automatic exterior doors
17 Passenger compartment
18 Inter-car bogie

Train à Grande Vitesse

The basic concept of the French "Train à Grande Vitesse" (TGV) is that of a high-speed train running on a traditional track modified to take account of the very high speeds envisaged. A major advantage of such a design concept is that during the many years that it takes to build a network of modern high-speed track the TGV can continue to run on existing lines, though not at full speed.

With the sudden large increases in the price of oil in the 1970s, the motive system originally planned – the gas turbine – was replaced by an electric system. The French national railways (SNCF) decided that their new high-speed trains would be powered from overhead cables carrying alternating current at 25,000 volts. A completely new track system was planned, to be used exclusively by the TGVs. Grades four times as steep as usual could now be allowed, with the result that track cost almost one-third less than a conventional rail link.

The first section of the new track was built between Paris and Lyon, a distance of 390km, and commercial service started in 1981. A new traffic control center in Paris was designed to be in continuous radio contact with every train in service.

For the planned top commercial speed of 270km/h, a highly efficient aerodynamic design was conceived to minimize power consumption. The result has reduced the drag of the TGV to the point where it consumes approximately the same amount of energy at 260km/h as a conventional train traveling at 200km/h.

One unique feature of the TGV, which also adds to its aerodynamic efficiency, is the positioning of the bogies between the coaches, giving an "articulated" design that requires only nine bogies to carry the eight bodies that make up the passenger part of the train. A conventional train has two bogies per coach. Mounting the traction motors in the body of the power-car with flexible drive to the axles has substantially reduced the mass of the bogie, and thus the forces exerted on the track. The difficult problem of picking up current from an overhead cable at high speed was resolved by using a special two-stage pantograph designed to maintain contact with the power cable. A new signaling system replaced all external signals with in-cab data, and the permitted speed is displayed in front of the driver who sets his controller to maintain that speed automatically.

The TGV is the world's fastest train, holding both the world speed record on rails of 380km/h and the record commercial speed of 270km/h. By comparison, the Japanese Shinkansen has a maximum permitted speed of 210km/h. The journey time from Paris to Lyon has been cut almost by half, and, through 1983, 97·5 percent of TGVs arrived on time.

See also
Introduction to Measuring 9–22
Engines 129–134
Electricity 135–140
Road Transport 199–208
Mining 229–236

Trains that float

The TGV probably approaches the highest speeds possible for trains with flanged steel wheels running on steel rails. For future advances in high-speed technology, attention may turn towards applying the principle of magnetic levitation (maglev), by which trains are suspended above the track by magnetic force, thus eliminating friction. The Japanese have already achieved a world track speed record of 517 kilometers per hour with their test vehicle ML-500 in June 1977, and both they and the West Germans aim to have high-speed commercial services in operation by the end of the 1980s.

Maglevs can be either "attractive" or "repulsive" In the first case, adopted by the German Transrapid system, a series of electromagnets are set in the train wings, which are slung below the guideway. When the current is switched on, these magnets are drawn up to the steel rail. A gap of 15 millimeters must be maintained to prevent the train sticking to the rail, and this is monitored by sensors which automatically switch the current to the electromagnet on and off.

Repulsive maglevs, such as the Japanese MLU 001, rely on superconducting magnets to induce eddy currents in coils embedded in the U-shaped guideway. These in turn create a magnetic field of the same polarity as that generated by the magnets in the train; the two fields repel each other and the train is levitated. With the magnets cooled by liquid-helium refrigeration units, the current continues to flow even after the power is cut off, resulting in a very economical operation. A wheeled suspension system is included in this design as the train has to be traveling at about 100 kilometers per hour to induce sufficient currents for total levitation to occur.

Both types of maglev are propelled by a linear induction motor, which works like a flattened electric motor (◀ page 136). The stator coil is "unwound" along the guideway (combined with the levitation coils in the German system, but placed separately in the Japanese) while the rotor becomes the reaction rail in the train. When current flows through the stator coils, alternating magnetic polarities are precisely controled to attract and repel the magnetized rotor in such a way as to push the train forward. Increasing the frequency accelerates the train; cutting the power serves to brake it.

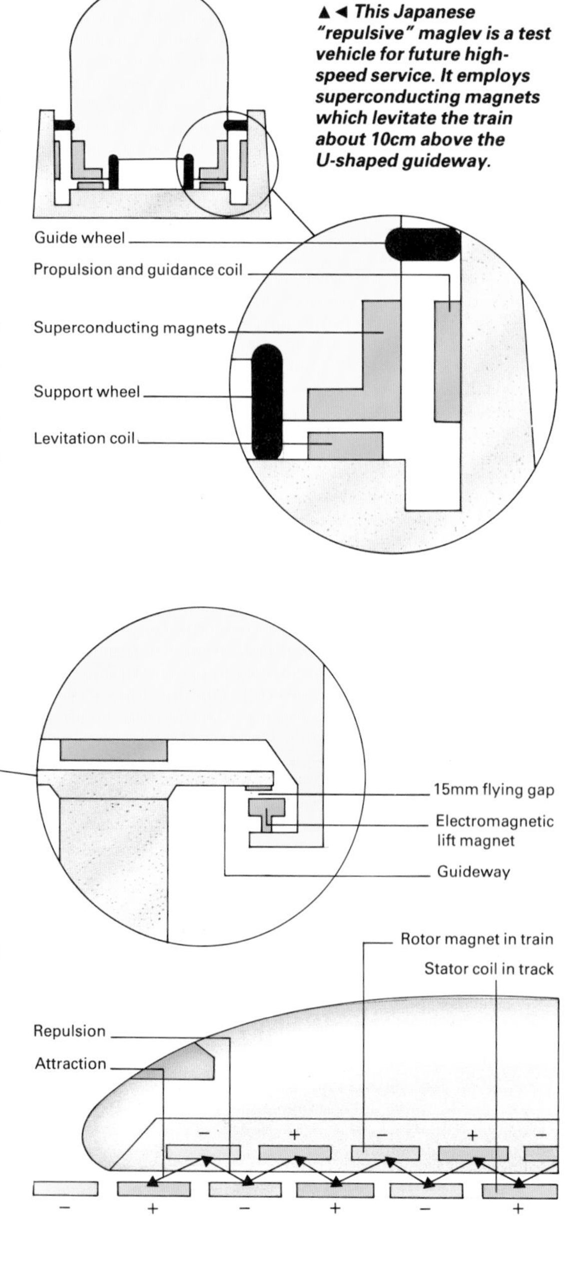

▲◀ This Japanese "repulsive" maglev is a test vehicle for future high-speed service. It employs superconducting magnets which levitate the train about 10cm above the U-shaped guideway.

▲◀ The Birmingham (UK) maglev became the world's first commercial service in 1984. An "attractive" version, it runs its 620m route levitated 15mm above the track by eight electromagnets.

► By controling the current in flat stator cells, unlike poles can be created just ahead of stator and rotor magnets (pulling the vehicle forwards) and like poles just behind (pushing it forwards).

Road Transport

The development of the horseless carriage...Designing for the checkered flag...Bicycles and human-powered vehicles...The car of the future: Nissan CUE-X... Automotive design...PERSPECTIVE...Features of the car... Monocoque and ground effects...Bicycle gear systems... In-car navigation systems...Turbochargers and superchargers

▲ *The 1893 "Viktoria" designed by Karl Benz.*

The automobile is a machine of the 20th century, but all its basic components of power and transmission had been invented by the end of the 19th. The twin giants of the automobile's earliest history were the German engineers Karl Benz (1844–1929) and Gottlieb Daimler (1834–1900), who independently and successfully applied a practical internal combustion engine (◀ page 132) to wheeled vehicles, in 1885 and 1886 respectively. Benz's first vehicle was little more than a motorized tricycle, whose stability he improved with the four-wheeled "Viktoria" which was in limited production by 1890. Daimler meanwhile had licenced the French firm Panhard et Levassor to manufacture his engines, resulting in a design that appeared in 1892 and has remained fundamental to road vehicle technology: a forward-mounted engine passing drive to the rear wheels via a clutch and gearbox.

By 1900 an automobile industry was taking shape in France – conservatism had cost Germany the lead that could have been built on the pioneering work of Benz and Daimler – and many other features of the modern car, including the electric starter and all-steel bodies, were to appear before the First World War. By that time the industry's leadership was moving to the United States, with the introduction in 1908 of moving-line production by the industrialist Henry Ford (1863–1947) for his Model T, which was to achieve sales of more than 15 million. In 1915 the more sophisticated but equally pioneering Packard Twin Six appeared at the other end of the scale. Meanwhile the needs of war gave impetus to truck development and introduced many young men to the design and use of motor vehicles.

▲ *Henry Ford in his 1905 Model F.*

Automobile development after 1918

Through the next two decades cars became more widely available throughout the Western world. The makeup of cars changed as steel bodies became the norm and four-wheel brakes were widely adopted. Refinements such as synchromesh gearboxes and independent front suspension spread in the 1930s.

After the Second World War the motor industry generally resumed with its pre-war range of models, and in part passed through periods of technical stagnation disguised by exaggerated styling. Companies were sometimes sidetracked into producing bizarre economy vehicles. The French company Citroën had for long concentrated on producing front-wheel drive models, but the appearance of the British Austin-Morris Mini in 1959 signalled a general trend towards that layout; soon rear-engined cars virtually disappeared.

By the 1980s, when the car was taken for granted as a means of independent transport in most countries, the main challenge facing the industry worldwide was not that of producing vehicles to meet the fuel crises that had been predicted in the previous decade, so much as of meeting the problems of manufacturing overcapacity.

▲ *The Austin Morris Mini of 1959.*

Chronology

1885 Gas-driven engine tricar developed by Karl Benz
1886 Gottlieb Daimler's four-wheeled motor car appeared
1889 Pneumatic tire introduced by J.D. Dunlop in Ireland for bicycle
1890 Benz's "Viktoria" became world's first production car
1895 Pneumatic tires adapted for use on motor vehicles
1898 Appearance of shaft drive, introduced on the first Renault
1898 Independent suspension fitted to the French Deauville
1901 First "modern" car, the Mercedes, introduced by Daimler
1902 Internally-expanding drum-brake developed by Louis Renault
1903 Four-wheel drive introduced by Dutch firm, Spyker
1904 Pneumatic tyre with raised flat tread introduced by Michelin
1908 Model T Ford, first production-line car, built in USA
1909 Telescopic shock absorbers invented in France
1916 Windscreen wipers introduced
1921 Hydraulic brakes introduced on Duesenberg Model A
1921 Appearance of first cars equipped with supercharger
1923 All-steel bodies introduced by Studebaker in USA
1924 Electric fuel pump became regular feature on production cars in Wills Sainte Claire
1928 Freewheel differential developed by Brazilian D. de Lavaud
1929 Front-wheel drive introduced on American Cord L-29
1929 Synchromesh gearbox developed by Cadillac

Layout and steering

The principal components of a motor vehicle are its engine (➧ page 132), the powertrain (the means of transmitting the engine power to the driven wheels), suspension and brakes, as well as the chassis, the basic structure on which all the parts are assembled. The usual form of the chassis is now a unitary or "monocoque" construction (➧ page 202), enclosing engine, passenger and luggage compartments.

For almost 50 years the conventional layout placed the engine at the front, driving the rear wheels, and many automobiles (especially the larger-engined models) are still built on these lines. Front-wheel drive has now become the most widely used layout for smaller cars, as well as some medium-sized models. With manual and automatic gearboxes a form of clutch is required to engage the transmission to the engine flywheel. This normally takes the form of a friction disk which can be pressed against the flywheel face and withdrawn. The gearbox allows a driver to maintain optimum speed (and hence torque) regardless of road speed. A propeller shaft carries power from gearbox to the differential, which is a system of final drive gears that permits a wheel on the outside in a bend to rotate faster than the inside wheel.

Rack-and-pinion steering, which came into use in the 1930s, is simple and effective. The pinion (a gear wheel) at the outer end of the steering column meshes with a rack, which it moves to the left or right; tie rods transfer this movement to the wheels. Power steering, using hydraulic servos, became almost essential on the large and heavy automobiles of the late 1940s and 1950s in the United States, and it was later adopted for speed of response in other vehicles as well, although it can make steering too light. Some manufacturers recently have experimented with four-wheel steering systems to stabilize high-speed steering.

Suspension systems

The earliest motor vehicles used springs like those used on horse-drawn vehicles, but it was soon realized that a system interposing springs between wheels and chassis was needed. For many years leaf springs in a variety of forms, combined with rigid axles and sometimes with dampers to curb the tendency to bounce or oscillate, were almost universal. Independent front suspension came to mass-production vehicles in the 1930s, and more recently all-independent suspension became widespread, using systems such as trailing arms, double wishbone and MacPherson strut. There are also multilink systems and interconnected arrangements and now there is a move towards "active" systems (➧ page 208).

Allied to the springs are the dampers. Early versions relied on friction, then hydraulic types came in, followed by the telescopic types most widely used today.

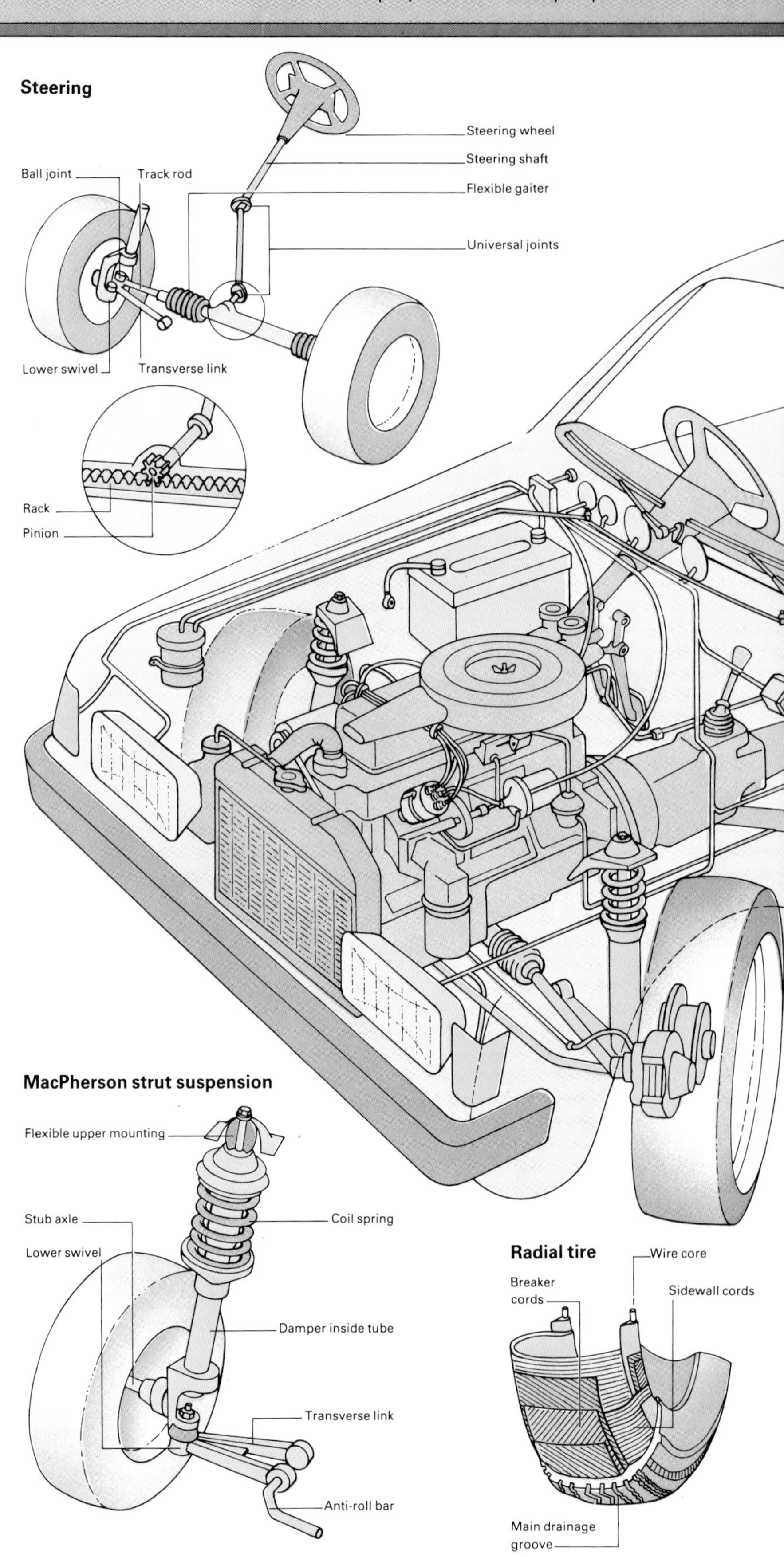

1934 Citroen made first mass-produced car with monocoque construction, front-wheel drive and independent front-wheel suspension
1934 Automatic overdrive developed by Chrysler
1936 Mercedes-Benz 260D became first diesel-engined production car
1937 Automatic transmission developed by Chrysler for Oldsmobiles
1939 Hydramatic transmission introduced
1947 Tubeless tire introduced by Goodrich of USA
1948 Radial tire introduced by Michelin
1948 Four-wheel drive introduced to Land Rover
1950 Power-assisted steering introduced for production cars by Chrysler
1953 Fiberglass reinforced plastic bodywork used on Chevrolet Corvette
1962 Monocoque chassis construction introduced on Lotus 25 Grand Prix car
1963 Wankel rotary engines introduced to production cars by NSU in Japan
1966 Electronic fuel-injection systems developed in Britain
1972 Dunlop introduced "safety" tires: punctures sealed by liquid compounds inside tire
1973 Hydrogas suspension, using pressurized nitrogen, introduced in British
1977 Hydrogen gas used experimentally as fuel
1979 Low-pollution direct-injected Stratified Charge (DISC) engines developed in USA
1980 Four-wheel drive successfully applied to road vehicles, with Audi Quattro

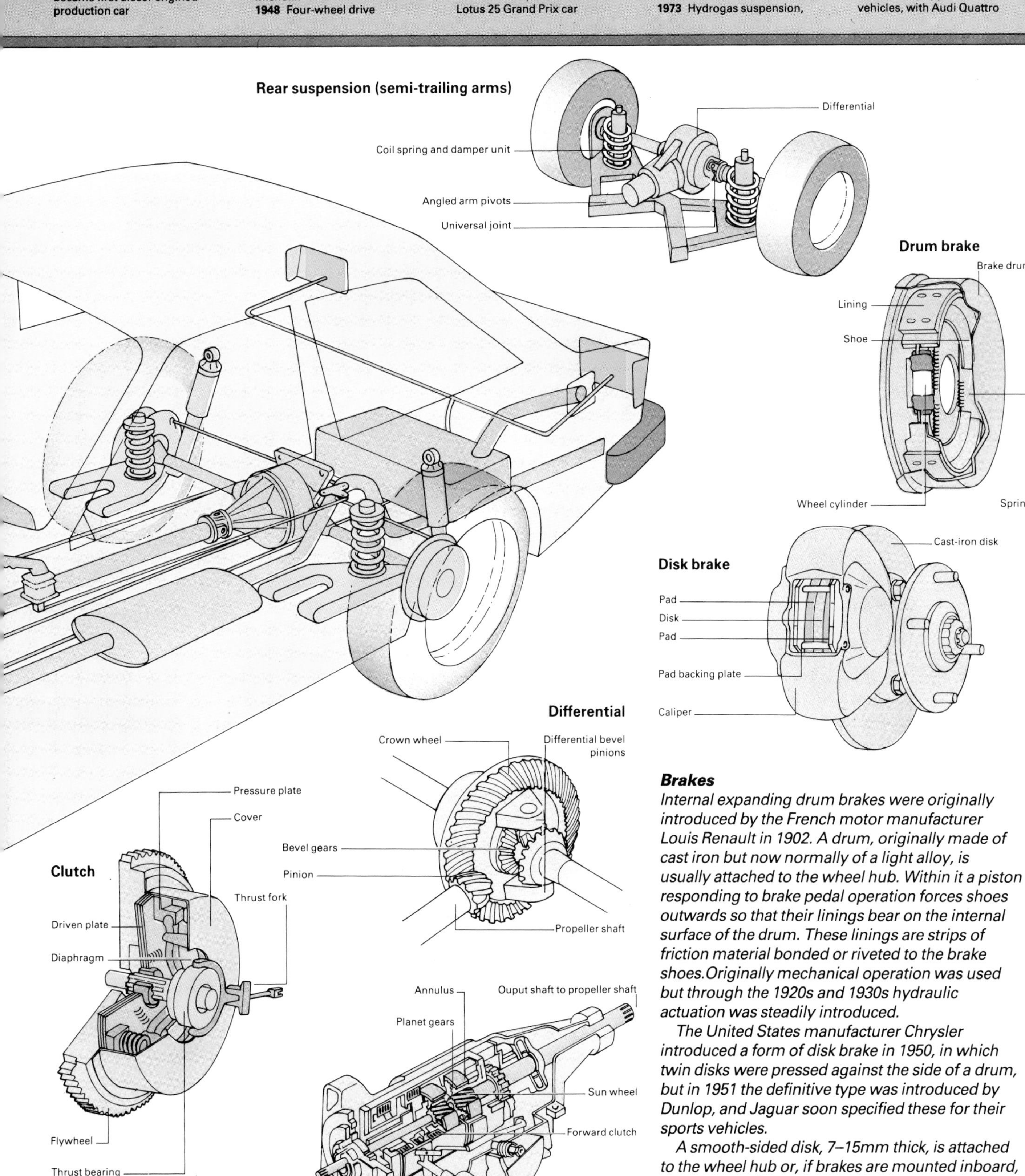

Brakes

Internal expanding drum brakes were originally introduced by the French motor manufacturer Louis Renault in 1902. A drum, originally made of cast iron but now normally of a light alloy, is usually attached to the wheel hub. Within it a piston responding to brake pedal operation forces shoes outwards so that their linings bear on the internal surface of the drum. These linings are strips of friction material bonded or riveted to the brake shoes. Originally mechanical operation was used but through the 1920s and 1930s hydraulic actuation was steadily introduced.

The United States manufacturer Chrysler introduced a form of disk brake in 1950, in which twin disks were pressed against the side of a drum, but in 1951 the definitive type was introduced by Dunlop, and Jaguar soon specified these for their sports vehicles.

A smooth-sided disk, 7–15mm thick, is attached to the wheel hub or, if brakes are mounted inboard, to a drive shaft from the hub. An external caliper houses lined pads, which are forced onto machined surfaces of the disk by hydraulically operated pistons.

Monocoque chassis and ground effects

In 1962 the British designer Colin Chapman introduced monocoque chassis construction on his Lotus 25 Grand Prix car. The idea was not completely original but earlier designs, from the American Cornelian racing car of 1914 to the monocoque center section of the Jaguar D-type sports-racing car of 1954, had not been followed through. Within ten years, Chapman's lead had been followed by every other manufacturer of front-rank racing vehicles, and the tubular space-frame chassis which it superseded was used only in cars for secondary categories.

In the Lotus 25, the monocoque, or single-shell hull, consisted of two hollow side members containing the fuel tanks and linked by the floor and bulkheads. It proved lighter, more rigid (a factor that improved road-holding) and more crash-resistant than a space frame. It contributed to the reduction in vehicle frontal area, and was ideally suited to the practise of mounting the engine behind the cockpit, which had by then become universal. In 1967 Chapman introduced the modern practise of using the engine as a stressed chassis member, linking rear suspension and monocoque hull, in the classic Lotus 49.

The simplicity of the 1962 monocoque has given way to the complexity and cost of one-piece shells using carbon-fiber materials that have been "baked" in autoclaves – in part a spinoff from space technology (➧ page 246).

Colin Chapman was also responsible for one of the great racing innovations of the 1970s, "ground effects", a design which gave phenomenal cornering power. In ground-effect cars, the fuel tank and engine were on the center line of a slim monocoque, leaving the broad side pods clear of all ancillaries except radiators so that they could be used as inverted wing sections. These created low pressure areas which pulled the car down onto the road. But the rigid suspension systems necessary imposed enormous stresses and loadings, and this safety hazard led to the banning of ground effects in 1983.

Lola Ford Formula 1 car

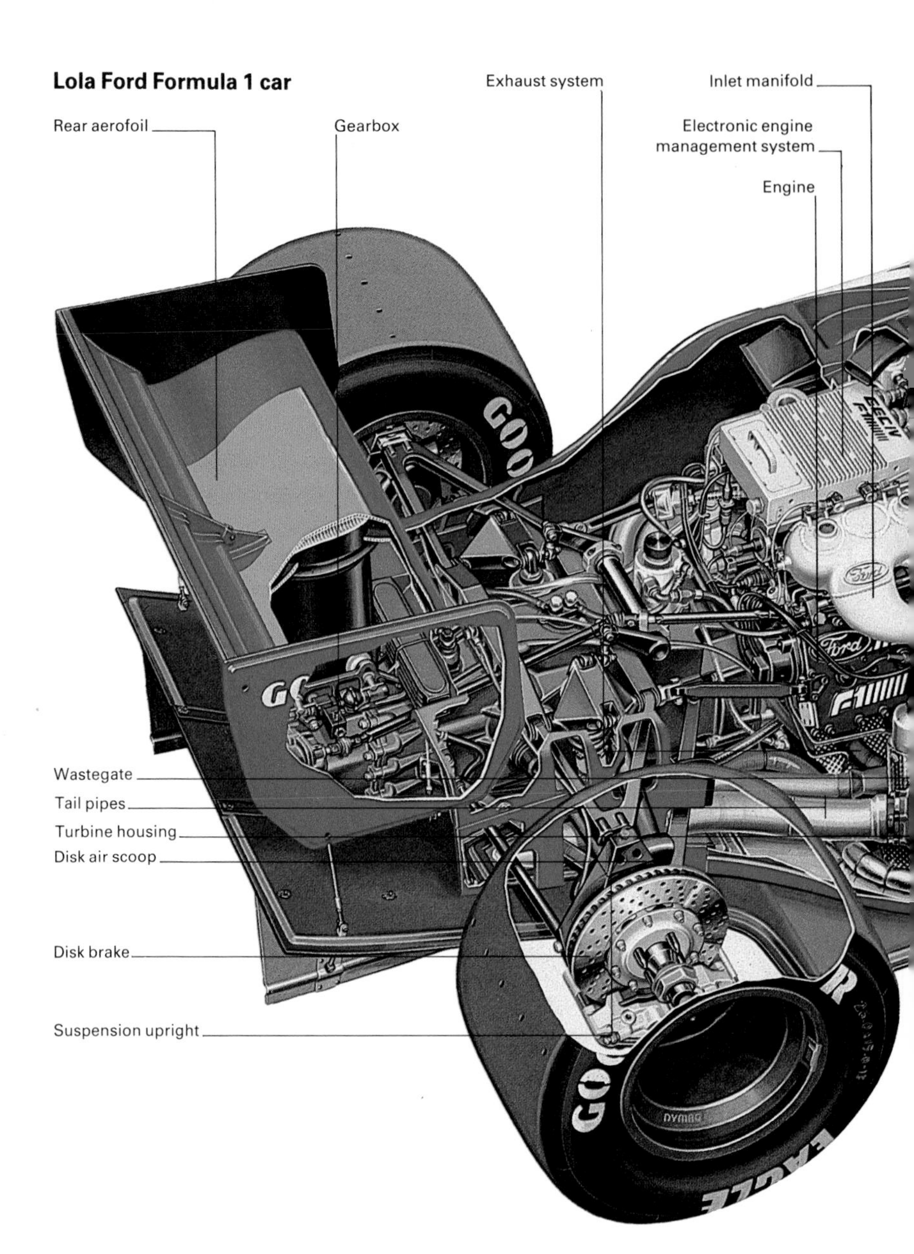

***►* In 1962 the Lotus 25 revolutionized Grand Prix car design, with a monocoque construction borrowed from aviation engineering. The superior stiffness of the chassis meant that more supple suspension systems could be incorporated than previously, which gave it an advantage over its space-frame rivals on tight corners.**

***◄* The Lotus 78, which appeared in 1977, was the first ground-effects racing car. Inverted fiberglass side wings, initially sealed by brush-lined side plates but soon more effectively plated with ceramic rubbing strips, created a downforce that sucked the car onto the ground.**

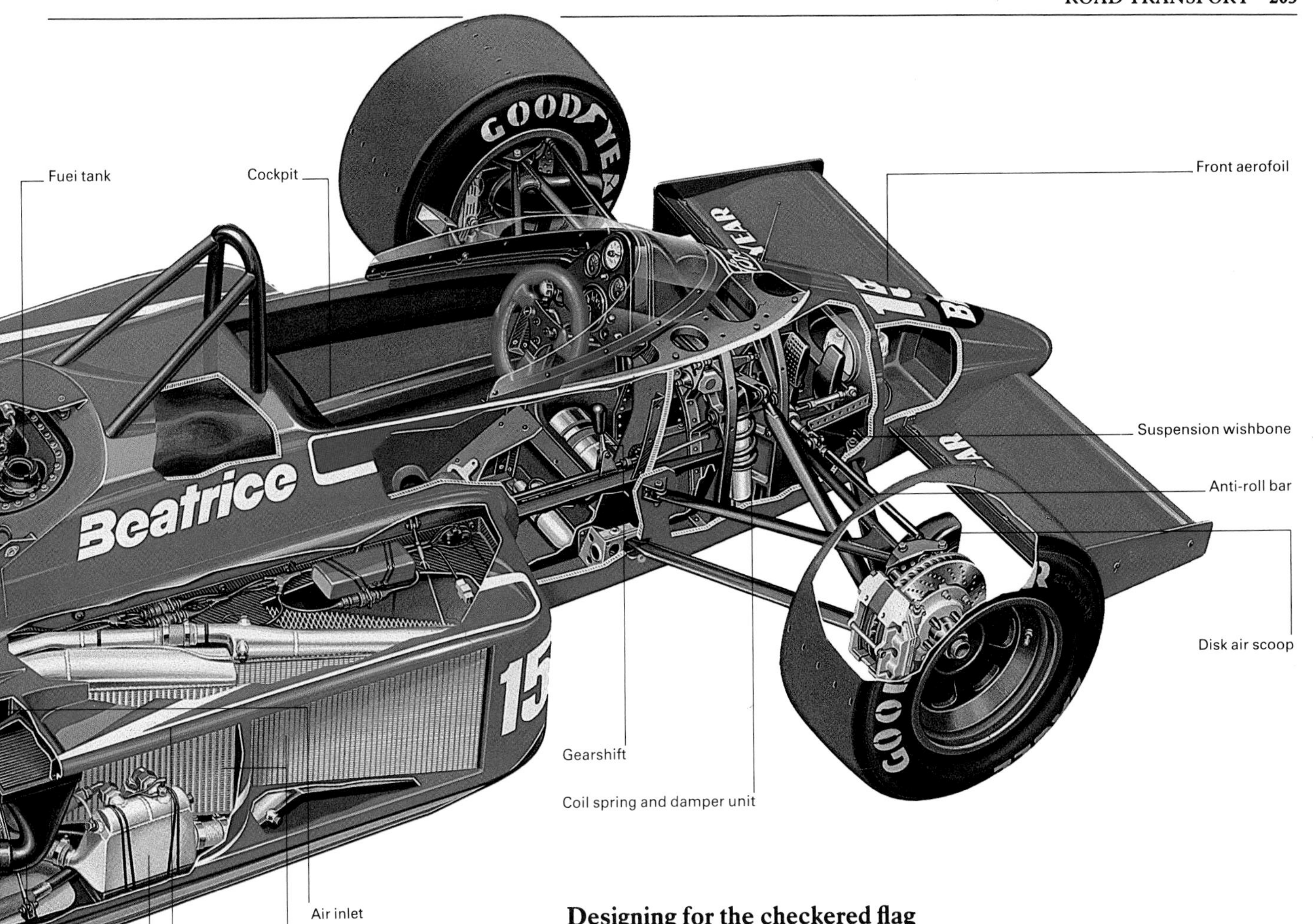

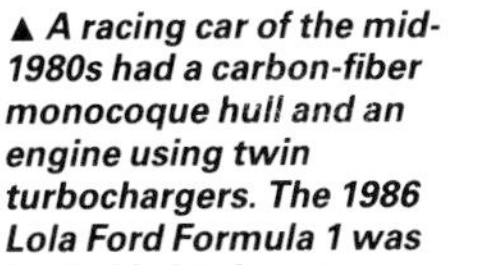

▲ *A racing car of the mid-1980s had a carbon-fiber monocoque hull and an engine using twin turbochargers. The 1986 Lola Ford Formula 1 was typical in having an electronic engine management system linked to a digital display, which controls fuel injection, ignition and turbocharger operation.*

Designing for the checkered flag

Motor sport has often forced the pace of automotive development, although some of the "novelties" have been adopted for cars only after having first been proved on other vehicles. Disk brakes, for example, were used on commercial vehicles and aircraft before they appeared on a racing car in 1938 and on a road car in 1952. Racing of all classes has been a great stimulus to road vehicle technology, although not all the innovations tried out on the track have found a wider application. Grand Prix cars before the First World War were the first to incorporate the efficient overhead camshaft engine; those of more recent vintage pioneered carbon-fiber monocoque hulls.

Apart from road-holding and handling, where the virtues of all-independent suspension were first appreciated, aerodynamics and engine power have been high racing priorities. For decades aerodynamics meant reducing frontal area wherever possible (though this was complicated by regulations that called for the wheels to be outside the bodywork), and perhaps "streamlining". In 1967 airflow was positively exploited for the first time as "wings" appeared on Grand Prix cars. These rapidly outgrew genuine knowledge of their application, and they were soon banned.

Ground effects, the introduction of "inverted wings" to hold the car on the road, was also curbed by regulations in 1983. In consequence there was a return to large aerofoils, and, to overcome the drag penalty that these imposed, designers turned to a search for more power and found it in turbocharging the engine (➧ page 208). Turbocharging in its turn was restricted by Grand Prix regulations after 1986 and from 1988 only atmospheric engines will be permitted. Advanced suspension which can be "tuned" through minute adjustment, sophisticated aerodynamics and very high power-to-weight ratios are now features of all racing cars built for road circuits or high-speed tracks.

The bicycle

The excellent power-to-weight ratio of the bicycle means that for many years it has been the most efficient form of regular road transport. Derived from the "hobby-horses" of the early 19th century (two-wheeled vehicles propelled by being "walked" along the road), true bicycles powered by pedals directly driving the wheels or linked to them by a flexible chain became popular in the last decades of that century. By 1900 the now-conventional diamond frame, with equal-sized wheels, chain-drive and crankshaft at the bottom of the seatpost had become commonplace. The invention of internal hub-gears and external derailleur gears made the bicycle suitable to hilly terrain, and inflatable tyres made the ride more comfortable.

This design of bicycle has proved perennially popular and adaptable to many purposes, but recent years have seen many new developments. Racing bicycles have adopted aerodynamic modeling, whether in the "cow-horn" low-profile handlebars or the oval tubes of the frame, and in 1984 carbon-fiber solid wheels, designed to reduce drag, were introduced in place of the conventional spoked variety. Expensive materials, notably titanium, are used to reduce the overall weight. Light and strong alloys, often of chrome-molybdenum, provide components that can be trusted over the roughest terrain, and specialist "mountain-bike" models have been produced for off-road riding, with strengthened and very stable frames, low gearing and broad wheels and tyres. The bicycle motocross (BMX) machine is a smaller version of the mountain-bike, adapted for racing over rough ground or for stunts which require both stability at low speeds and resilience to flying jumps. Many other designs have been tried, from the small-wheeled, folding bikes (some of which, like those built by the British firm Moulton, may match conventional styles in speed and handling characteristics) to those with a "slung-back" style on which the rider takes up a recumbent position – excellent for long distances but undesirable in traffic. Even more unusual designs have been tried out on machines designed to maximize aerodynamic efficiency, while some designers have experimented with reviving designs from the early history of the bicycle, using modern lightweight components.

▲ **This 1818 cartoon depicts the introduction of the "Draisienne" to the French mail service. Equipped with a steerable front wheel but still propelled by foot, this machine was demonstrated by the German Baron von Drais in that year.**

◄ **Between the arrival of pedal propulsion with Michaux's "Vélocipede" in 1863 and chain drive to the back wheel in 1879, various systems for attaching the pedals were tried. For this "high" or "standard" bicycle ("pennyfarthing") they were fitted to cranks on the front hub.**

▼ **Bicycle motocross (BMX) machines first became popular in the United States in the 1970s. Skillful maneuvers over tough courses are made possible by a light but strong frame backed up by wide handlebars, studded tyres and back-pedaled brakes.**

Gearing for bicycles

The use of a chain to transmit drive from pedals to wheels was introduced in 1879 with H.J. Lawson's "Bicyclette". As well as allowing greater flexibility in frame design, this arrangement meant that gears could now be added. Gears allow the rider to select how much forward motion should result from a single turn of the pedals, thus allowing effort to be matched to road conditions.

Derailleur gears, with a selection of cogs onto which the chain is fed by a guide mechanism, were first introduced in 1909; they are easy to access but require regular maintenance. Using derailleur gears on both bottom bracket and rear wheel allows a choice of up to 15 gears. The most common arrangement offers the rider a forward movement of 76–250cm for each turn of the pedals, according to the gear selected. Internal hub gears are simpler to operate but offer less scope; the usual mechanism, which was introduced in 1938, has up to five gears. The chain drives a sun-wheel, with which planet-wheels of various sizes engage. These are selected by the gear control, and slip into position as the freewheel turns.

▲ *The United States team brought new technology to Olympic racing in 1984. Fitted with lightweight frame, "cowhorn" handlebars, slimline wheels of unequal size and helium-filled tyres, their machines were built for lightness and aerodynamic efficiency.*

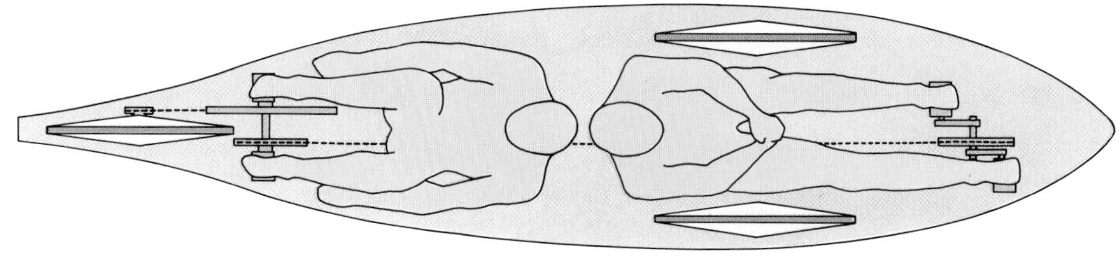

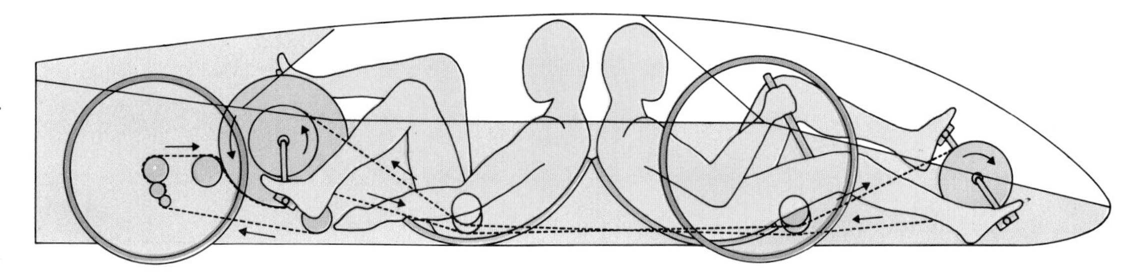

► *Shown here in plan and elevation, the Vector Tandem, a streamlined two-person recumbent, is the world's fastest human-powered road vehicle. Designed with scientific attention to aerodynamics, it set the speed record of 102km/h over 200m in 1980.*

A laser radar system automatically maintains a safe distance from the vehicle in front

Today's automobile of the future
In 1985, the Japanese firm Nissan unveiled the CUE-X (Concept for the Urban Executive). A working prototype automobile which may never reach full production, it is nevertheless built with a range of state-of-the-art technology that promises to set the trend in road vehicle development for the immediate future. The CUE-X incorporates many features already widely adopted, such as turbocharging, four-wheel drive, active suspension and anti-lock brakes. It is also equipped with an array of sensors, computers and electronics designed to optimize the response of working parts to the driver's will, by monitoring both the changing road circumstances and the condition of the vehicle itself.

*The operating elements of the turbocharger engine, including the cooling system, fuel injection, idling speed and valve timing, are all electronically controlled by a 16-bit microcomputer. The driver issues commands to the power unit through depression of the gas pedal. The throttle opening, linked to the pedal electronically rather than mechanically, is controlled to respond best to the driver's wishes while taking into account driving conditions. This "drive-by-wire" system is equivalent to the "fly-by-wire" facility being introduced to air transport with the A320 (**▶ page 222**). In the CUE-X, this electronic control is extended to the transmission and the brakes. Optimum traction to all four wheels is delivered via automatic distribution of engine torque. In braking, inputs such as rate of deceleration, payload changes and brake wear are accounted for in the response made to pedal effort.*

Incorporated within the electronic control system is one of the CUE-X's most innovative features. A laser radar system installed in the nose calculates the relative speed of the vehicle in front and determines whether it is a potential hazard. If necessary, it then slows the CUE-X by automatic control of throttle, transmission and brakes. Another novel facility is four-wheel steering. The power steering hydraulic system is extended to include the rear wheels. At speed, they are turned fractionally in the same direction as the front set, which enhances cornering and lane stability; at low speeds they are turned in the opposite direction for maximum maneuverability.

Nissan CUE – X

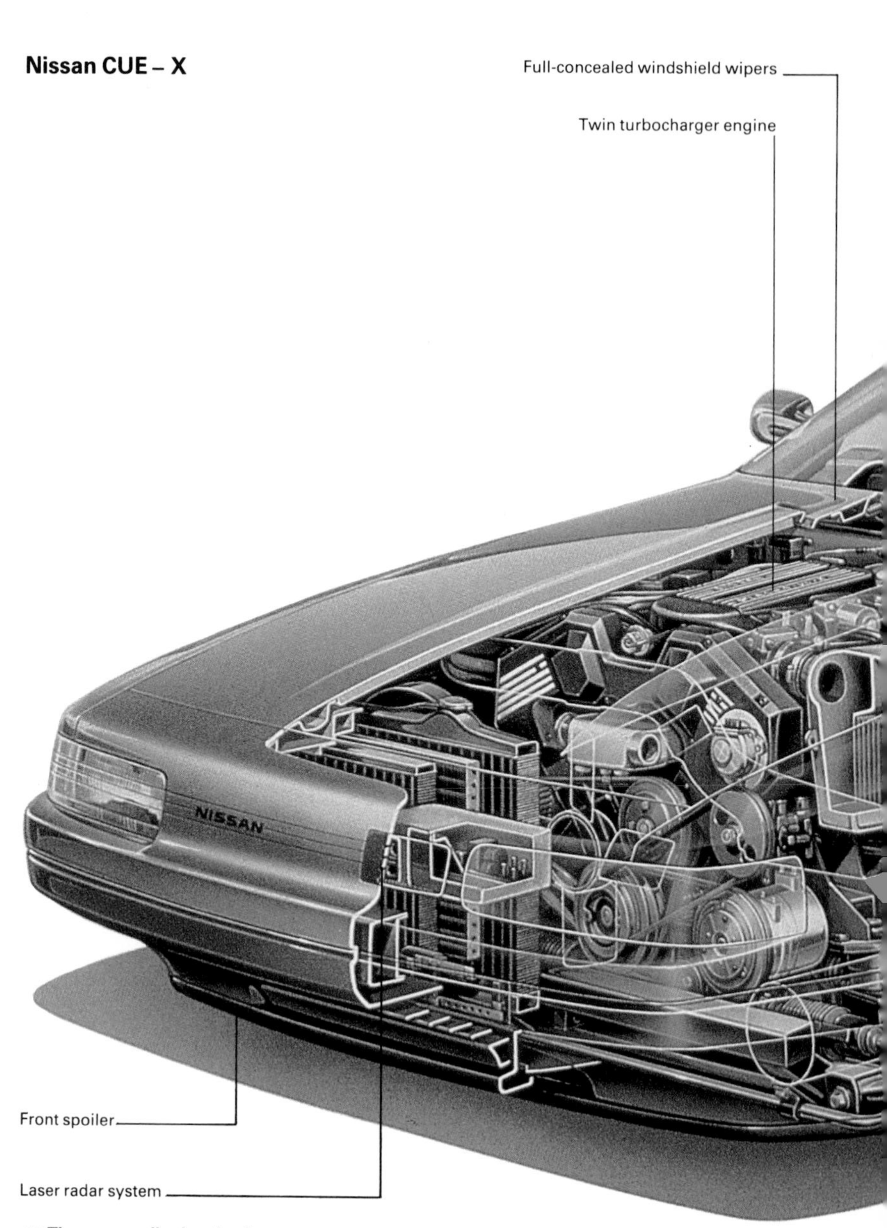

▼ The rear spoiler is raised automatically to reduce aerodynamic lift at high speed or to limit the possibility of aquaplaning in wet weather.

"Drive-by-wire"

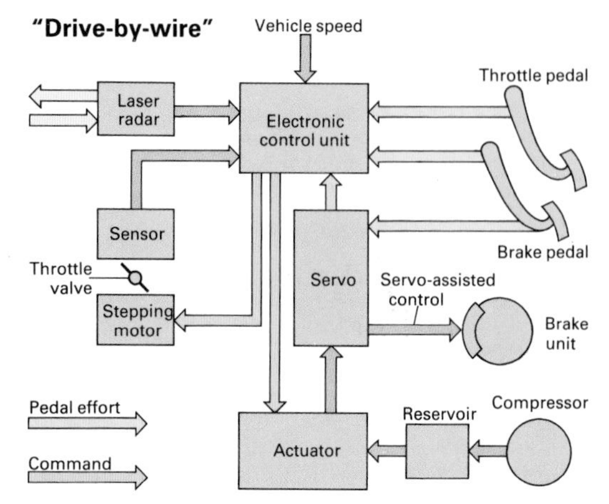

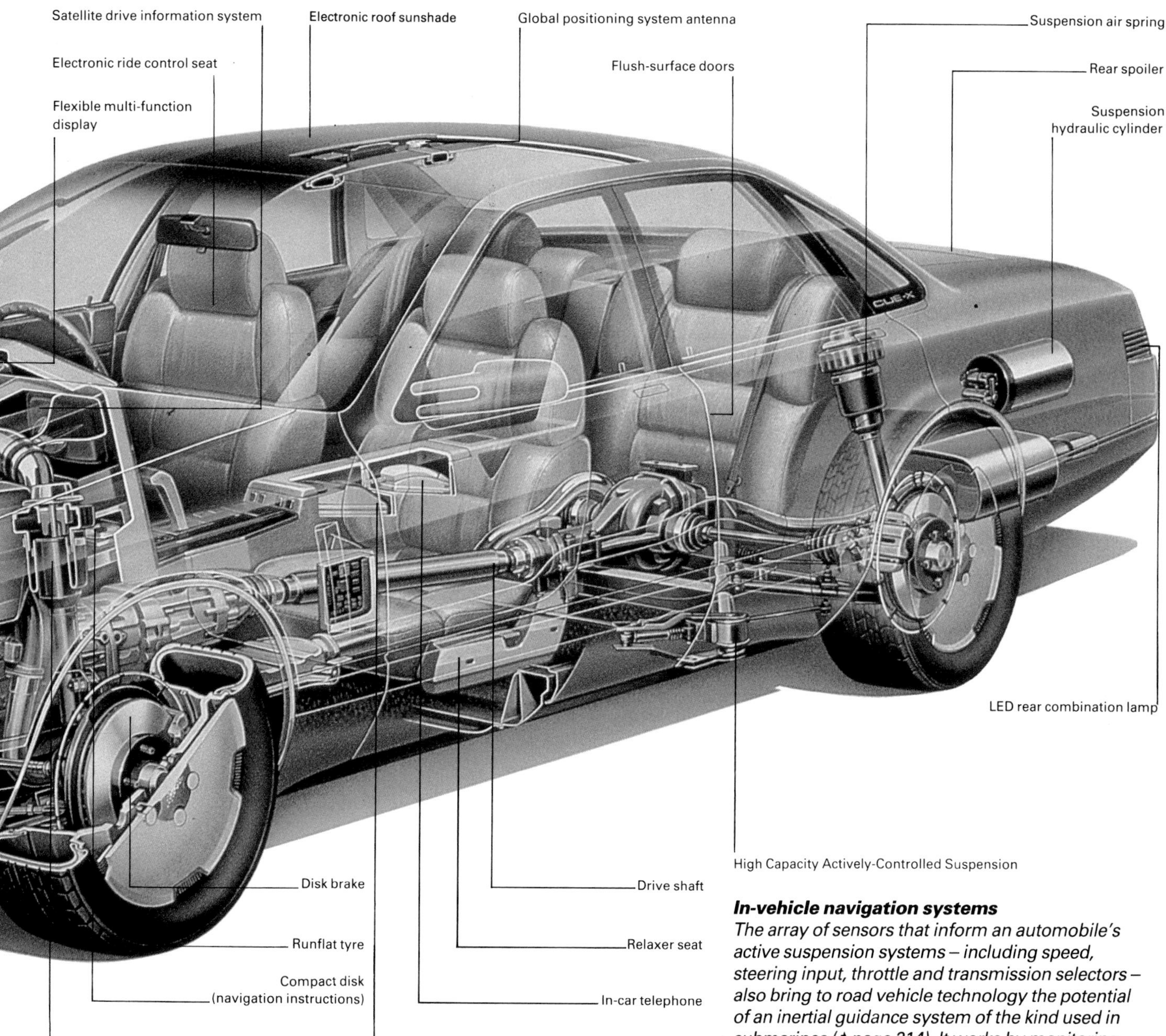

◄ ***The CUE-X drive-by-wire system centers on a microcomputer. Throttle and brakes (via a servo system) are operated electronically. The computer can also take instructions from a laser radar system, by slowing the vehicle automatically.***

► ***The CUE-X dashboard features two cathode ray tube display panels. Behind the wheel, the flexible multifunction display can be switched between speed, warning and service information modes; the screen on the left gives route guidance instructions.***

In-vehicle navigation systems

*The array of sensors that inform an automobile's active suspension systems – including speed, steering input, throttle and transmission selectors – also bring to road vehicle technology the potential of an inertial guidance system of the kind used in submarines (**➧ page 214**). It works by monitoring movement in two dimensions and relating back the information received to a known starting point. To avoid the possibility of accumulated error, the system installed in the CUE-X operates in conjunction with an in-car telephone system, available to give updates on current traffic conditions. Also planned is a Global Positioning System, a satellite facility originally developed for military use but which, if adapted for civilian purposes, would enable a driver to fix the position of the car to an accuracy within 10m. A computer correlates the positional data acquired in transit and compares it with street maps contained in digital form on compact disk (**➧ page 98**) and displayed as images on a cathode ray tube screen in the dashboard. At road intersections, the screen display gives guidance instructions to the driver, indicating which way to turn for the best route to the preselected destination.*

See also
Introduction to Measuring 9–22
Computer Software 123–128
Engines 129–134
Air Transport 217–224
Processing Raw Materials 237–244

New trends in automotive design

In the second half of the 1980s it became obvious that oil reserves were adequate for the foreseeable future, so the industry relaxed its efforts to develop such alternatives to the internal combustion engine as electric cars. These have appeared spasmodically since the 1890s, but have never become a practical means of personal transport. Like the gas turbine car, the electric car remains no more than a possibility for the future.

Completely new car engines are rare, since a design has to be competitive for ten years or more after its introduction, and legislation on engine emissions may be changed or brought into effect in a major market at any time during this expected life-span. There are two principal forms of exhaust emission control. Catalytic converters are favored in the United States and some European countries (▶ page 108). These components work at high temperatures; they are expensive and impose a fuel consumption handicap of as much as 20 percent. Lean-burn engines are a more positive alternative, and are already available on some production cars. The aim is to use a lean air/fuel mixture (18:1 or more by mass) in a speeded-up combustion phase. This calls for careful design of the combustion chamber.

Manufacturers have turned increasingly to the turbocharger to raise the power outputs of existing gasoline or diesel engines, at some cost in economy, or to diesel units where consumption is paramount (typically 25 percent better with a diesel, a fact which offsets the higher initial costs; ◀ page 133). Electronic engine management systems to control complementary variables such as throttle, engine speed, ignition timing, temperature, boost (if a turbocharger is fitted), and so on will contribute to efficiency, and innovations such as the compressed-air valve systems introduced by Renault on their Grand Prix engines may also find their way onto road vehicle power units.

Apart from the mechanical and marketing aspects of modern design, research continues into reducing the "injury-producing potential" of automobiles, and in the second half of the 1980s work is concentrated on projects for a "soft" car body.

▲ Incorporated within the Nissan CUE-X twin turbocharger system are ceramic turbine rotor blades. Just as strong as the more usual nickel alloy, the material's lighter weight cuts the moment of inertia (the tendency of a body to resist acceleration) by 35 percent, allowing a more flexible response from the engine.

Turbocharging and supercharging

The turbocharger was a refinement introduced in the mid-1920s to aircraft and diesel engines, which has recently been adopted by automobile designers as a means of substantially increasing the power output of existing engines. It sometimes provides an alternative to using engines of larger capacity.

A turbocharger is driven by exhaust gases that can otherwise be regarded as energy going to waste, unlike a supercharger which absorbs power in its drive. Both are forms of compressor, forcing into an engine more fuel/air mixture than it could otherwise take. In a supercharger this process is directly powered by the engine itself, usually from the crankshaft.

In a turbocharger a turbine in the exhaust system is used to drive a centrifugal compressor in the intake system. This forces mixture into the induction system under pressure, each cylinder burns more with each combustion stroke and, to a degree related to the boost used, engine power increases. Boost can be regulated simply, being proportional to exhaust gas flow, and in normal road car engines the amount of boost is modest. It can be maintained at a pre-set maximum level regardless of exhaust gas flow by a wastegate between engine and turbine, and this arrangement has been used to restrain engine power outputs in some forms of racing. In Grand Prix cars drivers can control boost to meet speed or fuel consumption requirements.

◀ A turbocharger produces little boost at low engine speeds, although it is much more powerful than a supercharger which functions throughout the range of the engine. This rally car is fitted with a supercharger, and others may have both turbos and superchargers in order to exploit the respective benefits of each form of assisted power.

Water Transport

Evolution of the ship...The fall and rise of sail...Power by steam, diesel and nuclear energy...Transport under the water: submarines and submersibles...Transport over the water: hovercraft and hydrofoils...PERSPECTIVE... Chronology...Computerized sailing ships...Screw propulsion...How a submarine works

The waterborne craft was among the first ventures of humanity into technology. Evidence from the earliest civilizations dates the first reed boats and canoes made from dug-out tree-trunks to about 8000 BC, but sea voyages were possibly made many thousands of years earlier. The boat evolved separately in different parts of the world, and some ancient forms, such as the Arab dhow, still survive.

In 16th-century Europe, the full-rigged, three-masted ocean-going ship became the standard design, surpassed in seaworthiness only by the Chinese junk. The design was a marriage of traditions: the *cog*, a broad, deep-hulled version of the Viking longship with stern rudder, square sail, bowsprit and built-up "castle" at the stern; and the *nef*, a carvel or smooth-sided vessel of Byzantine origin, which inherited the ribbed frame of the ancient Phoenician-type round ship and incorporated lateen rigging first developed by the Arabs and gave great maneuverability. The resulting hybrid was of sturdy but streamlined construction, with a mix of square and lateen sails which allowed the ship to sail close to the wind as well as running with it. Although ships became larger, faster and more fortified, the pattern was kept until the advent of iron construction, steam and screw propulsion revolutionized water transport more than three hundred years later.

▲ Marco Polo's description of a junk made in 1298 – the earliest known account in the West – could equally be applied to one in use on Chinese waters today. Steered by a hinged sternpost rudder and maneuverable sails, the hull is subdivided into watertight compartments by transverse bulkheads, allowing some parts to be flooded without sinking the ship.

◄ The "Great Harry" was an English warship dating from the 1550s. Typically, her forecastle was set lower than on earlier ships, presenting less wind resistance at the bow, thus improving handling and speed, particularly to windward. A frame construction with hull planking and a longitudinal beam known as a keelson fastened to the keel were essential structural supports to this new, larger, breed of fighting galleons.

Chronology

Rowing and sail power
8th millennium BC Reed and dugout boats used
3rd millennium BC Sailing boats used in Egypt
c.1200 BC Ocean-going "roundships" used by Phoenicians
5th century BC Age of Greek trireme warship
3rd century AD Fore and aft sails introduced by Arabs
8th century Viking longship, with hinged sternpost rudder and several masts developed, capable of crossing Atlantic
15th century Full-rigged ocean-going ship introduced in Europe
16th century Appearance of galleon style
18th century Naval frigates streamlined
19th century Clipper cargo ships built, designed for minimum resistance to water on cargo route from Far East to Europe
1980 "Shinaitoku Maru", first sail-assisted merchant ship for 50 years, built in Japan

Steam, diesel and nuclear power
1783 First working steamboat, "Pyroscaphe", built by Jouffroy d'Abbans in France
1802 First commercially-successful paddle-steamer launched in Scotland
1821 First iron-hulled merchant-ship launched, the "Aaron Manby"
1836 Screw propeller patented by Francis Pettit Smith in Britain, and by John Ericsson in USA
1838 First ocean crossings under steam power, in "Great Western", built by I.K. Brunel, and the tug "Sirius"
1839 "Archimedes", first sea-going screw ship, built by American Joseph Ressel

The sailing ship

Sail is one of the most ancient, most durable, and, in its essentials, one of the simplest technologies invented to convert a natural resource into power for transport. Not seriously challenged until the mid-19th century for sea-going vessels and still widely used in recreation, the sailing boat has retained many of its fundamental features for thousands of years.

Around 3000 BC, the Egyptians found that they could travel up the Nile by fitting a mast and square sail to their boats, and thus make use of the prevailing up-river winds. Extra sails – the topsail and the foresail – were added by merchants during the Roman period. These could be adjusted to take the wind from different quarters; but it was not until the fore-and-aft rig and lateen sail were introduced on the Arab dhows of the 3rd century AD that it became possible to sail into the wind. The Chinese junks had a similar ability, and despite their rather ungainly, multimasted appearance they demonstrated supreme aerodynamic efficiency. The full-rigged three-masted ship became standard in Europe in the 15th century, evolving in many forms through refinements to the sail combinations and streamlining of the hull to become the high-speed tea-trade clippers of the last years of the 19th century.

Despite the supremacy of the metal, engine-powered ship in the 20th century, sail is far from finished. New materials and designs in hull manufacture have developed sailing as a leisure and competitive sport. And sail is beginning to make a come-back in commercial areas as well: most notably in Japan, where sails have been fitted to cargo ships to assist engines and save fuel on long voyages. Such ships resemble the 19th-century hybrid sail-and-steam ships such as the British engineer I.K. Brunel's *Great Britain* (➧ page 212), except that today's versions have sails of steel and plastic as well as canvas, and are computer-controlled to take optimal advantage of the wind.

▲ ***Model of an ancient Egyptian boat used on the Nile.***

Wind direction and velocity meter
Cross-section – unfurled
Cross-section – furled
Steel frame
Canvas
Revolvable mast
Turntable
Fixed mast

▲► ***Launched in 1980, the Japanese tanker "Shinaitoku Maru" is fitted with two sets of rectangular sails constructed of canvas stretched across steel frames. In the first year of service, the sail rig was used in conjunction with the main diesel engine about 60 percent of the time, reducing fuel consumption by around 10 percent. A computer, monitoring wind speed and direction, automatically controls the sails, by opening, closing and rotating them in the manner shown in the diagram. The sails can thus derive maximum benefit from a favorable wind or, conversely, offer the least amount of resistance to a headwind. A microprocessor is linked to the power system, which adjusts the engine output according to whether wind assistance can be utilized by the sails.***

◄ ***This late 19th-century "windjammer" belongs to the last generation of sail trading ships. Spacious rather than swift, they were built to run routes around Cape Horn.***

1845 "Great Britain" launched, built by Brunel: first iron passenger liner with screw propulsion
1897 "Turbinia", first turbine-driven steamship, built by C.A. Parsons
1902 First marine diesel engine installed, on a French canal boat
1906 First Dreadnought battleship built
1937 Commercial hydrofoils introduced, on the Rhine
1953 First hovercraft, SRN1, built in Britain
1958 "Savannah", an American merchant ship, launched, the first nuclear-powered ship

Submarines
1620 First known submersible craft invented by Cornelius van Drebbel
1776 Bushnell's "Turtle" built: first submarine to use bouyancy tanks
1800 Fulton's "Nautilus" built, with hydroplanes to direct craft up or down when submerged, and carrying compressed air supply
1863 Hand-operated propeller-driven craft used in American Civil War to sink warships
1863 First mechanically-driven submarine built in France
1900 J.P. Holland builds first modern submarine, with internal combustion engine for surface travel, and electric motor for use when submerged
1944 Air-intake tube raised above water introduced on German U-boats, permitting use of diesel engine when submerged to periscope depth
1954 USS "Nautilus", first nuclear-powered submarine, launched

Australia II keel

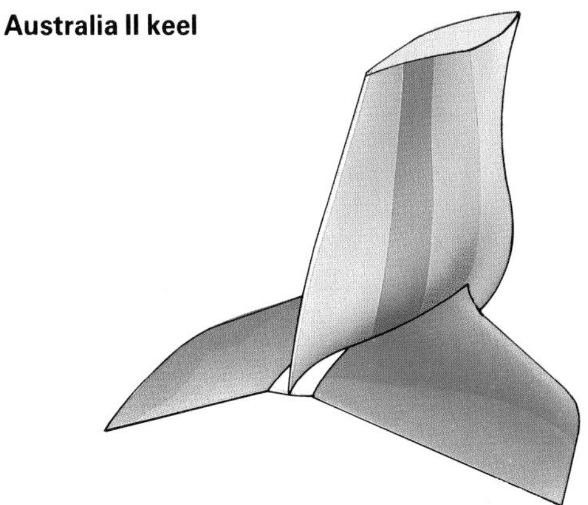

▲► Secrecy surrounding the design of modern racing yachts is a testimony to the contribution of technology. The success of "Australia II" (right) in the 1983 America's Cup race was partly attributable to the configuration of its keel, which was fitted with an underwater foil for improved stability. Design innovations for "Australia III" (left) for the race in 1987 were literally kept under wraps.

Engine-powered ships

Driving a boat by steam was first proposed in the mid-17th century by the Frenchman Denis Papin (page 129), but he was unable to develop the scheme. It was not until the late 18th century that serious experiments in steam-powered boats were undertaken, with the most notable advances achieved by Marquis Jouffroy d'Abbans in France in 1783 and John Fitch in the United States in 1787. They experimented with using a chain to transmit drive from engine to paddles, but it was not until 1803, when the Scotsman William Dundas incorporated a crankshaft in the *Charlotte Dundas*, thus producing a more efficient rotary motion, that the first successful steam-driven vessel was seen.

Important developments followed in the 1830s. In 1834, the English inventor Samuel Hall patented the surface condenser, a method by which fresh water could be reconstituted from steam and reused. Four years later the survival in a storm of the grounded iron ship *Garry Owen* ended fears that such ships would sink if their hulls were damaged. The screw propeller, soon to outclass the paddlewheel in propulsive efficiency over high seas, was invented almost simultaneously by Englishman Francis Pettit Smith and Swedish-American John Ericcson in 1836. With the voyage of the *Great Western* in 1838 the British engineer I.K. Brunel (1806-59) demonstrated that the addition of a huge fuel capacity could increase a ship's range without affecting performance.

Despite this, lack of fuel over the longest voyages still gave sail the edge over steam. Brunel attempted to overcome this problem with the *Great Eastern*, at its launch in 1858 by far the largest ship yet built and a technical marvel well ahead of its time. It was equipped with sail as well as paddles and screw propeller, but remained plagued with problems and never sailed the route from Britain to the East Indies, for which it was intended.

Rivalry for the fastest Atlantic crossing spurred technological development. Hulls were redesigned to minimize resistance and maximize buoyancy, and iron plates held together by rivets gave way to welded steel. The high-pressure triple-expansion engines (page 129) provided a light but efficient power source, though in 1897 the British engineer C.A. Parsons (1854-1931) demonstrated the first turbine-driven boat (page 131). Later developments in gearing technology enabled the turbine to run at high speeds while the propeller turned at its optimal low speed.

In the 20th century, military needs generated unprecedented building programs and technological innovation, while the diesel engine took over from steam as the universal power source. Commercial interest in efficiency and economy produced still larger ships, both in the luxury liner market – particularly between the two world wars – and in the industrial sector, including the oil tanker and the freight container ship.

In the 1950s there were hopes that nuclear energy (page 141) would prove an important new power source for long-distance ships of all types. However, in all types other than submarines and aircraft carriers, the advantages of the vastly increased range offered by nuclear power have been outweighed by high costs and problems of safety. In the civilian sector, only Soviet icebreakers still find advantage in this source of power generation, while the United States, West Germany and Japan have all converted their nuclear-powered vessels to conventional fuels in recent years.

The screw propeller

First perfected in its bladed form by the Swedish-American engineer John Ericcson (1803-1889) in 1836, the screw propeller remains the most efficient instrument for converting a ship's power to motive thrust under water.

The blades are mounted radially on a central hub in a spiral configuration. When rotated, water is accelerated through the spiral, thus increasing its momentum. The thrust achieved is proportional to the mass of the water it is acting upon and its rate of acceleration. For the most efficient propulsion the mass of water should be large and acceleration small; this optimum is obtained by a large slow-running propeller. Wide blades are designed with the intention of restricting the phenomenon known as cavitation, whereby pressure reduction forward of the blades can result in cavities of vapor being created in the water, impairing the efficiency of the propeller.

Enclosing the propeller in an open-ended nozzle helps to accelerate water flow into it and increase thrust by up to 25 percent. Variable-pitch propellers allow thrust to be controlled as required, simply by altering the pitch of the blades in the water.

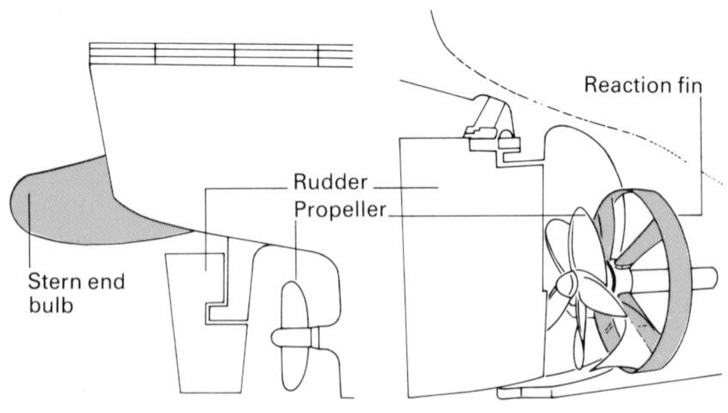

▲ ***The stern end bulb is designed to smooth out waves at the stern of the ship. This reduces eddy-making resistance and thus the power required for the design speed.***

▲ ***A reactor fin installed ahead of the propeller creates a prewhirl of water flowing in the opposite direction to the rotation of the propeller. Efficiency is increased by 4-8 percent.***

▲ ***The largest oil tankers, such as this Iranian vessel, are steam-driven. Displacing up to a quarter of a million tonnes, they rely on efficiently-designed turbines and screw propellers to drive them through the water.***

◄ ***In 1845 the "Great Britain" became the first steamship driven by a screw propeller to cross the Atlantic. In her iron hull, Brunel introduced watertight bulkheads and girders to give longitudinal strength.***

► ***The Finnish icebreaker "Otso" was launched in 1986. An air bubbling system creates a current of water and air between the hull and the ice, minimizing friction and resistance to the forward motion of the vessel.***

Nuclear power permits the submarine to travel more than half a million kilometers without refueling

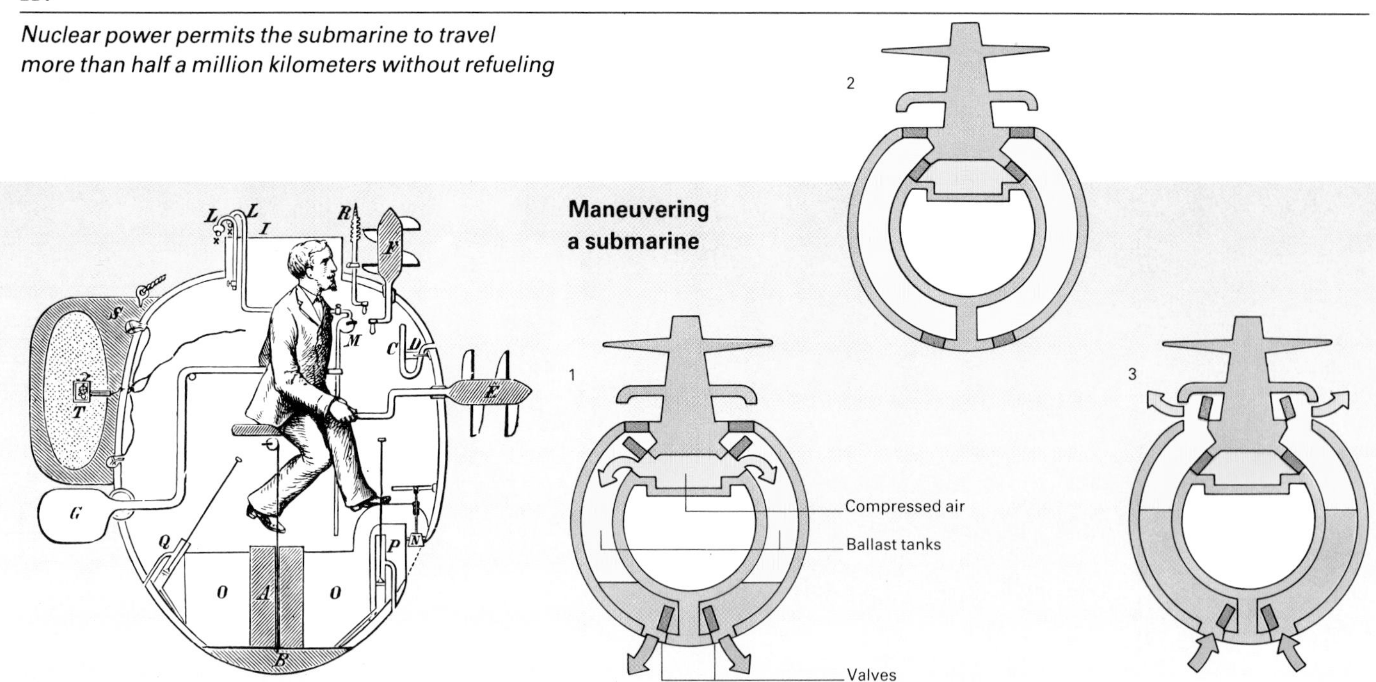

▲ *An 1885 drawing of Bushnell's "Turtle".*

Undersea craft

Transport below the surface has been dictated primarily by military requirements since its earliest developments. The first practical submarine, the *Turtle*, was designed by the American David Bushnell in 1776 as a means of screwing explosive charges to the hulls of British ships during the War of Independence. Although it proved ineffective, its mechanism of propulsion and the use of buoyancy tanks for submerging and surfacing set the standard that others followed. Robert Fulton's three-man *Nautilus* of 1800 introduced a metal construction; the French *Plongeur* of 1863 was powered by a piston fed by compressed air. Experiments with steam and electric power did not produce an effective long-distance submarine, but in 1900 the Irish-American John P. Holland devised a vessel with an internal combustion engine for use on the surface, and an electric motor while submerged. The application of the diesel engine, with a snorkel tube providing fresh air to the engine while submerged to periscope depth, led to the very successful development of submarines during the Second World War, culminating in 1954 with the world's first nuclear-powered submarine, the USS *Nautilus*.

The nuclear submarine

Before the advent of nuclear propulsion, the submarine's potential was limited by the need for oxygen both for the crew and for the engine. A nuclear reactor, which requires no oxygen and produces no fumes, offers a range of over 600,000 kilometers on a single fueling. New equipment for extracting oxygen and fresh water from the sea allows the submarine to remain submerged for long periods.

A pressurized water reactor (♦ page 144) delivers steam to drive the vessel's main propulsion turbine. This is coupled to the screw propeller either through reduction gearing or by turbo-electric drive. Ballast tanks control the submarine's buoyancy. When submerged, a state of neutral buoyancy is maintained by transferring water to and from the vessel's trim tanks. A submarine's course through the water is controlled on its yaw axis by a rudder, and on its roll axis by small adjustable fins or hydroplanes. Navigation is determined by an inertial guidance system of gyroscopes and accelerometers, combined with sophisticated sonar and radar tracking devices that are also essential to the submarine's role in detection and attack.

▲► *To surface (1), compressed air forces water from ballast tanks through valves beneath the vessel. At the surface (2), the ballast tanks are completely evacuated. To dive (3), water is admitted through the lower valves while air escapes through the upper set. The course is set by the movement of hydroplanes mounted fore and aft (4). Rudder control changes the submarine's course (5).*

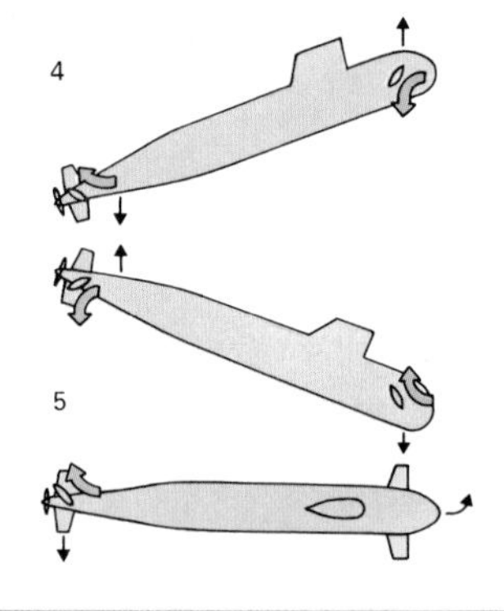

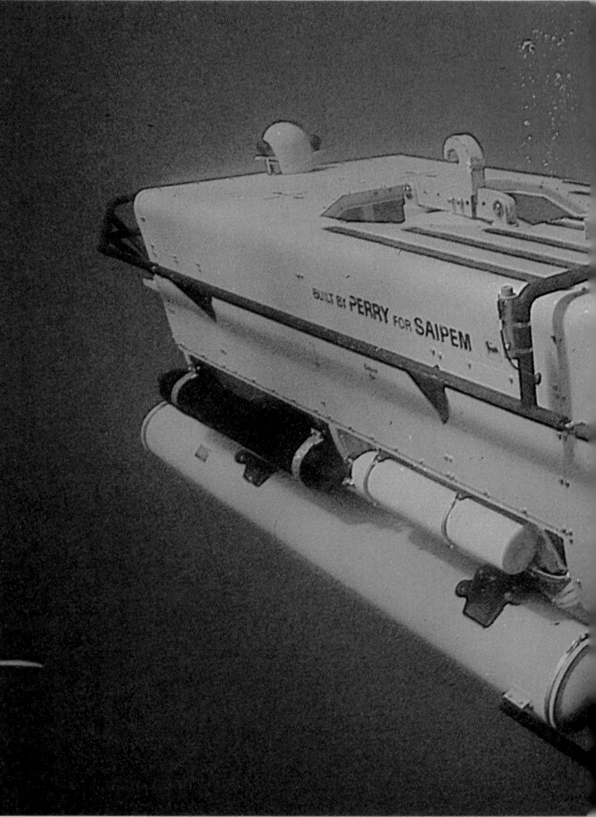

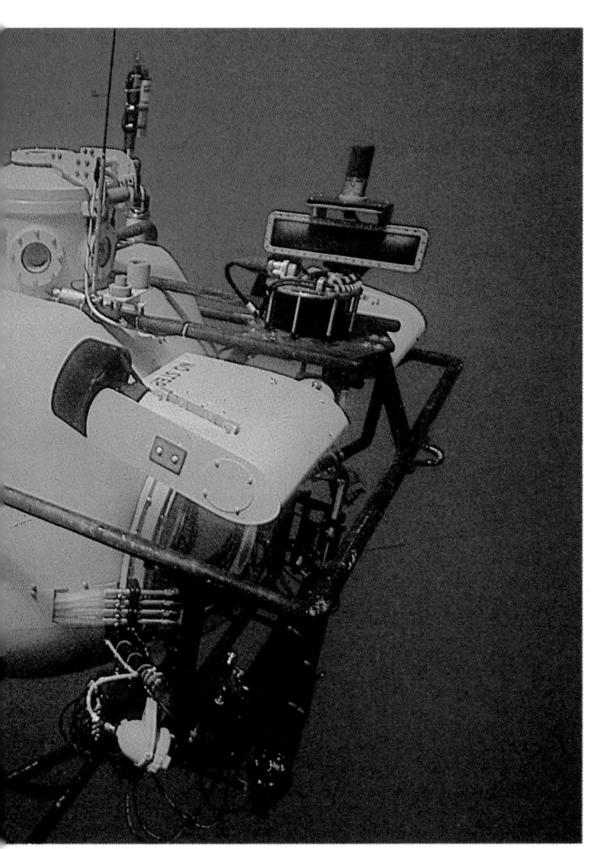

▲ The USS "Will Rogers" is a nuclear-propelled submarine carrying nuclear missiles able to strike targets within a range of 4,000km. It is linked to a satellite communication system and can launch the missiles while submerged.

◄► Not all submersible craft are built for military purposes. Oceanographic research, the search for minerals and oilrig maintenance all make use of bathyscaphes, small free-moving submarines designed to withstand high pressures at extreme depth. The Perry submersible (left) has a hull made of plastic, and is propelled by a small electric motor. Its cylinders, mounted outside the vessel, contain enough air for its occupants to survive one week. In a "lockout" chamber, divers are conditioned for the pressure before and after work outside. The Spider one-man submersible (right) enables the diver to work entirely contained within his "vehicle" – a similar arrangement to Bushnell's "Turtle".

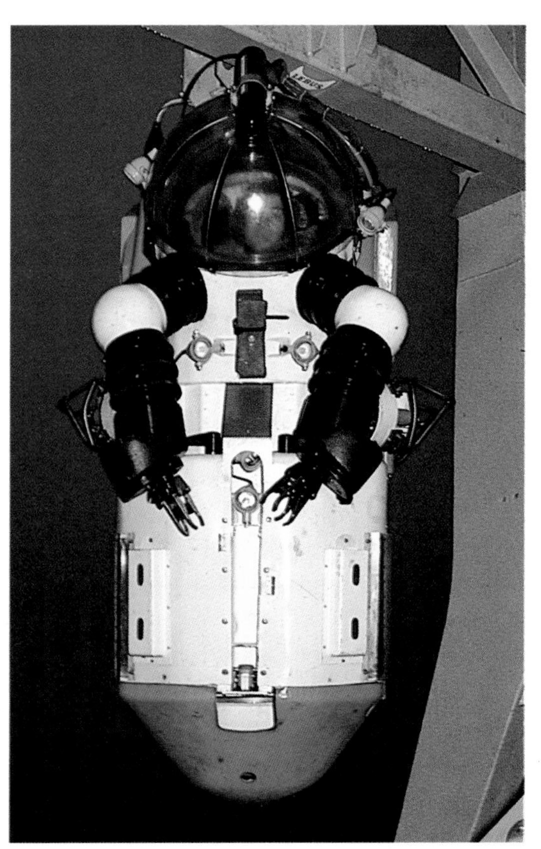

See also
Computer Hardware 115–122
Engines 129–148
Nuclear Power 141–148
Defense and Attack 165–170
Air Transport 217–224

▲◀ The hovercraft is powered by gas turbine engines that drive variable-pitch aircraft propellers to provide forward movement. Geared to these are lift fans which suck in the air, and force it into the plenum chamber within the flexible skirt. From there, the air is directed inwards and downwards by "fingers" to form a buoyant air cushion beneath the craft.

▼ Stilts supported by underwater wings lift the hydrofoil clear of the water's surface and thus minimize the drag that hinders speed. As with the hovercraft, high fuel consumption restricts long-distance operation, but various types of power unit and transmission designs have been tried, such as this modern jet-propelled hydrofoil built by Boeing.

A powerful, aerodynamically-designed speedboat is able to achieve speeds of up to 500 kilometers per hour. To do so the hull must rise out of the water to minimize contact with the surface and reduce drag. For passenger-carrying transport operating on routes where speed and convenience are essential, such as short sea crossings or island hops, exploitation of this principle is clearly an advantage.

A practical air-cushion vehicle (ACV, or hovercraft) was invented in 1953 by the British boat designer Christopher Cockerell (b. 1910). The simple notion that a buoyant air cushion produced by compressed air could support a substantial vessel has been applied to hovercraft making scheduled commercial crossings of the English Channel, but the principle is also ideal for amphibious craft operating in swamps or icebound waters throughout the world.

First demonstrated by the Italian airship designer Enrico Forlanini (1848-1930) between 1905 and 1911, the hydrofoil uses the same aerodynamic principle that produces lift in aircraft (➧ page 218) to raise the hull out of the water when traveling at speed. Water is more resistant than air, so a much smaller lifting surface is required on the "sea wing" than for an aeroplane. Hydrofoils are used on ferry services throughout the world but have proved most popular in the USSR.

Air Transport

The first powered flight...How an aircraft flies...Military and civil aviation...The jet airliner...New dimensions in flight...PERSPECTIVE...Chronology...Aerodynamics... Helicopters...Boeing 707...The survivable crash...The A320 and "fly-by-wire"

On 17th December 1903 the American brothers Wilbur and Orville Wright achieved the first controlled, powered flight in their "Flyer I" at the Kill Devil Hills, North Carolina, USA. The last of the four short flights covered 260 meters against a stiff breeze. This triumph followed three years of experiments with gliders during which they had learned to obtain three-axis control in the air (with pitch, roll and yaw). By 1905, the "Flyer III", a biplane with a span of over 12 meters and powered by a 16 horse-power engine driving two pusher propellers, could bank, turn and make figures-of-eight, and fly for 30 minutes at between 50 and 60 kilometers per hour. Ninety years on, the same configuration of unstable, "tail-first" flight and pusher propeller power is making a reappearance in advanced light aircraft design, and may be seen on the airliners of the 1990s.

Since 1903, the boundaries of wing-borne flight have been pushed back. The top speed has risen from around 50 to 3,520 kilometers per hour, height from a few meters to 35,000 meters altitude, and duration from 35 meters to 20,000 kilometers nonstop. The time it takes to cross the Atlantic has been cut from days to hours, and in the process a worldwide, multibillion-dollar, multimillion-passenger air transport industry has developed. The Wright brothers carried their first passenger in 1908. Within 11 years, the first fare-paying passenger flights were inaugurated between London and Paris. That first scheduled daily international airline flight across the English Channel took 2½ hours.

▲ Powered by two sophisticated turboprop engines and constructed using the latest composite materials, the 1986 Beech "Starship 1" is nevertheless laid out not unlike the "Flyer I". Designed by computer for optimal stability and performance, the propellers are rear-mounted while "tailplanes" are fitted to the nose.

▼ The "Flyer I", seen here piloted by Orville Wright on its maiden flight in 1903, was a descendant of earlier gliders tested by the Wright brothers. Successful methods of control were obtained by elevator, rudder and wing-warping, combined with a home-made internal combustion engine of a suitable power-to-weight ratio.

Chronology

1783 First manned balloon flight, by Montgolfier brothers in hot-air balloon in Paris
1783 First hydrogen-balloon flight, by J.A.C. Charles in Paris
1852 First flight of mechanically-propelled airship by Henri Giffard of France
1891 Controlled glider flight experiments begun by Otto Lilienthal in Germany
1900 Prototype Zeppelin large-volume airship with internal-combustion engine and aluminum frame
1903 First sustained, controlled, powered flight in heavier-than-air machine, by Orville and Wilbur Wright in USA
1907 First free vertical flight by Paul Cornu's twin rotor helicopter in France
1908 First cross-country flight, from Mourmelons to Rheims in France, by Henry Farman, in machine using ailerons
1909 First oversea flight, from France to England, by Louis Blériot
1910 Airship entered commercial service with Zeppelin in Germany
1912 Monocoque construction introduced, for French Deperdussin racer
1915 German Junker J1 built, first all-metal cantilever-wing aircraft
1919 First direct non-stop transatlantic crossing, by Britons John Alcock and Arthur Brown
1927 Charles Lindbergh made first solo transatlantic flight, from New York to Paris
1930 Jet engine patented by Frank Whittle in Britain, with gas turbine for jet propulsion in aircraft
1935 Douglas DC-3 "Dakota" made its maiden flight

How an aircraft flies

An aircraft has three control axes: pitch, roll, and yaw. Pitch control is provided by elevators on the tailplane which cause the nose to rise or fall, so increasing or decreasing wing angle of attack and, with it, lift. Pitch control therefore enables the aircraft to climb or descend.

Roll control is provided by ailerons on the outermost part of the wing. These are deflected asymmetrically – one up, one down – to cause one wing to rise while the other drops. At high speed, devices called spoilers are sometimes used to "dump" lift on one wing, so achieving the same effect with less wing bending.

To turn, the pilot first rolls, or banks, the aircraft then pitches the nose up into the turn. Yaw control, provided by a rudder on the fin, is used sparingly to keep the plane pointing in the right direction. The fin itself provides "weathercock" directional stability.

The aircraft moves through the air under the power of its engines either converted to drive one or more propellers or supplying jet propulsion (◀ page 134). Propellers give forward thrust by accelerating air around rotating blades, aerodynamically shaped and twisted to maintain a favorable angle of attack. Variable-pitch propellers can vary their angle to maintain efficiency at different speeds.

The pilot does not fly the aircraft unaided. Autostabilization keeps it steady by sensing and cancelling out disturbances automatically. The autopilot flies a course programmed into the navigation computer, while an autothrottle controls engine thrust. Autoland enables the aircraft to land automatically in zero visibility by homing in on intersecting radio beams (◀ page 13).

Aerodynamics

The principle of flight relies on the fact that as air speeds up its pressure drops. A wing is shaped so that air flowing over it is forced to accelerate, creating a low-pressure region on top of the wing. The difference in pressure between the upper and lower surfaces of the wing generates an upward force – lift – which supports the aircraft in flight.

To generate more lift, the angle at which the wing attacks the air is increased, forcing the air flowing over the wing to accelerate even more. A point is reached, however, when the air can no longer negotiate the leading edge of the wing. It separates and the wing begins to lose lift. This is the stall.

A wing works hardest during take-off and landing, when airspeed is slow but maximum lift is required. There are several ways to increase wing lift at low speeds. Lowering trailing-edge flaps increases the curvature of the wing, generating additional lift. Flaps which extend as well as lower increase both wing curvature and area. Leading-edge slats can be extended to delay the stall.

With lift comes drag – drag due to frontal area, skin friction, and to lift itself. Drag is energy wasted on the air. In order to minimize the amount of fuel burned to propel an aircraft through the air, therefore, designers strive to maximize lift while minimizing drag. One way is to sweep the wing but, to generate the same amount of lift as a straight wing, a swept wing has to be bigger or use lift-improving devices such as flaps and slats.

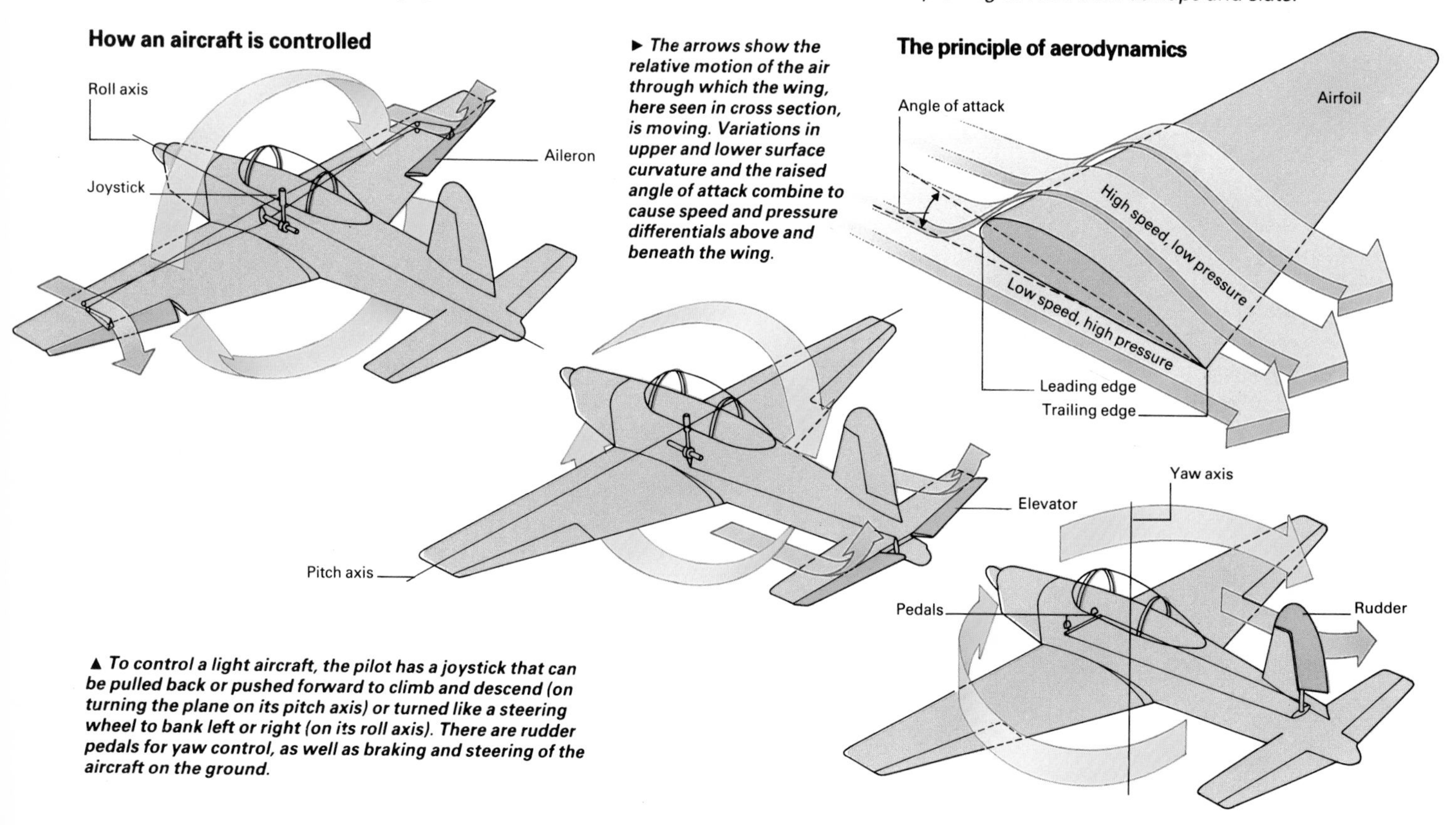

▶ The arrows show the relative motion of the air through which the wing, here seen in cross section, is moving. Variations in upper and lower surface curvature and the raised angle of attack combine to cause speed and pressure differentials above and beneath the wing.

▲ To control a light aircraft, the pilot has a joystick that can be pulled back or pushed forward to climb and descend (on turning the plane on its pitch axis) or turned like a steering wheel to bank left or right (on its roll axis). There are rudder pedals for yaw control, as well as braking and steering of the aircraft on the ground.

1936 Heinrich Focke's Fa-61 helicopter airborne
1937 US Lockheed XC-35 built, the first fully pressurized aircraft
1937 World's largest and newest airship, "Hindenburg", burst into flames in New Jersey, ending age of airship travel
1939 German Heinkel He-178 built, first turbojet aircraft
1939 Prototype modern helicopter VS-300 designed by Igor Sikorsky, with single main rotor and small tail rotor
1941 First flight the Gloster Meteor, powered by Whittle's jet engine,
1947 US Bell X-1 exceeded speed of sound in level flight
1949 De Havilland Comet, first jet airliner, introduced into service in Britain
1954 First experimental flight of vertical take-off aircraft, in Britain
1958 Boeing 707, first United States jet airliner, entered commercial service
1967 Computer guidance systems installed in aircraft for first time. Autoland blind landing system introduced in British airliner Trident 2
1969 Anglo-French supersonic airliner Concorde made maiden flight
1970 Boeing 747, first wide-bodied "jumbo" jet entered service
1977 Successful flight by heavier-than-air human-powered aircraft, Gossamer Condor, in USA
1979 First man-powered flight across the English Channel, by Gossamer Albatross
1981 Solar Challenger crossed English Channel, powered only by solar cells
1981 Superplastic metal alloys introduced on Airbus A310

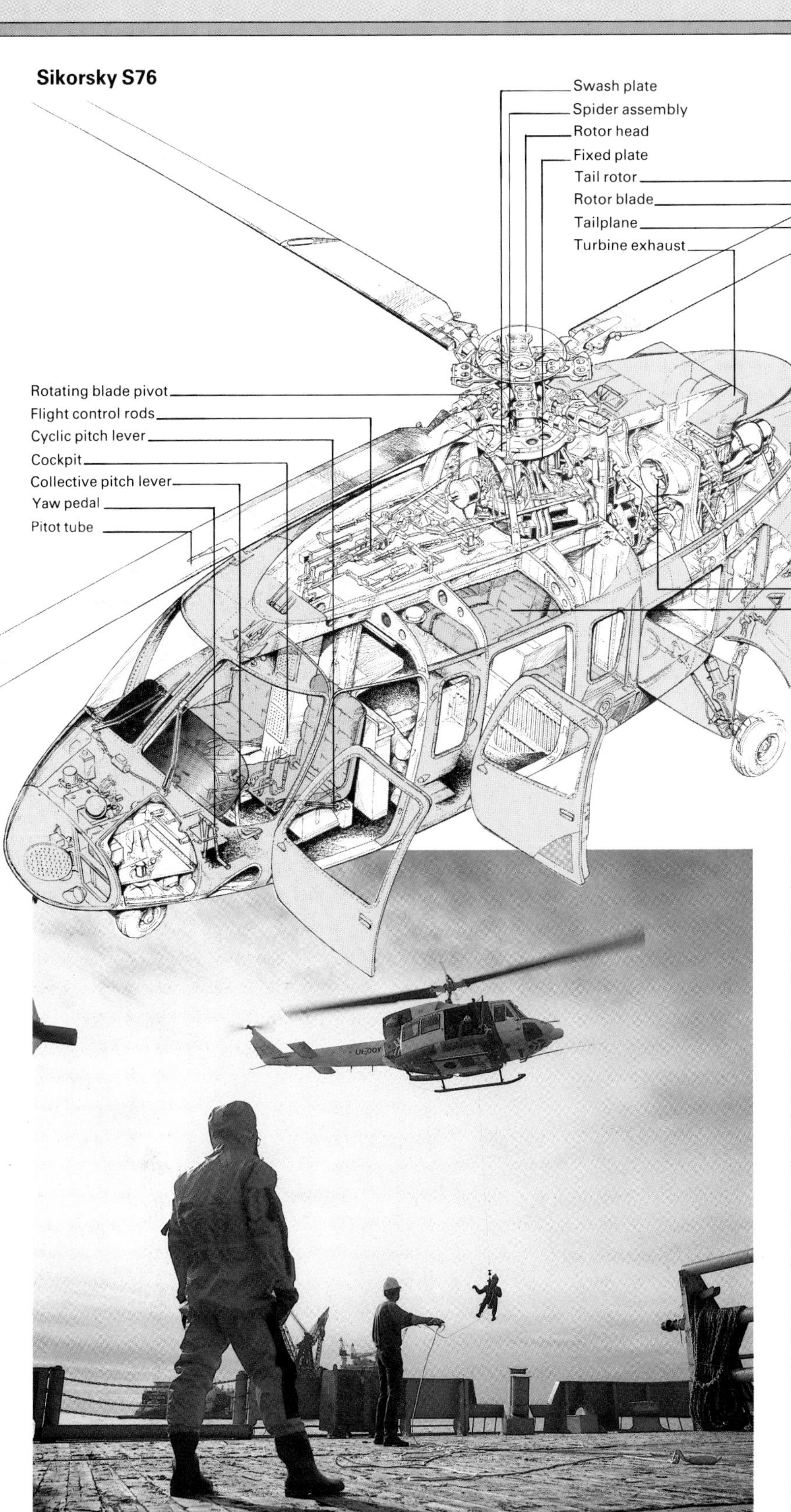

◀ The internal mechanism of a modern helicopter is revealed in this drawing of the Sikorsky S76. Twin turboshaft engines transmit drive to main and tail rotors, while a system of rods and linkages, a movable swash plate and "spider assembly" beneath the main rotor head link the pilot's control to the pitch of the rotor blades and thus determines the helicopter's lift and flight pattern.

Helicopters

A helicopter rotor blade operates on the same principle as an aircraft wing except that, because it rotates, it can generate lift when the aircraft is stationary. This enables the helicopter to take off and land vertically and to hover.

To increase the lift, or thrust, generated by a blade, the blade's angle of attack, or pitch, is increased. To climb or descend in the hover the pitch of all the blades is varied collectively. To make the helicopter fly forwards, backwards or sideways, the rotor disk, and with it the thrust, has to be tilted in the direction required. This is done by varying the pitch of each blade cyclically as it rotates.

In a helicopter with only one main rotor, the tendency of the fuselage to rotate in the opposite direction to it is counteracted by a tail-mounted anti-torque rotor.

To control the helicopter the pilot is provided with a handbrake-like collective pitch control lever to climb or descend, a cyclic pitch control joystick for pitch and roll control, and rudder pedals controlling tail-rotor thrust for yaw control.

◀ The helicopter is an invaluable craft where tight maneuverability is a priority. Able to land or take off from the deck of a ship, the helicopter is used by civil airlines, the military and oil companies.

Influences on aircraft development

The civil and military spheres of aerospace are inextricably linked. Developments in one inevitably lead to advances in the other. Usually it is the military sphere that leads, but this is not always so.

Some of the most important airliners of the age have owed their prosperity, if not their existence, to military contracts. The Douglas DC-3 was conceived as an airliner, yet 90 percent of the 13,000 built were ordered by the military. In return, surplus DC-3s released for civil use after the Second World War played a key role in the post-war rebirth of air transport.

Airliner design was profoundly changed by two inventions given impetus by the Second World War and the military desire for faster fighters – the jet engine and the swept wing. In the course of developing the new jet fighters the speed, smoothness, and quietness advantages of the jet for airline passengers first became apparent.

The Second World War had another profound effect on air travel. The need to ferry bombers to Europe across the inclement Atlantic led to advances in communications, navigation, and weather-forecasting that after the war enabled landplanes to operate routinely over routes previously thought suitable only for flying boats. This, coupled with the wartime development of long-range landplane transports, contributed to the rapid post-war expansion of civil air travel.

Examples of military programs spinning off into the civil arena abound. The Boeing 707, perhaps the most influential of post-war airliners, owes its existence to a US Air Force requirement for an air-refueling tanker able to match the performance of its new B-52 bombers. Boeing's 747, the first "widebody" airliner and still the world's largest, resulted from a USAF competition to design a massive cargo aircraft. Boeing lost to Lockheed's C-5, but the turbofans developed for their prototype were successfully applied to the new 747 (◀ page 134). In the fuel crisis of the 1970s, the "big fan" was the savior of the airline industry. In the continuing war against noise and fuel consumption, the big fan remains the airline's primary weapon.

Examples of the civil world leading the military are less common, but do occur. Development of the Anglo-French Concorde supersonic airliner came at a crucial time when all of Britain's supersonic combat aircraft programs had been cancelled. The contribution Concorde made to keeping technology moving, and to development of the European Tornado fighter-bomber, is only now being appreciated.

The 707 story

Nowhere is the symbiosis between civil and military aerospace more evident than in the story of the Boeing 707, the United States airliner that contributed more than any other to the expansion of air passenger transport.

The now-familiar configuration with swept wing and podded jet engines had evolved in the early 1950s. The 707 was conceived as the Model 367, an improved, jet-powered, version of Boeing's C/KC-97 air-refueling tanker/transport for the US Air Force. The US Air Force, however, did not have funds to spare, so in 1952 Boeing took the bold step of spending $16 million of its own money on building and flying a prototype airline version, known as "Dash 80".

This aircraft flew in July 1954, five years after the world's first jet airliner, the Comet. Later that year the USAF relented and bought the first handful of an eventual 800-plus C/KC-135 tanker/transports based on the Dash 80.

Then in 1955 Boeing received USAF permission to sell an airliner version, the Model 707. This, the first US jet airliner, entered service in 1958, by crossing the Atlantic nonstop.

◀ Following independent research by Britain and Germany in the years leading up to the Second World War, it was a German aircraft, the Heinkel He-178, that made the world's first jet-powered flight, on 27th August 1939. The He S-3b turbojet engine, developed from the model pictured here, was designed by Dr Hans von Ohain. The rival British engine of Frank Whittle took to the air later, but became the prototype power unit for the world's first jet airliners, thanks to the postwar technology exchange between the victorious Allies.

▶ Modern turbofan engine in production.

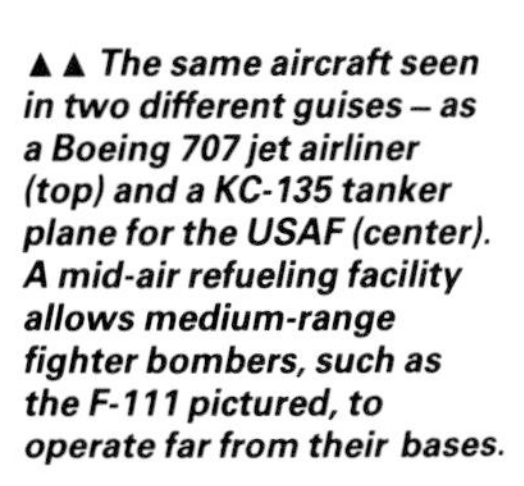

▲▲ *The same aircraft seen in two different guises – as a Boeing 707 jet airliner (top) and a KC-135 tanker plane for the USAF (center). A mid-air refueling facility allows medium-range fighter bombers, such as the F-111 pictured, to operate far from their bases.*

► *The Panavia Tornado is one of the latest generation of multi-role fighter-bombers. Its "variable geometry" wings are swept back to restrict drag at supersonic speeds in imitation of Concorde's "delta" configuration, but returned to an extended or conventional position for maximum lift at take-off and landing.*

▲ A Douglas DC-3 "Dakota". Many are still flying after 50 years' service.

The age of the airliner

The forerunner of the modern airliner is widely regarded to be Boeing's Model 247, which entered airline service in 1933. Carrying 10 passengers, the 240-kilometer-per-hour aircraft pioneered the familiar low-wing, all-metal layout. This design was most successfully adapted by Douglas for its 21-seater DC-3, which first flew in 1935, on the 32nd anniversary of the Wright brothers' first flight. When production ended in 1946 some 13,000 had been built in the United States, Soviet Union and Japan. The DC-3 is still in service 50 years after its introduction.

Although aircraft like the Model 247 and DC-3 helped to promote air travel within the United States, they were not intended to cross the Atlantic. In the 1920s, it was thought that airships would perform that role, but the Hindenburg inferno in 1937 prematurely ended their popularity. Their place was taken by flying boats, but their reign ended almost as soon as it began, cut short by the outbreak of war in 1939. By 1945, surplus long-range landplane transports were being pressed into airline service, beginning routine nonstop flights across the Atlantic. The age of passenger air transport had begun.

Before the war, only civil servants and other privileged individuals traveled by air, but now the existence of large numbers of servicemen and women, accustomed to wartime air travel, inspired an expansion of the air transport industry. This expansion coincided with the introduction of fast, comfortable jet-powered airliners. The first was the de Havilland Comet, but the most successful was the Boeing 707. Together they inaugurated scheduled nonstop transatlantic services within weeks of each other in 1958.

A decade later, 1969 saw the first flight of very different, but equally influential airliners – the supersonic Concorde and the wide-body Boeing 747. Never a commercial success, the Concorde remains today the only supersonic airliner in service, having carved out a profitable niche as a transatlantic executives' express. The 747, by contrast, has been an astounding commercial success – almost 700 aircraft have been sold. With a passenger capacity of nearly 400, its technology has contributed greatly to the new era of low-cost mass transportation between North America and Europe.

The survivable crash

In August 1985 a Japan Airlines Boeing 747 crashed into a mountainside after its rear pressure dome cracked in midflight, causing the tail to fall off. With a death toll of 520, it was the world's worst air disaster to date, yet, miraculously, four passengers survived. Advances in technology have improved aircraft efficiency and economies, but crashes, malfunctions and acts of sabotage still occur. Indeed, as air transport grows in popularity, more and more lives could be at risk. There are a number of measures known to improve safety, some already standard in airlines, that could lead to many more survivals from otherwise fatal accidents.

The body of the aircraft itself provides the first line of resistance to impact – and it is the newer, larger widebodies that prove the safest in this respect. Less flammable fuels – kerosine rather than gasoline – or chemical additives could minimize the risk of explosion that leads to so many deaths even if the impact of a crash is survived. Inside the aircraft, the spread of fire could be checked by supplying fire-blocker covers to the seats, but if a fire should break out, smoke detectors and cabin alarms could expedite action to extinguish it. The passengers could be protected by smoke hoods and guided to the exits by emergency lighting, especially if positioned near the floor and illuminating escape direction arrows. Easy-to-open exit handles and plentiful escape hatches are an obvious precautionary measure. Even the way you sit could save your life: head down between the knees with arms clasped over the head is safest. Seats facing backwards, as installed in many military transporters improve your chances too.

Where all these safety measures exist, information to passsengers on how to use them, instructions to staff in simulated training to put them into effect and regular operational checks all make vital contributions. Avoidable practises such as permitting luggage to block gangways and ignorance of the nearest fire exit claim lives unnecessarily.

▲ A helicopter flies over the debris which was all that remained of the Japanese Airlines Boeing 747 after it had crashed into Mt Sangoku, about 80km northwest of Tokyo on 12 August 1985. The bigger and more impact-resistant widebody airliners are more likely than conventional aircraft to protect their occupants in the event of a crash. Four passengers survived this one.

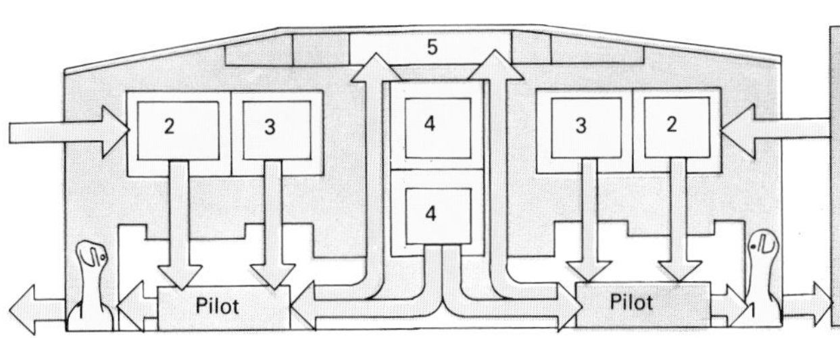

Electronic Flight Control System (EFCS)

SEC 3 computer systems operate the spoilers and elevators, controlling pitch and roll			**ELAC** 2 computer systems operate the ailerons and elevators, controlling pitch and yaw		**FAC** 2 computer systems aid mechanical rudder operation, controlling yaw (also tailplane trim)	
1	2	3	1	2	1	2

▲ ***The EFCS controls the A320's track and attitude. The flight pattern, as well as navigation and engine data are monitored by the pilot who uses the sidestick to input orders to the EFCS. The pilot retains thrust control throughout, and can take over from the autopilot via the flight control unit.***

1 Sidestick
2 Primary flight display
3 Navigation data
4 Engine warning and systems display
5 Flight control unit

The A320 and "fly-by-wire"

The A320 is the latest product from the successful European Airbus consortium, and represents the state-of-the-art in airliner design in the 1980s. Seating 150 passengers, it is conventional in layout, with swept wings and podded turbofan engines, but it incorporates many design innovations. Most significantly, it is the first "fly-by-wire" airliner. The pilots have miniature sidestick control columns and the traditional mechanical linkages to the control surfaces have been replaced by electrical signals traveling along wires. Effectively a flight control computer interprets and implements a pilot's commands based on an "intimate knowledge" of aircraft behavior.

A computer also navigates the A320 from climbout through to touchdown, following a fuel-saving profile more accurately and consistently than could a human pilot. Pilot and copilot both monitor color television screens which replace the traditional dials and gauges while still more computers monitor the engines and other systems, replacing the flight engineer.

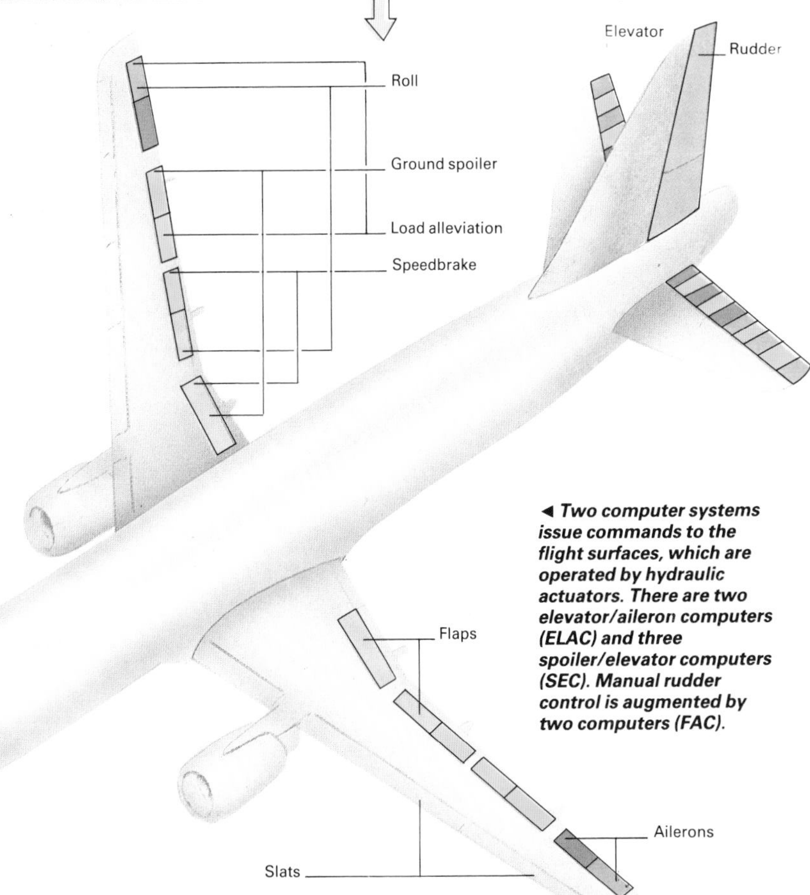

◀ ***Two computer systems issue commands to the flight surfaces, which are operated by hydraulic actuators. There are two elevator/aileron computers (ELAC) and three spoiler/elevator computers (SEC). Manual rudder control is augmented by two computers (FAC).***

See also

▼ The success of the Gossamer "Albatross" was due to its lightness and stability. Built of lightweight plastics and with a wingspan of 29m, the pilot powered the aircraft by pedaling but needed to apply only modest effort to keep aloft.

▲ The Rutan "Voyager" was designed to fly around the world without refueling. The aircraft carries two people, and the journey around the world in December 1986 took nine days. This craft is typical of the alternative designs of the mid-1980s.

Alternative flying machines

The Second World War saw the end of the era when light aircraft flying was the exclusive preserve of the rich. In recent years, however, the escalating cost of fuel and aircraft has plunged the industry into recession. Many have seen technology as the only way forward – not the high technology of the large airliners, but the "alternative" technology of lightweight, cheap-to-build and easy-to-fly aircraft.

Lightest and cheapest of them all is the ultralight, a hanglider with an engine attached. In pursuit of safety and performance, the ultralight is evolving into a true miniature airplane with a lightweight plastic airframe. Indeed, plastic construction holds the key to the resurgence of private flying in all its forms. In the United States, the recent innovative ideas of designer Burt Rutan for low-cost, high-performance homebuilt planes have had the most far-reaching influence. Rutan has championed the return to the "tail-first" configuration used by the Wright brothers. The result is an aircraft that is difficult to stall and is therefore safe for the amateur weekend pilot. The use of reinforced plastics in place of metal has cut the real cost of building private aircraft (➧ page 246).

What of human-powered flight? The English Channel has been conquered by the Gossamer Albatross, the ultralight design of Dr Paul McCready, in 1979. Another design from the same stable made the first sustained solar-powered flights. Again plastics held the key, but the widespread use of large, unwieldy human-powered or solar-powered craft at present seems unlikely.

Space Transport

The Shuttle program... PERSPECTIVE...History of space travel ...Chronology...Space Shuttle launch sequence ...Inside the Shuttle...Spacesuits...Life support systems

When, in the late 1960s, the United States National Aeronautics and Space Administration (NASA) sent giant Saturn rockets into space carrying the Apollo spacecraft to the Moon and back, the cost was enormous because each booster rocket was used only once. To make space travel routine, it seemed essential to develop some form of reusable system. While the Apollo program was in progress, engineers in the United States were working at a spacecraft with wings that could fly into space, return and land like an aircraft.

The Shuttle was first seen in 1977 when an unpowered prototype was launched from the top of a Boeing 747. In five flights it was made to glide down to the runway, proving that it could return safely through the atmosphere. By 1981, the Shuttle "Columbia" was ready for space, its inaugural flight marred only when a few non-essential heat-insulating tiles fell off in the vibration of the launch (➧ page 248). By 1985 it was flying nine flights a year, and had taken laboratories into space manned by crews from the United States and Europe and launched satellites for many countries and commercial organisations. However, on the Shuttle's 25th mission, in January 1986, a fault in the booster-rocket seals caused a leak of fuel. The Shuttle "Challenger" disintegrated into a fireball 73 seconds after launch, a disaster that grounded the Shuttle program while a full enquiry was undertaken and engineers sought to change the design of the booster rockets. Even so, the Shuttle will be the launch vehicle for lifting satellites and modules for a space station planned for the mid-1990s.

Early space travel

More than a thousand years ago the Chinese used powder rockets in battle. Yet only since 1926 has the dream of space travel seemed real. In that year, the American Robert Goddard (1882-1945) became the first person to fire a liquid-fueled rocket into the air. Little more than 10 years later the German Wernher von Braun (1912-1977) designed the first rocket missile. Called V-2, it was a flying bomb that was fired against London in 1944 and 1945, providing a very effective demonstration that rocket power had arrived. Following the work of Sergei P. Korolev (1906-66) in bringing rocket propulsion together with space flight, the Russians took the lead in 1957 when they launched the first artificial satellite into orbit. This was followed in April 1961 with the first manned space flight, by Yuri Gagarin (1934-68).

For 20 years "cosmonauts" flew dramatic missions around the Earth in cramped conditions aboard tiny spacecraft. Some spent nearly two weeks in space to study the body's responses to weightlessness, some practiced docking two spacecraft together, and some went on "spacewalks" lasting several hours, drifting at the end of a lifeline. A select and carefully chosen group walked on the Moon, blasted from Earth on giant rockets 100 times more powerful than the German V-2.

▲ On 28 January 1986, Shuttle "Challenger" was destroyed and its seven occupants killed when the right-hand solid-fuel rocket booster leaked flame through a joint, igniting fuel in the external tank. It happened 73 seconds into the flight when "Challenger" was 13km above Cape Canaveral.

◀ In the years following his first successful rocket launch in 1926, Robert Goddard (seen here in 1931, second from right) carried out numerous tests with the aim of developing a craft capable of sustained flight into space.

Chronology

1926 First liquid-propellant rocket, built by Robert Goddard, lifted off in USA
1942 First successful launch of German A-4 rocket by Wernher von Braun's unit at Peenemünde, Germany; reached altitude of 85km
1944-5 4,320 V-2 rocket-powered bombs fired against British targets by Germany
1947 Corporal E surface-to-surface missile became first rocket to carry a ground-controlled radar system
1957 Sputnik 1, world's first satellite, launched from a Soviet R.7 rocket
1959 Soviet space probe Lunik 2 became first artificial object to reach far side of Moon
1961 Soviet astronaut Yuri Gagarin became first human in space, completing one Earth orbit in Vostok 1
1961 Alan Shepherd in United States Mercury 3 spacecraft reached 185km and controlled craft manually
1962 First TV pictures from manned spaceflight transmitted by Soviet astronaut Andryan Nikolayev in Vostok 3
1962 John Glenn became first United States astronaut to orbit Earth in Mercury 6
1963 Soviet Valentina Tereshkova became first woman in space
1965 Soviet A. Leonov completed first spacewalk, lasting 10 minutes
1965 First United States two-man crew to be maneuvered from one orbit to another in Gemini 3
1966 First docking between a manned and unmanned spacecraft achieved by United States Gemini 8

The Shuttle's dimensions

The Shuttle is equipped with a combination of solid and liquid propellant motors. In the tail are the three main liquid-fuelled engines which generate a combined thrust of more than 450,000kg, fed with almost four million liters of liquid hydrogen and oxygen from the external tank to which the orbiter is attached. Two solid rocket boosters – essentially huge fireworks – each generate more than 1·3 million kg thrust. These are jettisoned two minutes after liftoff, with the three main liquid engines continuing to burn for another six minutes, by which time the Shuttle is in orbit. Additional thrust from small maneuvering motors put it into a path 240km high. The external tank is jettisoned and falls back through the atmosphere where it is burned up and destroyed, unlike the boosters which parachute down into the ocean for retrieval and reuse. The orbiter returns by firing small braking rockets, using the atmosphere to slow it through friction. More than 30,000 thermal tiles, all installed by hand, protect its aluminum structure from the heat of re-entry (about 1,500°C). The orbiter carries a maximum of eight people and can remain in space for about 10 days.

Space Shuttle

Space Shuttle flight profile

1 Liftoff
2 Ascent stage – booster rockets and main engines
3 Burn-out and separation of boosters
4 Main engines continue to power Shuttle
5 Main engines cut off
6 External tank separates
7 Maneuver motors steer Shuttle into orbit
8 Deorbit retroengines fire
9 Communications blackout as heat builds up
10 Attitude adjusted for reentry
11 Communications restored
12 Glide approach with S-turns
13 Automatic landing phase
14 Steered towards runway
15 Final approach with split rudder acting as airbrake
16 Touchdown

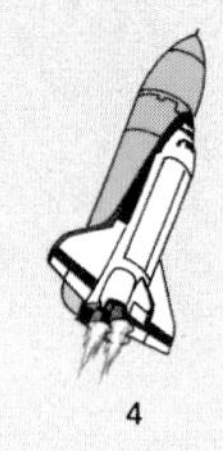

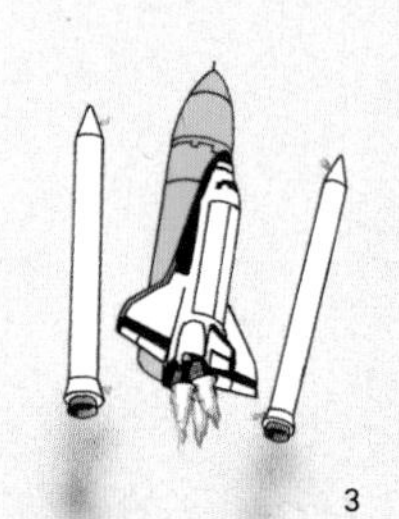

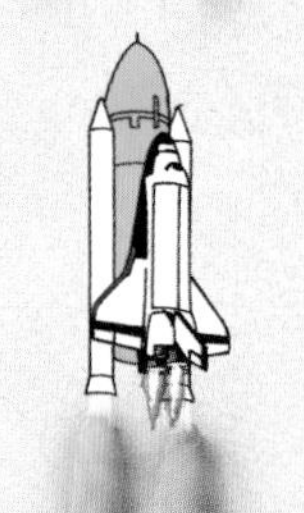

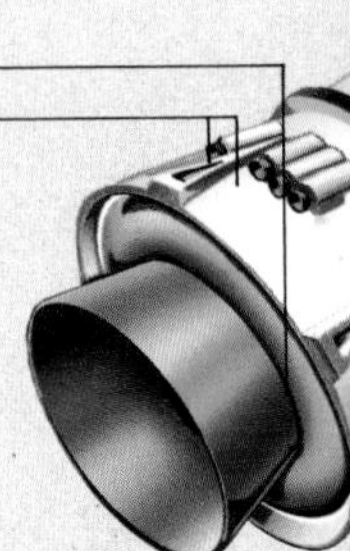

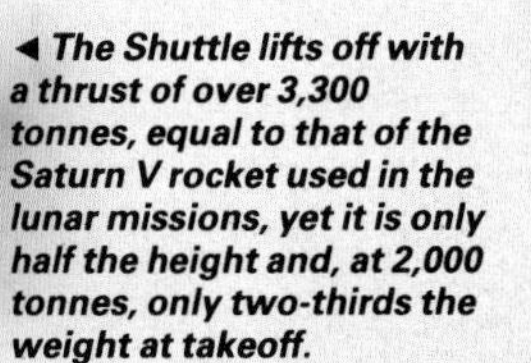

◂ The Shuttle lifts off with a thrust of over 3,300 tonnes, equal to that of the Saturn V rocket used in the lunar missions, yet it is only half the height and, at 2,000 tonnes, only two-thirds the weight at takeoff.

1966 Soviet Luna 9 relayed first TV pictures back from surface of Moon
1967 United States Saturn 5 rocket launched for first time
1967 Soviet spacecraft Venera 4 softlanded on Venus
1968 United States Apollo 8 completed first manned flight around Moon and back
1969 Soviet Soyuz 4 and 5 achieved first docking between two manned spacecraft
1969 United States astronauts Neil Armstrong and Edwin Aldrin became first humans on Moon, during Apollo 11 mission
1971 Soviet Salyut 1 became first manned orbiting scientific laboratory
1972 Launch of Pioneer 10, United States probe intended to fly-by Jupiter and become first artificial object to escape Solar System
1973 United States space station Skylab 1 launched
1975 United States Apollo spacecraft docked with Soviet Soyuz 19
1976 United States probe Voyager 1 soft-landed on Mars
1977 United States Space Shuttle Enterprise made first free flight from a Boeing 747
1977 Voyagers 1 and 2 launched on "grand tour" of giant planets
1979 European Space Agency three-stage satellite vehicle Ariane L01 launched
1981 (April) First ground launch of the United States Space Shuttle Columbia
1981 (November) Columbia made second flight, becoming first spacecraft to make more than one flight into space
1986 Space Shuttle Challenger exploded shortly after takeoff, killing all members of crew

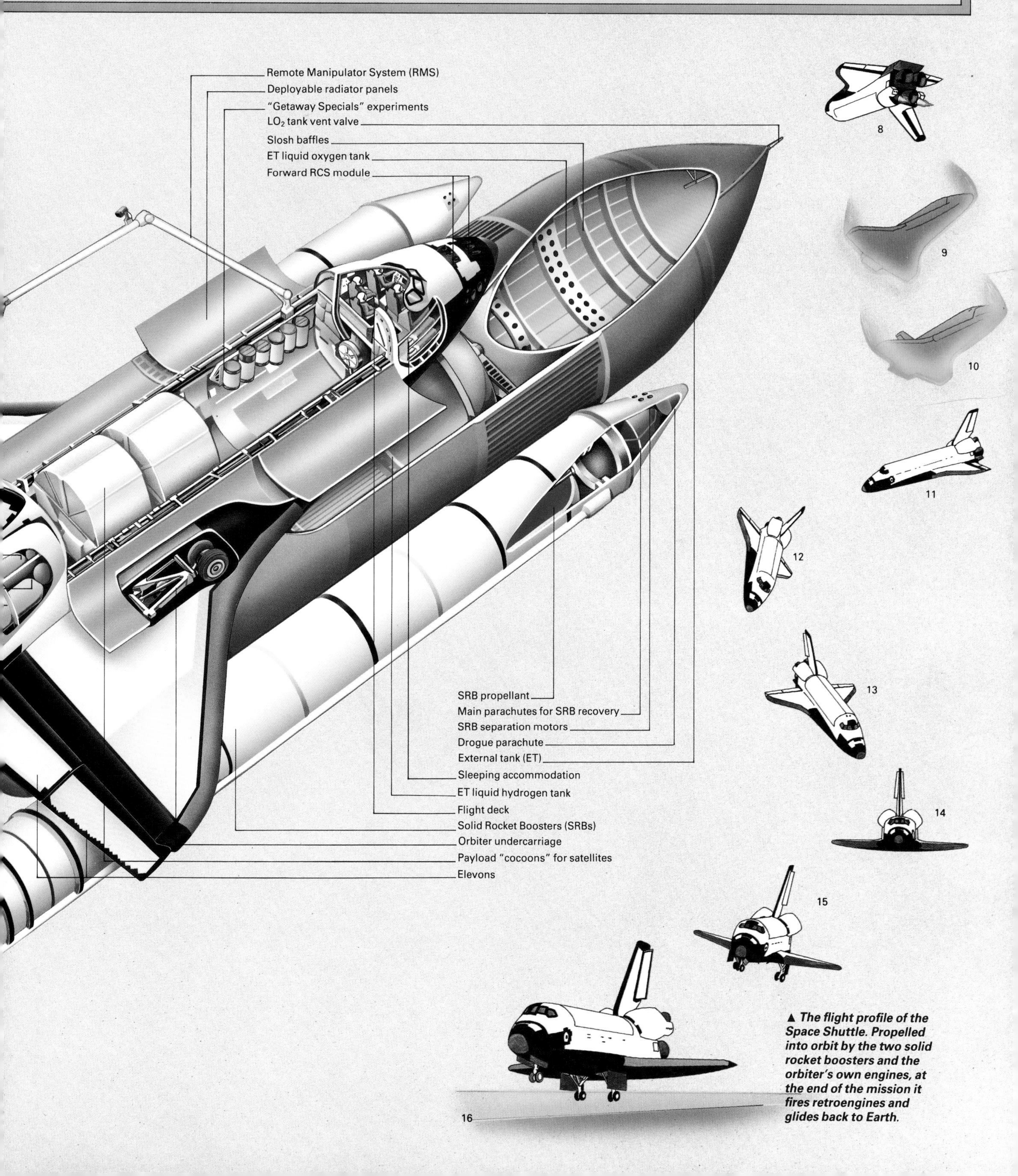

▲ ***The flight profile of the Space Shuttle. Propelled into orbit by the two solid rocket boosters and the orbiter's own engines, at the end of the mission it fires retroengines and glides back to Earth.***

See also
Optical Telescopes 57–64
Engines 129–134
Defense and Attack 165–170
Radio and Television 185–192
New Materials 245–248

Life support systems

Because space is a vacuum, space travellers must take with them an environment in which to live and breathe. This includes at least 0·24 atmospheres of pure oxygen (or, preferably, a nitrogen-oxygen atmosphere at 1 atmosphere); a temperature of 17–25°C, humidity of between 10 percent and 50 percent and a very low level of potentially toxic gases such as carbon dioxide and natural body excreta. The earliest United States manned capsules (Mercury and Gemini) carried replenishing oxygen, filters to remove exhaled carbon dioxide and, in the two-man Gemini, a recirculated supply of purified water from the cooling system. Water was produced as a byproduct of the fuel cells (♦ page 135) in which hydrogen and oxygen were brought together over a catalyst to produce electricity. This system was also used in the Apollo spaceships, and is found in the Shuttle, too, which is the first United States spacecraft to adopt a sea-level atmosphere of oxygen and nitrogen.

All these systems are "open-loop", using up limited stores of nitrogen and oxygen that are carried in tanks. Tests have been carried out on "closed-loop" systems in Soviet spacecraft, which show promise that atmospheric gases can be produced in quantities sufficient to support human life in a permanent system of oxygen and nitrogen with water as a byproduct. Only with such a system, providing the travelers with a miniature ecosphere, could really long space flights take place.

Travelers may be permitted to remain in weightless conditions for no more than a few months, since the human heart deteriorates unless it is loaded by gravity. Bones lose their mineral content and limbs weaken. United States and Soviet scientists are experimenting with modules of a space station rotating to provide artificial gravity. Again, this is an essential requirement for manned space journeys lasting several years.

Clothing for space

When outside the life-support system of a spacecraft, astronauts on spacewalk or lunar walk missions have to take their own environments with them. In designing an effective spacesuit a formidable range of criteria have to be met. The suit must protect the wearer from the unfiltered heat of the Sun by incorporating a water circulation-tube network. The suit must be completely airtight, tough enough to resist accidental damage and to protect against bombardment by micrometeoroids. Flexibility, however, is essential to allow freedom of movement outside the spacecraft.

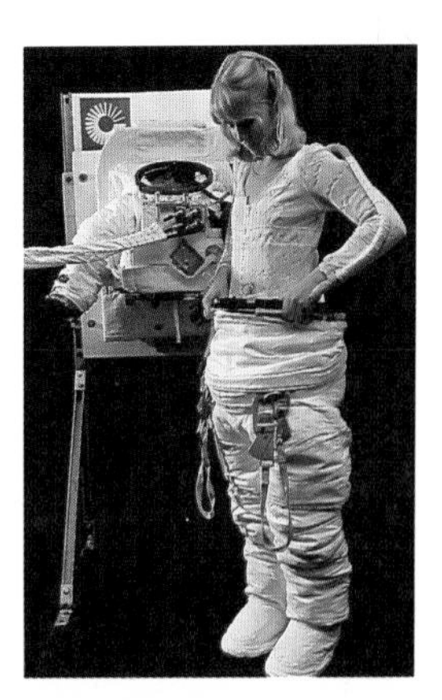

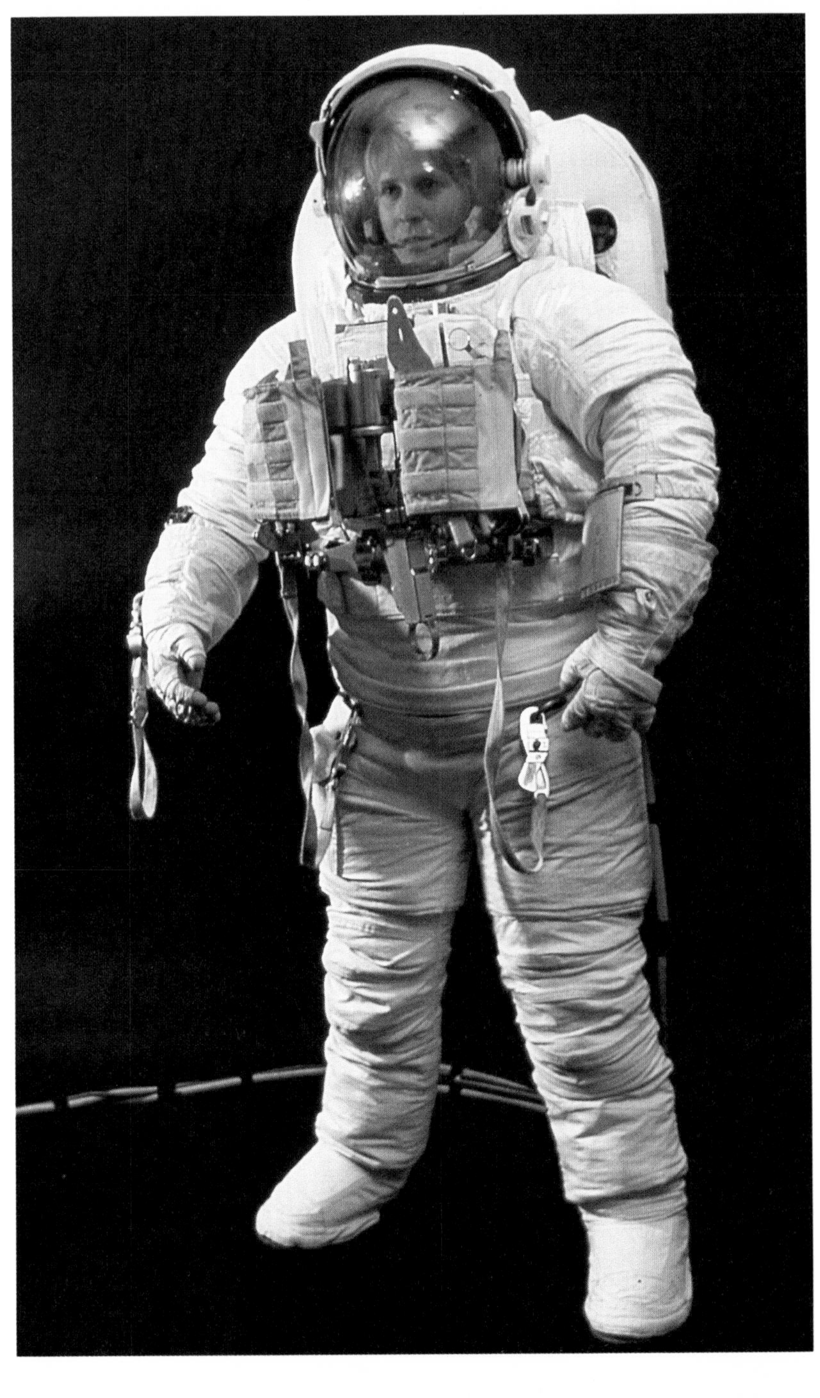

►** A spacesuit is more than just protective clothing to be worn in space. Termed by NASA an "Extravehicular Mobility Unit", the modern spacesuit is a complete life-support system made up of several layers. On top of the long underwear with cooling circulation tubes comes a neoprene-coated pressure garment with flexible joints, then a layer designed to withstand radiation and meteroids, and finally two layers of non-flammable, abrasion-resistant Teflon fabric. The helmet contains three protective visors, and the boots are made of tough composite material. An assemblage of tubes connects this astronaut with her oxygen supply, cooling liquid pump and humidifier.

Mining

The origins of mining...Mining coal underground...Oil drilling...Mining without miners...PERSPECTIVE... Mechanization in mining...Underwater mining... Open-cast mining...Strip mining

Mining, the process by which useful materials are obtained from or beneath the Earth's surface, is one of humankind's oldest activities. In prehistoric times, precious metals were obtained from "placer" deposits and flint mined using primitive tools such as pieces of antler.

Through mining we obtain not only precious substances such as gold, silver, diamonds and opals, but many essential metal ores and chemicals such as sulfur, soda ash and phosphate rock. Mining also provides building materials and fossil fuels.

Mining exploits non-renewable resources. Over the centuries the more obvious deposits have become exhausted. Prospecting for new sources of mineable material now involves sophisticated geophysical and geochemical techniques. Computer expert systems (◀ page 122) are also being developed to predict valuable mineral sites. Once a potentially useful deposit has been found, it has to be explored thoroughly to establish its economic potential (◀ page 68). At this stage, when test drillings are carried out, mining proper may be said to begin.

In addition to the traditional underground mine, mining is carried out in several different ways. Open-cast mining is used widely to obtain mineral ores and coal (▶ page 233). The majority of mineral mining now takes place above ground, with lead, antimony, tungsten, molybdenum and precious metals being the major substances still obtained from underground mines.

Solution mining, in which salts are dissolved in water pumped into mineral deposits, is an important source of chemicals, while microbial leaching is being developed by biotechnologists for obtaining valuable elements from low grade ores that have previously been unworkable.

▲ An illustration from Agricola's treatise on mining of 1556, showing shafts, tunnels and the hand-powered winches used to haul the minerals to the surface.

▼ Panning for gold in the traditional manner, in north-west Pakistan. Gold may be found in placer deposits (mixed in superficial gravel), and can be separated out by sieving.

Mechanization and modern mining

The major change in mining technology since 1945 has been in the extent of mechanization. This has affected underground mining through the introduction of large new machines, particularly for longwall mining, and surface mining, where very large-scale earth-moving machinery is now used. Open-cast mines, which are least common in Europe, tend to account for most of the world's most productive mines for this reason.

There are limits to mechanization, forced on mining engineers by the geology of particular sites. Sometimes the adaptability of human labor makes it more valuable than a machine in exploiting an awkwardly shaped deposit. Recently development has been aimed at mechanization for difficult deposits. A Dutch-designed "push-button miner" can exploit seams of coal of 60–160 cm thick and follows contours of the seam by using sensors which measure its natural radioactivity and compare this with that of the surrounding rock.

Longwall mining techniques may advance 1,000 meters at a time along a seam

Underground mining

The first materials to be "mined" came from naturally-occurring outcrops of ore. These arose where folding of the Earth's surface led to ore deposits being buried at an angle, in a hillside for example. As the surface eroded, the ore deposit was exposed.

As such deposits are exploited, material is removed from the outcrop, and then from further into the ground, thus creating a "drift" or "adit" mine. Despite their historic origins, such mines can still be developed today. The giant coalfield discovered in Selby, England, in the early 1980s will be exploited in part by drift mining.

The conventional image of a mine is one in which a vertical shaft is sunk into the ground until it reaches a buried body of ore. Horizontal galleries are then dug into the deposit and useful material is taken to the surface through the shaft. Modern mines have more than one shaft; these are lined with material such as steel to maintain stability.

The deeper a deposit lies, the greater the weight of "overburden", or rock and other material between the deposit and the surface. Consequently, the difficulty and danger of mining increase with depth.

In the traditional method of underground mining, rooms are cut out of the deposit, with pillars of material being left as roof supports. The number and position of these pillars depends on site geology. Originally, the material removed was loosened with hand-tools, but subsequently explosives came into use. In current practise, a section of material to be mined is undercut by machine at the bottom of the working face. Explosive charges are then placed higher up the face and detonated. The material expands and breaks into fragments which can be removed by conveyor belt. An explosive charge of 5–7 kilograms is sufficient to shatter 50 tonnes of coal.

As work proceeds and the room grows larger, long steel bolts are driven into the roof to strengthen it. In some cases, when mining in a particular area is finished, the pillars themselves are mined and the roof collapses, leading to surface subsidence.

As depth increases, room and pillar mining becomes less feasible. Longwall mining permits exploitation of deeper deposits. With this technique, material is removed along a length of the deposit. In a modern coal mine, the face is often 200 meters long, but in exceptional circumstances may exceed one kilometer. Giant mechanical cutters fitted with steel bits tipped with silicon carbide remove up to one meter thickness of deposit in each pass along the wall.

At each end of the face are tunnels through which men and equipment reach the face and through which the mined material is taken away. The roof is held up by hydraulic roof supports which move forward as the face is removed. The area behind the supports usually caves in rapidly. This has the advantage that surface subsidence is more immediate and predictable than in the case of room-and-pillar mining. Longwall mining was pioneered in Europe, which has deeper coal deposits than those generally found in America and Australia. The technique was not introduced seriously into the USA until the mid-1970s but is now becoming increasingly important there.

Underground mining has always been hazardous, but mechanization has introduced its own hazards. The high-speed cutting of rock increases the quantities of dust in the air. In the case of coal mining, cutting the rock can release methane gas. If this ignites in a dust-laden atmosphere, it can cause the coal dust to explode. High-speed mining machinery is fitted with methane sensors, so that the machines stop if the gas reaches a dangerous level.

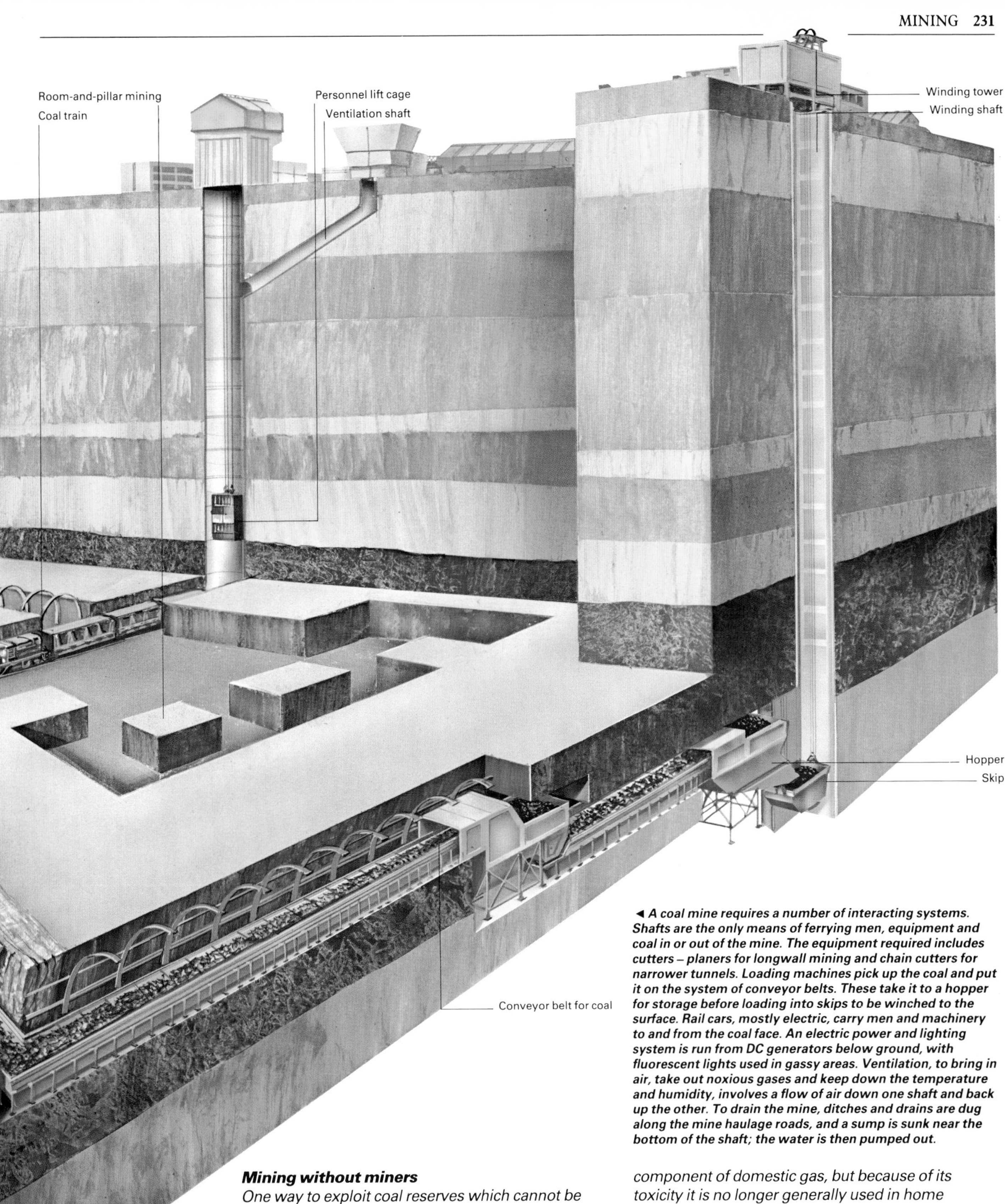

◄ ***A coal mine requires a number of interacting systems. Shafts are the only means of ferrying men, equipment and coal in or out of the mine. The equipment required includes cutters – planers for longwall mining and chain cutters for narrower tunnels. Loading machines pick up the coal and put it on the system of conveyor belts. These take it to a hopper for storage before loading into skips to be winched to the surface. Rail cars, mostly electric, carry men and machinery to and from the coal face. An electric power and lighting system is run from DC generators below ground, with fluorescent lights used in gassy areas. Ventilation, to bring in air, take out noxious gases and keep down the temperature and humidity, involves a flow of air down one shaft and back up the other. To drain the mine, ditches and drains are dug along the mine haulage roads, and a sump is sunk near the bottom of the shaft; the water is then pumped out.***

Mining without miners

One way to exploit coal reserves which cannot be mined conventionally is in situ gasification. If the coal deposit is overlain by impervious rock, two shafts are drilled into it. Air, sometimes enriched with oxygen, is pumped down one shaft and the coal is ignited. Initially it burns to carbon dioxide, but as it passes over the unburned coal towards the exit shaft, it is partially reduced to carbon monoxide. Although carbon monoxide does not have as much energy in it as elemental carbon, it can be used as an energy source by industry in some parts of the world. It used to be a significant component of domestic gas, but because of its toxicity it is no longer generally used in home supplies. Carbon monoxide toxicity also has to be considered in choosing suitable sites for in situ gasification, as leakage to the surface in inhabited areas could cause an environmental disaster.

As a result the technique is most applicable to deep coal deposits such as those found in Europe. Belgium, France, Germany and Great Britain are all investigating this technique, with the most advanced project being undertaken in Belgium with financial support from the European Economic Community.

The largest open-cast mines are up to 1,000 meters deep

◄ *Firing at the copper mine of Chuquicamata in northern Chile, one of the largest open-cast mines in the world.*

► *A unit train at the Valdivia nitrate mine in northern Chile, removing the mineral for processing. Such trains may be more than 1,000m long, and are used where large quantities of material have to be moved considerable distances; over shorter distances large trucks, up to 350 tonnes, are cheaper.*

► *The Great Hole at Kimberley, South Africa, the product of a large number of individual diamond mines sunk independently in the early 1870s.*

▼ *An excavator at a strip mine in Illinois. Despite the huge volumes of material moved, strip mining can permit the land to be restored after the ores have been removed.*

Open-cast mining

The world's largest mines are open-cast mines, in which mineral deposits near the surface are exploited. In addition to coal and iron and copper ores, chemicals such as phosphate rock (used in fertilizer manufacture) and construction material such as sand and gravel are obtained by surface mining. Open pit mines are usually inverted cones, consisting of many terraces. The conical shape is dictated by the need to bank the mine walls at an angle at which the slope will be stable, so that material does not come away and fall onto the next terrace. Stability depends not only on the type of rock, but also on the effect that rain or floodwater may have on it. These mines may reach a depth of one kilometer, but the cost of exploiting the ore increases with depth.

Once the overburden – the soil and other material between the deposit and ground level – has been removed, the mineral is broken up by drilling and blasting in the case of hard rock or by mechanical cutting. Usually, layers of 5-15m in depth are removed at a time.

When the rock or coal has been loosened it is excavated by large-scale equipment, such as power shovels and bucket-wheel excavators, and then transported by truck or train. The buckets used may hold as much as 180 cubic yards of material and more than a quarter of a million tonnes per day may be taken from such a mine.

Where the amount of overburden is too great for economical exploitation of the underlying deposit, auger mining can be used to get more material from a mine. A giant drill or auger is used to recover material from beneath the overburden. This type of mining was used to produce a small percentage of US coal in the 1950s and 1960s, but has been little used since for economic reasons.

Strip mining is similar to open-cast mining, but used for shallower deposits. A slice of overburden between 10 and 50m wide is removed with earth-moving equipment. The underlying mineral is then removed. When the next strip of overburden is taken off, it is used to cover the previous strip.

In recent years, strip mining in developed countries has been rigorously controlled by environmental authorities. In some cases, mining companies have to preserve the topsoil separately from the rest of the overburden and make sure that it goes back onto the surface, so that plant growth can re-establish itself quickly.

An oil well may be up to 8,000 meters deep. The borehole is produced with rotating drill bits, which are connected to the surface by 9-meter lengths of drill pipe, the whole being called the drill "string". The bits wear out rapidly, so the string has to be dismantled and reassembled at regular intervals.

The borehole into which the drill cuts is filled with a viscous liquid called the drilling mud. This circulates through the drill string and brings rock fragments to the surface where they are filtered out before the mud is recycled, helping to cool and lubricate the drill. The mud also helps to control any flow of oil or gas from the well.

After oil is found in sufficient quantity to justify production, production tubing is put in place, and the wellhead topped with a "Christmas tree", a complex of valves and tubing. Gas and water are removed from the oil, and it is transported, by pipeline or by tanker, to the refinery (➧ page 241).

Offshore drilling

About 25 percent of the world's oil now comes from offshore fields. Offshore drilling for oil began in the late 1940s, off the Louisiana coast of the USA, but it is mainly since the mid-1960s that the technology has developed. Offshore drilling is mostly carried out from platforms sitting on the seabed in the shallow waters (usually not more than 200 meters) of continental shelves. However, deep-sea drilling is also possible from floating platforms.

Platforms which sit on the seabed are built to withstand the highest wave likely to occur once in 100 years. In the North Sea between Britain and continental Europe, this means that the lowest deck is at least 25 meters above sea level. The platforms also have to be able to withstand constant battering by smaller waves. The Stratfjord "B" oil platform, 160 kilometers off the Norwegian coast, weighs almost 900,000 tonnes. Most of this weight is in the concrete used as the platform's base. Other types of platform use piles driven into the seabed to give the support, or a series of anchor cables like the guy ropes of a tent. By drilling at different angles, a single platform can drill a large number of boreholes.

▲ An offshore oil drilling platform in the Cook Inlet of Alaska.

◄ The drilling bit, which may be tipped with diamonds, used for deep drilling. The earliest drills used a pounding action rather that a rotary one; this technique was liable to produce an uncontrollable "gusher" when it found oil.

Offshore drilling

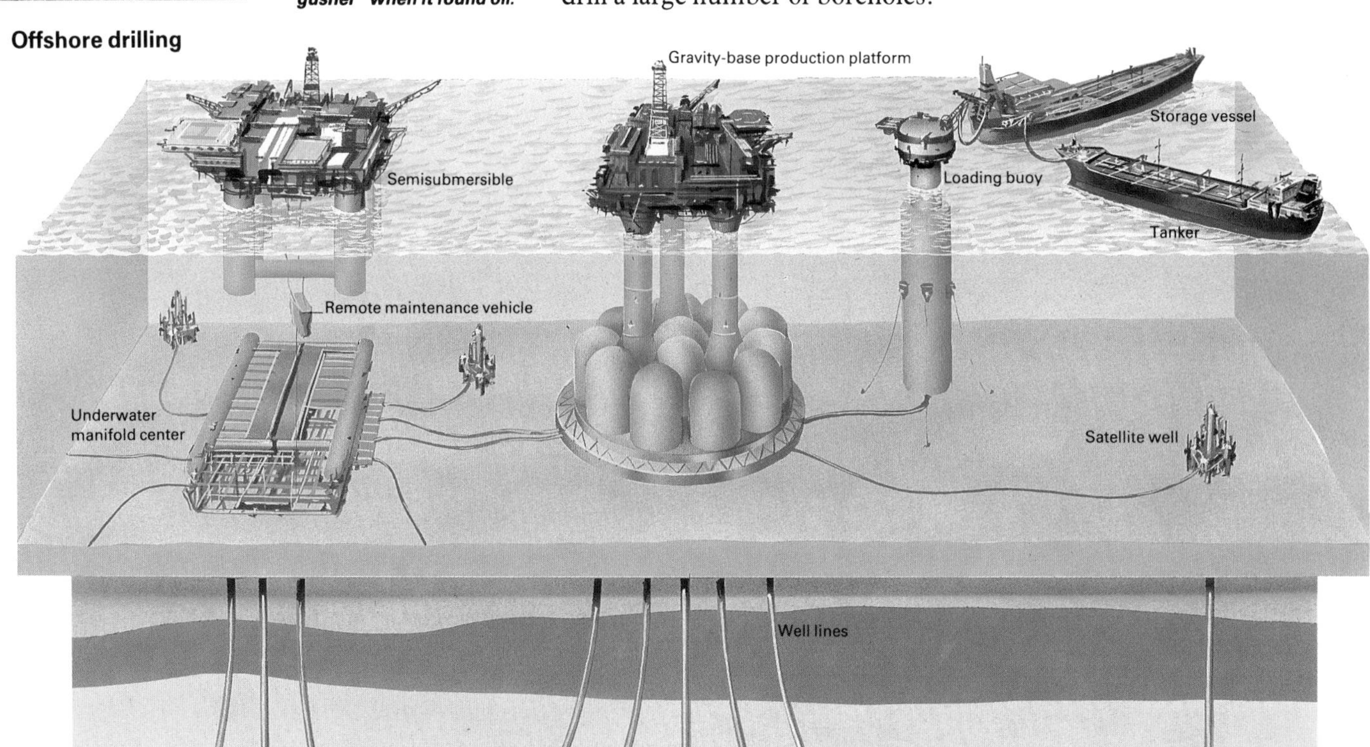

▲ **The Alaskan pipeline and (inset) the Trans-Siberian pipeline, up to 1·22m in diameter, provide an efficient and cheap method of transporting crude oil. Pumping stations are placed along the route to maintain the flow.**

◄ **More and more activity on an offshore oil field takes place under the water. The platform itself may be semi-submersible, or may sit on the bottom with massive concrete supports. An underwater manifold center, serviced by remote-controlled submersibles, acts as the head of a number of wells, sunk at angles into the oil field, while satellite wells have separate wellheads. The oil is passed to the surface and stored in a permanently moored tanker before being taken by ship to a refinery.**

Underwater mining

In some mining operations, the overburden is liquid rather than solid. Floating barges called dredges are used to obtain valuable materials from the beds of both lakes and seas. Material can be recovered by draglines, endless belts of buckets or even by suction. Tin ore is one of the major materials obtained commercially by this method, with more than 10 percent of world production coming from offshore operations in Thailand and Indonesia.

Increasingly, as existing land-based mineral deposits are depleted, the potential of offshore deposits is being reevaluated. For many years, useful materials have been recovered from the continental shelf – the coastal waters up to 200m deep. Here, environmental conditions are usually not too severe and large quantities of material are obtained – such as lime, sand and shells for Portland cement, siliceous sand and ore-bearing sands such as chromite (from which chromium can be refined), ilmenite and rutile (titanium), magnetite (iron) and monazite (thorium and other rare earth elements).

Deep-water mining

Recovering materials from deeper waters has proved difficult so far. A scheme to recover diamonds offshore in South Africa was started in the 1960s, but it failed a few years later, on economic grounds.

A major resource of interest is the nodules which occur on the deep ocean floor bearing manganese (a metal that is of considerable importance when alloyed with copper and aluminium or added to steel to strengthen it). It has been suggested that these nodules could be recovered by sucking them up to the surface. However, not only is the technology insufficiently developed, but also there is no international agreement yet about how the seabed outside territorial waters should be exploited.

Seawater itself is probably the most valuable resource mined offshore at present. More than 10 million tonnes of salt are obtained from seawater by evaporation each year. Magnesium is also present as dissolved salts and more than three-quarters of magnesium production in the United States is obtained from seawater (➧ page 242).

See also
Seeing with Sound 65–72
Chemical Analysis 105–108
Electricity 135–140
Processing Raw Materials 237–244
New Materials 245–248

Mining without miners

In various parts of the world deposits of water-soluble minerals have been protected from dissolution. The most famous are perhaps the Chilean nitrate beds, which were virtually the only inorganic source of nitrogen fertilizer until the development of the nitrogen fixation process by Fritz Haber (1868-1934) and Carl Bosch (1874-1940) in the early 20th century. These deposits remained intact largely because of arid local conditions. However, even in wetter parts of the world such as western Europe, deposits of soluble minerals have been preserved by being overlain with impermeable rock.

In the past, the salt and soda ash mines of Europe were worked by miners in conventional shaft mines. Now, such materials are more frequently obtained by solution mining. Hot water is piped into the soluble mineral deposit. Here it forms a solution which is pumped back to the surface, where the mineral can be recovered by evaporation or by chemical conversion to an insoluble derivative.

Where a mineral is insoluble, it may be converted into a soluble form and then be leached from the ore deposit with water. More than 10 percent of the world's copper is now produced in this way. Although this technique can be described as biotechnology, it has been used for at least 3,000 years.

Various bacteria can derive metabolic energy from oxidizing either iron or sulfur. As copper often occurs as a sulfide in conjunction with iron sulfide, these bacteria have been naturally producing soluble copper salts for millions of years. It was this that first led to the recovery of copper from mine drainage water so many years ago.

Now, however, the process has been studied in detail and applied more efficiently. In dump leaching, ore dumps of up to four billion tonnes are treated with acidified water. This helps the bacteria, such as *Thiobacillus ferro-oxidans*, to convert copper sulfide to soluble copper sulfate.

Biomining has also been used to recover uranium by converting it to a soluble ion. In future, perhaps with the help of developments in genetic engineering, it may be extended to other metals, such as lead, molybdenum and zinc. The advantage of bacterial processes is that they usually require less energy than conventional mining and also cause less environmental damage.

Mining sulfur with water

Water is generally a problem in mines, and elaborate arrangements are needed to provide drainage. However, it can actually be used to extract minerals in difficult circumstances. One of its earliest applications was in the sulfur recovery process, developed at the end of the 19th century by Herman Frasch (1851-1914) and still used.

In various parts of the world, notably around the Gulf of Mexico, large deposits of elemental sulfur occur in limestone-capped salt domes. These deposits were probably formed by microbial action which reduced sulfate rock to sulfur. Such deposits are frequently hazardous to exploit, because of the poisonous hydrogen sulfide gas which accompanies the element.

Sulfur does not dissolve in water, but it has a relatively low melting point. The Frasch process uses three pipes of different diameters, arranged concentrically. These are placed down a borehole into the sulfur deposit, usually some 200m under the surface. Initially superheated water is pumped down the outer two pipes under pressure (at high pressure the boiling point of water exceeds the melting point of sulfur). The sulfur melts and forms an underground pool. After 24 hours, sufficient sulfur has usually melted to make it worthwhile to start forcing it to the surface.

The supply of water to the middle pipe is stopped and compressed air is blown down the inner pipe. This causes the hot water and molten sulfur to form a frothy liquid which rises up the middle pipe as more pressurized hot water is pumped down the outer pipe. The mixture is transported through heated pipes at the surface to settling tanks where the sulfur separates out. The purity of sulfur obtained in this way is 99·5–99·9 percent.

▼ The Dead Sea in Israel contains many valuable minerals, including potash, bromine and common salt. These are extracted commercially by solar evaporation in chemical works such as the one shown here. The brine is concentrated in large ponds of about 20 hectares to separate out impurities before passing to crystallization pans where salt of various grades is deposited.

Processing Raw Materials

Refining raw materials...Refining metals and non-metallic substances by heat...Electrolysis: aluminum and copper...Processing by water and solvents... PERSPECTIVE...Recycling scrap metal...Iron and steel making...Gold recovery...Aluminum processing... Recycling road surfaces

Most raw materials processing is the production of metals from ores, in which the metal exists in combination with other elements. Metallic ores usually need to be concentrated by physical processes before further treatment can take place. This often involves crushing or grinding of the ore, followed by flotation. Metal compounds are frequently more dense than the rock particles in which they are embedded, but when agitated in water containing a foaming agent they concentrate in the bubbles, while the rock particles sink. Many ores, particularly those containing metal sulfides, can often be concentrated tenfold in this way.

After being concentrated, the ores are normally processed by one of three major metallurgical techniques: pyrometallurgy (involving heat ◀ page 238), electrometallurgy (involving electricity ▶ page 242) and hydrometallurgy (involving water ▶ page 243). Extracting pure metal from the ore is known as smelting.

Recycling

Depletion of basic mineral sources has led to increased interest in recycling waste materials. In practise, a considerable amount of recycling occurs during the processing of raw materials.

Only about half the steel produced each year is made from new iron. Much of the remainder comes from two types of scrap steel. Some steel is "scrapped" without ever leaving the steel works, because it does not reach the required specification. In addition to this "revert" scrap, a considerable amount of industrial scrap is produced by manufacturers of steel articles. If a complicated shape, such as a car wing panel, is cut from a steel sheet, small pieces are left over. These can be returned to the steel mill for reprocessing.

The third form of scrap is the most difficult to handle. This is steel in articles which have come to the end of their useful life, such as broken washing machines. To recycle this material, the component parts have to be separated easily and the quality of the material used must be known. As steels may be of widely differing composition they cannot be mixed together and reprocessed easily.

The problem is even more acute if the steel is intimately associated with another metal. Tin cans consist of sheet steel coated with a very thin layer of tin. Unless the tin is removed, it forms iron-tin compounds if the metal is heated strongly. If a tin can has been used to hold food, there are probably residues adhering to its inside. The only economic way to remove this organic matter is to burn it off, but that causes migration of tin into the steel.

Many recycling processes have been developed, for example for the silver from photographic materials. Some uses to which materials are put seem likely to ensure that recycling will never be feasible. For example, the materials used in a telephone contain more than 40 different elements. Nevertheless a process has now been developed to separate the metals from electronic circuitry.

Recycling is most successful when materials can be designed with recycling in mind. Until recently, many soft drink containers were made primarily of tinned steel, but ring pull caps were aluminum. Before these could be reprocessed, the aluminum had to be separated from the ferrous metal. Now, a large number of drink cans are made entirely from aluminum and can be reprocessed more easily. More than 50 percent of the aluminum cans used in Australia are recovered and reprocessed.

In addition to the possibility of future shortages of particular materials, the rise in energy costs in recent years has increased interest in recycling some materials, such as aluminum. This is because the energy required to reprocess the metal is only a fraction of that needed to win it from its ore. For aluminum, the energy needed to reprocess the metal is only 5 percent of that required to produce new aluminum from bauxite.

◀ Old car bodies contain recoverable copper and nickel as well as steel. The bodies are shredded and the ferrous parts removed magnetically; other metals are extracted by flotation. Another process sprays the bodies with liquid nitrogen and then fragments them. At very low temperatures the steel is brittle and breaks like glass, while the other metallic elements remain malleable.

Refining by heat

Pyrometallurgy may involve one or more steps. Some ores, such as galena (lead sulfide) and sphalerite (zinc sulfide), are roasted to convert them to oxides which are then smelted. This process also produces sulfur dioxide, which can be used to make sulfuric acid.

Metal oxides, including those produced by roasting and those found naturally – as in iron ores (➧ page 240) – can be heated with a reducing agent such as carbon (often in the form of coke). The carbon serves a dual function, not only acting as a fuel source to generate intense heat, but also combining chemically with the oxygen to release free metal.

Another form of heat treatment of impure metals is distillation, in cases where the metal has a sufficiently low boiling point. In some smelting processes, that for zinc for example, smelting and distillation take place simultaneously.

Some metals form compounds with carbon monoxide. These carbonyls can be used to produce very pure metals, because they vaporize easily at fairly low temperatures. Nickel, iron and cobalt all react directly with carbon monoxide, and both silver and copper also can be purified in this way via their halide salts.

Oil products

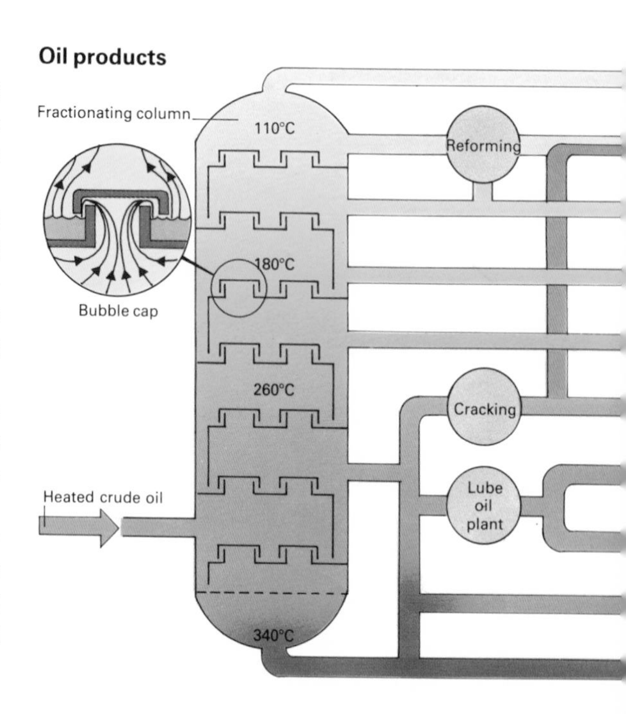

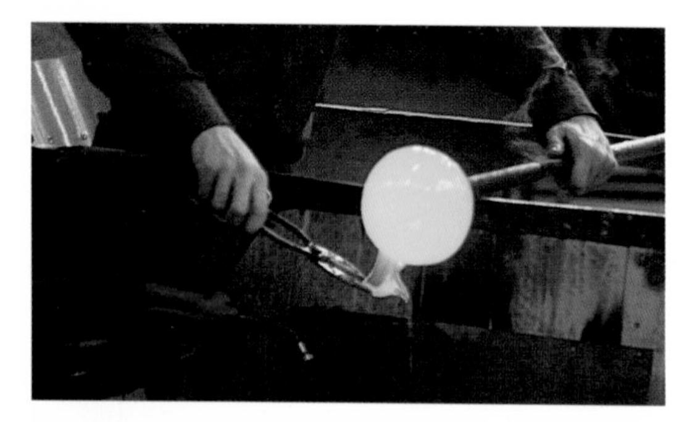

◄ ***Traditional glassblowing is done by gathering molten glass on a hollow rod, blowing a bubble which is then shaped by rolling. The base is formed, then the top snipped off and shaped.***

▼ ***Zinc is derived by making a zinc-lead concentrate from which other waste materials have been removed, and smelting this in a blast furnace; pure lead is also produced.***

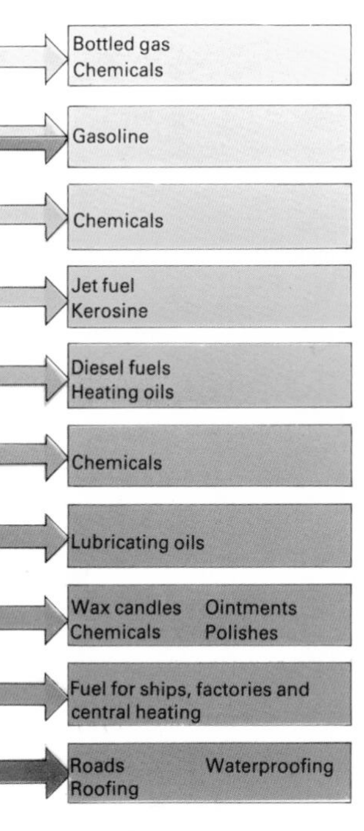

◀ *Crude oil is heated and passed into a fractionating column, a distillation plant made up of a number of different levels. The temperature is highest at the base of the column, and the lighter hydrocarbon vapors can rise up the column. Each grade of oil has its own boiling point, and so it condenses at a particular level; rising vapors bubble through it as it condenses.*

▼ *An oil refinery involves several complex processes. After fractioning, some of the heavier oils are passed to the catalyst cracking plant to be broken into lighter grades. Sulfur may be removed from diesel fuel. The heavier fractions may be further processed, and a vacuum distillation plant produces lubricating oil and paraffin wax.*

Heat treatments for non-metallic substances

Non-metallic raw materials may be processed by similar technologies to those used for metals. The active ingredients of cement and concrete are obtained by mixing clay and limestone or chalk to form a powder, a slurry, or, in the most fuel-efficient process, a semi-wet "cake". This is then passed through a rotary kiln at high temperature. This forms a "clinker" which is ground to a powder. A small amount of gypsum (calcium sulfate) is added to control the setting rate and the result is a hydraulic cement which, on mixing with water, forms a strong solid. Most cement is used in the form of concrete, which contains other non-reactive bulk materials. Slag from iron smelting is used in some concretes.

The various petroleum products are produced by distilling the crude oil in a fractioning column and separating it out into its various hydrocarbon components; the process is made more flexible by introducing catalysts to break up the organic compounds.

Another example of raw material processing which uses heat is the preparation of quicklime (calcium oxide), needed for the production of one of the raw materials for glass manufacture. This is made by heating natural limestone (calcium carbonate) in coke-fired kilns.

Modern furnaces are so efficient at removing impurities that carbon has to be replaced in the steel to prevent brittleness

Iron and steel

The pyrometallurgical processes most widely used in the world today are iron and steel production. Most of the iron ore converted to metal each year is made into steel. World steel production is about 700 million tonnes per annum.

Iron has been produced for centuries by smelting iron ore (usually an iron oxide, like rust) with a source of carbon (originally charcoal, now coke or a mixture of coke and fuel oil) and limestone. Most iron ores smelted today contain between 60 and 80 percent iron oxide. The remainder is often a mixture of earth, clay, stones and sand. The limestone forms a slag with these silicon-containing impurities and helps to purify the metal.

At high temperatures, the oxygen in the ore combines with carbon to form carbon oxide gases and free metal. Unfortunately, molten iron dissolves about 4 percent of its own weight of carbon, making the metal – called pig iron – very brittle when it solidifies.

To produce steel, the carbon content of crude iron has to be reduced to less than 1 percent. The first large-scale production of steel followed the invention of the steel converter by Henry Bessemer (1813-1898) in 1856.

Bessemer discovered that passing a blast of air through molten iron was sufficient to burn off the impurities, including the carbon. The energy given off by formation of oxides of these impurities keeps the metal molten.

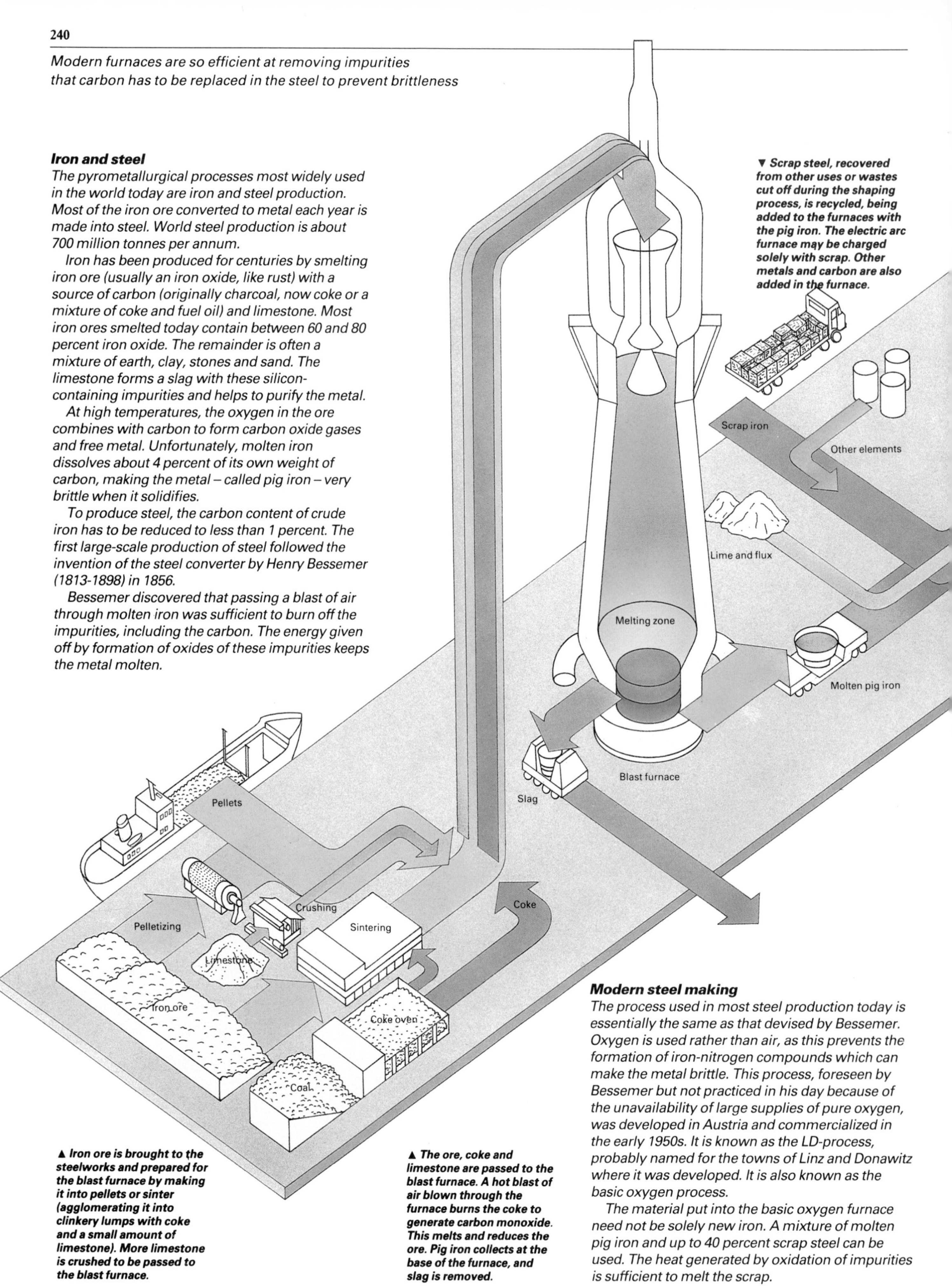

▼ **Scrap steel, recovered from other uses or wastes cut off during the shaping process, is recycled, being added to the furnaces with the pig iron. The electric arc furnace may be charged solely with scrap. Other metals and carbon are also added in the furnace.**

▲ **Iron ore is brought to the steelworks and prepared for the blast furnace by making it into pellets or sinter (agglomerating it into clinkery lumps with coke and a small amount of limestone). More limestone is crushed to be passed to the blast furnace.**

▲ **The ore, coke and limestone are passed to the blast furnace. A hot blast of air blown through the furnace burns the coke to generate carbon monoxide. This melts and reduces the ore. Pig iron collects at the base of the furnace, and slag is removed.**

Modern steel making

The process used in most steel production today is essentially the same as that devised by Bessemer. Oxygen is used rather than air, as this prevents the formation of iron-nitrogen compounds which can make the metal brittle. This process, foreseen by Bessemer but not practiced in his day because of the unavailability of large supplies of pure oxygen, was developed in Austria and commercialized in the early 1950s. It is known as the LD-process, probably named for the towns of Linz and Donawitz where it was developed. It is also known as the basic oxygen process.

The material put into the basic oxygen furnace need not be solely new iron. A mixture of molten pig iron and up to 40 percent scrap steel can be used. The heat generated by oxidation of impurities is sufficient to melt the scrap.

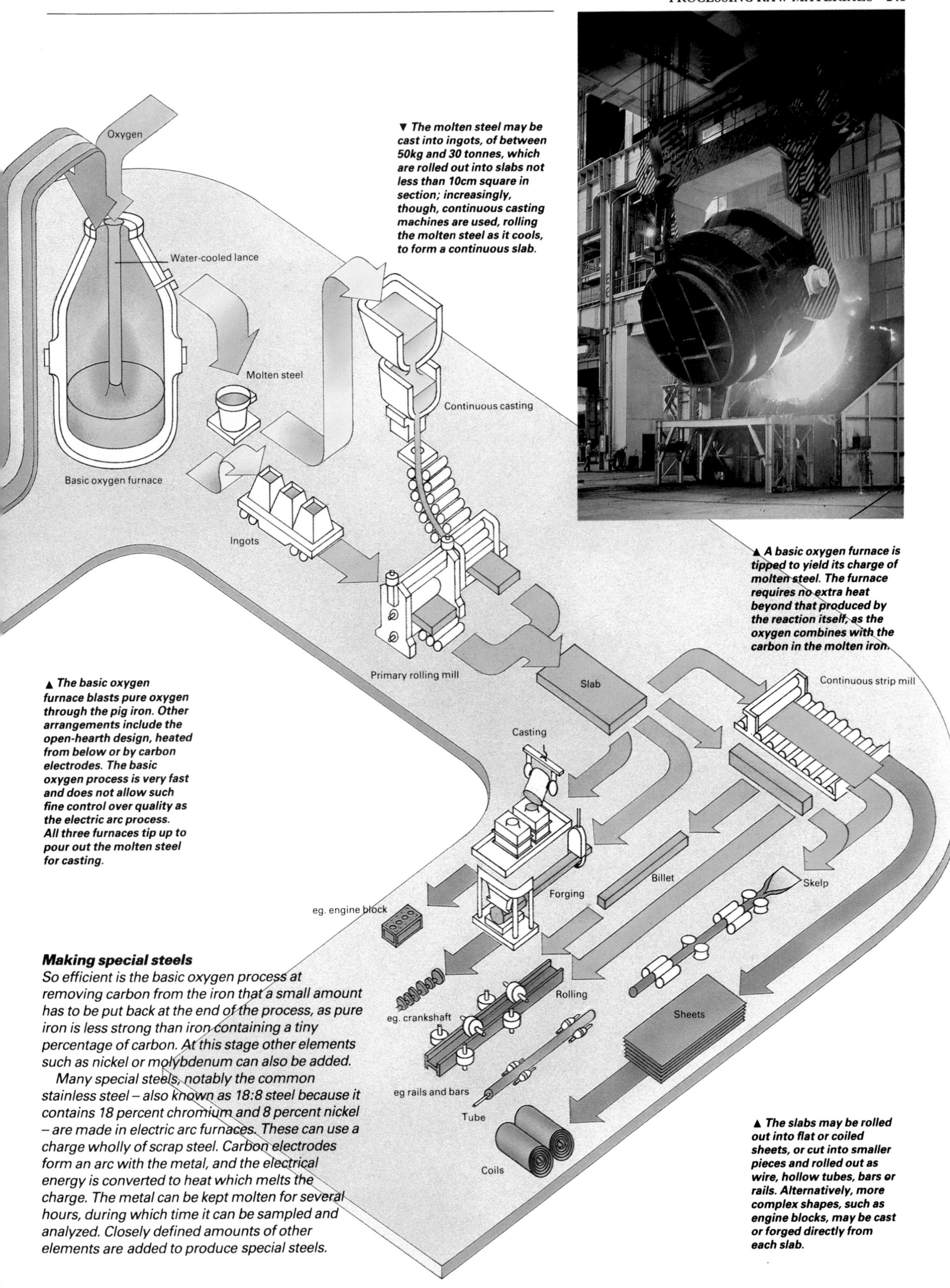

▼ ***The molten steel may be cast into ingots, of between 50kg and 30 tonnes, which are rolled out into slabs not less than 10cm square in section; increasingly, though, continuous casting machines are used, rolling the molten steel as it cools, to form a continuous slab.***

▲ ***A basic oxygen furnace is tipped to yield its charge of molten steel. The furnace requires no extra heat beyond that produced by the reaction itself, as the oxygen combines with the carbon in the molten iron.***

▲ ***The basic oxygen furnace blasts pure oxygen through the pig iron. Other arrangements include the open-hearth design, heated from below or by carbon electrodes. The basic oxygen process is very fast and does not allow such fine control over quality as the electric arc process. All three furnaces tip up to pour out the molten steel for casting.***

Making special steels

So efficient is the basic oxygen process at removing carbon from the iron that a small amount has to be put back at the end of the process, as pure iron is less strong than iron containing a tiny percentage of carbon. At this stage other elements such as nickel or molybdenum can also be added.

Many special steels, notably the common stainless steel – also known as 18:8 steel because it contains 18 percent chromium and 8 percent nickel – are made in electric arc furnaces. These can use a charge wholly of scrap steel. Carbon electrodes form an arc with the metal, and the electrical energy is converted to heat which melts the charge. The metal can be kept molten for several hours, during which time it can be sampled and analyzed. Closely defined amounts of other elements are added to produce special steels.

▲ ***The slabs may be rolled out into flat or coiled sheets, or cut into smaller pieces and rolled out as wire, hollow tubes, bars or rails. Alternatively, more complex shapes, such as engine blocks, may be cast or forged directly from each slab.***

Gold is now derived from the dumps of waste thrown up at the South African mines of the early years of this century

Aluminum

Production of aluminum is the major electrometallurgical process. A nearly pure aluminum oxide is needed, which means that the ore – which is principally bauxite – has to be purified before processing.

The main impurities in bauxite are other oxides, notably those of iron, tin and silicon. The aluminum oxide is separated from these by treatment with sodium hydroxide. Unlike the other oxides, aluminum oxide dissolves in this and nearly pure aluminum oxide can therefore be recovered from the solution.

Aluminum oxide, or alumina, has a very high melting point (about 2,000°C), but this can be reduced to about 1,000°C if the alumina is mixed with cryolite (sodium aluminum fluoride). Electrolysis of the molten mixture then produces free metal.

In the Hall-Héroult process, the electric current provides sufficient energy to keep the bath of oxide-cryolite molten as well as to split the oxide into its elements. The aluminum metal is heavier than the molten salt mixture and sinks to the bottom of the bath. Oxygen is discharged at the anode. As the latter is made of carbon, it combines chemically with the oxygen and has to be renewed continually. Half a tonne of electrode is needed for each tonne of metal produced.

Concern has been expressed about pollution from aluminum smelters, as toxic fluoride salts may escape into the atmosphere. This has led to development of a process in which alumina is converted to aluminum chloride. This is then electrolyzed to give aluminum and chlorine gas. As well as eliminating fluoride pollution, this process reduces electricity consumption by 30 percent.

►▲ Copper is normally refined by smelting when found as a sulfide, but by electrolysis when found in its oxide form. In this process, the ore is leached with sulfuric acid to form copper sulfate (lower photograph). This is then electrolyzed, using an insoluble anode. Very pure copper deposits at the cathode.

The electrolysis of aluminum

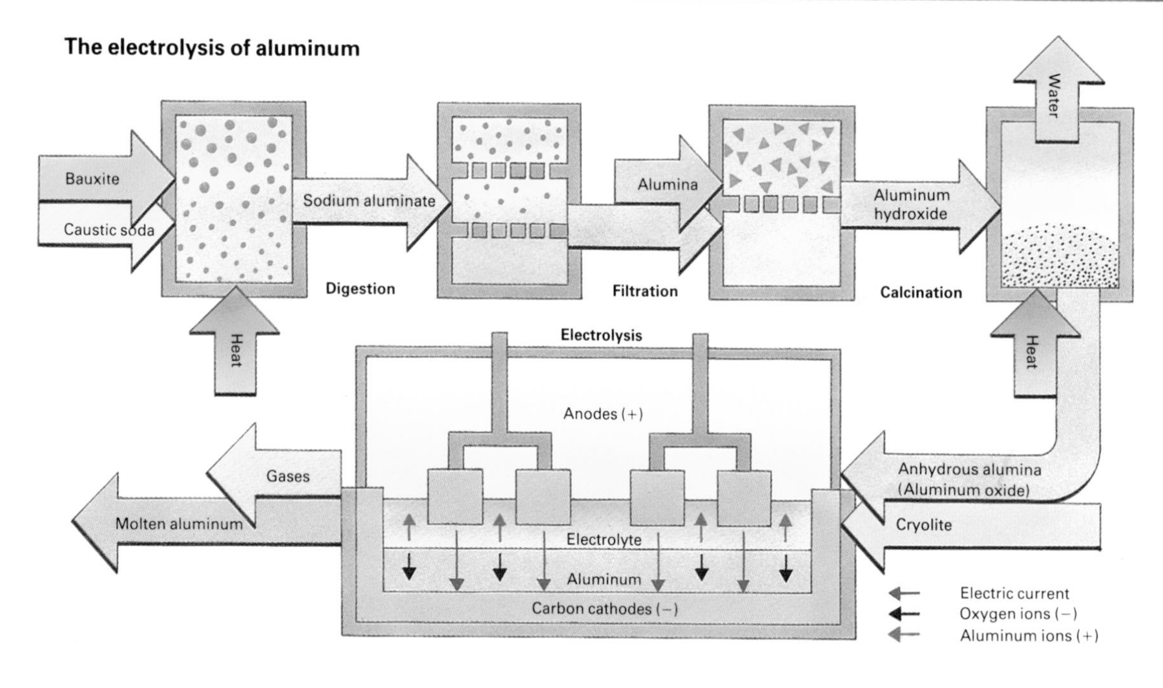

◄ The Bayer process for extracting aluminum is one of the most widely used methods of metal refining. Alumina (aluminum oxide) is first produced from bauxite by "digesting"; the alumina dissolves in caustic soda to form sodium aluminate and impurities are filtered out. On cooling, crystals of aluminum hydroxide are formed, which are heated and give up their water to become anhydrous alumina. This is mixed with cryolite or artificial aluminum sodium fluoride for use as the electrolyte in the electrolysis; aluminum is deposited at the cathode while the oxygen unites with the carbon anode to form carbon monoxide. Half a tonne of anode is used for every tonne of aluminum produced.

Electrometallurgy

Another way to purify impure metals is by electrometallurgy. This technology can also be used to win metals from ores. Aluminum, which is regarded as the second most important metal after iron, is the most common example of an element produced from its ore by an electrical process. More than 13 million tonnes are produced annually worldwide. Magnesium is also produced electrolytically. Magnesium chloride is obtained from seawater or magnesium oxide is extracted from the earth. The oxide is converted to magnesium chloride by heating with chlorine. The magnesium chloride is transferred to electrolytic furnaces where the metallic magnesium is produced.

Different products may be obtained by passing electricity through a molten metallic salt and a solution of that salt. For example, the highly reactive element sodium can be obtained by electrolysis of fused sodium chloride. The metal is formed at the cathode (negative electrode) and chlorine gas is discharged at the anode (positive electrode). If a solution of sodium chloride is electrolyzed, however, hydrogen from the water is discharged at the cathode. Chlorine is still obtained from the anode and the overall result is that the solution is changed from sodium chloride to sodium hydroxide (caustic soda). This process is used industrially to manufacture sodium hydroxide, an important industrial raw material.

Solution electrolysis is widely used to purify metals, such as copper, which are less reactive than sodium. To do this impure copper is used as a sacrificial anode in an electrolysis circuit, a piece of pure copper is the cathode, and the electrolyte is a solution of copper sulfate. As electric current passes through the circuit, copper dissolves from the anode and plates out from the solution onto the pure cathode. The impurities either stay in solution or form a sediment, but in neither case do they transfer to the cathode.

In some processes, the impurities, or "anode slimes", are processed separately to purify the metals in them. Selenium and tellurium, two rare metals, are obtained commercially in this way.

Some important chemical raw materials, such as sodium hydroxide and chlorine, are obtained by the electrolysis of solutions, paralleling the production of metals from fused salts by electrometallurgy.

Gold: from fleece to carbon-leaching

The ancient Greek story of Jason and the golden fleece probably has its origins in primitive technology. Gold is one of the few metals found in its elemental state rather than as a chemical compound. As gold-bearing rock erodes, tiny particles of metal are washed away, together with the valueless rock particles. In Classical times, one way of collecting gold was to peg fleeces in the beds of rivers which had their sources in or passed over gold-bearing rock. The heavy particles of metal became entangled in the hairs of the fleece, while the unwanted rock carried on to the sea.

During the great gold rushes of the 19th century, the traditional panning for gold represented little improvement on this ancient technology. One attraction of gold as a metal is its unreactivity, but this makes its separation by chemical means difficult. It will form an amalgam – a sort of alloy – with mercury and the first technological advance in gold processing was a technique in which mercury was mixed with finely crushed gold ore in order to extract the gold. This process was inefficient and also left behind vast sand heaps.

Towards the end of the 19th century, a chemical technique for gold separation was developed. Although generally unreactive, gold does form chemical complexes with cyanide ions. The cyanide treatment required the ore to be more finely ground and suspended in water so that the gold could be dissolved. This led to the development of huge heaps of spoil called slimes.

The sand heaps and slime dams of South Africa are now themselves a major source of gold, for they are being reprocessed. In the Witwatersrand area around Johannesburg, there are 250 slime dams and 100 sand dumps covering 8,000 hectares.

▼◄ Dumps from the early days of mining in South Africa are now a major resource for gold. They are washed down with high-pressure water cannon, and the material concentrated. This is leached to remove uranium salts and roasted to convert pyrites to sulfur dioxide. The gold is dissolved by cyanide and absorbed onto a special form of carbon.

See also
Chemical Analysis 105-108
Nuclear Power 141–148
Road Transport 199-208
Mining 229–236
New Materials 245–248

Using solvents for metal processing

A fast-growing area of metal processing is hydrometallurgy. This involves purification via solution of the required material. In order to form a solution effectively, the substance to be dissolved needs to be in small particles, so grinding is a prerequisite for most hydrometallurgy. In some cases, the ground ore is also roasted before contact with water.

The production of some metals, such as copper, from leach waters has been carried out for centuries. What has made hydrometallurgy grow in the past quarter of a century has been the development of organic solvents which will dissolve metals selectively, so that they can be extracted easily from aqueous solution.

Most metallic salts only dissolve easily in so-called polar solvents, such as water. As most organic solvents, which are usually based on petroleum, are non-polar, metal salts do not usualy dissolve in them. However, while the atomic bomb was being developed, it was discovered that the uranium salt, uranyl nitrate, would dissolve in diethyl ether (anesthetic ether). This led to a process for purifying uranium by solvent extraction; the uranium is removed from the water by the organic solvent while the impurities remain behind (♦ page 142).

Solvent extraction metallurgy depends on two factors. The first is dissolution of the required metal, as a salt in water. The second is the availability of an organic chemical compound which will dissolve in a solvent that does not mix with water, such as kerosene, and which will also react with the required metal ion to form a complex or "chelate" that is still soluble in the non-aqueous solvent. The use of such compounds overcomes the difficulty that most metal ions are not soluble in organic solvents.

In the case of copper, chemicals called hydroxyoximes are used as metal-chelating agents. These were introduced in the 1960s and have revolutionized copper production. A variety of other substances have been developed which have an affinity for a particular metal ion and solvent extraction is now widely used.

The solubility of the metal in water can be altered by the presence of other substances, such as acids and alkalis. By altering the acidity of a solution, often coupling this with the use of different organic solvents, it is possible to separate complex mixtures of metals.

In July 1983, Matthey Rustenburg Refiners Ltd opened the world's most advanced platinum metals refinery in Royston, England, to exploit solvent extraction technology. In the first stage of their process, a concentrated raw material is dissolved in an acid solution. An organic solvent called methylisobutylketone removes gold, iron and tellurium. Under slightly different conditions, palladium is then extracted into a different organic solvent. An oxidation process removes ruthenium and osmium from solution, after which platinum is extracted by an organic solvent containing an ammonia-based chelating compound. After this, the aqueous solution is concentrated. Organic solvent extraction removes iridium. Finally the aqueous solution is passed through an ion-exchange column which takes out rhodium.

Solvent extraction technology is particularly important in the nuclear industry, in recovering active materials from nuclear fuels. "Breeder" reactors produce more fuel than they consume (♦ page 147). Uranium is converted to plutonium during the breeding process and the plutonium can then be used in a different type of reactor, once it has been purified by solvent extraction.

▲ *Platinum has very high resistance to corrosion and a high melting point; it is useful for applications where oxidation or corrosion must be minimal. It can also be used in treating cancer; its preparation for this purpose is shown here.*

Hydrometallurgy for non-metals

Many chemical raw materials are also obtained by solution processes, similar to those used in hydrometallurgy. Table salt (sodium chloride) has been obtained from seawater and other brines for centuries. In addition to sodium chloride, most brines also contain other important minerals. If the water is evaporated slowly in a series of pans, it is possible to separate different substances, such as calcium carbonate, calcium sulfate, sodium chloride, magnesium sulfate and magnesium chloride. A chemical works on the Dead Sea in Israel carries out these separations using solar energy and recovers bromine from the residual solution by electrolysis.

Recycling road surfaces

Among non-metallic substances the recycling of paper and glass are well known; a relatively new development is the recycling of constructional materials. Around 1970 the reuse of asphalt and concrete, especially from road surfaces, became a large-scale activity. These surfaces consist of crushed stone aggregate bound together with petroleum bitumen; both elements are increasingly expensive and finite resources. The road surface is planed off, sometimes by burning but more often by cold milling. If required, the concrete base can be broken out as well. The recovered material is heated, mixed with a little additional aggregate and bitumen, and then relaid for resurfacing.

New Materials

Artificial diamonds...Hard artificial materials... Alloys...New composites...Plastics...New uses for ceramics...PERSPECTIVE...Building better icebergs... Glassy metals

As exploitable resources of traditional materials are used up, incentives grow to develop new materials or modify existing materials which are not in short supply. For many years diamond, a natural but rare substance, was the hardest material known. In addition to their decorative function, diamonds are important industrially in grinding and cutting operations. A diamond is composed solely of carbon atoms, linked together in the tightest possible configuration of chemical bonds. Softer forms of carbon also exist: in graphite, for example, the carbon atoms are linked in planes which can slide over one another. This makes graphite useful as a solid lubricant, as well as in pencil "leads".

Experimenters tried to convert graphite into diamond by subjecting it to very high pressures, mimicking the processes believed to have taken place geochemically. After many years of development, the first artificial diamonds were made in 1953. Initially this was done because of the potential value of synthetic gems; eventually, however, it was to supply the need for industrial diamonds, and processes which do this have operated since the late 1950s. Today 50 percent of the world's industrial diamonds are synthetic.

Industrially, diamonds have a drawback: carbon burns. Consequently, when a diamond tool heats up from friction, it oxidizes away. Also in the 1950s, scientists at General Electric in the United States created by high-pressure processes a compound made of equal parts of boron and nitrogen. An atom of boron has one fewer electron than an atom of carbon, an atom of nitrogen one more. Consequently, a mixture of equal parts of the two elements has the same number of electrons as a collection of carbon atoms and can bond chemically to give a diamond-like structure. The resultant compound, boron nitride or borazon, is very hard but resists oxidation. It may be used, for example, as a superior steel grinding abrasive.

Surface alloys

Another way to make exceptionally hard materials is by metalliding, or creating a surface alloy. It has been known for centuries that metal alloys may be much harder than the major element of which they are composed. Bronze, for example, has been known since prehistoric times as an alloy in which a small amount of tin imparts considerable hardness to copper, the main element.

Until recently all alloys have been bulk alloys: the composition of the material has been the same all the way through. In metalliding, atoms of a "hardening" element are diffused into the surface layer of a piece of metal, thus creating a layer which has the properties of the alloy. Diffusion of boron atoms into molybdenum, for example, makes the normally soft molybdenum as hard as diamond. Boron can also be compounded with titanium and zirconium to similar effect.

▲ *A bronze jug from Cyprus, from about the 12th century BC. Bronze, the first alloy ever developed, made stronger and more durable utensils and weapons than iron or copper.*

▼ *A diamond synthesis press, and (inset) synthetic diamonds on the tip of a dentist's drill. Diamonds are made at a temperature of 1,650°C and a pressure of 10,000 tonnes.*

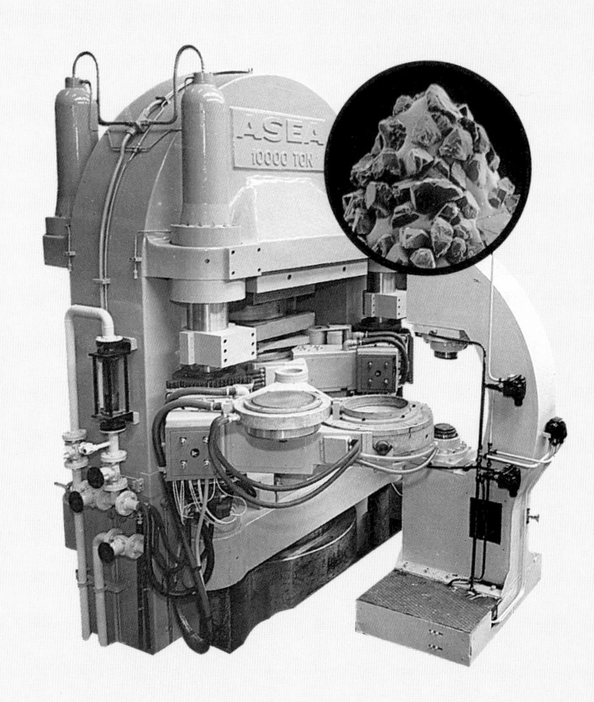

One major development in materials in recent years has been the introduction of new composites, that is combinations of materials with widely different properties. The idea of a composite is not new. Natural materials such as bone and bamboo are composites, while the ancient Greek builders used metal bars to strengthen marble more than 2,000 years ago.

Most modern composites use synthetic resins – polymeric materials made from oil or other organic materials – combined with glass, carbon or mineral fibers. The fibers stop or deflect cracks which would lead to breakage of the resinous material on its own, thus making it less brittle.

If only a very small percentage of fibers is needed (about 2 percent) they can be added to the liquid resin, which can then be molded into the desired shape. However, addition of fibers makes the resin more viscous and if a higher percentage of fiber is needed the resin has to be added to it. A mat of fibers may be blown onto a shaped gauze mold and resin then poured in. This is how complex shapes such as crash helmets are made.

Glass-reinforced plastics have been used to make a wide variety of objects that were previously made from metal. More recently, composites have been developed using carbon fibers, themselves made from a synthetic polymer, polyacrylonitrile, by controlled carbonization. It was thought once that carbon-fiber strengthened material would have many high-technology uses. However, although carbon-fiber composites are very stiff they are not strong in tension. Thus the carbon-fiber fan blades for the Rolls Royce RB211 jet engine, developed in the late 1960s, had to be replaced by the metal titanium. Subsequently carbon-fiber composites have been used widely in applications where their lack of tensile strength is not a drawback, such as sports equipment and artificial limbs.

Research aimed at improving the properties of composites continues not only because of the advantages they offer in substituting for expensive metals, but also because their lower density can offer other economies. Each kilogram that can be saved on the weight of a large aircraft will cut its fuel consumption by 15 barrels of oil over the craft's lifetime.

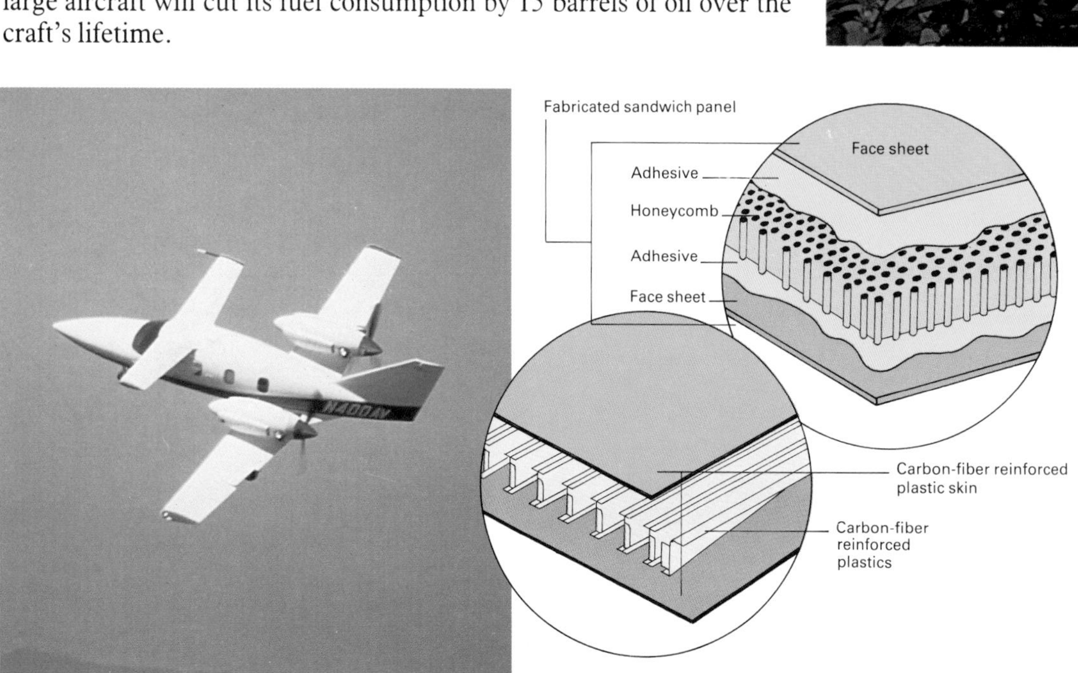

► A spray boom 24m wide, used for crop spraying, and made of lightweight composites. Such materials may be up to four times as strong for their density as high tensile steel. They can simplify design and are cheaper to produce than components of traditional materials.

◄ The Avtek 400, a revolutionary design of executive jet, with composites accounting for 71 percent of its structural weight. The fuselage has no metallic fasteners. Panels are made in a honeycomb-sandwich design 12mm thick, and held together with high-performance adhesives (inset above), and the airbrakes are solid monolithic laminates with skin and ribs of graphite-fiber/epoxy tapes (below).

◄ ***Plastic sheets can be used to cover large greenhouses preventing evaporation and, as here, permitting fruit to be grown in what were once desert regions of the Middle East.***

▲ ***Early plastics were inflammable and often brittle, but new types of plastic shield have been developed in the 1980s for riot police to resist missiles and firebombs.***

Building better icebergs

One of the most unusual materials developed during the 20th century is pykrete, a composite of wood pulp and ice. In 1942, the British government started a project, headed by Geoffrey Pyke, to see whether it was possible to construct aircraft carriers from ice. An early version of the idea was to hollow out part of an iceberg and level off the top to make a runway. At the time, there was a great shortage of traditional constructional materials and Pyke calculated that, even if one had to build an ice ship, the energy required to manufacture ice was only 1 percent of that needed to make steel. And, unlike steel, ice is unsinkable.

Ice was soon found to be unreliable; different samples had widely differing strengths. However, addition of a small percentage of wood pulp to water gave an ice composite which was not only much stronger than ice itself, but also far less variable.

Plans were devised for a pykrete aircraft carrier which would provide a runway 600m long and 6m wide. A drawback was that pykrete deformed unless kept at −15°C. The design therefore included refrigerating circuits as well as motors to propel the artificial iceberg.

The project was abandoned in 1944 for several reasons. An important one was that developments in aircraft design meant that a 600m runway was no longer adequate. Also, the need for such floating airstrips had decreased as the range of aeroplanes increased and it was now possible to patrol the Atlantic with land-based aircraft. Nevertheless, a similar material is now being used in Oslo Fjord, Norway, as the construction material for a wharf. And it is being studied as a construction material for offshore oil platforms.

See also
Shelter 157–164
Road Transport 199–208
Space Transport 225–228
Mining 119-236
Processing Raw Materials 237–244

◀ **The Space Shuttle (▸ page 226) is protected from the intense heat of re-entry into the Earth's atmosphere by more than 30,000 individually contoured silica fiber tiles, attached to the upper and lower surfaces by adhesive.**

▼ **Silicon nitride turbine blades made by Rolls-Royce for use in a helicopter engine. Other ceramic elements planned for this engine include coatings for the bearings and the flame tube, and the low-pressure nozzle. The engine is designed to operate at up to 40,000 rpm at 1,100°C. The individual blades are machined from blocks and fixed in a metal disk. The joint between ceramic and metal is potentially a weak link, and a technique to cast disk and blades together is being researched to overcome this problem.**

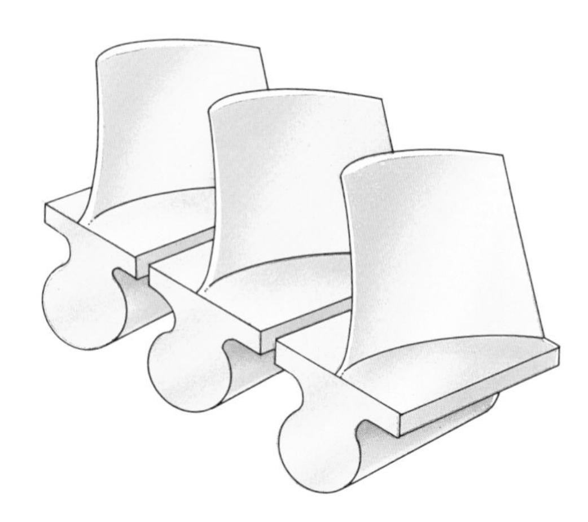

As metals become more expensive, interest grows in developing non-metallic materials to take their place. Millions of dollars have been spent on researching high-temperature ceramics for use in nuclear reactors and gas turbines. In addition to saving on cost and metal use, ceramic turbines and jet engines should be able to operate at higher temperatures, since ceramics cope well with temperatures of 1,000°C or more. Ceramic blades do not need internal air cooling and do not expand with heat, so ceramic turbines should be more efficient than conventional nickel-chrome alloy ones. Ceramics will bring also appreciable benefits to the design of internal combustion engines, if used for the cylinder walls.

Silicon nitride was chosen in 1971 as a potential turbine material, because it is very hard, thermally stable and resistant to oxidation. Its component elements, silicon and nitrogen, are both very common – sand is silicon oxide and nitrogen makes up 80 percent of the atmosphere. However, silicon nitride is too hard to be shaped by machining, and components must be finished accurately before firing, something difficult to achieve since a ceramic shrinks appreciably at this stage. To get round this problem, the ceramic may be mixed with plastic and baked, forming a spongelike material with holes where the plastic has vaporized. These holes can then be filled with graphite or nitrogen, and the resulting ceramic will shrink only minimally on firing. Even so, silicon nitride can be prepared in its most effective, maximum-density form only under conditions which spoil its properties.

An alternative family of compounds is the sialons, named after the symbols for the elements of which they are made: silicon, aluminum, oxygen and nitrogen. These can be produced by reacting silicon nitride with alumina, usually in the presence of yttrium oxide. The sialons do not have the processing drawbacks of silicon nitride. They have been used in cutting tools for machining metals and are candidates for ceramic turbines.

Glassy metals

New alloys can improve on existing materials. Adding 3 percent lithium to aluminum, for example, gives an alloy which is 10 percent stronger and nearly 10 percent less dense, thus offering the possibility of weight-saving in aircraft.

New ways of preparing alloys are also producing new commercial possibilities. By cooling a ferrous alloy very fast – at rates of up to 1,000,000°C per second – the formation of crystals of metal is prevented. The resultant amorphous alloys have a glassy structure which gives them unusual magnetic and corrosion-resistant properties. These alloys can be magnetized and demagnetized very easily. This means they could be used to reduce energy losses in electric motors and transformers (▸ page 140). If used in all the powerline transformers in the USA, they could reduce by two-thirds current annual heat losses of 30 billion kWh.

Credits

Key to abbreviations: SPL Science Photo Library
b bottom; bl bottom left; br bottom right; c center; cl center left; cr center left; r right; t top; tl top left; tr top right; l left.

9t Sally & Richard Greenhill **9b** J. Cochin/Explorer **10t** Robert Harding Picture Library **10b, 11** Science Museum/Michael Holford **12-13** Rex Features Ltd **12b** ZEFA **13bl** Grapes & Michaud/SPL **13br** Max-Planck Institute for Radio Astronomy/SPL **14t** Science Museum **14-15** Royal Ordnance Factory, Enfield **15tl** Ann Ronan Picture Library **15tr** Dick Luria/SPL **16t** Mary Evans Picture Library **16b** Ann Ronan Picture Library **17tl, 17tr** Explorer Archives **17cl** Ann Ronan Picture Library **17cr** BBC Hulton Picture Library **17b** Ann Ronan Picture Library **18l** Intergraph (GB) Ltd **18r, 19l, 19r** Applicon (UK) Ltd **20/21** Royal Greenwich Observatory **21t** British Museum/Michael Holford **21b** Rex Features Ltd **22t** Dr David Jones/ SPL **22b** Dr Gary Settles/SPL **23t** Michael Holford **23bl** Science Museum/Michael Holford **23br** Ann Ronan Picture Library **24** Michael Holford **25** The National Maritime Museum, London **26** Avia Watches **27** Science Museum/ Michael Holford **28** Royal Greenwich Observatory **29t** Michael Freeman **29b** British Museum/Michael Holford **30l, 30r** Jerry Mason/New Scientist **31t** Michael Freeman **31b** Bureau International des Poids et Measures **33** Science Museum **34-35** B. & C. Alexander **35t** Rex Features Ltd **35cl** Defiant Weighing Ltd **35cr, 35b** Paul Brierley **36t, 36b** Mettler Instruments Ltd **37t** National Maritime Museum, London/Michael Holford **37bl** British Museum/Michael Holford **37br** Woodmansterne/Clive Friend **38-39** Royal Geographical Society **39tl** The Royal Society **39br** Clyde Surveys Ltd **40l** Zeiss West Germany **40r** Clyde Surveys Ltd **42** Science Museum/Michael Holford **43** Thorn EMI **44-45** David Parker/SPL **44, 45** Thorn EMI **47t** Science Museum/Michael Holford **47b** Ann Ronan Picture Library **48t, 48b** Paul Brierley **49** Zeiss West Germany **50t** David J. Patterson/Oxford Scientific Films **50-51** Sinclair Stammers/SPL **51l, 51r** David J. Patterson/Oxford Scientific Films **52-53** CNRI/SPL **53t** Zeiss West Germany **53bl** SPL **53br** M. Serrailler/SPL **54l, 55tl, 54-55** IBM, Zurich Research Laboratory **56t** David Parker/SPL **56b** Professor Erwin Mueller/SPL **57t** Lowell Observatory Photograph **57b** Hank Morgan/SPL **58l** The Royal Society **58r** Daily Telegraph Colour Library **59** Anglo-Australian Telescope Board **60l** Ann Ronan Picture Library **60c, 60r** Royal Greenwich Observatory/SPL **61t** Los Alamos National Laboratory/SPL **61c, 61b** Royal Greenwich Observatory/SPL **62l** George Wright/Daily Telegraph Colour Library **62-63** Royal Observatory, Edinburgh **63tl** © by Scientific American, Inc. All rights reserved. **63tr** Perkin-Elmer **64t** Multi-Mirror Telescope **64c** G. Weigelt & G. Baier, University of Erlangen **64b** Professor A. Labeyrie, CERGA **65l** Crown Copyright **65r** Sensitron ab **66** R.B. Mitson **67** NASA/SPL **68, 69t** Shell Photo Service **69b** British Petroleum Co Ltd **70** CNRI/SPL **71t 71b** CEGB **72t, 72cl, 72cr** Nigel Burton, VG Semicon **74-75, 75t** Philips Industrial X-ray, Hamburg **75b** X-ray Astronomy Group, University of Leicester/SPL **76t, 76c** Tiede/Wells Krautkramer **76bl** Dr John Arens/SPL **76br** Royal Observatory, Edinburgh **77** AERE, Harwell **78** Sinclair Stammers **79** ERSAC Ltd **80t** Bell Labs **80b, 81** Max-Planck Institute for Radio Astronomy/SPL **82-83** Art Directors' Photo Library **82b** SERC/Rutherford Appleton Laboratory **83** Dr R.J. Allen et al/SPL **84-85** Crown Copyright **84b** SERC/Rutherford Appleton Laboratory **85b** Hunting Geology & Geophysics Ltd **86t** General Electric Medical Systems, USA **86b** Mary Rose Trust **87t** Royal Photographic Society **87c** Ann Ronan Picture Library **87b** Bayerisches Nationalmuseum **88-89** Dr Harold Egerton/SPL **89** Gamma/Frank Spooner Pictures **90t** Science Museum/Michael Holford **90** Minolta **91t** Ann Ronan Picture Library **91b** Peter Newark's Western Americana **92-93** Adam Hart-Davis/SPL **93t** Canon **94** Abbey Road Studios/John Price **96** The South Bank Centre **97t** Dr Jeremy Burgess/SPL **97b** Reproduced with permission Crown Copyright Controller, HMSO **98** John Monforte **99l** SONY **99r** Margaret Coreau **100** Paul Brierley **101** Hughes Research Laboratory, Malibu **102** John Walsh/SPL **104t** Lawrence Livermore Laboratory/SPL **104b** Alexander Tsiaras/SPL **105l** Leatherhead Food RA **105r** Phillip Hayson/Colorific! **106-107** Science Museum **106** AEI/John Price Studios **107** Michael Freeman **108t** AERE Harwell **108c, 108b** Johnson Matthey **109** Science Museum **110-111t** Fermilab **110c** Michael Freeman **110b** Heini Schneebeli/SPL **111** Jean Collombet/SPL **112-113** University of Cambridge, Cavendish Laboratory **113t** CERN **113b** Patrice Loiez, CERN/ SPL **114l** CERN/SPL **114r** Patrice Loiez, CERN/SPL **115c** Science Museum **115b** Division of Applied Sciences, Harvard University, Cruft Laboratory **116t** University of Liverpool **116b** The Mitre Corporation Archives **117** Manfred Kage/Oxford Scientific Films **118** Sherwood Data Systems Ltd **119** Tony Stone Worldwide **120-121** Nippon Steel Corporation **121c, 121b** Hank Morgan/SPL **122t, 122b** Tony Stone Worldwide **123** Sally & Richard Greenhill **124** Rediffusion Simulation Ltd **124-125, 125** Mapart-Forschungsgruppe "Komplexedynamik" Universität Bremen **126t** Sally & Richard Greenhill **126bl** Meta Machines Ltd **126br** Michael Lesk **127** Dan McCoy/Colorific! **128t** Intel **128b** John McGrail, Wheeler Pictures/Colorific! **129** Ann Ronan Picture Library **130t** BBC Hulton Picture Library **130-131** Popperfoto **131br** CEGB **132t** Mary Evans Picture Library **132b, 133** Ford Motor Company Ltd **134** Quadrant Picture Library **135l** Mary Evans Picture Library **135r** Ann Ronan Picture Library **136-137** Vautier-de Nanxe **137** Michael Freeman **138-139** NEI Parsons Ltd (Drawing courtesy of "Modern Power Systems") **138** Ann Ronan Picture Library **139** G.R. Roberts **140t** ZEFA **140b** Jim Howard/Colorific! **141** Gamma/Frank Spooner Pictures **142t** UKAEA **142c** British Nuclear Fuels Plc **142b** Vautier-de Nanxe **143t** British Nuclear Fuels Plc **143c** UKAEA **143b** COGEMA **144-145** Bill Pierce/Contact/Colorific! **145** UKAEA **146-147** Quadrant Picture Library/Nuclear Engineering **148** JET Joint Undertaking **149** ZEFA **150** John McGrail/Wheeler Pictures/Colorific! **151t** Gamma/Frank Spooner Pictures **151b** Daily Telegraph Colour Library **152t** James Sugar/Black Star/Colorific! **152c** Martin Bond/SPL **152-153** Gamma/Frank Spooner Pictures **153r** John Running/Black Star/ Colorific! **154t** Vautier-de Nanxe **154b** Mark Edwards/ Earthscan **155** Johnson Matthey **157c** Tor Eigeland/ Susan Griggs Agency **157b** Adam Woolfitt/Susan Griggs Agency **158** Michael Freeman **159t** Foster Associates **159b** Robert Harding Picture Library **160-161** All-Sport **161t** B. & C. Alexander **161c** Michael Freeman **161b** ZEFA **162-163** Rex Features Ltd **162bl, 162br, 163b** Tom McHugh/SPL **164** Alex Bartel/SPL **165** US Army/MARS, Lincs **166l** Rex Features Ltd **166-167** McDonnell Douglas Helicopters/MARS, Lincs **168** British Aerospace **169** Ford Aerospace & Communications Corporation **170t** Flight/ Quadrant Picture Library **170b** Rex Features Ltd **171** Mary Evans Picture Library **172t, 172bl** ZEFA **172-173b, 173br** Linotype Ltd **176** Art Directors' Photo Library **177l, 177r** Science Museum/Michael Holford **178** British Telecom **179** Plessey **180-181t** AT & T **180bl** Ann Ronan Picture Library 180-181b TCL/AEI/John Price Studios **181r** Art Directors' Photo Library **182c** TCL/AEI/John Price Studios **182b** Ann Ronan Picture Library **183** TCL/John Price Studios **184c** Philips **184b** ICL/Archie Miles **185t, 185b** The Marconi Company Ltd **186** Science Museum/Michael Holford **187** BBC Hulton Picture Library **190t** Images by Goodman/Rex Features Ltd **190bl** Quantel **190br** Art Directors' Photo Library **191** Pete Addis/New Scientist **192t** NASA/SPL **192c** Art Directors' Photo Library **193tl** Ann Ronan Picture Library **193tr** BBC Hulton Picture Library **193b** Ann Ronan Picture Library **194** Union Pacific Railroad Museum Collection **195** Union Pacific System **196tl, 196tr, 196c** French Railways Ltd **198t** Japan Railway Technical Service **198b** Martin Bond/ SPL **199t** National Motor Museum **199c** Mary Evans Picture Library **199b** Autocar/ Quadrant Picture Library **202-203t** Ford Motor Company Ltd **202b, 203b** LAT Photographic **204t** Ann Ronan Picture Library **204c** John Topham Picture Library **204b** Eamonn McCabe **205t** Gamma/Frank Spooner Pictures **205b** © by Scientific American, Inc. All rights reserved. **206-207t** Nissan Motor Co Ltd **206b** Kevin Radley **207b, 208t** Nissan Motor Co Ltd **208b** All-Sport/Vandystadt **209t** ZEFA **209b** Mary Evans Picture Library **210t** British Museum/Michael Holford **210b** ZEFA **211t** Sipa-Press/Rex Features Ltd **211b** Leo Mason **212-213b** Science Museum **213t** ZEFA **213br** Wärtsilä **214tl** Mary Evans Picture Library **214-215b** Dick Clarke/Planet Earth Pictures **215t** Dick Halstead/Contact/Colorific! **215br** John Menzies/Planet Earth Pictures **216t** British Hovercraft **216b** Boeing Marine Systems/MARS, Lincs **217t** Beech Aircraft Corporation **217b** Science Museum **219t** Flight/Quadrant Picture Library **219b** British Petroleum Co Ltd **220-221c** Adrian Balch **220bl** National Air & Space Museum, Smithsonian Institution **220br** Quadrant Picture Library **221t** Aviation Picture Library **221br** Jerry Young **222t** McDonnell Douglas Corporation **222b** Sipa-Press/Rex Features Ltd **223t** Airbus Industrie **224t** Doug Shane/Visions/Colorific! **224b** Du Pont (UK) Ltd **225c, 225b** David Baker **228** Anthony Wolf **229t** Mansell Collection **229b** Robert Harding Picture Library **230** National Coal Board **232l** Vautier-Decool **232-233b** Jim Howard/Colorific! **233t** Vautier-Decool **233c** De Beers Consolidated Mining **234t** Joe Rychetnik/Black Star/Colorific! **234c** British Petroleum Co Ltd **235t** Tony Stone Worldwide **235tr** Gamma/Frank Spooner Pictures **236** ZEFA **237** Robert Estall **238c** B. & C. Alexander **238bl** RTZ **238-239** Tony Stone Worldwide **241** Nippon Steel Corporation **242t** RTZ **242c, 243l** Terry Spencer/Colorific! **243r** East Rand Gold & Uranium Co Ltd **244** Johnson Matthey **245t** British Museum/Michael Holford **245b** De Beers Industrial Diamonds **245inset** Manfred Kage/SPL **246-247** Daily Telegraph Colour Library **246b** Du Pont, Geneva **247tr** Rex Features **247b** Framers Weekly Picture Library **248t** Rockwell International/MARS, Lincs.

Artists
Robert and Rhoda Burns, Kai Choi, Simon Driver, Alan Hollingbery, Kevin Maddison, Colin Salmon, Mick Saunders

Indexer
Margaret Cooter

Additional research
Paul Holister

Typesetter
Peter Furtado

Further Reading

General

Armytage, W.H.V., *A Social History of Engineering* (Faber and Faber)
Derry, T.K. and Williams, Trevor I., *A Short History of Technology*, (Oxford University Press)
Encyclopedia of Science and Technology (5th edition, McGraw-Hill)
Pitt, Valerie H., *The Penguin Dictionary of Physics* (Penguin Books)
Williams, Trevor, *A Short History of Technology* (Oxford University Press)

Specific topics

Landes, David S., *Revolution in Time: Clocks and the Making of the Modern World* (Harvard University Press)
Dickinson, G.G., *Maps and Air Photographs*, (Edward Arnold)
Stock, John T., and Vaughan, Denys, *The Development of Instruments to Measure Electric Current* (Science Museum, London
Cohen, D., *In Quest of Telescopes* (Sky Publishing Corp)
Dent, D. and Young, A., *Soil Survey and Land Evaluation* (Allen and Unwin)
Darius, J., *Beyond Vision* (Oxford University Press)
Henbest, N., and Marten, M., *The New Astronomy* (Cambridge University Press)
Claugher, D., *Scanning Nature* (Cambridge University Press/British Museum (Natural History)
Driscoll, Roger, *Practical HiFi Sound* (Hamlyn)
Dalton, Stephen, *Split Second, the World of High Speed Photography* (J.M. Dent)
Matthewson, D., *Revolutionary Technology* (Butterworth)
Collier, R.J., Buckhardt, C.B. and Lin, L.H., *Optical Holography* (Academic Press, Inc.)
Jackson, K.G., *Newnes Book of Video* (Butterworth)
Sherwood, M., *New Worlds in Chemistry* (Faber and Faber)
Sutton, Christine, *The Particle Connection* (Hutchinson)
Atkins, P.W. *Physical Chemistry* (2nd edition; Oxford University Press)
Morgan, E., *Microprocessors, a Short Introduction* (Department of Industry, London)
Augarten, Stan, *Bit by Bit: An Illustrated History of Computers* (Allen and Unwin)
Patterson, Walter C., *Nuclear Power* (Penguin Books)
Brinkworth, B.J., *Solar Energy for Man* (Compton Press)
Armstead, H. Christopher, *Geothermal Energy* (F and F.N. Spon)
Gordon, J.E., *Structures* (Penguin Books)
Vale, Brenda and Robert, *The Autonomous House* (Thames & Hudson)
Kennedy, Col.W.V. *et al*, *The Intelligence War* (Salamander)
Astrua, Massimo, *Manual of Colour Reproduction* (Fountain Press)
Hutchings, Ernest A.D., *A Survey of Printing Processes* (Heinemann)
Povey, P.J., *The Telephone and the Exchange* (Pitman)
Owen, David, *The Complete Handbook of Video* (Marshall Editions)
Nock, O.S., *Encyclopedia of Railways* (Octopus)
Wise, D. Burgess, Boddy, W. and Laban, B., *The Automobile, the First Century* (Orbis)
Stinton, Darrol, *The Anatomy of the Aeroplane* (Collins)
Chant, Christopher, *Aviation, an Illustrated History* (Orbis)
Gatland, Kenneth, *The Illustrated Encyclopedia of Space Technology* (Salamander)
Alexander, William and Street, Arthur, *Metals in the Service of Man* (8th edition Penguin Books)
Gordon, T.E., *The New Science of Strong Materials* (Penguin Books)
Macdonald, E.H., *Alluvial Mining* (Chapman-Hall)
Williams, W. Randolph, *Mine Mapping and Layout (Prentice-Hall)*

Glossary

Acceleration
The rate at which the velocity of a body changes, usually measured in meters per second per second.
Accelerator
In particle physics, a research tool used to accelerate subatomic particles to high velocities.
Acoustics
The science of sound, dealing with its production, transmission, reflection and absorption.
Adhesives
Substances that bond surfaces to each other by the force of attraction between contacting surfaces of unlike substances.
Aerial
See ANTENNA.
Air-cushion vehicle
A marine, land or amphibious vehicle supported on a high-pressure cushion of air maintained by a system of fans. Also known as a hovercraft.
Algorithm
A set of mathematical operations which together, and in the correct order, constitute a complex operation. Algorithms are used in planning computer software.
Alloy
A material of metallic character prepared by combining metals with one another or with non-metals such as carbon or phosphorus.
Altimeter
A device that measures height. Measurements relative to sea level are made using pressure measurement and a local reference, or relative to local terrain using electromagnetic devices.
Amplifier
A device that multiplies a signal by a constant factor.
Amplitude
The maximum deviation from its mean value of a physical property which is subject to modulation.
Amplitude modulation
Common method of encoding a carrier wave in radio transmission, with the signal modifying the amplitude, rather than the frequency, of the wave.
Analog
A description of a signal that may vary continuously.
Anode
The positive electrode of a battery, cell or electron tube. The electrode at which electrons leave the system to an external circuit.
Antenna (or aerial)
A component in an electrical circuit that radiates or receives radio waves.
Atomic clock
A device used to CALIBRATE clocks using the constant frequencies of the electron spin reversals of the cesium atom to define an accurate and reproducible time scale.
Balance
An instrument used for measuring the mass of an object by comparison of its weight with objects of known weight.
Bandwidth
The difference between the upper and lower limits of the frequencies needed or available to transmit a signal.
Battery
A device for converting internally-stored chemical energy into direct-current electricity.
Binary code
Number system on the base 2, frequently used in computers, microprocessors and for digitized data of all kinds.
Biotechnology
The use of micro-organisms for industrial purposes.
Bolometer
An instrument used to measure radiant energy, notably in the infrared and microwave wavebands.
Bubble chamber
Device used to observe the paths of subatomic particles, by reducing pressure as the particles pass through so that bubbles form along their paths.
Calibration
The provision of a scientific instrument with a numerical scale in accordance with an agreed scale or by comparison with a standard measure.
Calipers
A device used for measuring length, usually made up of a pair of bowed metal legs pivoted at one end.
Cam
A mechanical device that, on rotating, gives a regular, repetitive motion to a member held in contact with it.
Camcorder
A VIDEO camera, or camera that records moving images on magnetic tape.
Catalyst
A substance which increases the speed of a chemical reaction but is itself chemically unchanged at the end of the reaction.
Cathode
The negative electrode of a battery, cell or electron tube. The electrode at which electrons enter the system from an external circuit.
Cathode ray tube
An evacuated glass tube containing at one end a CATHODE and ANODE; electrons from the cathode are accelerated through the anode to form a beam that hits a screen. It forms the principal component of an OSCILLOSCOPE and a TELEVISION set.
CCD
See CHARGE-COUPLED DEVICE
Central processing unit
The part of a computer that draws data from the memory, processes them in the arithmetic and logical unit and outputs or stores the results, following the instructions of the operator or program.
Charge-coupled device
A SEMICONDUCTOR device used in imaging systems, such as telescopes, to produce a VIDEO signal. A PHOTON striking one of an array of capacitors generates a charge which is amplified.
Chromatography
A technique of chemical analysis in which a sample is mixed with a carrier and passed over a porous surface. Components are adsorbed at different rates, and separation occurs.
Cloud chamber
A device for making visible the tracks of subatomic particles, by passing them through a saturated vapor; the passage of charged ions through the chamber causes the gas to condense into droplets.
Combined heat and power station
Electrical generating station in which exhaust heat from the turbines is passed to neighboring homes and factories.
Composite
Material made up of two or more substances, with properties different to those of the substances that make it up.
Computer
A device that performs calculations and stores their results, according to a PROGRAM.
Conductor
A substance capable of carrying electric current.
CPU
See CENTRAL PROCESSING UNIT
Cyclotron
A particle accelerator in which the particles travel in a spiral path in a strong magnetic field.
Database
A system of centralized information storage to which individual computer users may have access, often by telephonic links.
Data processing
All the operations performed by a computer on the information it receives.
Diffraction
The property by which a wave motion enters a geometrically shadowed region when passing an opaque object.
Digitization
The conversion of an ANALOG signal into a digital one, usually in BINARY code.
Dynamo
Also known as a generator, a machine for converting mechanical power into electrical power, normally by rotating a coil in a magnetic field.
Electrode
A component in an electrical circuit at which current is transferred between metal conductors and a gas or electrolyte.
Electrolysis
The technique of producing a chemical reaction by passing a current through an electrolyte, so that quantities of a substance are deposited at the CATHODE or ANODE.
Electromagnetic radiation
The form in which energy is transmitted through space or matter, using an electromagnetic field. Its wavelengths carry radio waves and infrared waves, visible light, ultraviolet, X-rays and gamma rays.
Electron
A subatomic particle of negative charge, commonly in orbit around an atomic nucleus.
Electron gun
The part of a CATHODE RAY TUBE that produces, accelerates and deflects the beam of ELECTRONS.
Electronics
A science dealing with SEMICONDUCTORS and devices where the motion of ELECTRONS is controlled.
Feedback mechanism
A device that controls the operation of a system by detecting the effect and regulating the output automatically.
Filter
An arrangement of components that transmits signals in a given frequency range and rejects others.
Fission
In nuclear physics, the changing of an element into two or more elements of lower atomic weight, with the release of energy.
Flotation
The process of recovering minerals by mixing the pulverizing ore with water, and pumping air through the mixture. The mineral particles form a froth which can be skimmed off.
Frequency
The rate at which a wave motion completes its cycle.
Frequency modulation
The encoding of a signal by varying the frequency of the carrier wave.
Fuel cell
Direct-current power source, in which a chemical fuel must be supplied to the cell while in use.
Fusion
In nuclear physics, the merging of two nuclei to form a new element of higher atomic weight, resulting in the release of energy.
Geiger counter
An instrument for detecting the presence of alpha particles, and beta-, gamma- and X-rays.
Gyroscope
A spinning disk mounted so that its axis can adopt any orientation; used in INERTIAL GUIDANCE systems.
Halflife
The period required for half the isotopes of any given radioactive sample to undergo decay.
Heat pump
A device for transferring heat from a cold region to a hotter region, using a working fluid such as ammonia.
Holography
The creation of three-dimensional images by photographing the subject when illuminated by a split laser beam, and reproducing the image by recreating the beam.
Hydroelectricity
The generation of electricity by using moving water to drive generator turbines.
Hydrofoil
A structure attached to a boat which generates lift from the water, raising the hull clear of the surface.
Hydrophone
An adaptation of the MICROPHONE for use underwater.
Induction
The phenomenon in which an electric field is generated in an electric circuit when the number of magnetic field lines passing through it changes.
Inertial guidance
An automatic navigation system using accelerometers and gyroscopes to sense and record changes in a vessel's velocity.
Information technology
The technology relating to the collection, processing, storage and transmission of information.
Integrated circuit
A structure (often a silicon chip) on which many individual electronic components are assembled.

Interferometer
An instrument employing interference effects, used for measuring wavelength or for detecting the direction of a radio source.
Ion
An atom that has become electrically charged by gain or loss of electrons.
Isotopes
Atoms of a chemical element with the same number of protons but different number of neutrons.
Jet propulsion
The propulsion of a vehicle by expelling a fluid jet backwards, whose momentum imparts a forward movement to the vehicle.
Laminate
Component in which several thin sheets of different substances are bonded together with resins.
Laser
Device producing an intense beam of parallel light with a precisely defined wavelength; laser is short for "light amplification by stimulated emission of radiation".
LCD
See LIQUID CRYSTAL DISPLAY.
Liquid crystal display
Form of displaying information using liquid crystals to block or allow the passage of light onto a reflective surface.
Lithography
Method of printing by treating the printing plate such that some areas attract ink while others repel it.
Maglev
"Magnetic levitation" train, in which electromagnets are used to lift the train off the track and move it along.
Magnetic resonance inspection
Also called nuclear magnetic resonance, the use of the spin of an atom when subjected to a strong magnetic field to form images of the interior of objects, including the human body, by detecting the presence of particular elements.
Mainframe
A large central computer, to which smaller computers and other devices may be linked.
Maser
A device used as a microwave oscillator or amplifier; "microwave amplification by stimulated emission of radiation".
Microcomputer
A computer with the CENTRAL PROCESSING UNIT contained on a single silicon chip.
Micrometer
Handheld device used to measure distance accurately.
Microprocessor
An electronic device that receives, processes, stores and outputs information according to a preprogrammed set of instructions.
Microscope
A device that produces images of small objects, by focusing light, electrons, sound waves or X-rays.
Moderator
In a nuclear reactor, a substance used to slow the neutrons released by fission.
Modem
"Modulator-demodulator", a device used to convert the output of a computer into a form suitable for transmission by telephone.
NDT
Nondestructive testing; the inspection of materials or machines without destroying or removing parts.
NMR
See MAGNETIC RESONANCE INSPECTION.
Noise
Unwanted sound, current or voltage that interferes with the signal.
Nuclear magnetic resonance
See MAGNETIC RESONANCE INSPECTION.
Optical fiber
Extruded glass fiber of high purity, used to transmit a light signal; used widely for telephone systems.
Oscilloscope
A CATHODE RAY TUBE in which a varying signal current creates a line graph on the screen.
Pantograph
Hinged parallelogram instrument. One version may be used in technical drawing to enlarge the movement of the pen; a sprung pantograph linkage is used to maintain contact between a train and overhead cables.
Photoconductive detector
Electric component whose conductivity rises with the amount of light falling on it.
Photogrammetry
The use of aerial photographs in mapmaking.
Photon
A quantum of electromagnetic energy, often thought of as the particle associated with light.
Photovoltaic cell
A device for converting light radiation into electricity.
Piezoelectricity
The relationship between mechanical stress and electric charge exhibited by certain crystals.
Pitot tube
A device for measuring air speed or water speed, by comparing the pressure in the flowing stream with that in the static.
Polarized light
Light in which the orientation of wave vibrations displays a definite pattern.
Program
The set of instructions to be followed by a COMPUTER or MICROPROCESSOR.
Proton
A stable particle found in the nucleus of an atom, carrying a positive charge.
Pyrometer
A device for measuring high temperatures.
Radar
"Radio detection and ranging": a technique that locates the position of distant objects by measuring the time taken for radio waves to travel to them, be reflected and return.
Radiation
The emission and propagation through space of subatomic particles or of waves forming part of the ELECTROMAGNETIC SPECTRUM.
Radio
Communication between distant points using radio waves, electromagnetic radiation of a certain frequency.
Radioactivity
The spontaneous disintegration of unstable nuclei, accompanied by the emission of RADIATION of alpha particles, beta particles or gamma rays.
Radiocarbon dating
The technique for dating ancient objects by measuring the radioactive decay of carbon-14, a long-lived ISOTOPE, in their material.
Raster
A series of parallel sweeps, such as the sweeps of an electron beam across the image on a television camera or receiver.
Refraction
The change in direction of energy waves as they pass from one medium to another.
Remote sensing
The use of distant sensing devices (normally carried on aircraft or spacecraft) to detect features on the Earth.
Resistance
The ratio of the voltage applied to a conductor to the current flowing through it.
Resonance
The large response of an oscillatory system when driven near its natural frequency.
Robot
Machine that does work automatically according to a PROGRAM. A robot may be able to detect limited changes in its environment and adapt its movements accordingly.
Rocket
A JET-propulsion engine in which all the substances needed to create the propellant gases are carried internally.
Scintillation counter
An instrument for detecting ionizing RADIATION, which causes a light flash in a phosphor.
Seismograph
A device used to detect seismic waves caused by earthquakes or explosions.
Semiconductor
A material whose electrical conductivity varies with temperature and impurity. By introducing impurities to different regions of a semiconductor, it can be modified for different electrical purposes.
Sensitivity
The ability of a measuring instrument to respond to small variations in the input signal.
Servomechanism
An automatic control device in which a command signal from a reference device uses FEEDBACK to control a higher-power output device.
SI units
"Système Internationale", the scientific units internationally agreed since 1960.
Signal
The variable impulse by which information is carried through a system.
Sintering
The bonding together of powder particles below melting point to consolidate ores.
Software
The elements of a computer system that are not intrinsic to the hardware; notably the PROGRAM.
Sonar
"Sound navigation and ranging", a technique used to locate objects underwater by detecting the time taken for sound waves to travel to, be reflected and return.
Spectroscopy
The production, measurement and analysis of spectra, much used by astronomers, chemists and physicists.
Strain gauge
A device for measuring strain at the surface of a material by determining variations in the current through a PIEZOELECTRIC device attached to it.
Stroboscope
A device that produces regular brief flashes of light.
Superconductor
Material cooled to extremely low temperatures, at which electrical resistance is zero; used in large electromagnets.
Synchrotron
A large ACCELERATOR in which the particles are accelerated around a circular path.
Telescope
A device for detecting ELECTROMAGNETIC RADIATION from very distant objects.
Television
An apparatus for communicating moving pictures by radio transmissions.
Thermocouple
An electric circuit involving two junctions between different metals or SEMICONDUCTORS; a current is generated if these junctions are at different temperatures.
Tokamak
A ring-shaped device in which a plasma is contained for experiments in FUSION reaction.
Torque
A measure of the effectiveness of a force in setting a body in rotation.
Transducer
A device for converting an input signal into an output signal of another energy mode, normally electrical.
Transformer
A device for altering the voltage of an alternating-current electricity supply.
Transistor
An electronic device made of semiconductors used in a circut as an amplifier, rectifier, detector or switch.
Turbine
A machine for directly converting the energy of a flowing fluid into rotational energy.
Ultrasonics
The use of high-frequency sound waves to detect the internal structures of solid objects.
Work
The result of a force moving its point of application.
Video
See TELEVISION.
Videotape
The magnetic tape on which a VIDEO signal can be recorded.
X-rays
Electromagnetic radiation with a wavelength between that of ultraviolet radiation and gamma rays.

Index

Page numbers in *italics* refer to illustrations, captions or side text.

A

B

C

D

E

F

G

H

I

J

K

L

M

N

P

U

V

W

Z